Active and Quasi-Optical Arrays for Solid-State Power Combining

WILEY SERIES IN MICROWAVE AND OPTICAL ENGINEERING

KAI CHANG, Editor
Texas A&M University

FIBER-OPTIC COMMUNICATION SYSTEMS • *Govind P. Agrawal*

COHERENT OPTICAL COMMUNICATIONS SYSTEMS • *Silvello Betti, Giancarlo De Marchis and Eugenio Iannone*

HIGH-FREQUENCY ELECTROMAGNETIC TECHNIQUES: RECENT ADVANCES AND APPLICATIONS • *Asoke K. Bhattacharyya*

COMPUTATIONAL METHODS FOR ELECTROMAGNETICS AND MICROWAVES • *Richard C. Booton, Jr.*

MICROWAVE RING CIRCUITS AND ANTENNAS • *Kai Chang*

MICROWAVE SOLID-STATE CIRCUITS AND APPLICATIONS • *Kai Chang*

DIODE LASERS AND PHOTONIC INTEGRATED CIRCUITS • *Larry Coldren and Scott Corzine*

MULTICONDUCTOR TRANSMISSION-LINE STRUCTURES: MODAL ANALYSIS TECHNIQUES • *J. A. Brandão Faria*

PHASED ARRAY-BASED SYSTEMS AND APPLICATIONS • *Nick Fourikis*

FUNDAMENTALS OF MICROWAVE TRANSMISSION LINES • *Jon C. Freeman*

MICROSTRIP CIRCUITS • *Fred Gardiol*

HIGH-SPEED VLSI INTERCONNECTIONS: MODELING, ANALYSIS, AND SIMULATION • *A. K. Goel*

HIGH-FREQUENCY ANALOG INTEGRATED CIRCUIT DESIGN • *Ravender Goyal (ed.)*

FINITE ELEMENT SOFTWARE FOR MICROWAVE ENGINEERING • *Tatsuo Itoh, Giuseppe Pelosi and Peter P. Silvester (eds.)*

OPTICAL COMPUTING: AN INTRODUCTION • *M. A. Karim and A. S. S. Awwal*

MILLIMETER WAVE OPTICAL DIELECTRIC INTEGRATED GUIDES AND CIRCUITS • *Shiban K. Koul*

MICROWAVE DEVICES, CIRCUITS AND THEIR INTERACTION • *Charles A. Lee and G. Conrad Dalman*

ANTENNAS FOR RADAR AND COMMUNICATIONS: A POLARIMETRIC APPROACH • *Harold Mott*

INTEGRATED ACTIVE ANTENNAS AND SPATIAL POWER COMBINING • *Julio A. Navarro and Kai Chang*

FREQUENCY CONTROL OF SEMICONDUCTOR LASERS • *Motoichi Ohtsu (ed.)*

SOLAR CELLS AND THEIR APPLICATIONS • *Larry D. Partain (ed.)*

ANALYSIS OF MULTICONDUCTOR TRANSMISSION LINES • *Clayton R. Paul*

INTRODUCTION TO ELECTROMAGNETIC COMPATIBILITY • *Clayton R. Paul*

INTRODUCTION TO HIGH-SPEED ELECTRONICS AND OPTOELECTRONICS • *Leonard M. Riaziat*

NEW FRONTIERS IN MEDICAL DEVICE TECHNOLOGY • *Arye Rosen and Harel Rosen (eds.)*

NONLINEAR OPTICS • *E. G. Sauter*

FREQUENCY SELECTIVE SURFACE AND GRID ARRAY • *T. K. Wu (ed.)*

OPTICAL SIGNAL PROCESSING, COMPUTING AND NEURAL NETWORKS • *Francis T. S. Yu and Suganda Jutamulia*

ACTIVE AND QUASI-OPTICAL ARRAYS FOR SOLID-STATE POWER COMBINING • *Robert A. York and Zoya B. Popović (eds.)*

Active and Quasi-Optical Arrays for Solid-State Power Combining

Edited by

ROBERT A. YORK
University of California, Santa Barbara

ZOYA B. POPOVIĆ
University of Colorado, Boulder

A WILEY-INTERSCIENCE PUBLICATION
JOHN WILEY & SONS, INC.
NEW YORK/CHICHESTER/WEINHEIM/BRISBANE/SINGAPORE/TORONTO

This text is printed on acid-free paper.

Copyright © 1997 by John Wiley & Sons, Inc.

All rights reserved. Published simultaneously in Canada.

Reproduction or translation of any part of this work beyond that permitted by Section 107 or 108 of the 1976 United States Copyright Act without the permission of the copyright owner is unlawful. Requests for permission or further information should be addressed to the Permissions Department, John Wiley & Sons, Inc., 605 Third Avenue, New York, NY 10158-0012.

Library of Congress Cataloging in Publication Data:
Active and quasi-optical arrays for solid-state power combining/
   edited by Robert A. York, Zoya B. Popović.
       p.   cm. – (Wiley series in microwave and optical engineering)
   Includes index.
   ISBN 0-471-14614-5 (alk. paper)
   1. Power electronics. 2. Integrated optics. 3. Power amplifiers.
4. Solid state electronics. I. York, Robert A. II. Popović, Zoya
B. III. Series.
TK7881.15.A37 1997
621.381'1—dc21   96-39031

Printed in the United States of America

10 9 8 7 6 5 4 3 2 1

Contents

Contributors		xiii
Foreword		xv
Preface		xix

1 Quasi-Optical Power Combining — 1
Robert A. York

   1 Applications Potential and Technological Constraints — 2

   2 Power-Combining Circuits — 5

   3 Spatial or Quasi-Optical Arrays — 9
      3.1 Maximum Power Density — 15
      3.2 Quasi-Optical Array Topologies — 17
      3.3 Planar Antennas for Arrays — 21

   4 Analytical Techniques for Quasi-Optical Arrays — 25
      4.1 Passive Array Analysis — 25
      4.2 Active and Nonlinear Circuit Analysis — 28
      4.3 Nonlinear Dynamics — 31

   5 Array Characterization and Figures of Merit — 33
      5.1 Aperture Efficiency and Polarizers — 34
      5.2 Dominant-Mode Coupling — 37
      5.3 Effective Radiated Power of Sources/Transmitters — 39
      5.4 Gain of Beam Amplifiers — 40
      5.5 Combining Efficiency — 42

   6 Conclusions — 42

   Acknowledgments — 44

   References — 44

2 Spatial Power Combining 49
Mark A. Gouker

  1 Feeding and Combining Approaches 50

  2 Trade-Offs for Spatial Power-Combined Amplifiers and Transmitters 53
    2.1 System Configurations 53
    2.2 Array Architecture (Tile Approach or Tray Approach) 55
    2.3 Circuit-Fed Versus Spatially Fed 58
    2.4 Grid Arrays Versus Distinct Component Arrays 61
    2.5 Monolithic Versus Hybrid Construction 63
    2.6 Trade-Off Summary 65

  3 Losses in Spatial Power-Combining Systems 66

  4 Circuit-Fed/Spatially Combined Amplifier Arrays 68
    4.1 Differences from Phased Arrays 68
    4.2 Components for Arrays 69
    4.3 Examples 72

  Acknowledgments 82

  References 83

3 Active Integrated Antennas 85
Siou Teck Chew and Tatsuo Itoh

  1 Design Issues 86
    1.1 Size of Antenna and Active Devices 86
    1.2 Surface Wave Excitation 86
    1.3 Heat Sinking 88
    1.4 Free-Space Mutual Coupling 88
    1.5 Unwanted Radiation from the RF Circuit 91
    1.6 Antenna as a Resonator in Oscillator Design 92
    1.7 Nonferrite Device Integration 93
    1.8 Antenna Dynamic Load 94
    1.9 Lack of Simulation Tool 94
    1.10 Testing in a Non-50 Ω Environment 95
    1.11 Others 96

  2 Review of the Field 96
    2.1 Amplifier Type 97
    2.2 Oscillator Type 98
    2.3 Frequency-Conversion-Type Circuits 100
    2.4 Optical-Integrated Type 102

  3 Choice of Planar Antennas 104
    3.1 Introduction 104

	3.2 Patch Antenna	105
	3.3 Slot Antennas	109
	3.4 Others	112
	4 FDTD Analysis and Visualization	112
	4.1 Introduction	112
	4.2 FDTD	113
	4.3 Integration of the Active Device in FDTD	114
	4.4 Simulation of Active Antennas	114
	5 Case Studies	115
	5.1 Gunn/Patch Oscillator [82,83]	115
	5.2 Active Slot Antenna [48]	118
	5.3 Noncontact ID Transponder [61]	120
	5.4 Monopulse Switch [84]	122
	5.5 Doppler Transceiver [87]	125
	References	128
4	**Coupled-Oscillator Arrays and Scanning Techniques**	**135**
	Jonathan J. Lynch, Heng-Chia Chang, and Robert A. York	
	1 Introduction	136
	2 Oscillator Modeling	138
	2.1 Injection-Locking	142
	2.2 Oscillator Noise	144
	3 Systems of Coupled Oscillators	145
	3.1 Derivation of the Dynamic Equations	147
	3.2 Stability of Solutions	149
	3.3 Broadband Coupling Networks	150
	3.4 Analysis, Synthesis, and Simplifications	151
	3.5 Linear Arrays with Nearest-Neighbor Bilateral Coupling	153
	3.6 Experimental Results	157
	3.7 Transient Response to Tuning Variations	163
	3.8 Phase Noise Analysis	166
	4 Scanning Oscillator Arrays	171
	4.1 Unilateral Injection-Locking	171
	4.2 Stephan's Scanning Approach	174
	4.3 Bilateral Coupling with Symmetric End Tuning	176
	4.4 Variations	180
	Appendix: Kurokawa's Substitution	181
	Acknowledgments	183
	References	183

5 Quasi-Optical Antenna-Array Amplifiers — 187
Zoya B. Popović, Robert A. York, Emilio A. Sovero, and Jon Schoenberg

1 Introduction — 188
  1.1 Quasi-Optical Amplifier Gain — 190
  1.2 Quasi-Optical Amplifier Array Feed — 190

2 Antenna Elements for Amplifier Arrays — 193
  2.1 Patch Antennas — 193
  2.2 Slot Antennas — 193
  2.3 Broadband Tapered-Slot Antennas — 201

3 Plane-Wave-Feed Amplifier Arrays — 205
  3.1 Polarization-Preserving Class-A 24-Element MESFET Patch Array — 205
  3.2 High-Efficiency Class-E MESFET Slot Array — 208
  3.3 C-Band Folded-Slot Array — 210
  3.4 X-Band Multiple-Slot Array with Commercial MMICs — 213
  3.5 X-Band Tapered-Slot Array in Waveguide — 214
  3.6 Ka-Band Quasi-Monolithic 2.4-W Array — 216
  3.7 Monolithic 42-GHz Quasi-Optic Amplifiers — 219
  3.8 60-GHz Monolithic Patch/Slot Amplifier Array — 223

4 Lens Amplifiers — 226
  4.1 Historical Development of Constrained Lenses — 227
  4.2 Two-Dimensional Patch Lens Amplifier Array — 231
  4.3 Low-Noise CPW Slot Lens Amplifier — 234

5 Conclusions — 240

References — 242

6 Multilayer and Distributed Arrays — 245
Amir Mortazawi, Carl L. Brockman, and John F. Hubert

1 Introduction — 245

2 A Multilayer Amplifier Array — 248
  2.1 Double-Layer Amplifier Architecture — 251
  2.2 Through Wafer Coupling Mechanism — 252
  2.3 The Amplifier's Unit Cells — 254
  2.4 A Unit Cell with High Active Device Density — 255
  2.5 Near-Field Excitation of Spatial Amplifiers — 256
  2.6 Excitation Using Hard Horn Feeds — 258
  2.7 Spatial Amplifier Measurements — 260

3 Multilayer Spatial Amplifier Arrays — 263
  3.1 Monolithic Design of Double-Layer Arrays — 264

    4 Spatial Power-Combining Oscillators Based on an Extended
      Resonance Technique   267
      4.1 Unit Cell Design   268
      4.2 Spatial Power-Combining Oscillator Array   270

    5 Discussion   272

    References   274

7 **Planar Quasi-Optical Power Combining**   277
  Michael B. Steer, James W. Mink, and Huan-Sheng Hwang

    1 Introduction   277

    2 Theory of Planar Quasi-Optical Waveguiding   280

    3 Planar Quasi-Optical Oscillator   284

    4 Rectangular Waveguide Transition   287

    5 Planar Quasi-Optical Amplifier   288

    Acknowledgments   292

    References   292

8 **Grid Oscillators**   293
  Zoya B. Popović, Wayne A. Shiroma, and Robert M. Weikle II

    1 Introduction   293
      1.1 What Are Grid Oscillators?   293
      1.2 Figures-of-Merit for Grid Oscillators   296

    2 Overview   298
      2.1 Two-Terminal Grid Oscillators   299
      2.2 Three-Terminal Grid Oscillators   301
      2.3 Comparison of Reported Grids   307

    3 Analysis Techniques   310
      3.1 The Induced EMF Method for Planar Grids   311
      3.2 Generalized Full-Wave Analysis   314
      3.3 Verification of the Grid Models   317

    4 Power Optimization   320

    5 Cascaded Grids   322
      5.1 Voltage-Controlled Grid Oscillator   322
      5.2 Dual-Frequency Grid Oscillator   325
      5.3 Three-Dimensional Grid Oscillator   325

    6 Conclusion   327

    References   327

9 Grid Amplifiers — 331
Michael P. De Lisio and Cheh-Ming Liu

1 Introduction and Background — 331

2 Modeling — 337
   2.1 Gain Modeling — 337
   2.2 Stability Modeling — 340

3 A 100-Element Hybrid pHEMT Grid Amplifier — 342
   3.1 Grid Construction — 342
   3.2 Gain — 346
   3.3 Angular Dependence — 353
   3.4 Noise — 356
   3.5 Power — 358

4 A Monolithic HBT Grid Amplifier — 361
   4.1 Grid Construction — 361
   4.2 Gain — 362
   4.3 Angular Dependence — 365
   4.4 Power — 365

5 A Monolithic pHEMT Grid Amplifier — 367
   5.1 Grid Construction — 367
   5.2 Gain — 368
   5.3 Tuning Range — 370

6 Conclusions — 373

Acknowledgments — 373

References — 373

10 Beam-Control Arrays — 377
Karl D. Stephan

1 Background — 377
   1.1 Passive Grids for Millimeter-Wave Beams — 378
   1.2 Active "RADANT" Grids for Microwave Beams — 380

2 Active Grid Arrays: Basic Principles of Operation — 381
   2.1 Passive Square Mesh on a Dielectric Interface — 381
   2.2 Active Mesh: Reflection Design — 383
   2.3 Active Mesh: Transmission Design — 385
   2.4 Active Mesh: Phase Shift Design — 386

3 Advantages and Limitations of Beam-Control Arrays — 387
   3.1 Electrical Limitations — 388
   3.2 Mechanical Limitations — 392

  4 Examples of Beam-Control Arrays   394
   4.1 Switching Hybrid PIN-Diode Array for 94 GHz   394
   4.2 Switching and Phase Shifting Monolithic Varactor-Diode Array at 60 GHz   399
   4.3 Phase-Shifting Monolithic Varactor-Diode Array at 94 GHz   401
  5 Conclusion   406
  Acknowledgments   407
  References   407

11 Frequency Conversion Grids   **409**
Jung-Chih Chiao

  1 Motivation   409
   1.1 Application   409
   1.2 Sources   410
  2 Waveguide Multipliers   411
  3 Quasi-Optical Grid Multipliers   415
   3.1 Concept   415
   3.2 Advantages   416
   3.3 Achievements   416
  4 66-GHz Frequency Doubler Grid   417
  5 99-GHz Frequency Tripler Grid   420
  6 THz Frequency Doubler Grid   422
   6.1 Planar Schottky Diodes   423
   6.2 The 6×6 Diode-Grid Arrays   425
   6.3 Design Approach   427
   6.4 Measurements   435
   6.5 Nonlinear Analysis   444
  7 Sideband Generator   446
  8 Conclusion   449
  Acknowledgments   450
  References   450

12 Quasi-Optical Subsystems   **455**
Zoya B. Popović and Gerald Johnson

  1 Introduction   455
  2 Transmitting Quasi-Optical Subsystems   458
   2.1 Two-Level Power Combining   458

		2.2 Beam Steering Using a Lens Amplifier	463
		2.3 Beam Forming Using a Lens Amplifier	469
	3	Receiving Quasi-Optical Subsystems	470
		3.1 Self-Oscillating Grid Mixer	471
		3.2 Receiving Lens Amplifier	473
		3.3 Quasi-Optical Receiver with Diversity	474
	4	Some Other Components for Quasi-Optical Subsystems	476
		4.1 A Quasi-Optical Linear-to-Circular Polarizer	476
		4.2 A Quasi-Optical Isolator/Directional Coupler	478
		4.3 Quasi-Optical Modulators	479
	5	What Needs To Be Done?	481
		5.1 Application Example—Space Communications	481
		5.2 Work To Be Done	482
		References	483
13	**Commercial Applications of Quasi-Optics**		**485**
	Richard C. Compton, Mehran Matloubian, and Mark J. Vaughan		
	1	Introduction	485
	2	Power Requirements	486
		2.1 Modulation Schemes	486
		2.2 Spectral Efficiency	487
	3	Device Technologies	488
		3.1 Diodes	489
		3.2 HBTs	491
		3.3 FETs	494
		3.4 Device Comparison	496
	4	Oscillator Arrays	498
		4.1 Array Phase Noise	498
		4.2 Modulation	502
		4.3 Omniazimuthal Arrays	505
	5	Amplifier Arrays	508
		5.1 Nonlinearities in Arrays	508
		5.2 Heat Dissipation	508
	6	Imaging and Receiver Applications	509
	7	Realization of Quasi-Optical Arrays	512
		References	514

Index   **523**

Contributors

Carl L. Brockman
Lockheed Martin Electronics and Missiles
Orlando, FL 32819-8907

Heng-Chia Chang
Department of Electrical and Computer Engineering
University of California
Santa Barbara, CA 93106

Siou Teck Chew
Department of Electrical and Computer Engineering
University of California
Los Angeles, CA 90024-1594

Jung-Chih Chiao
Department of Electrical Engineering
University of Hawaii at Mānoa
Honolulu, HI 96822

Richard C. Compton
Millimeter-Wave Wireless Laboratory
School of Electrical Engineering
Cornell University
Ithaca, NY 14853

Michael P. De Lisio
Department of Electrical Engineering
University of Hawaii at Mānoa
Honolulu, HI 96822

Mark A. Gouker
Lincoln Laboratory
Massachusetts Institute of Technology
Lexington, MA 02173-9108

John F. Hubert
Lockheed Martin Electronics and Missiles
Orlando, FL 32819-8907

Huan-Sheng Hwang
Department of Electrical and Computer Engineering
North Carolina State University
Raleigh, NC 27695-7911

Tatsuo Itoh
Department of Electrical and Computer Engineering
University of California
Los Angeles, CA 90024-1594

Gerald Johnson
Lockheed Martin Corporation
Denver, CO 80201

Cheh-Ming Liu
Rockwell International Science Center
Thousand Oaks, CA 91360

Jonathan J. Lynch
Hughes Research Laboratories, Inc.
Malibu, CA 90265-4799

Mehran Matloubian
Hughes Research Laboratories
Malibu, CA 90265-4799

James W. Mink
Department of Electrical
and Computer Engineering
North Carolina State University
Raleigh, NC 27695-7911

Amir Mortazawi
Department of Electrical Engineering
University of Central Florida
Orlando, FL 32816

Zoya B. Popović
Department of Electrical
and Computer Engineering
University of Colorado
Boulder, CO 80309-0425

Jon Schoenberg
Phillips Laboratory
Kirtland Air Force Base
Albuquerque, NM 87117

Wayne A. Shiroma
Department of Electrical
and Computer Engineering
University of Colorado
Boulder, CO 80309-0425

Emilio A. Sovero
Rockwell International Science
Center
Thousand Oaks, CA 91360

Michael B. Steer
Department of Electrical
and Computer Engineering
North Carolina State University
Raleigh, NC 27695-7911

Karl D. Stephan
Department of Electrical
and Computer Engineering
University of Massachusetts
Amherst, MA 01003

Mark J. Vaughan
Endgate Corporation
Sunnyvale, CA 94086

Robert M. Weikle II
Department of Electrical Engineering
University of Virginia
Charlottesville, VA 22903

Robert A. York
Department of Electrical
and Computer Engineering
University of California
Santa Barbara, CA 93106

Foreword

Quasi-optical components interact with free-space beams rather than waveguide modes or transmission-line voltages. In active quasi-optical circuits, diodes or transistors are distributed over a plane surface, and these devices interact with the beams. This is a radically unconventional approach to electronics, because there are no connecting wires or waveguides. The promise of active quasi-optics is greatly increased power from solid-state devices. Individually, solid-state devices may produce little power, but the outputs of many devices are combined with low loss in free space. In transmitters, this gives a large output power. In receivers, the input power is divided among many devices, increasing the saturation power and improving the dynamic range. Elegant quasi-optical approaches have been demonstrated for steering beams electronically. Quasi-optical amplifiers are multimode devices that can amplify beams at different angles, beams of different shapes, and even several beams simultaneously. The quasi-optical approach is particularly attractive at millimeter and submillimeter wavelengths, where many diodes and transistors can be fabricated simultaneously as a single monolithic circuit. This frequency range is a challenge for conventional integrated-circuit technology because the output power of solid-state devices decreases with frequency and the loss of conventional transmission-line combiners increases with frequency.

This is the appropriate time for a volume on quasi-optics. Many early problems with oscillator and amplifier stability and with monolithic fabrication have been overcome, and modeling has advanced sufficiently to allow a good comparison with experimental measurements of gain, radiation pattern, and power. At the same time, interest is developing in industry for commercial applications, as well as in the military for weapons systems. The editors, who have themselves made major contributions to the field of quasi-optics, have done an outstanding job of organizing the book and securing chapters from the leaders in the field.

The first experiments in active quasi-optics were carried out before the Second World War by Shintaro Uda, renowned as the inventor of the Yagi antenna. Uda was a professor of electrical engineering at Tohoku University in Sendai, Japan. In his book, *Short Wave Projector—Historical Records of My Early Days,* Uda describes many fascinating experiments. On pages 69–71, there is a 600-MHz oscillator consisting of seven vacuum tubes, a dipole antenna, and a reflecting mirror. In addition, he built a quasi-optical transmitter amplifier consisting of a linear array of nine vacuum tubes and eight dipole antennas alternately spaced along an open-wire transmission line. He showed that received power increased rapidly with the number of tubes and antennas, indicating both that transmitter power increased with the number of tubes and that directivity increased with the number of dipoles.

The roots of present active quasi-optics research go back to the early 1960s. In 1961, Georg Goubau and Felix Schwering at the U.S. Army's Fort Monmouth Laboratory showed that power could be transported in Gaussian beams with low loss by lenses and reflectors (G. Goubau and F. Schwering, "On the guided propagation of electromagnetic wave beams," *IRE Transactions on Antenna Theory and Propagation,* AP-9, pp. 248–255, 1961). In 1963, Roscoe George and E. M. Sabbaugh, at Purdue University, showed that closely spaced diodes in an overmoded waveguide could efficiently rectify a powerful microwave signal (R. H. George and E. M. Sabbaugh, "An efficient means of converting microwave energy to dc using semiconductor diodes," *IEEE International Convention Record, Electron Devices, Microwave Theory Techniques,* vol. 11, part 3, pp. 132–141, 1963). George and Sabbaugh produced 20 watts DC from 680 diodes with 30 watts input at 2450 MHz.

However, active quasi-optics has only recently become the dynamic research field it is today. In October 1993, the *IEEE Transactions on Microwave Techniques* published a special issue on Quasi-Optical Techniques. Starting the following year, the Microwave Symposium began devoting two sessions to quasi-optics. Most of the basic components in microwave receivers and transmitters have been demonstrated as quasi-optical circuits. These include amplifiers, oscillators, mixers, multipliers, and switches. Substantial powers in the 10-W range at 10 GHz have been achieved, and monolithic millimeter-wave amplifiers have been demonstrated up to 60 GHz. Quasi-optical oscillators and amplifiers have been tested in wireless communications circuits, transponders, and beam switching systems. Aggressive developers now have the opportunity to build quasi-optical systems for a new generation of communications and radar equipment.

We would like to express our particular appreciation to James Mink, formerly Director of the Electronics Division of the U.S. Army Research Office, and currently Professor of Electrical Engineering at North Carolina State University. Dr. Mink's paper, "Quasi-optical power combining of solid-state source," published in the *IEEE Transactions on Microwave Theory and Techniques* in February 1986, marks the beginning of the modern era of active quasi-optics. In addition, Dr. Mink developed and managed the U.S. Army Research Office program that has provided the major support and leadership for most of the significant advances in the field.

DAVID RUTLEDGE
JAMES HARVEY

California Institute of Technology
Pasadena, California

U.S. Army Research Office
Research Triangle Park, North Carolina

Preface

This book summarizes approximately one decade of work in the emerging field of quasi-optical power combining. During this time many researchers have contributed to the field, in the United States and abroad. Unfortunately, not everyone who has made a valuable contribution can be directly involved with a book project like this. We have chosen to focus on work carried out in the United States with which we are most familiar, and on a set of chapter topics which, while not exhaustive, represent most of the salient features of quasi-optical power-combining arrays and competing design philosophies. The individual chapter authors, all acknowledged leaders in the field and distinguished by their seminal work in various aspects of active and quasi-optical arrays, have made every effort to ensure that important contributions made by other groups have been well represented in the book and cited in the references. Our apologies for any inadvertent omissions; this is a rapidly changing and expanding field.

For those readers who have not closely followed developments in this area, we must emphasize that this book and the work contained herein would not have been possible without the vision, leadership, and mentoring provided by Professor David Rutledge at the California Institute of Technology (CalTech). Material from at least seven of the thirteen chapters can trace their origins back to his group at CalTech.

Several others and their associated organizations deserve recognition for further influencing, directing, and even subsidizing some of the seminal work. We would especially acknowledge Dr. Jim Mink and Dr. Jim Harvey, former and current directors, respectively, of the electronics division at the U.S. Army Research Office. Dr. Mink and Dr. Harvey have done an outstanding job of focusing much of the work effort while at the same time allowing the investigators considerable academic freedom to explore new avenues for research. Special thanks also go to Dr. J. Aiden Higgins of the Rockwell Science Center, Dr. Paul Greihling of the Hughes Research Laboratories, and Dr. Sander Weinreb at the

University of Massachusetts (formerly of the Martin Marietta Laboratories in Baltimore). In times of shrinking research budgets, the intellectual and financial support of these organizations has been invaluable.

ROBERT A. YORK
ZOYA B. POPOVIĆ

University of California
Santa Barbara, California

University of Colorado
Boulder, Colorado

CHAPTER ONE

Quasi-Optical Power Combining

ROBERT A. YORK
University of California, Santa Barbara

Active antenna arrays and active apertures, increasingly found in phased-array radar applications, are also potentially efficient power combiners. This observation is exploited in spatial or "quasi-optical" combining schemes for creating high-power, high-frequency components. These techniques are relatively frequency-independent, but are expected to have the greatest impact in the millimeter-wave to sub-millimeter-wave range, between 30 GHz and the low-terahertz region. Although this frequency regime is accessible with modern solid-state device technology, basic limitations of the devices present serious challenges to the development of millimeter-wave electronic systems. Spatial or quasi-optical power combiners provide a unique and compelling solution to these problems, but at the same time introduce new challenges. This chapter outlines motivations for power-combining and limitations of circuit-oriented techniques, attractive features of the quasi-optical or active array combiners, and various types of arrays and design issues that are subsequently the focus of later chapters. Commonly used analytical techniques and methods for characterizing the arrays are also provided.

Active and Quasi-Optical Arrays for Solid-State Power Combining, Edited by Robert A. York and Zoya B. Popović.
ISBN 0-471-14614-5 © 1997 John Wiley & Sons, Inc.

2 QUASI-OPTICAL POWER COMBINING

1 APPLICATIONS POTENTIAL AND TECHNOLOGICAL CONSTRAINTS

Motivations for exploiting the millimeter-wave frequency regime have long been recognized [1]. Millimeter-wave systems promise large bandwidths and significant reduction in size and weight compared to existing microwave electromagnetic systems. A variety of scientific, commercial, law-enforcement, and military applications exist or have been proposed which will depend upon or benefit from improvements in millimeter-wave technology. These include spectroscopic detection of atmospheric pollutants (ozone, carbon dioxide, methane, etc.) and chemical warfare agents, high-resolution imaging systems for low-visibility environments and optically opaque media (foul weather runway imaging, concealed weapons detection, low-altitude night-flying systems), and short-range covert communications systems and satellite cross-links. Some of these applications exploit the absorption characteristics peculiar to wave propagation in the atmosphere, shown in Fig. 1.1. Long-range communications and radar systems typically target the "windows" at 35, 94, 140, and 200 GHz, while short-range or covert systems make use of characteristic absorption peaks at 21, 60, 119, and 183 GHz.

The interests of the military have driven much of the millimeter-wave device and systems technology development in the United States. The U.S. Army, in particular, has a strong interest in millimeter-wave systems, projecting that approximately 80% of its future electronics systems will operate above 30 GHz

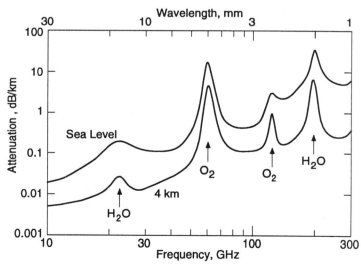

FIGURE 1.1 Measured atmospheric attenuation as a function of frequency. Attenuation "windows" are at 35, 94, 140, and 200 GHz, and absorption peaks at 21, 60, 119, and 183 GHz. The data represent attenuation in a typical atmosphere, and they can change significantly due to a number of environmental factors.

[2]. Military applications are many and include ground- or ship-based radars, missile seekers, and short-range battlefield communications, and they may operate at frequencies up to 94 GHz with required power levels ranging from tens to hundreds of watts.

Millimeter-wave frequencies have also been targeted for many civilian applications including automotive collision avoidance, blind-spot indicators, freeway tolling and surveillance, and other intelligent vehicle and highway systems. In some of these, available aperture is extremely limited (a few square inches), which requires operating frequencies up to or above 100 GHz for sufficient antenna gain. In addition to freeway tolling, many other commercial RF tagging or ID transponder concepts have been proposed to take advantage of the small wavelengths associated with millimeterwaves. These include grocery tagging systems for supermarkets, product tracking in manufacturing plants, and personnel monitoring in high-security areas. An example of such technology is discussed as a case study for integrated antennas in Chapter 3.

Millimeter-wave systems show strong potential for use in indoor wireless communications [3], or other environs subject to severe multipath distortion where the increased bandwidth is required to maintain robust communication links. In addition to the increased bandwidth and small size, such systems are preferred over hard-wired electrical or fiber-optic systems since they can be easily adapted to frequent changes in physical or geographical working environment, as is the case in modern manufacturing facilities, and are potentially more cost-effective when cable installation costs are considered. Overcrowding of the lower microwave spectrum also suggests a future demand for millimeter-wave communications links. Currently, cellular communications, satellite TV, police radar, and so on, are rapidly occupying the low microwave region. Chapter 13 of this book describes relevant technical issues associated with the role of active arrays in future wireless systems and possibilities for eventual commercial exploitation of the millimeter-wave regime.

These many possibilities have been slow to materialize, however, despite great advances in high-speed solid-state device and materials technology, much of which (in the United States) is a direct result of the (D)ARPA MMIC program [4]. GaAs PHEMT and InP-based HEMT technology is now relatively well established for applications up to 100 GHz, and improvements continue to be made. Above this frequency, two-terminal devices—IMPATT, Gunn, Schottky multiplier diodes and (possibly) RTDs—are potentially useful. Economies of scale are one reason for the slow growth of the millimeter-wave industry. However, the key impediment to solid-state-based millimeter-wave systems development is the limited power-handling capacity of semiconductor devices in this frequency range.

A comparison of average (CW) power for several common solid-state devices is shown in Fig. 1.2, alongside best available results from vacuum electronic devices. The shaded region corresponds to current or projected applications requirements for military and commercial systems. Based on Fig. 1.2, it is no surprise that vacuum electronics are still the dominant technology for power in

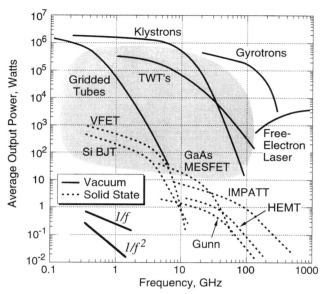

FIGURE 1.2 Power handling capacity (continuous) from a variety of millimeter-wave sources [6].

the millimeter-wave region. However, solid-state electronics are generally more desirable in terms of size, weight, reliability, and manufacturability. Economic considerations can also favor solid-state systems which can be mass produced using modern IC technology.

The limited power-handling capacity of the solid-state devices, along with the downward trend with frequency, is a fundamental limitation. The solid-state devices all exhibit an initial $1/f$ frequency dependence due to imperfect efficiency and subsequent generation of heat (the device active area shrinks with increasing frequency and therefore can dissipate less power for a given maximum allowable junction temperature), followed by a $1/f^2$ dependence which is essentially an electronic limitation imposed by impedance matching considerations [5]. The latter is a significant problem for modern three-terminal GaAs and InP-based devices operating in the millimeter-wave region. It is possible that wide band-gap semiconductors like silicon carbide (SiC) and gallium nitride (GaN), along with new device technologies such as vacuum microelectronics, could improve existing state of the art in power by an order of magnitude or more (vertical translation of the curves in Fig. 1.2). However, the downward trend with frequency will persist, and based on prior experience with compound semiconductors, it may take considerable time before the newer technologies reach the required level of maturity.

To satisfy system requirements, the power from many individual devices must be added coherently. Many circuit-based power-combining techniques have been devised, but these typically encounter difficulties at millimeter-wave frequencies where large numbers of devices must be combined and transmission losses are more conspicuous. Spatial or quasi-optical power-combining methods address these limitations. At the same time, spatial and quasi-optical techniques introduce new problems. To better appreciate the issues involved, it is useful to first understand some of the limitations of conventional power-combining techniques at millimeter-wave frequencies.

2 POWER-COMBINING CIRCUITS

At the device level, connecting a number of devices in parallel (or series, or both) is the most common and easiest way to increase power in a circuit. A multiple gate-finger HEMT is a simple example (Fig. 1.3). This lumped approach is broadband and essentially independent of the type of device (two terminal or three-terminal) and the manner in which the device is used (amplifier, oscillator, multiplier, etc.). Thermal and electronic (matching) considerations, as well as the influence of propagation delays that accrue as the device length or periphery increases, place limits on the number of devices that can be combined in this fashion.

Circuit or waveguide components can be added to more effectively parallel several devices with the correct phasing and impedance matching. There are many published techniques for doing so, and these have been extensively reviewed by Chang and Sun [7]; some of the representative concepts are reviewed here for convenience. At this level of combining it may become important to distinguish between the functional properties of the components—that is, amplifiers, oscillators, multipliers, and so on.

The "corporate" structure shown in Fig. 1.4a is a common combiner network found in planar circuits and antenna array feeds. The outputs from a number of circuits are successively combined using two-way adders such as Wilkinson combiners. The number of individual devices is $N = 2^K$, where K is the number of stages. The net output power is found to be $P_{\text{out}} = P_0 2^K L^K$ where P_0 is the output power per device and L is the insertion loss of each stage. The combining efficiency is therefore $\eta = L^K$. This is plotted in Fig. 1.4b for typical examples of insertion loss. As the frequency increases, the insertion-loss of circuit combiners typically increases at the rate of $f^{1/2}$ to $f^{3/2}$ depending on the transmission

FIGURE 1.3 An example of combining at the device level. This is a InP-based power HEMT with twelve 70-μm gate fingers. (Courtesy of Hughes Research Laboratories, Malibu, CA.)

6 QUASI-OPTICAL POWER COMBINING

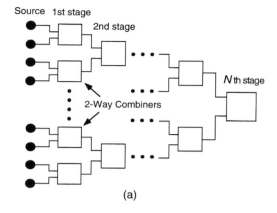

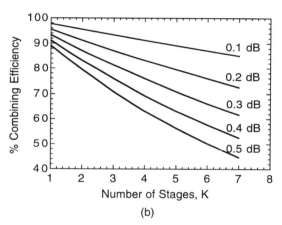

FIGURE 1.4 (a) A "corporate" combiner employing cascaded stages of two-way adders such as Wilkinson combiners. (b) Theoretical combining efficiency of the corporate structure for different levels of insertion loss for each two-way adder.

medium (see Fig. 1.6). The number of devices required to achieve a certain power objective also increases due to the $1/f^2$ limitation. As a result, the power-combining efficiency can become unacceptably low in the millimeter-wave region, and thus is limited to a small number of stages. A representative example combining four InP-based HEMT amplifiers is shown in Fig. 1.5.

Losses are therefore a key issue in large-scale power combining. Figure 1.6 contrasts losses associated with typical metallic and dielectric transmission-line structures and loss associated with propagation through the atmosphere [8]. Closed waveguide components typically have excellent performance in the low microwave and millimeter-wave region, and not surprisingly some of the most efficient combining strategies have been waveguide-based schemes. An example is the demonstration of a 400-W X-band combiner operating at 10–10.7 GHz, which combined forty 10-W GaAs MMIC power amplifier stages [9]. This

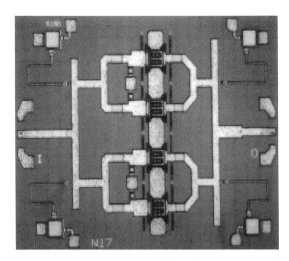

FIGURE 1.5 Example of a small power amplifier using "corporate" combiner/splitters based on Wilkinson circuits. (Courtesy of Hughes Research Laboratories, Malibu, CA.)

particular "chain" combiner is shown in Fig. 1.7; the operation is similar to a traveling-wave amplifier, where all of the amplifier input stages are fed by a common waveguide, and all of the output stages similarly feed a common waveguide. To ensure that there is an equal power division to all the amplifiers on the input waveguide, cross-guide couplers with adjustable power coupling ratios are required for each amplifier stage.

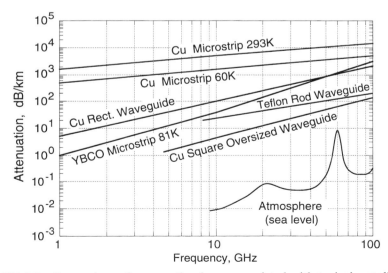

FIGURE 1.6 Comparison of propagation losses associated with typical metallic and dielectric transmission-line structures and loss associated with propagation through the atmosphere [8].

8 QUASI-OPTICAL POWER COMBINING

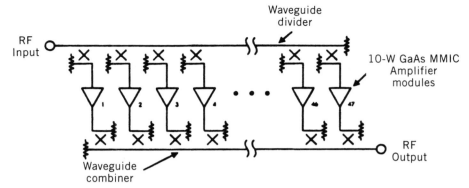

FIGURE 1.7 Waveguide-based chain combiner topology used successfully for a 400 W X-band power amplifier. (From reference [9], © 1995 IEEE.)

The above examples are most appropriate for power-combined amplifiers. Different schemes are required for combining multiple devices for high-power sources (oscillators), since the devices must be coupled in such a way to encourage and maintain mutual coherence. This is usually done by combining the devices in a common resonant cavity. A classic example is the Kurokawa wave guide cavity combiner [10], illustrated for a rectangular waveguide in Fig. 1.8.

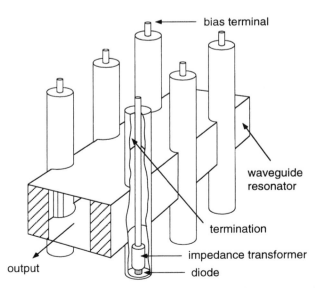

FIGURE 1.8 Kurokawa single-cavity multiple device combiner [10]. For large numbers of devices, multimoding problems can arise from distributing devices along the resonant dimension (see text).

The resonant cavity is periodically loaded with devices, each with its own impedance matching network and bias supply. The devices are terminated so as to present an equivalent negative resistance (or conductance) to form an oscillator or reflection amplifier, and they are typically spaced by $\lambda_g/2$ (where λ_g is the guide wavelength) to ensure the correct phasing and therefore constructive power summation. Combining efficiency for these configurations can be quite high, although ohmic losses are still an important limitation at very high frequencies. Excellent results have been at or below 44 GHz using variations of the cavity combiner and high-power IMPATT diodes on diamond heatsinks [11].

Manufacturing difficulties associated with tiny waveguide dimensions and a nonplanar hybrid topology make the Kurokawa combiner impractical for large-scale combining above 60 GHz. However, another important limitation arises with resonant combiners as the cavity dimensions are increased to accommodate more devices. The natural resonances of the structure can then become too close together, and "mode-hopping" or "multimoding" problems can develop. Suppression of undesired modes becomes increasingly difficult with increasing numbers of devices [12]. This problem is traced to distributing the devices primarily along a resonant dimension. Quasi-optical or spatial combiners address this problem, as well as ohmic losses, in a manner compatible with monolithic processing technology.

3 SPATIAL OR QUASI-OPTICAL ARRAYS

High losses associated with circuit combining schemes can be avoided using antenna arrays for power combining. An early demonstration of the strong potential for antenna-based power combiners was by Staiman et al. [13], who constructed a 100-W, 100-element amplifier array at 410 MHz (Fig. 1.9). Each amplifier fed a dipole antenna above a ground plane, with the dipoles interconnected and closely spaced, much like the Rutledge grids described later. This approach has also been employed at millimeter-wave frequencies with some success. Durkin et al. [14] described a 35 GHz "active aperture" using IMPATT amplifiers driving a printed slot array. Chang et al. [15] also reported a Ka-band array using GaAs MMIC amplifiers and tapered-slot antennas.

Figure 1.10 contrasts this "spatial" or "active array" approach (Fig. 1.10b) with the corporate transmission-line combiner (Fig. 1.10a). In principle, the power combining efficiency can be extremely high, since the energy is combined in a low-loss dielectric medium (air). Since combining losses are reduced, the technique is well suited to combining large number of devices or circuits, as required for millimeter-wave systems. Although the lossy feed network is retained, this is not disastrous, since any losses in the feed network can be compensated by preamplification of the input signal without seriously degrading overall system efficiency (Chapter 2). An added benefit of low-loss spatial combining is an overall noise reduction that improves with increasing numbers of devices. This reduction can be significant when large numbers of devices are combined, ranging from a dependence of $1/\sqrt{N}$ [16] to $1/N$ [10], where N is

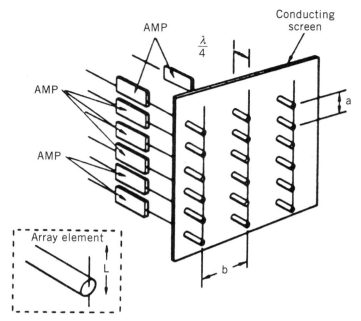

FIGURE 1.9 An early spatial combiner array topology using a short dipole array fed by solid-state amplifiers, which produced 100 W with 100 elements at 410 MHz. (From reference [13], © 1995 IEEE.)

the number of array elements. In addition, such arrays should degrade gracefully in the event of a device failure.

Some problems typically encountered with this approach are (a) imperfect radiation efficiency of planar antenna structures and (b) energy loss associated with the difficulty in collecting all of the generated output power. Chapter 2 discusses the circuit-fed spatial combiner in much greater detail and describes several case studies.

The spatial combiner concept can be further extended by replacing the circuit feed network by a spatial or "quasi-optical" feed, as shown in Fig. 1.10c. This approach exploits the tendency of any beam emerging from a source with a finite aperture to expand as it propagates (diffraction), thereby distributing the power in the beam over a larger surface. This approach is especially attractive at millimeter-wave frequencies where optical beam-guiding techniques become practical. In this case, lenses or reflective optics can be used to focus the input signal onto the array, and also focus the output energy to improve collection efficiency. The natural configuration for such a system is therefore a Gaussian-beam waveguide, with the array placed at a beam waist where the phase front is planar. An important challenge is then electromagnetic design of the array receiving and transmitting surfaces and associated active circuitry to couple

SPATIAL OR QUASI-OPTICAL ARRAYS 11

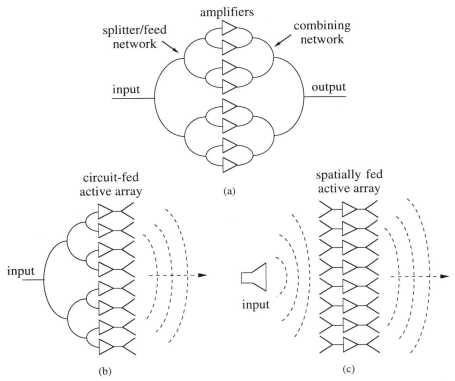

FIGURE 1.10 Evolution of spatial and quasi-optical combiners. (a) Circuit-fed/circuit-combined array using the corporate structure of Fig. 1.4. (b) Spatial combiner using a circuit-fed active array. (c) Spatially fed/spatially combined or "quasi-optical" combiner.

efficiently to the dominant mode. Another significant problem is associated with nonuniform excitation of the array. Chapters 5 and 6 address this problem further.

Several system functions can be accommodated using this combining scheme. Figure 1.11 illustrates some possibilities for quasi-optical combiners in a Gaussian beam environment (see Goldsmith [17] for background material on quasi-optical systems). Both transmissive and reflective arrays are shown. In most cases, isolating input and output signals can be a challenge since the receiving and transmitting sides of the array may be electromagnetically coupled. A common solution for transmissive arrays is illustrated in Fig. 1.11, where external wire-grid polarizers are placed on opposite sides of the array, and the input and output signals are orthogonally polarized. Reflective arrays are somewhat more complicated, since input and output signals must be coupled from the same side of the array. One possible solution would be to employ

12 QUASI-OPTICAL POWER COMBINING

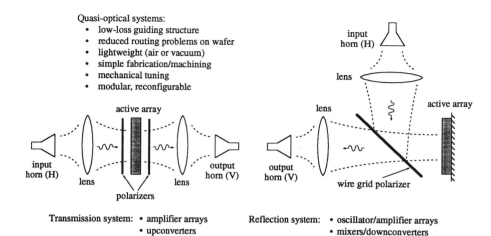

FIGURE 1.11 Methods for utilizing active array power-combiners in an open Gaussian-beam or quasi-optical system.

specular reflection at oblique incidence; this technique would permit reflection amplifiers to be cascaded for higher gain. Alternatively, use of orthogonal polarizations and a polarization diplexer (Fig. 1.11) or orthomode transducer [19] is possible. Faraday rotation devices [18] can also be used for both transmissive and reflective arrays if copolarized input and output beams are desired.

In later chapters the terms "spatial" and "quasi-optical" are used interchangeably. Some arrays very closely resemble ordinary antenna arrays, in which case the use of the term "quasi-optical" may seem gratuitous. In fact, the term has been somewhat abused. Though there is no formal definition, the term "quasi-optical" is usually understood to mean an electronic system that employs optical-beam-guiding components (lenses and/or shaped mirrors) for signal distribution/collection, like those of Fig. 1.11. It can certainly be argued that *any* antenna array, particularly if feeding a large lens or shaped reflector, is therefore quasi-optical. However, classical antenna arrays or spatial combiners differ from quasi-optical arrays in the use of circuit-based feed networks to ensure mutual coherence between the array elements. A quasi-optical array can therefore be defined as *an active antenna array that employs methods traditionally associated with optical systems for achieving mutual coherence between array elements, as well as optical methods for power collection*. Similarly, Gouker (Chapter 2) suggests distinguishing between "circuit-fed" and "spatially fed" combiners; a quasi-optical combiner is therefore a "spatially fed/spatially combined" array.

In either case, the distinction is still tenuous. Many of the quasi-optical array concepts described in the book can also be designed for use in closed metallic waveguide (Chapters 5 and 6) and surface waveguides (Chapter 7). Indeed, it is quite possible to employ optical techniques in a planar circuit environment like

SPATIAL OR QUASI-OPTICAL ARRAYS   13

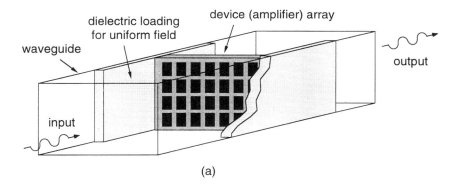

(a)

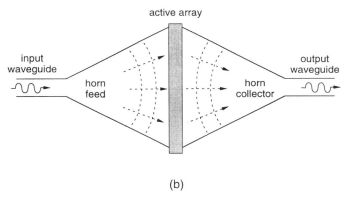

(b)

FIGURE 1.12 Active arrays can be used in closed waveguide structures. (a) Planar array in a standard rectangular waveguide with dielectric loading for uniform excitation. (b) Active array in a flared waveguide (horn) to accommodate more devices, which requires additional measures to compensate for spherical phase front.

microstrip or stripline, with the Rotman lens feed [20] being a perfect example. The term "quasi-optical" should therefore not be taken to simply imply the existence of dielectric lenses or curved mirrors, but rather a qualitative *methodology* based on wave interference and diffraction that is distinct from conventional transmission-line systems.

Closed waveguide-based array combiners (Fig. 1.12) may be attractive at frequencies below 100 GHz for several reasons: Diffraction losses and focusing errors are minimized or eliminated, since all of the energy is confined by the waveguide walls; the metal walls provide a convenient heatsink for the arrays; waveguide components are readily available in this frequency range and may be more economical than quasi-optical components; and using quasi-optical arrays in this way allows them to be retrofitted into existing waveguide systems. The nonuniform field profile in the waveguide means that edge elements on the array do not couple efficiently to the waveguide mode. This can be addressed by

14 QUASI-OPTICAL POWER COMBINING

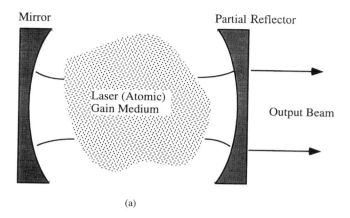

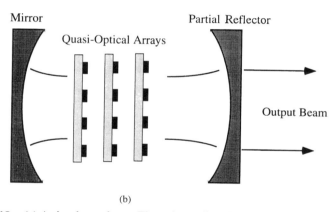

FIGURE 1.13 (a) A simple gas laser. The gain medium is a collection of tiny amplifying devices with integral antennas (excited gas molecules), with feedback provided by external mirrors. (b) Analogous quasi-optical oscillator using planar active arrays in a semi-confocal resonator.

using dielectric loading (Fig. 1.12a) or sidewall corrugations, as discussed in Chapters 5 and 6. Furthermore, standard waveguide dimensions may not accommodate a sufficient number of devices unless the waveguide cross section is enlarged. A simple illustration of this is shown in Fig. 1.12b. To compensate for the spherical phase front at input and output, either internal lenses may be employed, or the relative propagation delay through each array element can be manipulated. The latter is described in Chapter 5. Higher-order mode suppression techniques may also be necessary in oversized waveguides.

The quasi-optical approach to component design is quite logical when viewed from the perspective of optical systems. One would anticipate that as frequencies approach the infrared, systems would increasingly resemble their optical counterparts. In fact, this suggests a fruitful approach to millimeter-wave component development: extrapolate successful optical systems or techniques downward in frequency to address problems or difficulties arising from pushing conventional electronic circuit techniques upward.

For example, consider a simple gas laser system, as shown in Fig. 1.13a. Each gas molecule (when properly excited) delivers energy to an incident wave in much the same way as a negative resistance injection-locked amplifier. The individual contribution of each "device" is microscopic, but a large amount of power is obtained from the ensemble and carried in a propagating beam whose transverse dimensions are set by the external focusing elements (the mirrors). The active devices are densely distributed over the *entire cavity volume*, and the combined power is not continuously guided by lossy metallic boundaries. A quasi-optical version suitable for millimeter-wave systems is shown in Fig. 1.13b. One or more planar active arrays, fabricated by standard integrated circuit techniques, replace the gas lasing medium. Since the devices are spread out transverse to the direction of beam propagation, the resonator (if present) can be operated in a low-order mode to avoid axial multimoding problems (refer to difficulties associated with waveguide cavity combiners described above).

Compatibility with existing millimeter-wave integrated circuit fabrication technology makes the approach particularly attractive, although integration of active devices with planar antennas presents many challenges to efficient power combining. The example illustrated in Fig. 1.13 also suggests that the quasi-optical power-combining approach scales to the highest frequencies that can be accessed by electronic devices and fabrication technology. At the lower frequency limit, the size of the optical components becomes a practical limitation, at which point waveguide or planar transmission structures must be used. The frequency range between 30 GHz and 1 THz is most likely to benefit from quasi-optical combining concepts.

3.1 Maximum Power Density

In the laser system of Fig. 1.13a, the molecules interact with the propagating beam very efficiently. This occurs because (a) the spacing between the "devices," d, is a small fraction of a wavelength, and (b) the cross-sectional diameter, D, of the "array" is large compared with a wavelength. That is, $d \ll \lambda \ll D$. Ideally, an active or quasi-optical array would come as close to this optical limit as possible, but several other engineering issues must also be considered which are affected by the unit cell dimensions and overall array area. Desired output power and efficiency, thermal management, available input power, circuit functionality, and biasing are examples.

The unit cell area of an active array combiner is proportional to the square of the wavelength, $(\alpha\lambda)^2$ for a square unit cell, or equivalently $(\alpha c/f)^2$, where c is the speed of light in air, f is the frequency, and $\alpha \leq 0.8$ for a typical array. Therefore, as the frequency increases, the unit cell area shrinks at approximately the same rate as the solid-state power limitation described in Fig. 1.2. Consequently the available power/unit cell area is roughly independent of frequency for a given device technology and array topology.

It would appear that arrays with small unit cells (small α) would produce more power for a given aperture and device size, but this is misleading since heat removal considerations require the device size to be scaled with area. In other words, the maximum achievable power density will be dictated by heat removal, which in turn favors devices with high *efficiency* for increased packing density. Assuming that array area or available aperture can be chosen freely, the absolute power per device is much less important than the efficiency. Once thermal limits are reached, or the finite size of the device precludes tighter packing density, the output power can be increased simply by increasing the aperture area. What makes this plausible is the capability for combining very large numbers of devices and a flexible beam diameter afforded by the quasi-optical approach. Low-power/high-efficiency devices also tend to have better manufacturing yields and lower noise figures. In the latter case, the advantage is compounded by the use of larger numbers of devices, as described earlier.

If the available aperture is fixed by application requirements, this favors use of a device technology with the maximum power per unit area *for the given thermal management scheme and circuit configuration*. Note that many device technologists will quote a power per unit area for their device without specifying the class of operation or a heatsinking configuration. Heat removal considerations may also strongly favor *reflective* arrays which have an integral groundplane or heatsink on the backside of the substrate, rather than *transmissive* or bidirectional arrays for which the output beam can travel through the substrate.

In practice, it may be difficult to achieve the maximum available power density for a specific device technology, since each unit cell must accommodate not only the active device but also one or more antennas, matching circuitry, and DC bias circuitry. Array components described in later chapters integrate these components in many different ways, exploring a variety of trade-offs with power density and circuit complexity. An additional complication is nonuniform excitation of the array, which may require the device density to vary across the aperture for optimum efficiency. As shown in Fig. 1.14, a typical millimeter-

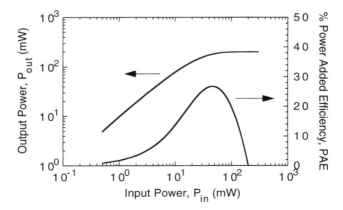

FIGURE 1.14 Power saturation and efficiency of a typical millimeter-wave amplifier.

wave amplifier has a well-defined optimum efficiency point which is achieved within a small range of drive power, so a Gaussian-illuminated array would require a Gaussian distribution of device density, or variable device size across the aperture. This in turn can exacerbate heat removal. In other types of waveguides it may be possible to create a more uniform excitation profile across the array using dielectric loading (Fig. 1.12a) or corrugated sidewalls. This scheme tends to be more narrowband, but is very attractive for quasi-optical power combining. This is discussed in Chapters 5 and 6.

DC biasing is particularly troublesome in large arrays, and it must be applied in such a way that a failure of one device or less-than-perfect fabrication yield does not lead to a catastrophic failure of all of the devices, as might be the case if the devices were biased in parallel. Independent biasing of each array element may be impractical in a large array unless a "tile" or "tray" approach is adopted (Chapter 2) which would provide additional "real estate" for bias networks. Bias isolation resistors must typically be avoided to maintain a high efficiency. Biasing is a difficult issue which is probably best considered on a case-by-case basis, and it is addressed in most of the chapters of this book.

3.2 Quasi-Optical Array Topologies

Integration of active devices with antennas has a long history, dating back to at least the mid-1940s (see Preface and Chapter 2). In the early 1960s, active dipole and slot antennas employing parametric devices ("PARANTs") and tunnel diodes were developed [21]. This approach was extended to spatial amplifier arrays, or "antennifiers" [22], and also to reflective arrays which used diode switches in a large pillbox array aperture for beam scanning [23]. Analytical work regarding scattering from active surfaces was also carried out during the early 1970s [24].

The development of planar antennas and high-performance microwave integrated circuit technology in the 1980s led to a strong interest in active arrays, primarily for compact and/or conformal military electronics. Rutledge et al. [25] published an important work on integrated antennas in 1983 (still widely cited) that includes several examples of planar antennas and arrays integrated with semiconductor devices, with an emphasis on devices for sub-millimeter-wave focal-plane receivers. Later papers by Wandinger and Nalbandian in 1983 [26], and by Mink in 1986 [27] suggested the marriage of active antennas and arrays with quasi-optical components for millimeter-wave power combining. This work emerged from the U.S. Army Research Office, which had previously carried out pioneering work on quasi-optical waveguiding systems in the 1960s [28]. Contemporaneously with Mink's classic paper, Camilleri and Itoh [29] also reported circuit-fed multiplier arrays, Lam et al. [30] reported quasi-optical diode grid arrays for frequency multiplication and beam-steering, and Stephan [31] reported coupled-oscillator arrays for power-combining. Subsequent work has largely originated from these early groups with seed funding from the U.S. Army Research Office. Significant industrial activity in this area began in the early 1990s.

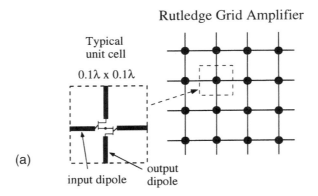

FIGURE 1.15 Comparison of amplifier arrays using (a) the Rutledge active grid concept (Chapters 8–12) and (b) active array concept (Chapters 2–7). Grids are formed from a tightly coupled metal mesh, with a small periodicity. Active arrays resemble conventional antena arrays.

A variety of array components have been devised, including amplifier arrays, oscillator arrays, frequency multiplier and mixer arrays, beam-switching arrays, and beam-scanning arrays. Most can be classified as a "grid" or an "active array"; the two approaches are contrasted in Fig. 1.15 for an amplifier configuration. The Rutledge grid concept is essentially an active Frequency-Selective Surface (FSS) formed from a tight metal mesh loaded with active or nonlinear devices, with a mesh period that is typically small compared to a wavelength. Alternatively, the active-array approach is an extension of classical antenna array systems, where each cell of the array is independent and contains a conventional planar antenna structure integrated with the appropriate circuitry, and the cell size is typically a half-wavelength square.

Both topologies could be equally well described as a grid or an active array, since they both involve periodic arrays of active devices in a radiating structure. The distinction is often made to distinguish between methods of analysis, or to

SPATIAL OR QUASI-OPTICAL ARRAYS   19

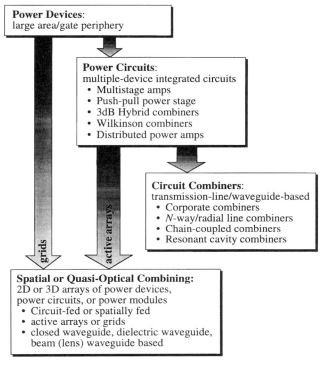

FIGURE 1.16   Power combining hierarchy.

emphasize the array periodicity or type of radiating structure. Grids typically involve elementary antennas with simple geometric shapes, analyzed using methods associated with infinite periodic structures (described in Section 4.1), whereas active arrays use common analytical techniques developed for planar circuits and phased-arrays. At the present time, no compelling motivations for choosing one approach over the other have been described; comparable results have been achieved using both types of arrays.

Similarities and differences are apparent in the examples of Fig. 1.15. Both configurations use orthogonally polarized input and output antennas for stability. Grid amplifiers typically use a differential amplifier circuit coupled to elementary dipoles (see Chapter 9). In the active array approach, a variety of planar antenna structures are possible, and there is more available area in the unit cell for on-wafer active circuitry. This does not necessarily imply a difference in power density (see previous section), but it does suggest greater compatibility with existing MMIC power circuits and simulation tools. Since active arrays accommodate somewhat larger circuits in a unit cell, this suggests the combining hierarchy shown in Fig. 1.16. Additionally, the phase delay between input and output antennas can be independently controlled within each array cell by increasing the path length. This raises the possibility of creating an "active lens" to compensate for the phase errors associated with feeding the array with a divergent beam (Chapter 5).

Biasing in both cases is difficult. For diode-multiplier (Chapter 11) or beam-control arrays (Chapter 10), the inherent interconnections of the grid approach lends itself to biasing. For other active-device arrays, however, out-of-band stability makes biasing equally complicated in both the grid and active array approach.

Although the grid unit cell contains an electrically small radiator, the disadvantages normally attributed to such antennas (poor radiation efficiency, difficult matching problems, limited bandwidth) are not relevant in this case, since the tight coupling to neighboring array elements (mutual coupling) strongly affects the circuit properties. An alternative viewpoint based on equivalent waveguide concepts is discussed later. Since the grid cell is small, the range of impedances that can be synthesized is typically somewhat limited, and there is little available room for additional matching elements on the substrate. However, external quasi-optical tuning elements (dielectric plates) can be used in place of on-wafer networks, and they can be adequately modeled in a single-mode system using transmission-line theory (Chapters 8–11).

The grid concept evolved from work on focal-plane detector arrays by Rutledge and Schwarz [32]. It is interesting to note that the concept was also independently conceived around the same time in connection with THz spectroscopy of semiconductor heterostructures and probing field-transport mechanisms in high-frequency devices [33]. In the limit of very small unit cells, the grid concept would more closely approximate the laser gain medium of Fig. 1.13a. If the devices are terminated in such a way as to create a negative resistance at the frequency of interest, the grid can then be viewed as a quasi-continuous conductive sheet with a negative sheet resistance, and hence it is capable of amplifying an incident electromagnetic beam. If the grid is placed before an external mirror to provide positive feedback, oscillation can occur. Grid oscillators based on this concept are discussed in Chapter 8.

Quasi-optical oscillators based on the active array approach have also been developed. The first example was suggested by Mink in his 1986 paper [27], consisting of an active array in a semiconfocal Fabry–Perot cavity. Chapter 7 describes a similar arrangement using a planar dielectric Fabry–Perot cavity. Stephan [31] suggested a more unusual approach using coupled-oscillators. Each oscillator feeds one element of the array, and mutual coupling between the elements (either through radiative interactions or on-wafer coupling networks) allows the oscillators to mutually synchronize through injection-locking [34]. Variations of this scheme are described in Chapters 4 and 6. Stephan also observed that phase relationships in coupled-oscillator systems can be manipulated in unusual ways that could be exploited for beam-scanning applications. This is also discussed in Chapter 4.

The organization of the book is illustrated in Fig. 1.17. Chapter 2 discusses spatial combining using circuit-fed architectures, which have a strong potential for the microwave and lower millimeter-wave regime. Chapters 4–7 cover active

SPATIAL OR QUASI-OPTICAL ARRAYS 21

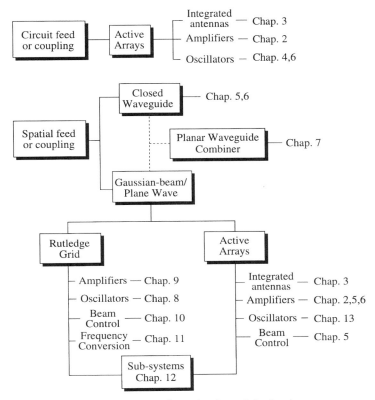

FIGURE 1.17 Organization of the book.

array oscillators and amplifiers, with Chapter 3 providing in-depth discussion of issues related to integrated antenna design and analysis for such arrays. Chapters 8–11 cover different types of grid arrays. Chapter 12 describes quasi-optical subsystem components which are created using combinations of grid and active array combiners. The last chapter of the book addresses issues related to fabrication of arrays and potential for commercial use in communications.

3.3 Planar Antennas for Arrays

An enormous variety of active antenna configurations have been developed which are suitable for active power-combining arrays. Many have been reviewed by Lin and Itoh [35]. Designs using both two- and three-terminal devices have been reported based on the patch antenna, coplanar-waveguide-fed slot antenna, tapered-slot antenna, and so on. Some common planar antenna structures are illustrated in Fig. 1.18; many more have been reviewed by Rebeiz [37]. Choice of antennas for active arrays is governed by many factors including ease of

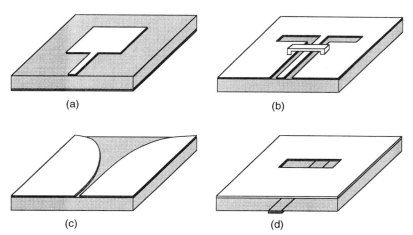

FIGURE 1.18 Common planar antennas for arrays: (a) patch antenna [36]; (b) CPW-fed slot [37]; (c) tapered-slot [38]; (d) microstrip-fed slot [39].

integration with active circuits, bandwidth, desired impedance for matching, input/output isolation, thermal management, and array topology ("tray" or "tile" approaches are discussed in Chapter 2).

The substrate plays an especially important role in planar antenna systems, particularly at millimeter-wave frequencies, markedly influencing the impedance, bandwidth, and radiation efficiency [39,40]. On electrically thick, high dielectric constant substrates (the typical situation for millimeter-wave antennas on semiconductor substrates), antennas suffer poor radiation efficiency as a result of surface-wave excitation. This is illustrated in Fig. 1.19a, showing the theoretical efficiency of a patch antenna on two different substrate materials as a function of substrate thickness. There are two competing effects in the figure: for patches on thin substrates, skin-effect losses limit efficiency, while thick substrates encourage substrate mode propagation. Even at the point of optimum efficiency, a significant amount of energy is lost to surface waves.

In an array, the surface-wave problem is further complicated by the relative phasing and spacing of the array elements. For certain phase distributions, substrate modes may not be strongly excited, while other phase distributions may resonantly enhance the substrate excitation; this leads to the phenomenon of scan-blindness in planar phased arrays [41], and would be disastrous in a power-combining array. Since the possibility of substrate mode excitation is nearly always present to some degree, there is a tendency to blame most problems with active arrays on surface waves or mutual coupling, but this is difficult to verify without careful electromagnetic analysis.

Two possible solutions to substrate mode problems are (1) to employ substrate alterations which inhibit propagation of surface waves and (2) to use a hybrid circuit fabrication technology (flip-chip, epitaxial transfer, etc.) whereby

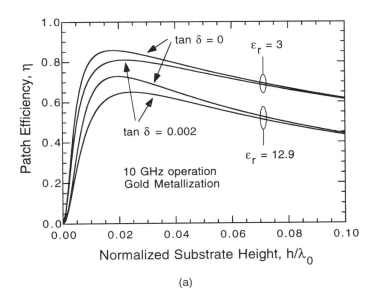

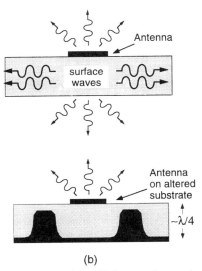

FIGURE 1.19 (a) Theoretical radiation efficiency of a patch antenna on two common substrate materials. For thin substrates, skin-effect losses limit the efficiency, while surface wave excitation limits the efficiency for large thickness. (b) Surface wave excitation might be controlled by altering the substrate [42].

the active devices on semiconductor substrates are integrated with antennas on a lower dielectric constant material. The first case is depicted in Fig. 1.19b. Photonic band-gap (PBG) structures might also be employed [42]. The flip-chip manufacturing approach is described in Chapter 13 and is attractive for several reasons. The antenna substrate can be chosen freely to optimize bandwidth, radiation efficiency, thermal conductivity, and so on. Furthermore, the approach can be more cost effective, since antennas typically occupy a large area on the semiconductor substrates that could otherwise be used for active devices.

Antennas on transverse planar arrays (tile approach) are constrained in size and therefore are usually standing-wave or resonant structures. This leads to a bandwidth limitation for canonical structures like the patch, dipole, and slot antennas. There are several possibilities for addressing the bandwidth limitation, some of which are illustrated in Fig. 1.20. These include using different modes of operation of conventional antennas (such as the second resonance of a slot [37]), circuit matching networks, adding direct- or parasitic-coupled radiators (shown in Fig. 1.20 coplanar and stacked patches [36] and folded-slots [43]), or using a tray architecture that facilitates the use of traveling-wave antennas like the tapered-slot [38] (see Chapter 2).

For oscillator arrays, bandwidth is generally not an important issue. The antenna itself is often used as the resonator or frequency-selective feedback element in the oscillator circuit to save space, although this typically results in a low-Q oscillator even for highly resonant antennas, since they are by definition lossy structures (radiation). Monolithic quasi-optical slot oscillators for arrays have been demonstrated at frequencies up to 155 GHz and 215 GHz using

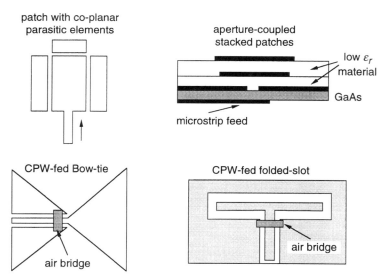

FIGURE 1.20 Some approaches to broadbanding planar antennas, involving use of parasitic or direct-coupled resonators or large traveling-wave structures.

pseudomorphic HEMTs [44]. Chapter 3 describes many active antenna designs and issues relevant to the design and analysis of integrated antennas for active arrays.

4 ANALYTICAL TECHNIQUES FOR QUASI-OPTICAL ARRAYS

Analysis of quasi-optical arrays involves (1) analysis of the passive array for a given excitation field and (2) an analysis of the device-circuit interaction. In both cases there are unique problems. Unlike ordinary antenna arrays, quasi-optical arrays may interact with Gaussian beams or other waveguide modes. Mutual coupling between array elements is significant in most designs and is influenced by a nonuniform excitation. Circuit design is complicated by possible strong electromagnetic coupling of the input and output beams, as well as by "load" impedances that are not broadband 50 Ω resistors but rather antennas with complex and highly frequency-dependent characteristics.

Additionally, coupling of nonlinear devices and circuits can lead to new (and interesting) nonlinear dynamical effects which are not easily modeled in conventional circuit simulators. Examples include mutual synchronization of array elements [45], spontaneous parametric oscillation [46], and chaotic dynamics [47]. Depending on the application, these may be desirable or undesirable effects, but in either case they must be carefully understood in order to exploit or avoid.

4.1 Passive Array Analysis

Finite active arrays and grids can be modeled using moment-method techniques developed for planar phased-arrays [41]. Alternatively, large transverse periodic structures can be approximated by infinite arrays, in which case analysis is reduced to a single unit cell in an equivalent waveguide, as shown in Fig. 1.21. For a certain class of symmetric structures excited by normally incident and vertically polarized plane waves, short-circuit boundaries (perfect electric conductors) on the top and bottom walls, along with open-circuit conditions (perfect magnetic conductors) on the side walls, properly image the unit cell to simulate the array. This approximation is suitable for large arrays in a wide-beam Gaussian system. Multimoding effects, which would be manifest as grating lobes and/or abnormally high sidelobe radiation, are prevented by keeping the cell and therefore the waveguide cross section significantly smaller than a wavelength. When only a single mode propagates, the array can be modeled as a shunt impedance across the equivalent waveguide. Once this array impedance is known, the equivalent circuit can be transferred to standard circuit design software (like EESof's LIBRA [56]) for simulating the behavior of active or nonlinear devices in the array.

26 QUASI-OPTICAL POWER COMBINING

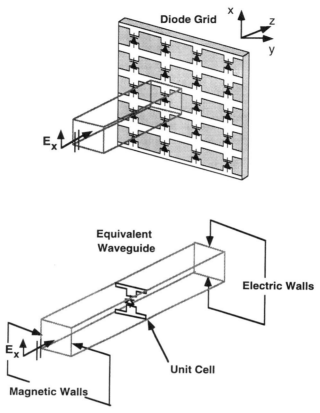

FIGURE 1.21 For uniform plane-wave excitation, array analysis can be reduced to an equivalent waveguide problem featuring a single unit cell. The symmetry in this case leads to a parallel-plate (TEM) equivalent waveguide.

A variety of analytical methods are available for treating the arrays using the unit cell technique. For electrically small unit cells with simple geometric shapes (grid arrays) where the current distribution can be guessed at, the induced EMF method is particularly attractive [48]. In this approach, the unknown transverse fields in the waveguide and the known currents on the grid are expanded in terms of the waveguide eigenmodes, $\hat{u}_{mn}$,

$$\bar{E}_t(x, y) = \sum_{m,n} e_{mn} \hat{u}_{mn}(x, y), \qquad \bar{J}_s(x, y) = \sum_{m,n} j_{mn} \hat{u}_{mn}(x, y) \qquad (1.1)$$

where e_{mn} are the unknown expansion coefficients (j_{mn} are known since $\bar{J}_s$ is known by definition). The magnetic fields are expressed similarly. The unknowns are then found by enforcing the boundary conditions at the array surface (standard mode-matching technique). The fields are then used to evaluate the

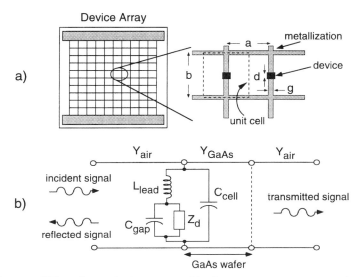

FIGURE 1.22 When the equivalent waveguide does not support propagation of high-order modes, equivalent circuits can be derived to characterize the arrays. (a) Unit cell of a simple diode grid. (b) Corresponding equivalent circuit for analysis.

induced EMF expression for impedance at some specified feed point,

$$Z = -\frac{1}{I_0^2} \iint_{\text{cell}} \bar{E}_t \cdot \bar{J}_s \, dS \tag{1.2}$$

where I_0 is the current at the feed point. As an example, the simple diode grid structure shown in Fig. 1.22a, which is assumed to have uniform current flow on the mesh for electrically small unit cells, has the following equivalent circuit parameters (defined in Fig. 1.22b):

$$C_{\text{cell}} = \frac{2a\epsilon_0}{b} \sum_{\substack{n=0 \\ n\,\text{even}}}^{\infty} \text{sinc}^2[k_y(b-g)/2](|k_z^{\text{GaAs}}| + |k_z^{\text{air}}|) \tag{1.3}$$

$$\frac{1}{C_{\text{gap}}} = \frac{2}{ab\epsilon_0} \sum_{\substack{m=0 \\ m\,\text{even}}}^{\infty} \sum_{n=0}^{\infty} \frac{(2-\delta_{m0})\sin^2(k_y d)\,\text{sinc}^2(k_x g/2)}{(k_x^2 + k_y^2)(\epsilon_r/|k_z^{\text{GaAs}}| + 1/|k_z^{\text{air}}|)} \tag{1.4}$$

$$L_{\text{lead}} = \frac{8\mu_0}{abg^2} \sum_{\substack{m=0 \\ m\,\text{even}}}^{\infty} \sum_{n=0}^{\infty} \frac{(2-\delta_{0n})(b\delta_{n0}-d)^2 \sin^2(k_x g/2)\,\text{sinc}^2(k_y d)}{(k_x^2 + k_y^2)(|k_z^{\text{GaAs}}| + |k_z^{\text{air}}|)} \tag{1.5}$$

where $k_x = m\pi/a$, $k_y = n\pi/b$, $k_z = \sqrt{\omega^2\mu\epsilon - k_x^2 - k_y^2}$ and where a is the cell width, b is the cell height, g is the lead width, d is the gap width, and δ_{mn} is the

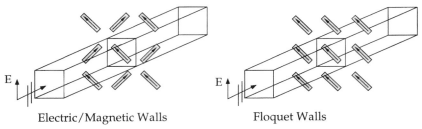

Electric/Magnetic Walls   Floquet Walls

FIGURE 1.23 Comparison of boundary conditions for array analysis.

Kronecker delta. These expressions are valid only when the equivalent wave guide supports TEM propagation, below the cutoff frequency $f_c = c/2a\sqrt{\epsilon_r}$. The active device is then inserted at the appropriate place, and the equivalent circuit of Fig. 1.22b is then analyzed using conventional circuit simulators. Chapters 8, 9, and 11 elaborate on this method for a variety of grid structures. Since the equivalent circuit model depends on the assumption of a uniform array excitation, the resulting circuit model cannot be expected to hold for self-excited arrays (i.e., oscillators) unless measures are taken to suppress undesired transverse modes. Substrate modes can be particularly troublesome for oscillator grids on high dielectric constant substrates [49].

Moment method and three dimensional finite difference time–domain (3D-FDTD) techniques have also been applied to analysis of passive grid structures. These are expected to be invaluable for active arrays with complicated unit cells and multilayer array geometries. In fact, only in special cases where the unit cells have the correct symmetry and the excitation conforms to this symmetry can the induced EMF method be applied. For arbitrary-shaped unit cells and excitation angles, Floquet boundary conditions must be used; the difference is illustrated in Fig. 1.23. Chapter 8 describes the application of moment-method techniques for parameter extraction in grids using such Floquet boundaries [50].

The FDTD method has been applied to both isolated antennas [51] and periodic grids [52]. An example demonstrating the potential of this technique for passive array analysis is shown in Fig. 1.24. A multilayer planar grid structure with an asymmetric unit cell, developed for possible use in an amplifier configuration [54], was analyzed using FDTD on an equivalent waveguide with Floquet boundary conditions [52]. Wire grid polarizers used in the measurement are incorporated into the FDTD analysis as shown in the unit cell of Fig. 1.24a. The longitudinal (z) direction is terminated in a Berenger PML absorbing boundary [53]. A comparison of the theory (no fitted parameters) and measurements in a Ka band (26–40 GHz) Gaussian-beam setup [54] is shown in Fig. 1.24b and shows excellent agreement over the band of operation.

4.2 Active and Nonlinear Circuit Analysis

Equivalent circuits can be extracted from passive array simulations for use in circuit simulators to facilitate the active component design. For simple grid

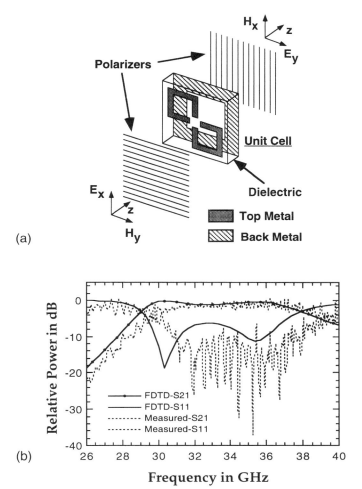

FIGURE 1.24 (a) Unit cell of the Kolias–Compton grid [54], and (b) preliminary comparison with an FDTD-grid code with no fitting parameters [52].

arrays analyzed using an equivalent waveguide model such as shown in Fig. 1.22b, the resulting active circuit can be easily analyzed using standard microwave CAD software such as EESof's LIBRA [56]. Cascaded grids can be similarly modeled, as described in Chapter 8. Chapter 9 describes interesting modifications of the transmission-line model that are required to simulate polarization-rotating amplifier grids like that depicted in Fig. 1.15. This is accounted for using two transmission-line circuits, one for each "mode" (polarization) of the system, which are coupled by the array.

Active array designs, on the other hand, are not always analyzed using an equivalent waveguide approach, since the larger array period reduces the influence of mutual coupling. A unit cell for an array amplifier, like those

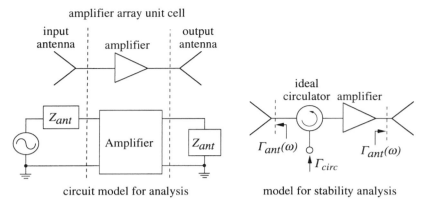

FIGURE 1.25 Typical active array amplifier unit cell, and equivalent circuits for analysis of frequency response and stability.

discussed in Chapters 5 and 6, is shown schematically in Fig. 1.25. Electromagnetic analysis of the antennas will yield information about self- and/or mutual impedances, labeled collectively as $Z_{ant}(\omega)$, which can be used for frequency-domain circuit analysis. The equivalent circuit is shown in Fig. 1.25, where the input antenna is replaced by a Thevenin equivalent, with an independent voltage source that is related to the incident field strength [64]. This example differs from standard microwave circuit analysis in that the source and terminating impedances are no longer broadband 50-Ω resistors but rather antennas with complex and highly frequency-dependent characteristics. Generalized scattering parameters [60], normalized to complex port impedances, could also be used.

In narrowband designs, the antenna can often be reduced to a simple series or parallel RLC network as depicted in Fig. 1.26a. In those cases, the reactive part of the impedance can be lumped in with the active circuit, and the terminating impedance (usually assumed to be real and frequency-independent) in the analysis software can be redefined as the radiation resistance of the antennas. This is shown in Fig. 1.26b. Even for narrowband designs, however, information about

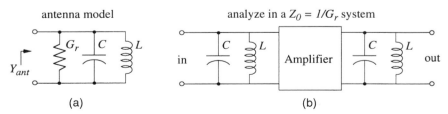

FIGURE 1.26 For narrowband applications the planar antenna can be adequately modeled as an RLC network. In this case, s-parameter analysis can proceed by adding the reactive part of the antenna impedance to the circuit and redefining the port impedance.

the out-of-band antenna impedance is required for stability analysis. This is especially significant because most planar antennas are resonant structures whose impedance trajectories can sample many regions in the complex impedance plane over a wide enough frequency range. A simple technique for stability analysis in these cases has been presented by Martinez and Compton [55], originally in connection with oscillator designs. This is shown in Fig. 1.25 and involves insertion of an ideal circulator function at the input or output of the amplifier. The antennas are modeled as arbitrary one-ports with a known reflection coefficient (or impedance) Γ_{ant}, which would come from an electromagnetic analysis. The input reflection coefficient, Γ_{circ} (the "circular" function [55]), is calculated over the entire frequency range where the active device has gain. A stable circuit requires that $|\Gamma_{circ}| < 1$ over this range of frequencies.

FDTD analysis is also attractive for analyzing active arrays because of the inherent ability to model nonlinear devices and the device-circuit interaction simultaneously. FDTD can handle difficult metallization geometries and dielectric inhomogeneities, while rigorously accounting for important effects of surface waves, ohmic losses, edge diffraction, and so on; active devices can be easily integrated into the analysis package using either a physics-based model or equivalent lumped circuits [57]; and FDTD is well suited to nonlinear array analysis since it is a time-domain simulator. FDTD is also an invaluable visualization tool. Rigorous analysis of two coupled integrated patch antennas using Gunn diodes has been simulated using FDTD [58], with excellent correlation between theory and experiment.

Another example is shown in Fig. 1.27, which shows an array of NonLinear Transmission Lines (NLTLs) using tapered-slot antennas to couple energy between the NLTLs and free-space. The array can be analyzed on the basis of a unit cell in a TEM waveguide (Fig. 1.21). An FDTD code was modified to model an NLTL comprised of 40 ideal reverse-biased abrupt-junction diodes in a 188.5-Ω slot transmission line, spaced 0.06 mm apart. A 10-GHz sinusoid was applied at the input of the equivalent waveguide, and the resulting waveform at the output is shown in Fig. 1.27. The waveform clearly exhibits shock wave formation, or a significant amount of energy at the higher harmonics. Also shown is a comparison with a SPICE simulation using an ideal transmission line, which verifies the FDTD implementation of the diode characteristics.

4.3 Nonlinear Dynamics

When nonlinear devices are coupled together in an array a number of unusual nonlinear phenomena are possible. These include mutual synchronization of array elements by injection locking [45], frequency multiplication, frequency division, or spontaneous parametric oscillation [46], and even chaotic dynamics [47]. Depending on the application, these may be desirable or undesirable effects, but in either case they must be carefully understood in order to exploit or avoid. The analysis of nonlinear dynamical phenomenon is still in its infancy,

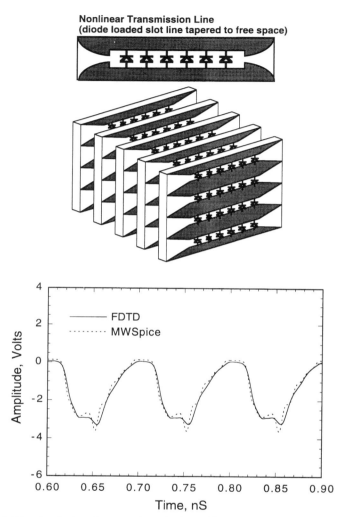

FIGURE 1.27 Analysis of a quasi-optical nonlinear transmission-line array using FDTD, and comparison to Microwave SPICE using a transmission-line model for the array.

but many interesting possibilities have been identified for electronic communication and control.

Oscillator array dynamics are one example. Mutual synchronization and/or external injection-locking of oscillator arrays was suggested for quasi-optical power combining by Stephan [31]. Coupled-oscillator arrays have been developed to demonstrate this concept at microwave frequencies, and some of this work is summarized in Chapter 4. Theoretical analysis of simple oscillator dynamics has revealed possibilities for simple beam-scanning technique, mode-locking for pulse generation analogous to a mode-locked laser, and chaotic communications. In addition, an understanding of oscillator dynamics is essential for predicting transient response of power-combining or beam-scanning arrays to determine bandwidth and/or scanning limitations, as well as noise properties. This work too is in its early stages, and practical issues such as amplitude dynamics and nonuniformities, frequency-dependent coupling networks, non-nearest-neighbor interactions, nonuniform tuning profiles of the oscillators, frequency-dependent device characteristics, and two-dimensional arrays have not been explored.

5 ARRAY CHARACTERIZATION AND FIGURES OF MERIT

An important but often neglected aspect of active array work is quantitative performance evaluation. Newcomers to the field may not appreciate the difficulties involved with active array measurements. These difficulties stem from a combination of (1) an inability to isolate radiative properties and active circuit performance from directly measurable quantities, (2) difficulties in directly relating individual unit cell performance to measured array (ensemble) properties, and (3) a lack of standardized evaluation criteria and calibrated measurement standards. These are further exacerbated by the high frequencies involved and, in the case of Gaussian beam systems, by problems with calibrating the measurement apparatus due to excitation of high-order beam modes. As a result, it has been difficult to draw meaningful comparisons between published active arrays and other types of circuit-based power-combiners, and also to diagnose problems if array performance objectives are not realized.

On the other hand, it is important to recognize that spatial combiners will often be utilized differently in systems than circuit-based counterparts—for example, as a focal plane feed for a reflector antenna (see Chapter 2)—and therefore may require a unique set of performance evaluation criteria that necessarily make comparison to conventional electronic combiners difficult. The primary goal of characterization is to assess performance relative to other active and quasi-optical arrays and to determine if the array fulfills larger system requirements when it is used correctly. Naturally, useful "figures of merit" should be easily determined from directly measurable quantities, should properly account for the loss mechanisms present in the arrays, and should reflect the manner in which the arrays are to be used.

Some work toward developing standard figures of merit for arrays has been published by Gouker [63], and important results from that work are included in this section. Several other important quantities, such as aperture efficiency and mode-coupling efficiency, are also described since they can impact the design of the array and the manner in which arrays are used and/or characterized. Two of the more common measurement techniques for sources and amplifiers, used in work described in later chapters, are also summarized here for convenience.

5.1 Aperture Efficiency and Polarizers

For most types of arrays, it is important to ensure that efficient use is made of the available aperture. This is characterized by the aperture efficiency, defined as the ratio of the effective area of the array to the physical area:

$$\eta_a = \frac{A_{\text{eff}}}{A_{\text{phys}}} \tag{1.6}$$

Normally it is desirable to maximize this efficiency, particularly in amplifier arrays where it is important to absorb all of the available incident power. It is instructive to examine this function quantitatively for a typical array. The effective area of any antenna is related to the directive gain, G, as [64]

$$A_{\text{eff}} = \frac{\lambda^2}{4\pi} G \tag{1.7}$$

where λ is the wavelength. The gain is in turn related to the directivity through [64]

$$G = \eta_r D \tag{1.8}$$

where η_r is the overall antenna efficiency including internal ohmic and dielectric losses, polarization and impedance mismatch losses, and so on. For simplicity it is assumed in the following that $\eta_r = 1$, and the antenna is oriented for beam incidence in the direction of maximum gain (typically the broadside direction). The gain is then equal to the maximum directivity, which is in turn given by

$$D = \frac{4\pi}{\oiint U_n(\theta, \phi)\, d\Omega} \tag{1.9}$$

where U_n is the normalized radiation pattern. For an array,

$$U_n = |F(\theta, \phi)|^2 \, |A(\theta, \phi)|^2 \tag{1.10}$$

where $F(\theta, \phi)$ is the element pattern, and $A(\theta, \phi)$ is the array factor. For an array in the x–y plane, with M antennas in the x direction spaced by d_x, and N antennas in the y direction spaced by d_y,

$$|A(\theta, \phi)|^2 = \left| \frac{1}{M} \frac{\sin(\frac{M}{2} k_0 d_x \sin\theta \cos\phi)}{\sin(\frac{1}{2} k_0 d_x \sin\theta \cos\phi)} \right|^2 \left| \frac{1}{N} \frac{\sin(\frac{N}{2} k_0 d_y \sin\theta \sin\phi)}{\sin(\frac{1}{2} k_0 d_y \sin\theta \sin\phi)} \right|^2 \tag{1.11}$$

Most planar antennas suitable for arrays have relatively small physical apertures and hence broad radiation patterns (low directivity). Consequently the directivity calculations are most strongly dependent on the size of the array, but the presence of external reflectors (polarizers, integral ground planes) also significantly influence the directivity and effective area. For illustrative purposes, consider an array of y-directed Hertzian dipoles, which are described by

$$F(\theta, \phi) = \sqrt{1 - \sin^2\theta \sin^2\phi} \qquad (1.12)$$

If the dipole is placed a distance h in front of an infinite ground plane so that it radiates into a halfspace, the element pattern is modified as

$$F(\theta, \phi) = \sin(k_0 h \cos\theta)\sqrt{1 - \sin^2\theta \sin^2\phi} \qquad (1.13)$$

The effective area of a square 10×10 array of Hertzian dipoles, computed from (1.6)–(1.13), is shown in Fig. 1.28 versus element spacing d, where $d = d_x = d_y$. Also shown for comparison are the aperture efficiencies for a uniformly excited aperture (here meaning an impressed current sheet) of the same physical size as the dipole array, which is taken to be $A_{\text{phys}} = (M d_x) \times (N d_y)$. Note that an ideal uniformly excited aperture has a 100% aperture efficiency ($A_{\text{eff}} = A_{\text{phys}}$) only when it is constrained to radiate into a half-space, as would be the case for an ideal reflector antenna.

For the dipole array radiating into freespace, the aperture efficiency is roughly 50% for the range of element spacings around $d = \lambda/2$, since half the

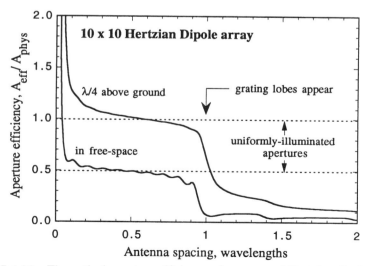

FIGURE 1.28 Theoretical aperture efficiency of an array of Hertzian dipoles in free space and above a ground plane (see Fig. 1.29), and comparison to uniformly illuminated apertures.

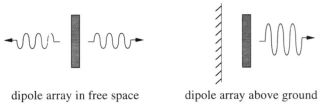

dipole array in free space   dipole array above ground

FIGURE 1.29 Direction characteristics of an array of dipoles.

available power is radiated into each half-space on either side of the array (Fig. 1.29). Consequently, if used as a receiver array, half of the incident power would be scattered away. When the array is spaced $\lambda/4$ in front of a mirror, however, the aperture efficiency effectively doubles, since the array is now constrained to radiate into (or receive from) a half-space. Figure 1.28 also emphasizes the requirement for element spacings less than one wavelength. The aperture efficiencies greater than 100% in the figure are not achieved in practice due to the physical size of most planar antennas, which would impose a lower limit on the range of antenna spacings. Furthermore, the curves represent an ideal situation in which additional loss mechanisms (polarization and impedance mismatch, dielectric and ohmic loss, spillover loss, etc.) have not been considered.

Based on these considerations, aperture efficiency is obviously linked to the choice of antenna in the unit cell. For certain types of antennas, like the dipole, slot, and the grid structures which radiate both above and below the array surface, external reflectors or lenses are *required* for efficient use of available aperture and efficient collection of output power. Figure 1.30 shows a few of the reported solutions to this problem. Figure 1.30a has been used in connection with patch antennas and microstrip circuits (see Chapters 5 and 6), and makes use of two microstrip antennas coupled through an aperture in the common ground plane. The ground plane otherwise decouples the incident and outgoing fields, which allows for copolarized input and output beams. Similarly, Fig. 1.30b uses a micromachined reflector antenna [65] fed by a printed dipole, with the two antennas linked by the active circuitry. A W-band monolithic amplifier cell has been demonstrated using this technique [66]. The primary tradeoff with these approaches is additional fabrication complexity associated with multilayer circuits, and also difficulties with modeling. Some prototype arrays using multilayer or double-sided circuits are described in Chapters 5 and 6.

Alternatively, many published results, including all of the grid amplifier work, has made use of orthogonal input and output polarization and external wire-grid polarizers for increasing the aperture efficiency. This is depicted in Fig. 1.30c. One useful feature of this approach is the allowance for mechanical adjustment of the polarizer location with respect to the array, which can be used for tuning (the polarizers are modeled as variable-length reactive stubs for each polarization; see Chapters 8, 9, and 11). A possible trade-off is bandwidth; the

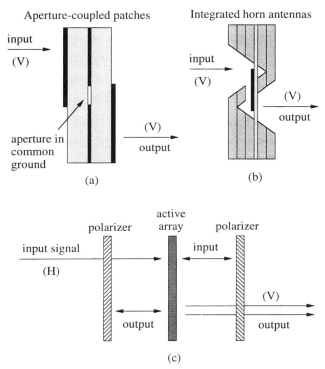

FIGURE 1.30 Published techniques for maintaining high aperture efficiency. (a) Aperture-coupled microstrip circuits (Chapters 5 and 6); (b) direct coupling using multilayer micromachining techniques [65,66]; (c) use of orthogonal polarizations and external polarizer components (Chapter 9).

polarizer-grid combination can create an undesirable resonance. This may not be a problem for narrowband applications (less than 10–15% fractional bandwidth) and can also be addressed using additional quasi-optical tuning elements or frequency-selective surfaces. It is important to emphasize here that the polarizers are not merely an option for arrays using bidirectional radiators, but are absolutely essential for obtaining the desired gain and combining efficiency.

5.2 Dominant-Mode Coupling

Many of the active arrays and grids described in the book are intended for eventual use in a guided-wave system, such as closed metallic waveguide, open dielectric waveguide, or more likely a Gaussian-beam waveguide. In each case, the array must be designed to couple effectively to the dominant mode of the system, which has important ramifications in the design and layout of the array. In a rectangular waveguide system, for example, the nonuniform field distribution in the TE_{10} mode may require nonuniform antenna spacing or tapering of device size to ensure maximum power-added efficiency and efficient coupling to the dominant mode. Alternatively, the guide can be dielectrically loaded or corrugated to create a uniform field, but coupling to higher-order modes is still a potential loss mechanism which must be addressed in the array design. Similarly in a Gaussian-beam system the array should couple efficiently to the lowest-order Gaussian beam mode, which would favor an array with an approximately circular aperture field or current distribution with a Gaussian amplitude tapering. Very few active array results to date have been specifically designed to optimize this coupling efficiency, but this is nevertheless an important practical objective.

In general the fraction of power, η_c, coupled from one beam to another or from a source distribution to a waveguide mode, is described by an overlap integral

$$\eta_c = \frac{|\langle \Phi | \Psi \rangle|^2}{\langle \Phi | \Phi \rangle \langle \Psi | \Psi \rangle} \qquad (1.14)$$

where Φ and Ψ are vectors describing the waveguide modes or aperture fields, and the inner product is defined as

$$\langle \Phi | \Psi \rangle = \iint_{-\infty}^{+\infty} \Phi \cdot \Psi^* \, dx \, dy \qquad (1.15)$$

in rectangular coordinates [59]. An arbitrary aperture field can be expanded into a set of the relevant waveguide modes. For closed waveguide systems, this is a standard textbook exercise [60]. For a beam waveguide, the relevant modes are the Hermite–Gaussian functions (see also Chapter 7). In this latter case each mode is doubly degenerate with $\hat{x}$ and $\hat{y}$ directed fields, assuming propagation in the $\hat{z}$ direction. Any two degenerate modes can be "combined" into one mode by rotating the x–y axes, leaving the field with components parallel to only one axis. This may not help if higher-order modes are important and are polarized in different directions. Usually we are interested only in the power coupled into the fundamental mode G_{00} with a well-defined polarization. The fraction of power coupled into G_{00} is then given by [61]

$$\eta_{00} = \frac{\left| \iint_{S_a} e^{jk(x^2+y^2)/2R_{ap}} e^{-(x^2+y^2)/w_{ap}^2} E(x,y) \, dx \, dy \right|^2}{\frac{\pi w_{ap}^2}{2} \iint_{S_a} |E(x_1 y)|^2 \, dx \, dy} \qquad (1.16)$$

where S_a is the aperture surface, $E(x, y)$ is the aperture field in the direction of interest and $\mathbf{E}(x, y)$ is the total field. R_{ap} is the radius of curvature of the equiphase surface at the origin (usually the center) of the aperture, where $x = y = 0$. For example, an array with the elements excited in phase has $R_{ap} = \infty$. R_{ap} for a conical horn is the distance from the imaginary apex of the horn to the

ARRAY CHARACTERIZATION AND FIGURES OF MERIT 39

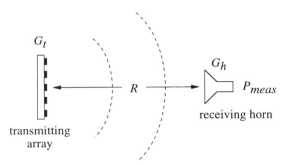

FIGURE 1.31 Measuring EIRP for transmitter arrays.

origin of the aperture. w_{ap} is the Gaussian beam radius at the aperture. This value is also an important design parameter for maximizing the coupling efficiency η_{00} [62]. Note that w_{ap} is *not* necessarily the beam waist (w_0) though it will be for most planar arrays designed for efficient coupling to a planar phase front.

5.3 Effective Radiated Power of Sources/Transmitters

The most basic, directly measurable figure of merit for transmitter arrays is the equivalent isotropic radiated power (EIRP, sometimes also written as ERP). The EIRP is is the product of the power generated by the transmitter, P_t, and the directive gain of the antenna array, G_t. The utility of this combination derives from its appearance in the Friis transmission equation [64], which is used for link calculations. For the transmitter configuration depicted in Fig. 1.31, the Friis transmission relates the measured power at the receiver under matched conditions, P_{meas}, to the transmitted power as

$$P_{meas} = P_t G_t G_h \left(\frac{\lambda}{4\pi R}\right)^2 \quad (1.17)$$

where G_h is the gain of the receiving antenna (usually a standard gain horn for measurements), R is the distance between the horn and the transmitting array, and λ is the operating wavelength. Naturally this formula is valid only when R is in the far-field of the transmitting array. Therefore a simple scalar measurement of received power determines the EIRP, given by

$$\text{EIRP} \equiv P_t G_t = \frac{P_{meas}}{G_h} \left(\frac{4\pi R}{\lambda}\right)^2 \quad (1.18)$$

The total RF power radiated by the array, P_{rad}, is also an important figure of merit, used in calculations of the DC–RF efficiency, system gain, and combining efficiency. In principle, this can be approximated by integrating over an experimentally determined 3D radiation pattern measurement. In some cases, good estimates can be made from principle pattern measurements. Note that the total radiated power is not the same as the total power generated by the array, P_t, since some energy is lost internally. This is accounted for in the previous results through the use of gain rather than directivity. Using (1.8) allows the radiated power to be computed in terms of the measured EIRP as

$$P_{rad} \equiv \eta_r P_t = \text{EIRP}/D_t \qquad (1.19)$$

where D_t is the directivity of the transmitter array. The directivity of the array, unlike the gain, can often be calculated to a high degree of accuracy from first principles. For a uniformly excited array, the directivity can be estimated from the physical aperture, as shown in Fig. 1.28. In addition, this formulation accounts for losses in the array such as unwanted phase and amplitude variations among the elements and the radiation efficiency of the antennas, since the directivity is calculated using the desired aperture distribution.

5.4 Gain of Beam Amplifiers

Beam amplifiers are to be used in guided-wave systems, and therefore they can (in principle) be characterized much like conventional two-port amplifiers. This is only possible when the arrays have been specifically designed to interact efficiently with a guided wave. Conventional antenna arrays can interact effectively with any beam cross section as long as the antenna spacing is sufficiently small, the layout reflects the symmetry of the beam, and the phase front is planar.

On the other hand, Gaussian beam measurement systems and other waveguide environments supporting multiple propagating modes can be difficult to calibrate. Scale models at lower frequencies are also difficult to characterize in guided-wave systems due to the large size of the components. Therefore, some of the experimental work described in later chapters makes use of a scalar far-field (plane wave) characterization setup, shown in Fig. 1.32. This technique is again based on the Friis transmission equation [equation (1.17)], and it requires two power measurements made with and without the amplifier array placed in the path of a simple point-to-point link. Using the notation in Fig. 1.32, the two measurements can be combined to give the result (for $R_1 = R_2 = R/2$)

$$\frac{P_{meas}}{P_{cal}} = \underbrace{G_r G_a G_t}_{\text{EIPG}} \left(\frac{\lambda}{4\pi R}\right)^2 \qquad (1.20)$$

where G_a is the power gain of the amplifier stage, and G_r and G_t are the directive gains of the receiving and transmitting surfaces of the array, respectively. The product of the power gain and array directive gains are analogous to the EIRP for transmitters, and have been identified in (1.20) as the effective isotropic power gain, (EIPG) [67]. This is an easily and directly measurable quantity, and

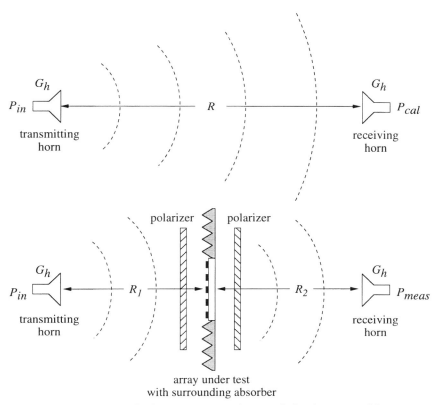

FIGURE 1.32 Far-field measurement of transmissive beam amplifiers.

hence a suitable figure of merit for comparison of arrays. The effective power gain of the array is naturally another important figure of merit. Like the effective power for transmitters [equation (1.19)], it is defined as

$$G_{\text{eff}} \equiv \frac{\text{EIPG}}{D_r D_t} \quad (1.21)$$

where D_r and D_t are the directivities of the receiving and transmitting arrays. This definition incorporates the effects of losses in the gain, and thus it is a more realistic basis for comparison to other arrays. In much of the grid amplifier work (Chapter 10), the directivities are estimated from the physical aperture of the array using (1.7) [68], giving

$$G_{\text{eff}} \approx \frac{\text{EIPG}}{A_{\text{phys}}^2} \left(\frac{\lambda^2}{4\pi}\right)^2 \quad (1.22)$$

This is a more conservative measure of the gain, since it effectively assumes 100% aperture efficiency. In connection with Fig. 1.28, this assumption is consistent with the use of polarizers surrounding the grid under test, as shown also in Fig. 1.30c. If polarizers are not used in the measurement, the effective gain computed from (1.22) would be too low by a factor of 6 dB.

5.5 Combining Efficiency

Combining efficiency describes the effectiveness of the combiner in summing all of the available power from the devices. The definition of combining efficiency should be consistent for active and quasi-optical arrays as well as for circuit combining approaches. In a quasi-optical array this is difficult to quantify since measurements of output power from each individual unit cell cannot be made separately. To this end, the combining efficiency is defined as the ratio of the total radiated power divided by the sum of the *maximum possible* power available from each of the devices in the array:

$$\eta_{\text{comb}} = P_{\text{rad}} \bigg/ \sum_{n=1}^{n} P_{n,\text{avail}} \qquad (1.23)$$

where N is the total number of power-generating stages in the array. The output power from any active device is dependent on the load impedance presented to it, and hence the maximum power available from each device, $P_{n,\text{avail}}$, is properly found through a load pull measurement, but idealized models for the active devices can also be used for useful estimates of combining efficiency [69]. This definition of the combining efficiency also correctly accounts for nonoptimum load impedances presented to the active device, and it precludes the possibility of combining efficiencies greater than 100% which can result when the available power is defined as the power radiated from a single antenna/active device pair (see Chapter 2, Section 3). Finally, the total efficiency of the systems is found by multiplying the combining efficiency by the device or circuit efficiency associated with each unit cell. In practice, this efficiency will most strongly determine overall system efficiency.

6 CONCLUSIONS

As the above discussions reflect, the design and analysis of active and quasi-optical arrays is an interesting and challenging multidisciplinary subject. Manufacturing considerations and device fabrication technology are additional topics of importance that have been left out of this chapter. Figure 1.33 underscores the challenges associated with quasi-optical array development. It is clear that a multidisciplinary team involving device, circuit, electromagnetic analysis, thermodynamic, and packing expertise is required for successful array development. The remaining chapters of this book each tackle a subset of these problems, but there is immense scope for future work in this area.

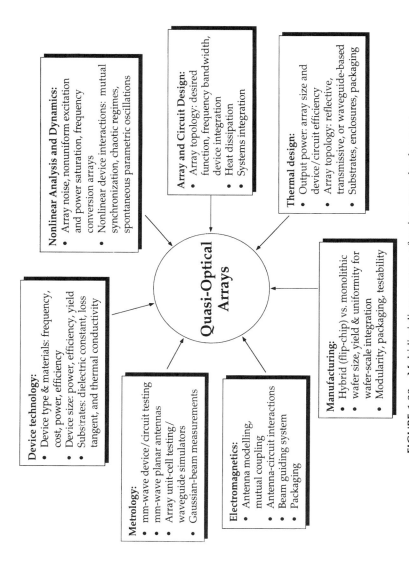

FIGURE 1.33 Multidisciplinary nature of active array development.

ACKNOWLEDGMENTS

The author thanks Dr. Mark Gouker at MIT Lincoln Laboratory (Chapter 2) for contributing material to the section on figures of merit, and he also thanks Eric Hufstedler at UC Santa Barbara for help with Gaussian beam coupling efficiencies.

REFERENCES

[1] P. Bhartia and I. Bahl, *Millimeter Wave Engineering and Applications,* Wiley, New York, 1984.

[2] Oral presentation by W. Gelnovatch, private (unclassified) quasi-optical workshop sponsored by U.S. Army Research Office, North Carolina State University, December 1995. This number is quoted with the permission of Dr. Gelnovatch.

[3] Y. Takimoto, "Recent activities on millimeter wave indoor LAN system development in Japan," *IEEE MTT-S Digest (Orlando),* pp. 405–408, 1995; H. H. Meinel, "Recent advances on millimeterwave PCN system development in Europe," *IEEE MTT-S Dig. (Orlando),* pp. 401–404, 1995.

[4] DARPA stands for the U.S. Defense Advanced Research Projects Agency (formerly ARPA). MMIC stands for Monolithic Microwave Integrated Circuits.

[5] S. M. Sze, *Physics of Semiconductor Devices,* 2nd ed., Wiley, New York, 1981, Chapter 10.

[6] Data adapted and updated from the following: K. J. Sleger, R. H. Abrams, and R. K. Parker, "Trends in solid-state microwave and millimeter-wave technology," *IEEE MTT-S Newsletter*, pp. 11–15, Fall 1990. Also S. Sze, ed., *High-Speed Semiconductor Devices,* 2nd ed., Wiley, New York, 1990; J. Browne, "Tubes continue the chase for power, gain, and bandwidth," *Microwaves & RF,* pp. 149–158, March 1990.

[7] K. Chang and C. Sun, "Millimeter-wave power-combining techniques," *IEEE Trans. Microwave Theory Tech.,* vol. 31, pp. 91–107, Feb. 1983. See also K. J. Russell, "Microwave Power Combining Techniques," *IEEE Trans. Microwave Theory Tech.,* vol. MTT-27, pp. 472–478, May 1979. M. Dydyk, "Efficient power combining," *IEEE Trans. Microwave Theory Tech.,* vol. MTT-28, pp. 755–762, July 1980.

[8] Data adapted from a variety of sources including: [1]; *Millimetre and Submillimetre Waves,* F. A. Benson, ed., ILIFFE Books Ltd., London, 1969; E. B. Ekholm and S. W. McKnight, "Attenuation and Dispersion for High-Tc Superconducting Microstrip Lines," *IEEE Trans. Microwave Theory Tech.,* vol. MTT-38, pp. 387–394, April 1990.

[9] M. Know et al., "400W X-band GaAs MMIC CW amplifier," *IEEE MTT Symposium Digest, (Orlando),* pp. 1605–1608, 1995.

[10] K. Kurokawa, "The single-cavity multiple-device oscillator," *IEEE Trans. Microwave Theory Tech.,* vol. MTT-19, pp. 793–801, Oct. 1971.

[11] J. W. McClymonds, Raytheon Research Division, Lexington, Massachusetts. Private communication, 1991.

[12] K. Fukui and S. Nogi, "Mode analytical study of cylindrical cavity power combiners," *IEEE Trans. Microwave Theory Tech.,* vol. MTT-34, pp. 943–951, Sept. 1986.

[13] D. Staiman, M. E. Breese, and W. T. Patton, "New technique for combining solid-state sources," *IEEE J. Solid-State Circuits,* vol. SC-3, pp. 238–243, Sept. 1968.

[14] M. F. Durkin et al., "35-GHz active aperture," *IEEE MTT-S Int'l Microwave Symp. Digest,* pp. 425–427, 1981.

[15] J. Chang et al., "Ka-Band Power-Combining MMIC Array," *15th International Conference on Infrared and Millimeter Waves (Orlando),* pp. 532–534, Dec. 1990.

[16] M. K. Sonmez, R. J. Trew, and C. P. Hearn, "Front-end topologies for phased array radiometry," *22nd European Microwave Cof. Digest,* Helsinki, Finland, pp. 1251–1256, Aug. 1992.

[17] P. F. Goldsmith, "Quasi-optical techniques at millimeter and sub-millimeter wavelengths," in *Infrared and Millimeter Waves,* vol. 6, Chapter 5, pp. 277–343, K. J. Button, ed., Academic Press, New York, 1982; P. F. Goldsmith, "Quasi-optics in radar systems," *Microwave J.,* pp. 79–100, Jan. 1991; P. F. Goldsmith, "Quasi-optical techniques," *Proc. IEEE,* vol. 80, pp. 1729–1747, Nov. 1992.

[18] S. J. Peters, "A compact Faraday rotator for beam waveguides," *International Symposium on Infrared and Millimeter Waves* (Orlando), pp. 699, 1991; also B. Lax, J. A. Weiss, N. W. Harris, G. F. Dionne, "Quasi-optical ferrite reflection circulator," *IEEE Trans. Microwave Theory Tech.,* vol. MTT-41, pp. 2190–2195, Dec. 1993.

[19] J. A. Benet, A. R. Perkons, and S. H. Wong, "Spatial power combining for millimeterwave solid state amplifiers," *IEEE MTT-S Dig. (Orlando),* pp. 619–622, 1993.

[20] W. Rotman and R. F. Turner, "Wide angle lens for line source applications," *IEEE Trans. Antennas Propag.,* vol. AP-11, pp. 623–632, Nov. 1963; see also D. H. Archer, "Lens-fed multiple-beam arrays," *Microwave J.,* vol. 27, pp. 171–195, Sept. 1984.

[21] A. D. Frost, "Parametric amplifier antenna," *Proc. IRE,* vol. 48, pp. 1163–1164, 1960; M. E. Pendinoff, "The negative conductance slot amplifier," *IRE Trans. Microwave Theory Tech.,* vol. MTT-9, pp. 557–566, 1961.

[22] C. H. Boehnker, J. R. Copeland, W. J. Robertson, "Antennaverters and antennafiers—unified antenna and receiver circuitry design," *10th Annual Symposium,* USAF Antenna Research and Development, Univ. of Illinois, Urbana, Illinois, 1960; J. D. Young, "Antennafiers for beam steering arrays," *14th Annual Symposium,* USAF Antenna Research and Development, University of Illinois, Urbana, Illinois, 1964; J. R. Copeland, W. J. Robertson, and R. G. Verstraete, "Antennafier arrays," *IEEE Trans. Antennas Propag.,* vol. AP-12, pp. 227-233, 1964.

[23] R. G. Malech et al., "The reflectarray antenna system," *12th Annual Symposium,* USAF Antenna Research and Development, University of Illinois, Urbana, Illinois, 1962.

[24] S. W. Lee and T. T. Fong, "Electromagnetic wave scattering from an active corrugated structure," *J. Applied Phys.,* vol. 43, no. 2, pp. 388–396, Feb. 1972.

[25] D. B. Rutledge, D. P. Neikirk, and D. P. Kasilingam, "Integrated circuit antennas," in *Infrared and Millimeter Waves*, vol. 10, K. J. Button, ed., Academic Press, New York, 1983, Chapter 1, pp. 1–90.

[26] L. Wandinger and V. Nalbandian, "Millimeter-wave power-combiner using quasi-optical techniques," *IEEE Trans. Microwave Theory Tech.*, vol. MTT-31, pp. 189–193, Feb. 1983.

[27] J. W. Mink, "Quasi-optical power combining of solid-state millimeter-wave sources," *IEEE Trans. Microwave Theory Tech.*, vol. MTT-34, pp. 273–279, Feb. 1986.

[28] J. R. Christian and G. Goubau, "Experimental studies on a beam waveguide for millimeter waves," *IRE Trans. Antennas Propag.*, pp. 256–263, May 1961; G. Goubau and F. Schwering, "On the guided propagation of electromagnetic wave beams," *IRE Trans. Antennas Propag.*, pp. 248–256, May 1961; G. Goubau, "Beam waveguides," *Advances in Microwaves*, vol. 3, L. Toung, ed. Academic, New York, 1968, pp. 67–126.

[29] N. Camilleri and T. Itoh, "A quasi-optical multiplying slot array," *IEEE Trans. Microwave Theory Tech.*, vol. MTT-33, no. 11, pp. 1189–1195, Nov. 1985; also S. Nam, T. Uwano, and T. Itoh, "Microstrip-fed planar frequency-multiplying space combiner," *IEEE Trans. on Microwave Theory Tech.*, vol. MTT-35, no. 12, pp. 1271–1276, Dec. 1987.

[30] W. W. Lam *et al.*, "Diode grids for electronic beam steering and frequency multiplication," *Int. J. Infrared Millimeter Waves*, vol. 7, no. 1, pp. 27–41, 1986.

[31] K. D. Stephan, "Inter-injection-locked oscillators for power combining and phased arrays," *IEEE Trans. Microwave Theory Tech.*, vol. MTT-34, pp. 1017–1025, Oct. 1986.

[32] D. B. Rutledge and S. E. Schwarz, "Planar multimode detector arrays for infrared and millimeter-wave applications," *IEEE J. Quantum Electronics*, vol. 17, pp. 407–414, March 1981.

[33] S. J. Allen, Jr. *et al.*, "Dispersion of the saturated current in GaAs from dc to 1200 GHz," *Appl. Phys. Lett.*, vol. 42, no. 1, pp. 96–98, January 1, 1983.

[34] K. Kurokawa, "Injection-locking of solid-state microwave oscillators," *Proc. IEEE*, vol. 61, pp. 1386–1409, Oct. 1973.

[35] J. Lin and T. Itoh, "Active integrated antennas," *IEEE Trans. Microwave Theory Tech.*, vol. 42, no. 12, pp. 2186–2194, Dec. 1994.

[36] J. R. James and P. S. Hall, eds., *Handbook of Microstrip Antennas* (two volumes), Peter Peregrinus Ltd., London, 1989.

[37] G. M. Rebeiz, "Millimeter-wave and terahertz integrated circuit antennas," *Proc. IEEE*, vol. 80, no. 11, pp. 1748–1770, Nov. 1992.

[38] K. S. Yngvesson *et al.*, "The tapered slot antenna—a new integrated element for millimeter-wave applications," *IEEE Trans. Microwave Theory Tech.*, vol. MTT-37, pp. 365–374, Feb. 1989.

[39] D. M. Pozar, "Considerations for millimeter-wave printed antennas," *IEEE Trans. Antennas Propagat.*, vol. AP-31, pp. 740–747, Sept. 1983.

[40] N. G. Alexopoulos, P. B. Katehi, and D. B. Rutledge, "Substrate optimization for integrated circuit antennas," *IEEE Trans. Microwave Theory Tech.*, vol. MTT-31, pp. 550–557, July 1983.

[41] D. M. Pozar, "Analysis of finite phased arrays of printed dipoles," *IEEE Trans. Antennas Propagat.*, vol. AP-33, pp. 1045–1053, Oct. 1985.

[42] E. R. Brown, C. D. Parker, and E. Yablonovitch, "Radiation properties of a planar antenna on a photonic-crystal substrate," *J. Opt. Soc. Am., B*, vol. 10, no. 2, pp. 404–407, Feb. 1993.

[43] H.-S. Tsai and R. A. York, "Multi-slot 50 Ω antennas for quasi-optical circuits," *IEEE Microwave Guided Wave Lett.*, vol. 5, no. 6, pp. 180–182, June 1995.

[44] S. E. Rosenbaum et al., "155 GHz and 215 GHz AlInAs/GaInAs/InP HEMT Oscillators," *IEEE Trans. Microwave Theory Tech.*, pp. 927–932, April 1995.

[45] R. A. York, "Nonlinear analysis of phase relationships in quasi-optical oscillator arrays," *IEEE Trans. Microwave Theory Tech.*, vol. MTT-41, pp. 1799–1809, Oct. 1993.

[46] R. J. Ram and R. A. York, "Parametric oscillation in nonlinear dipole arrays," *IEEE Trans. Antennas Propag.*, vol. AP-42, pp. 406–411, March 1994.

[47] R. J. Ram, R. Sporer, H.-R. Blank, P. Maccarini, H.-C. Chang, and R. A. York, "Chaos in microwave antenna arrays" (invited paper), presented at 1996 IEEE MTT-S International Microwave Symposium (San Francisco).

[48] R. M. Weikle II, "Quasi-Optical Planar Grids for Microwave and Millimeter-Wave Power Combining," PhD. dissertation; California Institute of Technology, Nov. 1991. Also see Chapter 8.

[49] D. W. Griffin, "Monolithic active array limitations due to substrate modes," *1995 IEEE Antennas Propaga. Soc. Symp.* (Newport Beach, CA), pp. 1300–1303.

[50] S. C. Bundy and Z. B. Popovic, "A Generalized Analysis for Grid Oscillator Design," *IEEE Trans. Microwave Theory Tech.*, vol. 42, no. 12, pp. 2486–2491, Dec. 1994; S. C. Bundy, W. A. Shiroma, and Z. B. Popovic, "Analysis of Cascaded Quasi-Optical Grids," *IEEE MTT Symposium Digest (Orlando)*, pp. 601–604, May 1995.

[51] H.-S. Tsai and R. A. York, "FDTD analysis of folded-slot and multiple slot antennas on thin substrates," *IEEE Trans. Antennas Propga.*, vol. AP-44, pp. 217–226, Feb. 1996.

[52] A. Alexanian, N. J. Kolias, R. C. Compton, and R. A. York, "FDTD analysis of quasi-optical arrays using cyclic boundary conditions and Berenger's PML," *IEEE Microwave Guided Wave Lett.*, vol. 6, no. 3, pp. 138, March 1996.

[53] J.-P. Berenger, "A perfectly matched layer for the absoption of electromagnetic waves," *J. Comput. Phys.*, 114, pp. 185–200, 1994.

[54] N. J. Kolias and R. C. Compton, "A microstrip based quasi-optical polarization rotator array," *IEEE MTT-Symposium Digest (Orlando)*, pp. 773–776, June 1995.

[55] R. D. Martinez and R. C. Compton, "A general approach for the S-parameter design of oscillators with 1 and 2-port active devices," *IEEE Trans. Microwave Theory Tech.*, vol. 40, pp. 596–574, March 1992.

[56] LIBRA and Microwave SPICE are are registered trademarks of EEsof, Inc., Westlake, CA.

[57] V. A. Thomas et al., "The use of SPICE lumped circuits as sub-grid models for FDTD analysis," *IEEE Microwave Guided Wave Lett.*, vol. 4, no. 5, pp. 141–143, May 1994; G. Massobrio, and P. Antognetti, *Semiconductor Device Modeling with Spice*, 2nd Edition, McGraw-Hill, New York, 1994.

[58] B. Toland, B. Houshmand, and T. Itoh, "Modeling of nonlinear active regions with the FDTD method," *IEEE Microwave Guided Wave Lett.,* vol. 3, no. 9, pp. 333–335, Sept. 1993.

[59] J. Lesurf, *Millimetre-Wave Optics, Devices & Systems,* Adam Hilger, London, 1990, p. 13.

[60] R. E. Collin, *Foundations for Microwave Engineering,* 2nd ed., McGraw-Hill, New York, 1992.

[61] G. V. Eleftheriades, "Analysis and design of integrated-circuit horn antennas for millimeter and submillimeter-wave applications," Ph.D. dissertation, The University of Michigan, pp. 78–80, 1993.

[62] R. J. Wylde, "Millimeter-wave Gaussian beam-mode optics and corrugated feedhorns," *IEE Proc.,* vol. 131, pt. H, no. 4, pp. 258–262, Aug. 1984.

[63] M. A. Gouker, "Toward standard figures-of-merit for spatial and quasi-optical power-combined arrays," *IEEE Trans. Microwave Theory Tech.,* vol. 43, pp. 1614–1617, July 1995.

[64] R. E. Collin, *Antennas and Radiowave Propagation,* McGraw-Hill, New York, 1990.

[65] C. Y. Chi and G. M. Rebeiz, "A quasi-optical amplifier," *IEEE Microwave and Guided Wave Lett.,* vol. 3, pp. 164–166, June 1993.

[66] T. P. Budka, M. W. Trippe, S. Weinreb, and G. M. Rebeiz, "A 75 GHz to 115 GHz quasi-optical amplifier," *IEEE Trans. Microwave Theory Tech.,* June 1993.

[67] H. S. Tsai, M. J. W. Rodwell, and R. A. York, "Planar amplifier array with improved bandwidth using folded slots," *IEEE Microwave Guided Wave Lett.,* vol. 4, pp. 112–114, April 1994.

[68] M. Kim et al., "A grid amplifier," *IEEE Microwave Guided Wave Lett.,* vol. 1, pp. 322–324, Nov. 1991.

[69] L. J. Kushner, "Output performance of idealized microwave power amplifiers," *Microwave J.,* vol. 32, pp. 103–116, Oct. 1989.

CHAPTER TWO

Spatial Power Combining

MARK A. GOUKER
Lincoln Laboratory, Massachusetts Institute of Technology

The preceding chapter introduced the circuit and spatial power-combining techniques to construct moderate power solid-state oscillators and amplifiers. Amplifier arrays require (1) a "feed" or splitter network to distribute the input signal to the individual amplifiers and (2) a combiner network to collect the power. The distribution and combining approaches are categorized here as either the *circuit* or *spatial* variety. This chapter will take a closer look at two of the four possible permutations of these approaches: (1) circuit-fed/spatially combined arrays and (2) spatially fed/spatially combined arrays. Various array architectures are described, and design trade-offs are discussed. The latter part of the chapter considers circuit-fed/spatially combined arrays in detail, with specific examples. Spatially fed/spatially combined arrays are discussed further in Chapters 5, 6, and 9.

Spatial power-combining was reported as early as 1968 [1] with the construction of a 100-element spatially fed/spatially combined array for operation at 410 MHz. In that work, one-watt bipolar transistors provided amplification between a pair of electrically short monopole antennas (one for receive and one for transmit). One of the first circuit-fed/spatially combined arrays was a 35-GHz radar transmitter reported in 1981 [2]. In this work, injection-locked pulsed IMPATT oscillators were used to drive a 32-element array (one oscillator per quadrant). Recently, quasi-optical approaches have dominated the research on spatial power combining and thus receive much attention in this book.

Active and Quasi-Optical Arrays for Solid-State Power Combining, Edited by Robert A. York and Zoya B. Popović.
ISBN 0-471-14614-5 © 1997 John Wiley & Sons, Inc.

However, more traditional antenna and microwave techniques are gaining interest in designs for spatial power-combining systems.

The circuit-fed/spatially combined amplifier arrays discussed in the last half of this chapter utilize traditional microwave practices. These arrays are characterized by distinct feed networks, amplifiers, and antennas. In many ways they resemble active aperture phased arrays, and in fact the architecture, methodology, and even components are taken from the modern phased array field. However, there is one important difference: The spatially combined array does not need electronic phase shifters. This seemingly small difference has tremendous impact in the simplification of the design and construction of the array.

1 FEEDING AND COMBINING APPROACHES

Four possible power-combining architectures are illustrated in Fig. 2.1. The circuit-fed/circuit-combined approach is the conventional approach for solid-state amplifiers. At microwave and millimeter-wave frequencies the two most common circuit types are microstrip and waveguide transmission lines. Microstrip transmission media offers the advantage of a smaller, more compact amplifier, but at the cost of lower combining efficiency. The reactive T-junction or Wilkinson-type dividers and combiners have greater loss than a waveguide T-junction. Waveguide dividers and combiners, on the other hand, offer lower loss but result in bulkier, heavier structures. In some applications a compromise is obtained by first combining in microstrip, forming a several-watt module, and then in waveguide through a multiport binary or radial combiner.

The circuit-fed/spatially combined design is the configuration of typical antenna arrays. In the spatial combining application, the array has a large number of solid-state amplifiers placed in circuit feed network, and their outputs are combined after conversion to free-space propagation via the antennas. Typically, every antenna is driven by an amplifier; however, there are viable system configurations where a single amplifier might feed several antennas. Under this definition an active aperture phased array is a spatial power combiner. While this is true, active aperture phased arrays are not considered in this chapter. These arrays are generally constructed for their ability to electronically steer the antenna pattern and thus have other priorities in their design. Furthermore, the unburdening of the design by eliminating the electronic phase shifters results in a number of advantages that make this configuration very attractive for power combining.

The spatially fed/circuit-combined approach is generally not used. This mixture of feeding and combining techniques pairs the disadvantages of both techniques. It is more difficult to achieve a uniform input signal to the amplifiers with the spatial feeding technique than with circuit feeding, and circuit combining has higher loss than spatial combining. Thus, this combination would result in

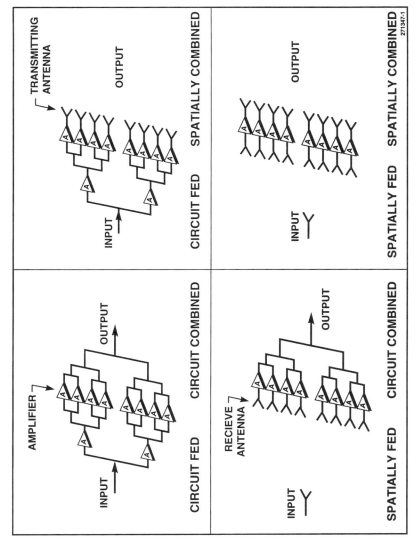

FIGURE 2.1 Four system configurations possible with the various combinations of circuit and spatial approaches for feeding and combining the signal in an amplifier array.

a system that is more difficult to build since both spatial and circuit components are necessary, and it would have poorer performance than other approaches.

Spatially fed/spatially-combined arrays are the most common type of spatial power-combined array. Quasi-optical power-combining falls into this category; however, the Fabry–Perot cavity approaches are somewhat a degenerate case since both the feeding and the combining take place in the electromagnetic mode of the cavity. Spatially fed/spatially combined arrays using distinct antenna elements and amplifiers are a relatively new development, but they are receiving a lot of attention. This configuration, while new to power-combining, has a proven track record in large-scale phased arrays (e.g., see reference [3]). If this configuration is used as a feed for a larger reflector antenna, the disadvantages of larger volumes associated with the spatial techniques may not be an issue. It is also possible to configure this approach as a stand-alone amplifier. A feed antenna illuminates the first array, and a collection antenna gathers the amplified signal and directs it to an output port.

The general attributes of the feeding and combining approaches can be summarized as follows:

Circuit-Fed

- Simple to obtain equal amplitude and phase to all of the elements
- Loss in the feed network but can be overcome using driver amplifiers
- Small volume

Spatially Fed

- Simpler system structure
- Requires larger volume to accommodate free-space fields
- Nonuniform amplitude and/or phase
- Spillover loss

Circuit-Combined

- Single output port for integration into system
- Loss is a function of the number of combining levels, thus the number of devices in the system

Spatially Combined

- Loss primarily due to phase and amplitude variation among the elements and is relatively independent of the number of elements
- Requires more volume for the free-space fields
- Output is in free space

2 TRADE-OFFS FOR SPATIAL POWER-COMBINED AMPLIFIERS AND TRANSMITTERS

There are a number of possible system configurations, array architectures (tray or tile), and construction methods (monolithic versus hybrid) for implementation of both circuit-fed and spatially fed combiners. Each has certain advantages and disadvantages for a given application. The disadvantages usually can be overcome but at the cost of impacting other aspects of the system design. These trade-offs are discussed below. Generally, there is some overriding system requirement, such as reliability, thermal management, or cost, that will guide the implementation decisions. When the design issues are evaluated, a few attractive system combinations emerge which are summarized at the end of this section.

2.1 System Configurations

A spatial power combiner can be operated as a stand-alone amplifier or as a transmitter. A stand-alone amplifier implies a distinct input and output port and insertion in a system with other waveguiding components. Spatial combiniers as stand-alone amplifiers are complicated by the basic attribute that the combined signal is in free space, unless a Gaussian beam-guiding system is employed. A more natural application of the spatial power combining technique is as a transmitter which takes advantage of the radiating aperture.

Transmitter arrays can be used alone, or can be designed to feed a reflector antenna, as shown in Fig. 2.2. The former case favors applications where the transmitter aperture must be kept small. In Fig. 2.2a, only the output array is visible with the rest of the system components concealed inside a housing. This array may be the output stage of a near-field cascade or a beam waveguide configuration, or it could be the face of a circuit-fed/spatially combined array. In system configurations where the output aperture is not restricted, there is strong motivation for using a reflector antenna as the output device. A reflector antenna is a relatively easy and cost-effective way to increase greatly the equivalent isotropic radiated power (EIRP) of the transmitter. Examples of such applications include satellite communications and microwave relay links.

Stand-alone amplifiers can be either circuit-fed or spatially fed. However, circuit-fed/spatially collected configurations are less likely since they involve using techniques and construction methods of both circuit and spatial approaches which will complicate the design and fabrication. For an amplifier with distinct input and output ports, it is more natural to feed and combine using the same approach. As mentioned above, the circuit-fed/circuit-combined configuration is the established approach for microwave solid state amplifiers.

Spatially fed/spatially combined amplifiers can be arranged in two different configurations: in a near-field cascade or in a beam waveguide, as shown in Fig. 2.3. The near-field cascade can be further divided into free-space systems (Fig. 2.3a), and guided wave systems (Fig. 2.3b). The beam waveguide

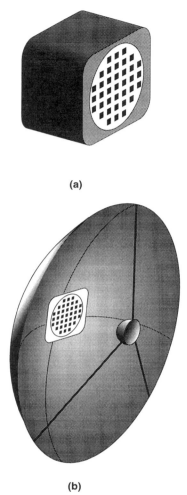

FIGURE 2.2 Transmitter system configurations: (a) Direct radiator configuration and (b) reflector antenna feed configuration.

configuration is a quasi-optical approach that uses lenses to form a Gaussian beam waveguide (Fig. 2.3c). The spatial amplifier arrays are placed at the beam waist. In designing a system with any of these configurations, one must bear in mind spillover loss, interaction of standing waves between arrays, and nonuniform amplitude and phase of the feeding signal. The interaction between multiple arrays placed in a near-field cascade/free-space configuration (Fig. 2.3a) is being studied and is discussed in Chapters 5 and 6. In the near-field cascade/guided-wave configuration (Fig. 2.3b), the intent is to use the mode of the waveguide to eliminate spillover loss and provide a more uniform illumina-

TRADE-OFFS FOR SPATIAL POWER-COMBINED AMPLIFIERS AND TRANSMITTERS

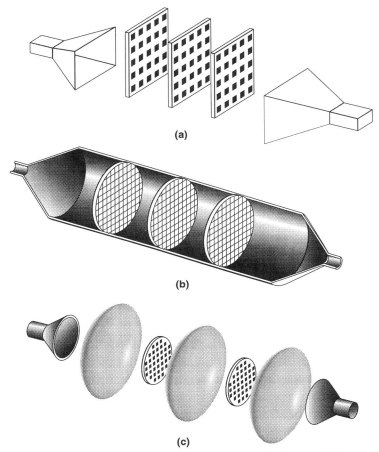

FIGURE 2.3 Amplifier system configurations: (a) Near-field cascade/free-space configuration, (b) near-field cascade/guided-wave configuration (cut-away view), and (c) beam waveguide configuration.

tion wave. The beam waveguide configuration (Fig. 2.3c) reduces spillover loss through the use of the lenses, and it provides a uniform phase front at the beam waist for input into the array. However, there is an inherent nonuniform amplitude distribution. This is discussed further in Section 2.3.

2.2 Array Architecture (Tile Approach or Tray Approach)

For a majority of applications it makes sense to divide the array into a number of subarray modules. At microwave and millimeter-wave frequencies the individual elements are too small to effectively construct them individually (there are always exceptions [4]). At the other extreme, there are very few arrays that

56 SPATIAL POWER COMBINING

can be effectively constructed as a single unit. Generally, imperfections in the manufacturing process or complexity of the array are the motivation for dividing the array construction into smaller units. The exception for construction as a single unit is the pursuit of wafer-scale integrated circuits for the grid array configuration and some other spatially fed/spatially combined configurations. At the present time, however, current manufacturing yields do not make wafer-scale integration practical for GaAs circuits operating in the upper microwave and millimeter-wave region.

The movement toward dividing the array into a number of subarray modules has spawned two different approaches. The tile approach denotes configurations which use relatively thin modules where the RF circuitry and active devices are placed on circuits parallel to the face of the array. They use integrated circuit

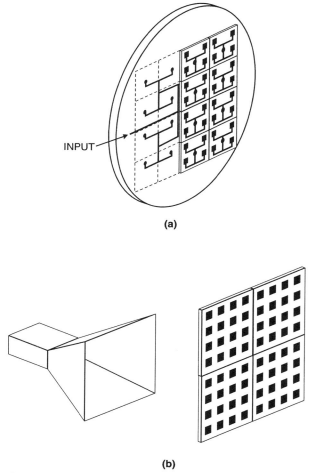

FIGURE 2.4 Illustration of the tile approach architecture: (a) Circuit-fed (subarrays on the left side are absent revealing the feed network to the subarrays) and (b) spatially fed.

antennas that radiate broadside from the antenna surface in the module (such as a microstrip patch). Figure 2.4 illustrates a circuit-fed and a spatially fed tile approach architecture. The other major array architecture is termed the tray or brick approach. This technique has traditionally been called the brick approach, but the term tray approach is more descriptive of the architecture. In this design the RF circuitry and active devices are placed on circuits perpendicular to the face of the array, and they usually contain end fire antenna elements (such as a linearly tapered slot or Vivaldi antenna). Figure 2.5 illustrates a circuit-fed and a spatially fed tray approach architecture.

The tile approach uses a fewer number of subarrays to populate the array; however, there is less space at each element to place the components. In a spatially fed tile approach configuration, one can generally fit the required components on one or two sides of the array, but heat removal or thermal management becomes the biggest design challenge. The design can be modified to contain a hea t-conducting layer to channel the heat from the center array to

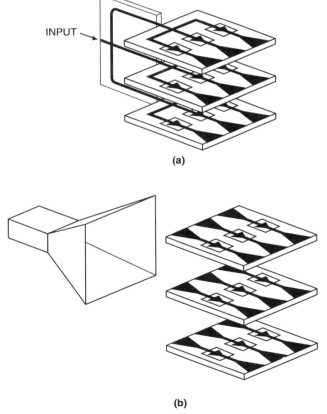

FIGURE 2.5 Illustration of the tray approach architecture: (a) Circuit-fed and (b) spatially fed.

the edges, but this heat-conducting layer will be relatively thick to obtain adequate thermal conductance. This thick layer complicates the transfer of the signal from the receive side of the array to the transmit side. A reflection configuration is also possible, so that the back side of the array can be for thermal management, but this complicates the propagation path of the signal through the system. A circuit-fed tile approach architecture has the advantage that the back side of the array can be used for thermal management, but the input feed network increases the amount of RF circuitry, generally leading to the need for a three-dimensional RF path.

The tray approach architecture provides more space for the RF circuitry and active devices. The availability of space should allow the RF circuitry to reside on a single layer, resulting in a less complicated design. Thus the larger number of required modules may be offset by simpler and less costly construction of the modules. Another advantage of this configuration is that the space between trays should permit a reasonably thick plate for heat conduction. The tray approach allows both circuit feeding and spatial feeding. Circuit feeding will result in a physically shorter system, but at the price of constructing a network to distribute the signal to each of the trays. The biggest disadvantage of the tray approach is the length of the system. There are many applications where this is not an issue. There are some uses, however, such as conformal arrays, where the dimension behind the face of the array is a critical system parameter.

In a related issue, the thermal management problem is a function of the output power and efficiency of the amplifiers in the array as well as the array configuration. Generally speaking, it is easier to maintain a lower junction temperature for lower-output power amplifiers. The thermal resistance path is characterized in units of °C/watt, so less waste heat results in a lower temperature rise at the MMIC attachment location. There are also advantages in the ease of dissipating the heat from a larger number of smaller-output power devices as opposed to a smaller number of higher-output power devices. The thermal energy is more uniformly spread over the heat exchanger, making the heat removal task easier.

2.3 Circuit-Fed Versus Spatially Fed

The main issues impacting a decision to use a circuit feed or spatial feed are (1) the feed losses and (2) the uniformity of the amplitude and phase. Isolation between the output and the input signals is also important. If the gain in the system is greater than the isolation, there is a possibility that the amplifier will oscillate. Even if the amplifier is stable, if the isolation is only a few decibels greater than the gain, there may be ripple in the output of the amplifier caused by destructive interference of the feed signal at the inputs of the amplifiers.

The major advantage of the circuit-fed approach is that it is relatively straightforward to obtain equal amplitude and phase at the input of the amplifiers. At millimeter-wave frequencies the construction details become more challenging, but good engineering practices can overcome these difficulties. Another advantage when circuit feeding is used for the tile approach is that it allows the back side of the array to be used for thermal management.

In the circuit-fed approach there is no spillover loss, but there is resistive loss in the feed network. The fan-out and resistive loss can be overcome by placing driver amplifiers at appropriate points in the feed network. The extra amplifiers add further complication to the feed network, but it is an effective solution. The number of driver amplifiers required to overcome the feed loss is insignificant compared to the number of amplifiers in the output stage, and thus the power consumed by these amplifiers has little impact on the overall system efficiency.

The biggest disadvantage of a circuit feed network is that it significantly complicates the construction compared to spatially fed arrays. In a tile approach array, at least three RF circuitry levels are required: one to distribute the RF to each of the subarray tiles, and two on each tile to control oscillations. The tray approach only requires one RF level on each tray, but the circuitry to feed each tray is still fairly complex. The key to unlocking the benefits of circuit feeding is to use modern packaging techniques and clever circuit design to simplify the complexity and construction of the subarray.

The advantage of the spatially fed approach is the simplicity of the feed. This approach, however, mandates a necessary trade-off between spillover loss and uniformity of amplitude and phase (similar to feeding a reflector antenna). On one extreme, it is possible to accept a large spillover loss to achieve uniform illumination, and then overcome the loss by designing more gain into the system. In a sense, this is the approach of circuit feeding; however, the spillover loss is likely to be higher than the losses in a corporate feed network.

The other extreme is to minimize the spillover loss by using a beam waveguide and accommodate the nonuniform amplitude distribution. The beam waveguide configuration (Fig. 2.3c) reduces spillover loss through the use of the lenses, and it provides a uniform phase front at the beam waist for input into the array. However, there is an inherent nonuniform amplitude distribution. (The amplitude distribution from the center of the beam to edge follows a Gaussian profile, hence the name Gaussian beam.) Nonuniform amplitude illumination presents a difficulty because it is more susceptible to distortion by the spatial amplifier. If the system is being utilized for power generation, the highest DC-RF efficiency is obtained when the amplifiers are operated in (or near) the gain compressed region. A system that inherently contains nonuniform amplitude input will suffer distortion because the amplitude at the edges of the beam increase disproportionately compared to the amplitude in the center of the array. This could have advantages in the output stage of a transmitter configuration, but attention to phase compression or distortion must also be given in the design.

There are two options for handling the nonuniform amplitude illumination that still allow the devices to operate under compression. The first is to use different-size amplifiers in the array. The second is to adjust the number of antenna elements feeding the amplifiers. The amount of adjustment under either option is guided by the rate of change in the power density of the wave incident on the array. The first approach requires more control over the gate periphery of the amplifiers. This approach is probably better suited to a monolithic fabrication since it is fairly straightforward to alter the size of the transistors when a custom mask set is designed expressly for the spatial amplifier. The second approach is better suited to a hybrid fabrication since a large quantity of similar amplifiers can be utilized, and their operation can be optimized by suitably designing the passive circuitry in the array (receive antennas and feed network to the amplifiers).

A third spatial feeding technique is to place the array inside a rectangular or circular waveguide, as shown in Fig. 2.3b [1,5,6]. The motivation for this is that intentionally excited higher order modes or dielectric loading along the walls of the waveguide can provide a more uniform illumination. The biggest difficulty with this approach is how large the waveguide can be made and still maintain the desired uniformity. If the array dimensions are going to be at least one wavelength or larger, an overmoded waveguide will be used. Unless closely controlled, the reactive fields of the antenna array will excite undesired higher-order modes making maintaining a uniform field distribution a formidable task. A variation on this is to use a waveguide structure whose fundamental mode is large compared to a wavelength and thus can handle reasonable-size arrays. A well-suited structure for this task is a Fabry–Perot cavity which is the corner stone of quasi-optical power combining.

Isolation between the output wave and the input feed is also an issue in spatially fed arrays. In some ways the spatial-fed approach is more vulnerable than circuit-fed arrays because there are two possible feedback paths. The first is directly through the array structure, and the second is via spurious reflections from objects in the vicinity of the arrays. The second path is an issue only for arrays that have appreciable gain. Generally speaking, spatially fed/spatially combined arrays are designed with the polarization of the input antennas orthogonal to that of the output antennas. This improves the isolation in the array.

A fourth possibility is to form a hybrid spatial, and circuit-feed system. In this scheme a spatial approach is used to feed the subarrays, and a circuit approach is used to feed the amplifiers on the subarray. Such a scheme has been reported for a tray approach architecture at 60 GHz [7]. The input port for the corporate feed network on each tray is an end fire dipole [8] as shown in Fig. 2.6a. When the trays are stacked in the array configuration, the input antennas form a vertical stack. Then a sectoral waveguide horn is used to feed the input to each tray, shown in Fig. 2.6b. The hybrid feed has great utility at these frequencies because it obviates constructing circuitry and the connectors to feed the tray stack, but it still maintains many of the benefits of the circuit-fed approach.

The issues concerning the feeding of the amplifiers in the array can be summarized as follows. The circuit-fed approach, in theory, provides a uniform feed signal to all the elements in the array. The price of the uniform feed is the increased circuitry of the feed network. This circuitry must be placed in the subarray module—which can be challenging, particularly for the tile approach

TRADE-OFFS FOR SPATIAL POWER-COMBINED AMPLIFIERS AND TRANSMITTERS 61

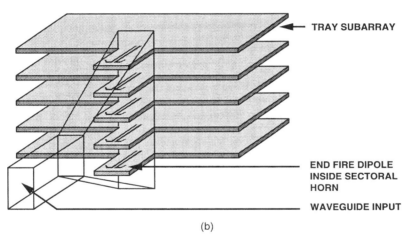

FIGURE 2.6 Hybrid spatial/circuit feed for a 60-GHz phased array: (a) Photograph of a single tray (courtesy of the Lockheed Martin Corporation) and (b) illustration of the assembled feed.

architecture. The spatially fed approach greatly simplifies the subarray construction, but it necessitates adjustments in the array design to overcome the nonuniformities of the feed.

2.4 Grid Arrays Versus Distinct Component Arrays

There are two fundamentally different approaches to configuring the active and passive elements in the design of amplifier arrays. The first is the traditional microwave approach of treating the transistors, antennas, and other passive circuitry as distinct elements and designing the elements using scattering parameter analysis. The second approach uses an integrated pairing of the transistor (or diode) and antenna to achieve oscillation or amplification. The device resulting from this pairing is referred to as an active antenna (Chapter 3) and forms the basic building block for quasi-optical arrays and grids discussed in later chapters. In these designs the effect of external polarizers and partial reflectors on the impedance presented to the transistor must often be considered. Scattering parameter design techniques are covered in many standard microwave texts, and the details of the active element and grid array design are covered in other chapters of this book. In this section we will consider the larger question of how well suited these two different approaches are to solving the system-level performance objectives. The three most prominent issues are the ability to provide the active device with the optimum load impedance, the amount of amplification that can be designed into the array, and the ability to choose an antenna that fits the needs of the system.

The distinct operation of the amplifier and antenna allows more control in optimizing the amplifier performance. To obtain maximum output power from the active device, one has to present the optimum load impedance. This includes engineering the load seen by the harmonics of the signal [9,10]. Modeling efforts for grid arrays are maturing and address issues like resonant frequency and the load seen by the active device, but they are not as mature as the models and the design techniques available for traditional microwave circuits. The optimum load impedance for an active device is found through a (harmonic) load-pull measurement. In traditional microwave practice, the load impedance is realized by proper amplifier circuit design. In monolithic arrays that can afford the space and in hybrid arrays that will make use of MMIC amplifiers, it makes sense to include the harmonic termination and impedance matching (nominally 50-Ω) circuitry. The circuit outside the amplifier can use impedance transformers or an appropriately designed antenna to provide an impedance match to the output of the amplifier.

With more traditional microwave designs, multistage amplifiers are possible which have more gain than active antenna or grid array designs. Higher-gain amplifiers enable array designs which fulfill many of the system requirements with a single array. This is opposed to active antenna and grid arrays, where a number of amplifier arrays are needed to achieve the required system gain and output power.

The final prominent issue in the decision between a grid array or a traditional array is the ability to choose an antenna to meet the system requirements. In the end, the antenna is simply a transducer to convert free-space propagation to guided-wave propagation or vice versa, so there is no single antenna that is best. There are simply antennas that make the system work and antennas which do not. The most important attribute of the antenna is bandwidth: impedance bandwidth, radiation efficiency bandwidth, and purity of polarization (axial ratio) bandwidth. Other attributes that must be considered are the sense of polarization, the physical size, and difficulty of construction. The ability to choose the antenna is most limited in the grid approach. Tile approach arrays which use either an active antenna or a distinct amplifier and antenna approach have more choice, but will probably tend toward antennas which are space efficient (resonant antennas or smaller). Arrays constructed with a hybrid approach have the most flexibility and can include cavity-backed (tile) and traveling wave antennas such as tapered slots (tray), spirals (tile), and horns (tray/tile).

2.5 Monolithic Versus Hybrid Construction

The circuit-fed/tile approach array is not well suited for monolithic construction. First, the input feed network takes a significant amount of space in the array, and routing the feed lines must be given a high priority, so generally there is not enough room in the array for the feed network, the amplifiers, and the antenna on a single layer. Second, even if space for all of the components can be found, the feedback path from the signal radiated by the antennas to pick-up on the input feed line is likely to cause oscillations. A multi-RF-level architecture constructed as a hybrid circuit is a better choice and is discussed in more detail below.

Spatially fed/tile approach arrays, however, are well suited to monolithic construction. Grid amplifiers easily fit all of the required components on one side of a GaAs substrate [11], and clever designs of more traditional approaches with a distinct receive antenna, amplifier, and transmit antenna have also been constructed on a single GaAs wafer [12]. The motivation for using a monolithic design is that the array can be fabricated in the production line along with, or in a manner very similar to, other GaAs substrates. At the end of the production line, the wafers would need very little postprocessing before inserting into a system. The goal of this approach is reduced system costs.

The biggest challenge for spatially fed/tile approach arrays is the thermal management issue. Removal of the waste heat produced by the active devices is a significant problem for the monolithic approach and will have to be overcome before practical systems can be constructed in this fashion. Some of the solutions being explored are liquid cooling [12], reflection configurations, thicker ground planes on the back side of the wafer, and even the use of thick diamond films on the wafer to conduct away the heat. While these approaches hold promise, they start to complicate the monolithic approach and will lead to higher system costs.

A spatially fed/tray approach and even circuit-fed tray approach arrays are also reasonable candidates for monolithic construction. The ability to place the wafers on a reasonably thick metal plate for thermal management permits the goal of having the wafers fabricated in the standard GaAs production line. There are disadvantages, and the immediate objection is the significant amount of wafer surface area that is wasted by the antennas and other passive circuitry. The retort to this objection is as follows: If the exotic efforts to address thermal management issues in the tile approach double or triple the cost of the processing, then one can afford to use two or three times the wafer area in a tray approach. Use of end fire dipole radiators [8] instead of long traveling-wave slot antennas can help reduce the length of the trays. To date, the tray approach has been largely overlooked, for monolithic as well as hybrid construction. However, there are many benefits to the tray approach, and it should receive more attention once the benefits are recognized.

Most of the arrays constructed so far have claimed to be hybrid-circuit frequency-scaled models of future monolithic millimeter-wave implementations. While it is true that the first few monolithic arrays have recently been constructed [5,11,12], the fact remains that most of the arrays constructed so far are hybrid. Once the realization is made that there is nothing wrong with a hybrid array, it allows the possibility of taking full advantage of the hybrid approach and many beneficial attributes can be designed into the system. In a properly executed design, the hybrid construction can incorporate features such as excellent thermal management, known good die, and optimization of the antenna radiation efficiency. However, an additional burden comes with having more choices and more latitude in the design afforded by the hybrid approach: These arrays must have increased performance to justify the additional construction effort.

Thermal management is more readily accomplished with a hybrid construction. A circuit-fed tile approach permits the entire back side of the array to be used for thermal dissipation. In a spatially fed/tile configuration a thick thermal conductor can be sandwiched between a receive array and an amplifier/transmit array. In the tray approach the thermal management issues are similar for either a monolithic or a hybrid construction.

One obvious advantage with a hybrid construction is the ability to use known good die. But more than this, the amplifiers can be screened according to their S_{21} performance, so the array is populated with amplifiers with similar phase and amplitude characteristics. The hybrid approach also directly takes advantages of the substantial work on MMIC power amplifiers in the past several years. In particular, systems can be designed that use commercially available amplifiers or amplifiers "as is" which have been developed for other communication and radar projects.

The hybrid approach relieves the pressure to use the minimum amount of wafer area; thus there is maximum flexibility in the antenna design. There is room to use a resonant or a traveling-wave antenna or a circularly polarized antenna with its more complicated and more space hungry feed. The designer also has the flexibility to choose the dielectric constant of the circuit board. The circuit board can be chosen for either its low dielectric constant, which improves radiation efficiency, or its thermal or manufacturing benefits.

The packing issues are the biggest challenge for the hybrid approach, but at the same time these issues provide the opportunity to achieve better performance and to achieve a design that more easily satisfies the system requirements. The emerging focus on packaging is well timed for the development of spatial power combined arrays. All of the multichip module techniques, MCM-C, MCM-D, and MCM-L, should find applicability in these arrays.

2.6 Trade-Off Summary

There are many options and trade-offs in the design of spatial power-combined arrays. There is no single best choice for a given application. Rather there are a number of ways to configure the design, each with its own advantages and disadvantages. The decision among the various choices should be guided by the overriding system goals (i.e., reliability, cost, etc.), by the talents of the engineer or team of engineers designing the system, and by the available manufacturing facilities. In this subsection we will reiterate some of the major points of the preceding discussion and make some suggestions about pairings of the various techniques and approaches that produce an attractive combination.

The major points covered in this section were:

- Use a reflector antenna if possible. It is an easy and cost-effective way to increase the EIRP of the transmitter.
- It is easier to construct systems designed with uniformly illuminated arrays. Either a circuit feed or a spatial feed with high spillover loss can be used to uniformly illuminate the array. The loss is readily overcome by adding extra gain in the system.
- The spatially fed/tile approach array is conceptually the simplest design, but it comes at the expense of more complicated thermal management schemes.
- The circuit-fed/tile approach arrays adequately address thermal management, but at the expense of increased construction complexity.
- Tray approach arrays, either hybrid or monolithic construction and either spatially or circuit-fed, have been overlooked but offer some significant advantages.
- The thermal management is easier for an array built with a larger number of lower-output power amplifiers than for an array built with a smaller number of higher-output power amplifiers.

Three particularly attractive combinations of design choices are listed below.

1. The first is a spatially fed/tile approach array using a monolithic construction. The array can be constructed with suitably sized subarrays ($1/4$ to 1 in^2) until wafer scale integration is feasible. The subarrays could be thinned as much as possible (2 to 4 mil) and epoxied to a dielectric substrate. The thermal conductivity of the substrate should be as high as possible—for example, AlN, BeO, or perhaps a diamond-coated substrate. The substrate would house the DC distribution. The thickness of the substrate should be $\lambda/2$ at the frequency of operation so the field will pass through the substrate with minimal reflection. Dipole (grid) or slots are the best candidates for the antennas since the radiation efficiency will be high if they are placed on the $\lambda/2$-thick substrate.

2. The hybrid-circuit/tile approach using conventional microwave techniques has many beneficial attributes: excellent thermal management, moderate bandwidths, excellent graceful degradation performance, and compact design. The only disadvantage of this technique is the complexity of the packaging. The subarray tiles required for this approach are arguably the most complex of any of the spatially combining techniques. There is limited area and the need for multiple RF and dc layers. The biggest challenge in this combination of approaches is using innovative circuit designs and three-dimensional packaging techniques to simplify the packaging complexities.

3. The final attractive combination is a spatially fed/tray approach array using a hybrid construction. This configuration has many advantages. It can handle the thermal dissipation from reasonably high-output power amplifiers, can be designed to have as much gain as necessary either with a single multistage MMIC or several MMIC amplifiers (input/output isolation is required), and can overcome the phase and amplitude nonuniformity by phase trimmers and compressing some of the intermediate stages. The hybrid construction would be relatively simple: a two-layer circuit board, one layer for RF and the other for DC, with cutouts for insertion of the MMIC amplifiers. The MMICs are attached directly to the carrier and connected by ribbon bonds to the RF circuit. The circuit board would permit use of traveling-wave antennas for broadband performance.

3 LOSSES IN SPATIAL POWER-COMBINING SYSTEMS

In a spatial power-combined array there are both losses that one normally associates with antenna arrays and losses one normally associates with amplifier circuits. If enough gain can be designed into the array, only the losses that occur after the output amplifier are significant. The loss of the input corporate feed network in circuit-fed systems or the spillover loss in spatially fed system can be hidden by adding enough amplification. The power consumed by the these extra driver amplifiers is usually minimal compared to the other amplifiers in the array.

Losses normally associated with antenna arrays include the metallic and dielectric dissipation loss and the loss to guided waves in the substrate ("surface waves"). These can be lumped in the radiation efficiency of the antenna. Variations from the desired phase and amplitude (usually uniform) of the antennas will appear as reduced directivity of the field patterns of the array—that is, wider main beam and higher side-lobe levels.

The most important loss normally associated with amplifier circuits is not presenting the amplifier with the optimum load impedance. This loss can be more significant than the normal mismatch loss which arises from not matching the active impedance of the antenna to the transmission line characteristic impedance. The output power of transistor amplifiers is dependent on the load connected to their output. Proper MMIC amplifier design will transform the characteristic impedance of the output port of the MMIC (usually 50 Ω) to this optimum load impedance. So if there is a mismatch at the antenna, the degradation to the system performance can go beyond the normal loss of the reflected energy at the antenna port. The reflected signal alters the load seen by the output transistor in the MMIC amplifier and can cause it to operate in a region of diminished performance.

As an aside, this phenomena is also responsible for the erroneous claims of combining efficiencies greater than 100%. These claims occur because the EIRP from a single amplifier–antenna combination (either an active antenna configuration or distinct component configuration) is measured as the baseline output power of the element. Then these elements are placed in an array, and the mutual coupling in the array causes the amplifier to see the active impedance of the antenna instead of the self-impedance. This change in load impedance as seen by the amplifiers is closer to the optimum load impedance than the isolated element. Thus, the amplifier produces more output power, resulting in apparent combining efficiencies greater than 100%.

Another important loss mechanism associated with the amplifier is feedback loss. This loss is caused by insufficient isolation between the output and input of the array. Suppose the gain of the amplifiers in the array is 20 dB, and the isolation between the output and the input is 21 dB. This feedback path is not sufficient to cause oscillations, but it is sufficient to cause significant destructive interference at the input port of the amplifiers. This condition manifests itself by ripple in the output power of the array. To minimize this ripple the isolation should be at least 10 dB, preferably 20 dB greater than the gain of the array.

While it is possible to enumerate the various loss mechanisms in the array, it is difficult to correctly identify their relative prominence in an operating array. The only unambiguous measurement that can be made for a transmitter array is the EIRP, and the only unambiguous measurement that can be made for an amplifier array is the EIPG. The figures of merit described in Chapter 1 may not provide a means to identify the specific loss mechanisms, but they do correctly account for total of losses present in the spatial power-combined array.

The final topic of this section is the expression for loss in directivity of an array caused by phase and amplitude variations and failure of elements in the array [13]. This expression is useful for predicting the performance of an array with known variations as well as providing a means to specify the allowable variations in the amplifiers and the point at which too many elements in the subarray have failed. It is given by

$$\frac{D}{D_0} = \frac{P_e}{1 + \Delta^2 + \delta^2} \qquad (2.1)$$

where D_0 is the is the no-error directivity, P_e is the probability that the element is functional, Δ^2 is the variance (mean-square) of the normalized amplitude error (amplitude of ith element $= 1 + \Delta_i$), and δ^2 is the variance of the phase error in radians squared. In the derivation of this expression the following assumptions are made: The phase variations, amplitude variations, and element failures are all random and independent of each other; the phase errors are reasonably small ($\delta^2 \ll 4$); and amplitude errors are reasonably small ($\Delta^2 \ll 4$). The effective transmitter power of the array, and therefore the EIRP, is also effected by failed elements. The expression for the EIRP in the presence of element variations is

$$\frac{\text{EIRP}}{\text{EIRP}_0} = \frac{P_e^2}{1 + \Delta^2 + \delta^2} \tag{2.2}$$

where EIRP_0 is the no-error equivalent isotropic radiated power.

4 CIRCUIT-FED/SPATIALLY COMBINED AMPLIFIER ARRAYS

In the circuit-fed/spatially combined approach, the feed is compact and can provide a uniform phase and amplitude drive signal to the amplifiers. Driver amplifiers can be added to overcome feed losses without significantly affecting overall system efficiency. The approach can be configured for excellent thermal management; thus high-output amplifiers can be used if desired. It provides the most flexibility in the choice of antenna. The most significant disadvantage is the packaging complexity. In tile approach arrays, space is limited, and a three-dimensional RF circuit is usually required. In tray approach arrays, the packaging issues are easier, but the overall size of the system will be larger. Even though the packaging issues are the biggest challenge in this approach, they also present an opportunity. If the array is constructed as a hybrid-circuit, the array performance can be enhanced. The amplifiers can be pretested to ensure not only that they are functional but that the phase and amplitude characteristics are similar. The dielectric constant of the circuit board material can be chosen to increase the radiation efficiency of the antennas which has a direct impact on the overall efficiency of the system.

4.1 Differences from Phased Arrays

There is an obvious similarity between the circuit-fed/spatially combined approach and modern phase array architecture. Indeed, every phased array system *is* a spatial power combiner. There are, however, significant differences between phased arrays and spatial power-combined arrays when the goal of the system is solely power combining. Since spatial combining arrays do not require an electronically scanned beam, a larger spacing between elements is permitted and no phase shifters are required. Not only are there fewer components by elimination of the phase shifters, but the data path to control the setting of the phase shifter is also eliminated. Providing the data for the phase shifter can be a bigger issue than the RF aspects of a phased array.

This combination of fewer components in a larger space greatly simplifies the layout and construction of the array. It also enables more flexibility for arranging the components for better thermal management. In particular, it allows the amplifiers to be mounted directly to low-thermal-resistance heat paths [14] where more complex phased arrays sacrifice the heat path for inclusion of phase shifter data control components [15].

The larger element spacing and fewer components per element can be exploited for more complex antennas such as an orthogonally fed circularly polarized patch antenna. The extra space also allows fabrication of the antenna on a lower dielectric constant circuit board. The lower dielectric constant increases the radiation efficiency. Finally, the combination of larger space for fewer components permits tile approach designs at higher frequencies where phased arrays must be constructed with a tray approach.

4.2 Components for Arrays

One of the major differences with the circuit-fed approach compared to the grid arrays or active antenna approaches is that the components are treated as separate entities. A disadvantage is that the components and the circuits composed from them are larger, but if enough space is available, this approach has many advantages, all of which lead to higher system performance. Using traditional components and techniques allows the designer to take advantage of the substantial work invested in the microwave field. While this statement is obvious, this fact is often overlooked in the enthusiasm to create something new. There are relatively new computer-aided design tools, both electromagnetic and circuit simulations, that can be used to optimize component and circuit designs. Each component has a well-defined input port(s) and/or output ports(s) and a well-defined function. The separation of functionality aids the ability to optimize the component, and the input and/or output ports allow characterization of the component and allow placement in the circuit with appropriate impedance transformer to optimize the circuit design.

Feed Networks and Transmission Lines There are a variety of transmission line configurations that can be used to construct the input feed network. Among these are rectangular waveguide, coaxial cable, microstrip, and stripline. It is reasonable to use a mixture of transmission lines in the feed network. For example, in one tile approach design [16], a rectangular waveguide corporate feed was used in the back plane or motherboard and a microstrip was used on the subarrays. It is also reasonable to use a mixture of spatial and circuit feeding. In a tray approach design, a spatial feed via a sectoral waveguide horn was to distribute the input signal to each of the trays, and a microstrip feed was used on each tray [7] (see Section 2.3).

In a corporate feed network based on microstrip or stripline, one has to decide which type of dividers to employ. Wilkinson dividers are a better choice than reactive T-junction since they provide some isolation between imbalances of the load connected to the output ports. This isolation is advantageous both to increase the uniformity of the input signal to all of the amplifiers and to enhance the graceful degradation performance if one of the amplifiers fails. If there is not enough room for Wilkinson dividers at all junctions, placing them at some of the junctions as close as possible to the amplifiers is still useful. The added benefits of these dividers warrant the extra construction complexity.

The ability to adjust the phase of the signal through the subarray is also a desirable feature to incorporate into the design of the feed network. Trimming the phase at every element would be too time-consuming even if a practical technique were found. A more appropriate strategy is to characterize the amplifiers before inserting them into the subarray and sort them into bins with similar S_{21} characteristics. Then populate the subarray with amplifiers from the same bin and trim the phase of each subarray to some nominal value. There are several ways the phase can be trimmed in the subarray. If a microstrip feed network is used on the subarray, one can design a series of hairpin line segments; and through a combination of laser ablation and ribbon bonding, one can modify the length of the microstrip line. A second technique is to use a small dielectric overlay on the microstrip line to adjust the phase delay. By choosing the height, length, and dielectric constant of the overlay, it is possible to obtain the desired phase delay through the line [17]. Another approach is to construct the feed network from the coaxial transmission line and use coaxial phase trimmers to compensate for both variations in the feed and variations among the subarrays.

Amplifiers The significant investment in GaAs MMIC technology in the late 1980s and early 1990s has pushed practical amplifier designs well into the millimeter-wave region. One-watt class output power amplifiers have been constructed which operate at 45 GHz [18,19], and useful multistage amplifiers have been constructed for operation at 94 GHz [20,23]. The maturing of MMIC amplifier technology is one of the primary motivators for spatial power combining. Certainly all array designs, monolithic and hybrid, will benefit from this recent progress; however, there is a growing pool of existing MMIC designs and amplifiers that are capable of being utilized "as is" in hybrid circuits.

Amplifiers designed for higher microwave and lower millimeter-wave frequencies tend to come in two types. The first type is driver amplifiers which typically have three stages, 20 dB of gain, and output powers in the range of 50–200 mW. The second type is power amplifiers which typically have only two stages, about 10 dB of gain, and output power in the range of 500–1000 mW. Most of these amplifiers are designed with 50-Ω input and 50-Ω output impedances. The driver amplifiers generally have higher manufacturing yields because the Q (quality factor) of the on-chip matching networks is lower, making them broader band and therefore less susceptible to process variations. The converse generally holds for the power amplifiers.

For tile approach architectures, space in the subarray is limited, and relatively high-gain amplifiers are needed to overcome the input feed losses. Thus, driver amplifiers are more likely to fit the design needs. The choice of driver amplifiers aids the thermal management issues since dissipating the heat from a larger number of smaller heat sources is easier than dissipating the same heat from a smaller number of larger heat sources. The higher manufacturing yields may actually result in lower system cost despite the fact that a larger number of amplifiers is used. In the tray approach configuration, more space for the amplifiers can be designed into the system. Thus, it is possible to cascade multiple MMIC amplifiers to achieve both higher gain and higher output power. One should be cautioned that designing too much gain into the system will only exacerbate the tendency to oscillate which is an inherent attribute of spatial power combined arrays. Further steps to increase isolation should be taken if multiple amplifiers are cascaded.

Antennas Some of the primary issues to bear in mind when choosing the antenna are input impedance, gain, radiation efficiency, polarization, size, and fabrication complexity. Using a distinct antenna and amplifier permits optimizing the antenna in test circuits before integrating it with the amplifier.

When characterizing the input impedance, it is the active impedance rather then the self-impedance that is of interest [22]. The active impedance is found from the ratio of the reflected signal to the input signal when all the elements in the array are being fed. This differs from the self- (or isolated) impedance of the antenna because mutual coupling in the array will modify the apparent reflected signal. The active impedance can be found in several different ways. It can be measured directly using bidirectional couplers, power splitters, phase trimmers, and so on, in conjunction with a frequency converter test set of a vector network analyzer. It can be found by measuring the self-impedance and mutual coupling between elements and using the data in a straightforward calculation [22]. It can also be found through calculation or electromagnetic simulation. Infinite array analysis will provide the active impedance. It is also possible to calculate the active impedance based on results of electromagnetic simulations for the self-impedance and mutual coupling. (Note that the simulation software should use a Green's function for an open structure to more accurately simulate the antennas' environment.) The author has observed excellent agreement for such calculations and measured results on proximity coupled microstrip antennas.

The radiation efficiency is another important parameter that should be considered carefully. The radiation efficiency is a direct factor in the overall combining efficiency of the system. A substantial loss to surface waves or use of a lossy substrate or circuit-board material will over shadow the benefits of the spatial power combining technique. The hybrid-circuit approach offers some advantages when considering the radiation efficiency. A lower dielectric constant circuit board can be used to reduce the amount of power loss to guided

72 SPATIAL POWER COMBINING

waves in the circuit board. Another technique is to surround the antenna with a metallic cavity, thus preventing the guide waves. One can place a dielectric puck with the antenna printed on top into a cavity machined from a metal carrier [23], or one can form the cavity using plated-through holes available in standard multilayer circuit-board processing [24].

Tile approach configurations are likely to use resonant antennas because of space limitations. Normally one wants an antenna whose main beam is broadside from the substrate or circuit board. Microstrip patch and resonant slot antennas are popular choices. In tray approach architectures, one usually desires antennas that have end fire radiation characteristics. Traveling-wave slot antennas [25,26] and endfire dipoles [8] are the most common choices. The traveling-wave antennas have the benefit of broad bandwidth; 6:1 bandwidths with acceptable radiation patterns have been reported [27].

4.3 Examples

In this subsection three examples of circuit-fed/spatially combined arrays are given. The first example is a tile approach 35-GHz monopulse active aperture for use as a radar transceiver. The second is a tray approach 32-GHz transmitter array with limited scan capability. Finally a tile approach 45-GHz transmitter array example is given. These arrays have been chosen to illustrate different array architectures. The first two examples also illustrate that functionality beyond straightforward power combining is relatively easy to incorporate. Adding this extra functionality, however, increases the complexity of the array and can result in compromises in the design from the power combining point of view. The last example illustrates a design for the sole purpose of power combining.

Motorola 35-GHz Monopulse Radar Array   One of the first arrays to recognize the benefits of circuit-fed/spatial-combining was a 35-GHz monopulse radar active aperture transceiver [2]. The basic array configuration, shown in Fig. 2.7a, divides the array into quadrants for the monopulse function. In each quadrant the signal is amplified by an IMPATT diode which in turn feeds eight antennas. The feed network is constructed with a balance stripline. The antennas are formed by exciting slots etched in one of the stripline ground planes, and the coupling ports for the IMPATT oscillators and comparator ports for the monopulse antenna are etched in the other ground plane. The feed network layout is shown in Fig. 2.7b, and a photograph of both sides of the array is shown in Fig. 2.8. In Fig. 2.8 the slot radiators are visible in the front side of the array on the left, and the IMPATT (injection locked) oscillators are seen on the back side of the array on the right.

The measured radiation patterns for the transmitted signal are shown in Fig. 2.9. There is a few-decibel irregularity in the side lobes caused by phase and amplitude nonuniformity in the array, but in general the patterns are typical of a uniformly illuminated circular aperture. The nominal peak power of the

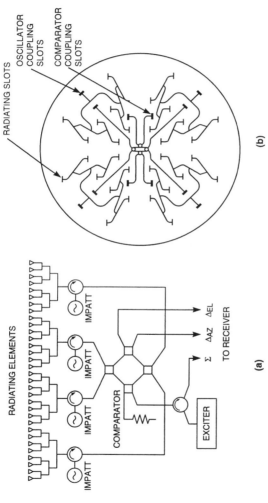

FIGURE 2.7 Design of a 35-GHz monopulse radar transmitter: (a) System configuration and (b) array layout. (From reference [2], © 1981 IEEE.)

FIGURE 2.8 Photograph of the 35-GHz monopulse radar transmitter. (From reference [2], © 1981 IEEE.)

oscillators under optimum conditions is 9 W; however, problems with the injection signal level feeding the oscillators resulted in an effective transmitter power of 25 W. This represents a 70% combining efficiency.

Jet Propulsion Laboratory 32-GHz Tray Approach Array The Jet Propulsion Laboratory's 32-GHz transmitter array was designed to feed a Cassegrain

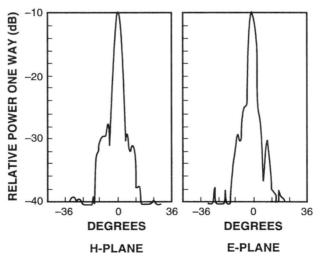

FIGURE 2.9 Measured sum pattern of the 35-GHz monopusle radar transmitter. (From reference [2], © 1981 IEEE.)

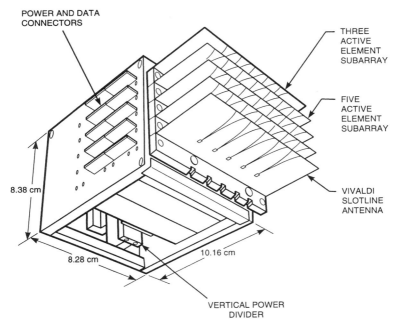

FIGURE 2.10 Ilustration of the 32-GHz tray approach, transmitter array, bottom side showing housing, slot antennas, and vertical power divider. (From reference [29], © 1990 IEEE.)

reflector antenna [28]. The transmitter design, illustrated in Fig. 2.10, is a tray approach configuration, consisting of 21 elements placed on five trays. The first and fifth trays have three elements, and the center three trays contain five elements. All of the trays have two dummy antenna elements, one on either side of the active elements. The array is designed for limited scanning, ±10°. Once the array is placed at the feed of a 10:1 magnification Cassegrain system, the reflector antenna will be capable of fine-tuning its pointing direction inside a 1° cone.

The input signal is distributed to the trays via a circuit-based, vertical power divider (Fig. 2.10) [29]. A photograph of the array is shown in Fig. 2.11. The input to the top tray is in the upper left portion of the photograph. The signal flows through a 1:3 power divider and into the input of each of the three channels. Each channel consists of a 3-bit phase shifter with associated ASIC controller, a driver amplifier, a power amplifier, and a Vivaldi slot antenna. The microstrip feed to the slot is visible in Fig. 2.11, and the actual slot antenna is illustrated in Fig. 2.10. A combination of amplifier stability problems, MMIC availability, and time constraints dictated that the final array be populated without the power amplifiers and that two different types of driver amplifiers be used: a 20-dB gain, 100-mW output power and a 12-dB gain, 25-mW output power.

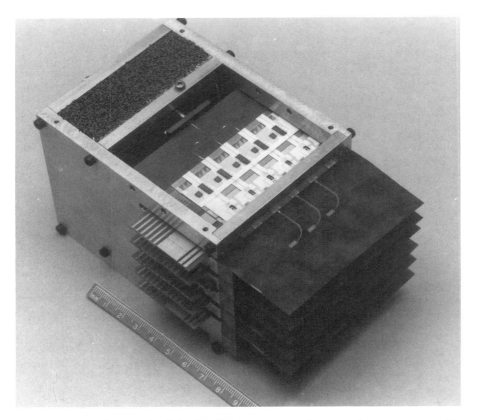

FIGURE 2.11 Photograph of the 32-GHz tray approach, transmitter array, top side showing details of tray. (Courtesy of the Jet Propulsion Laboratory, California Institute of Technology, Pasadena, CA.)

The measured E- and H-plane antenna patterns for the array pointing broadside are shown in Fig. 2.12. The modeled patterns in the figure are calculated from the measured phase and amplitude of each of the elements in the array. The reported array performance, which is based on an extrapolation where all elements use a 100-mW output power amplifier, is an EIRP of 26 dBW and an effective transmitter power of 1.6 W.

Lincoln Laboratory 45-GHz Tile Approach Array Lincoln Laboratory's spatial power-combined transmitter is also intended to feed a reflector antenna. The transmitter is composed of a number of subarray modules mounted on a common base plate as shown in Fig. 2.13. The base plate contains the DC power and RF input signal distribution to the subarray modules. The back side of the base plate is reserved for thermal management.

The design of a 4-by-4 element subarray is shown in Fig. 2.14. Each element consists of an MMIC amplifier and a cavity-backed, proximity-coupled microstrip antenna. There is a corporate feed network that distributes the input

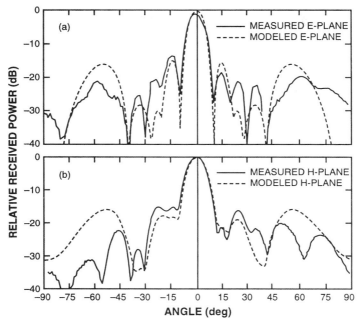

FIGURE 2.12 Measured field patterns for the 32-GHz tray approach, transmitter array. (Courtesy of the Jet Propulsion Laboratory, California Institute of Technology, Pasadena, CA.)

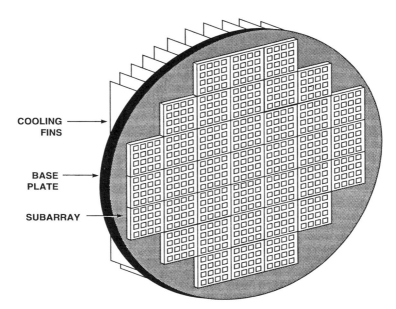

FIGURE 2.13 Illustration of the 45-GHz tile approach transmitter array.

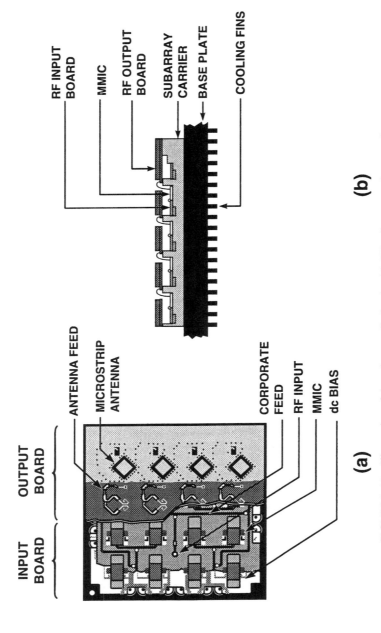

FIGURE 2.14 Illustration of the subarray tile for the 45-GHz tile approach transmitter array: (a) Top view and (b) side view.

signal to each of the MMIC amplifiers in the lower RF input layer. The output of the amplifiers is taken by ribbon bonds to the upper RF output layer that contains impedance matching circuitry and the patch antennas [24]. The ground plane of the RF output layer acts as a shield to reduce the feedback from the field radiated by the microstrip antennas to the RF input network. The entire subarray is integrated on a single metal carrier. The MMIC amplifiers are attached directly on the subarray carrier to provide a low thermal resistance path to the heat sink.

The subarray is a hybrid construction. The RF input layer and DC bias layer are constructed on 5-mil alumina substrates. The RF output layer is constructed with Duriod 6002 and standard circuit-board techniques to permit the fabrication of the plated-through holes which form the cavity-backed patch antenna. The dividers in the RF input feed network are Wilkinson dividers to provide isolation between the elements and improve the graceful degradation performance. The antennas in this array are circularly polarized, constructed by feeding the patch on two orthogonal sides and in phase quadrature. The patches are placed in the diamond set arrangement as a space reduction technique so that the subarray can be tiled into a larger array. Photographs of a brassboard version of the subarray are shown in Fig. 2.15. The brassboard is larger than the final subarray configuration because of the inclusion of individual bias lines to each amplifier.

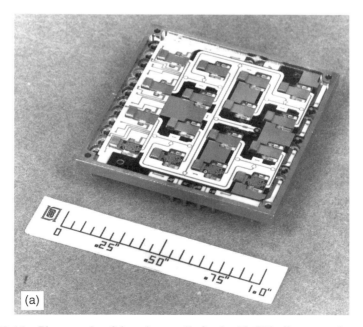

FIGURE 2.15 Photographs of the subarray tile for the 45-GHz tile approach transmitter array: (a) Partially assembled, showing RF input board and MMIC amplifiers and (b) fully assembled, showing cavity-backed patch antennas. (*Continued*)

80 SPATIAL POWER COMBINING

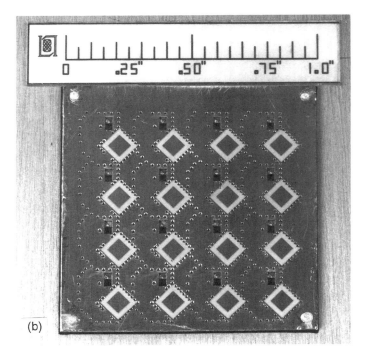

FIGURE 2.15 *(Continued)*

The brassboard subarray was characterized from 42.5 to 46.1 GHz. The measured EIRP, effective transmitted power, and DC-RF efficiency are shown in Fig. 2.16. The ripple in the measured results is caused by insufficient isolation between the field radiated by the patch antennas and the input feed network. The averaged results across the 43.5–45.5 GHz band are: 18.3-dBW EIRP, 530-mW effective transmitter power, 10.3% DC-RF efficiency, and 46.2% combining efficiency (not shown). Note that the loss of the antenna feed to create circular polarization and the RF layer to layer transition are included in the above figures of merit. The circular polarization axial ratio was 1.2 dB at 43.8 GHz, and the subarray showed a 3-dB axial ratio bandwidth of 3%. The narrow bandwidth was due largely to the reactive T-junction power divider in the antenna feed. A Wilkinson divider in the feed would improve both the axial ratio and the axial ratio bandwidth. Typical measured field patterns for the subarray are shown in Fig. 2.17, and the subarray exhibits theoretical maximum graceful degradation performance.

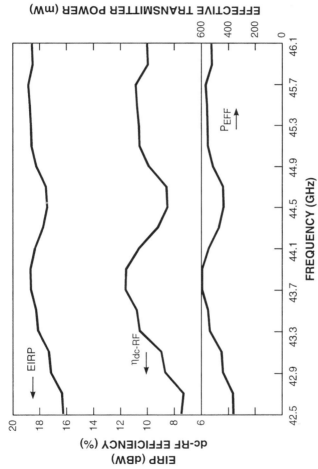

FIGURE 2.16 Measured EIRP, effective transmitted power and DC-RF efficiency for the 45-GHz tile approach subarray.

82 SPATIAL POWER COMBINING

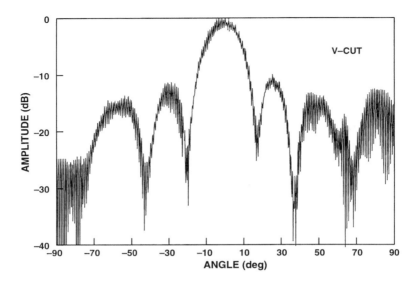

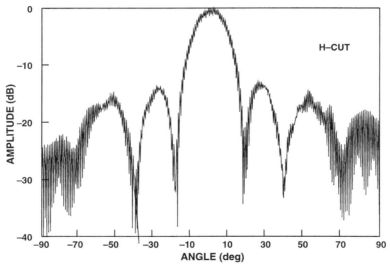

FIGURE 2.17 Measured spinning linear field patterns for the 45-GHz tile approach subarray.

ACKNOWLEDGMENTS

The author would like to thank Karen Lee of the Jet Propulsion Laboratory for providing figures on the 32-GHz tray approach array, and he would also like to thank John Windyka of Lockheed Martin Corporation for providing photographs of the 60-GHz phased array. The work on the 45-GHz tile approach arrays performed at the Lincoln Laboratory, Massachusetts Institute of Technology was sponsored by the Advanced Research Project Agency under Air Force contract F19628-95-C-0002.

REFERENCES

[1] D. Staiman, M. E. Breese, and W. T. Patton, "New technique for combining solid-state sources," *IEEE J. Solid-State Circuits,* vol. SC-3, pp. 238–243, Sept. 1968.

[2] M. F. Durkin et al., "35 GHz active aperture," *IEEE MTT-S Int. Microwave Symp. Dig.,* pp. 425–427, 1981.

[3] D. R. Carey and W. Evans, "The patriot radar in tactical air defense," *Microwave J.,* vol. 31, no. 4, pp. 325–332, May 1989.

[4] D. E. Riemer, "Packaging design of wide-angle phased-array antenna for frequencies above 20 GHz," *IEEE Trans Antennas Propag.,* vol. 43, pp. 915–920, Sept. 1995.

[5] J. A. Higgins, E. A. Sovero, and W. J. Ho, "44 GHz monolithic plane wave amplifiers," *IEEE Microwave and Guided Wave Lett.,* vol. 5, pp. 347–348, Oct. 1995.

[6] J. A. Benet, A. R. Perkons, and S. H. Wong, "Spatial power combining for millimeter-wave solid state amplifiers," *IEEE MTT-S Int. Microwave Symp. Dig.,* pp. 429–432, May 1994.

[7] D. McPherson et al., "Active phased arrays for millimeter wave communications applications," *Milcom 95 Conf. Record,* pp. 1061–1065, Nov. 1995.

[8] D. M. Snider, "A theoretical analysis and experimental confirmation of the optimally loaded and overdriven RF power amplifier," *IEEE Trans. Electron Devices,* vol. ED-14, pp. 851–857, Dec. 1967.

[9] L. J. Kushner, "Output performance of idealized microwave power amplifiers," *Microwave J.,* vol. 32, no. 10, pp. 103–116, Oct. 1989.

[10] C. M. Liu et al., "A millimeter-wave monolithic grid amplifier," *Int. J. Infrared Millimeter Waves,* vol. 18, pp. 1901–1909, Nov. 1995.

[11] J. Hubert, J. Schoenberg, and A. B. Popovic, "High-power hybrid quasi-optical Ka-band amplifier design," *IEEE MTT-S Int. Microwave Symp. Dig.,* pp. 585–588, Orlando, FL, May 1995.

[12] J. A. Kinzel, B. J. Edward and D. Rees, "V-band, space-based phased arrays," *Microwave J.,* vol. 30, No. 1, pp. 89–102, Jan. 1987.

[13] M. I. Skolnik, "Nonuniform arrays," in *Antenna Theory,* Part 1, R. E. Collin and F. J. Zucker, eds. McGraw-Hill, New York, 1969, p. 233.

[14] M. A. Gouker, R. G. Beaudette, and J. T. Delisle, "A hybrid-circuit tile approach architecture for high-power spatial power-combined transmitters," *IEEE MTT-S Int. Microwave Symp. Dig.,* pp. 1545–1548, May 1994.

[15] S. Sanzgiri, D. Bostrom, W. Pottenger, and R. Q. Lee, "A hybrid tile approach for Ka Band subarray modules," *IEEE Trans Antennas Propag.,* vol. 43 pp. 953–959, Sept. 1995.

[16] H. Wong et al., "An EHF backplate design for airborne active phased array antennas," *IEEE MTT-S Int. Microwave Symp. Dig.*, pp. 1253–1256, May 1991.

[17] M. A. Gouker and L. J. Kushner, "A microstrip phase-trim device using a dielectric overlay," *IEEE Trans. Microwave Theory Tech.*, vol. 42 pp. 2023–2026, Nov. 1994.

[18] T. H. Chen et al., "One watt Q-band class A pseudomorphic HEMT MMIC amplifier," *IEEE MTT-S Int. Microwave Symp. Dig.*, pp. 805–808, May 1994.

[19] W. Boulais, "A high power Q-band GaAs pseudomorphic HEMT monolithic amplifier," *IEEE MTT-S Int. Microwave Symp. Dig.*, pp. 649–652, May 1994.

[20] H. Wang et al., "A W-band monolithic 175 mW power amplifier," *IEEE MTT-S Int. Microwave Symp. Digest,* pp. 419–422, May 1995.

[21] S. Weinreb et al., "W-band 0.3 W PHEMT MMIC power amplifier module," *IEEE 1995 Microwave and Millimeter-Wave Monolithic Circuits Symp. Dig.*, pp. 33–36, May 1995.

[22] A. A. Oliner and R. G. Malech, "Mutual coupling in infinite scanning arrays," in *Microwave Scanning Antennas,* vol. 2, R. C. Hansen ed., Peninsula Publishing, Los Altos, CA, 1985, pp. 209–216.

[23] J. A. Navarro et al., "A 29.3 GHz cavity-enclosed aperture-coupled circular-patch antenna for microwave circuit integration," *IEEE Microwave and Guided Wave Lett.,* vol. 1, pp. 170–171, July 1991.

[24] S. M. Duffy and M. A. Gouker, "Experimental comparison of the radiation efficiency for conventional and cavity backed microstrip antennas," *1996 IEEE Antennas Propag. Int. Symp. Dig.,* pp. 196–199, June 1996.

[25] K. S. Yngvesson et al., "The tapered slot antenna—a new integrated element for millimeter-wave applications," *IEEE Trans. Microwave Theory Tech.,* vol. 37, pp. 365–374, Feb. 1989.

[26] Y. H. Choung and C. C. Chen, "44 GHz slotline phase array antenna," *1989 IEEE Antennas and Propag. Int. Symp. Dig.,* pp. 1730–1733, June 1989.

[27] P. J. Gibson, "The Vivaldi aerial," *Proc. 9th Eur. Microwave Conf.,* pp. 101–105, 1979.

[28] K. A. Lee et al., "A 32 GHz phased array transmit feed for spacecraft telecommunications," *TDA Progress Report 42–111,* vol. July–Sept. 1992, pp. 310–324, Jet Propulsion Laboratory, Pasedena, CA, Nov. 15, 1992.

[29] D. Rascoe et al., "Ka-Band MMIC beam steered planar array feed," *IEEE MTT-S Int. Microwave Symp. Dig.,* pp. 809–812, June 1990.

CHAPTER THREE

Active Integrated Antennas

SIOU TECK CHEW
TATSUO ITOH
University of California, Los Angeles

This chapter discusses design issues in active integrated antennas (AIAs). Some of the considerations are unique only to AIAs, while others are generic in any passive phased array. Unlike most of the chapters in the book where the focus is on active *arrays,* this chapter focuses on design of a single array element, as well as design of "intelligent" AIAs. Intelligent AIAs integrate multiple RF functions (e.g., frequency mixing and signal processing) in addition to power generation and amplification. In AIA designs, the choice of planar antenna affects the design topology and system performance directly, due to the direct integration of the active devices and antennas. Different planar antennas and their advantages, when used in AIA systems, will be discussed. An important area that is not extensively dealt with is the simulation tool for analyzing the AIA system as an entity. Current designs treat the electromagnetic radiation and the linear/nonlinear circuit separately or model the electromagnetic characteristics with simple equivalent circuit. This has a significant impact when both the antenna and circuit are integrated. Measured system amplifier gain or oscillating frequency will be different from initial design specifications. In this chapter, the simulation tools for such circuits will be dealt with, in particular those making use of Finite Difference Time-Domain (FDTD) methods. Case studies of several circuits will be highlighted to show how certain constraints, layout and electrical in nature, are resolved.

Active and Quasi-Optical Arrays for Solid-State Power Combining, Edited by Robert A. York and Zoya B. Popović.
ISBN 0-471-14614-5 © 1997 John Wiley & Sons, Inc.

1 DESIGN ISSUES

1.1 Size of Antenna and Active Devices

As the frequency of operation scales up to millimeter-wave spectrum, the size of the antenna is reduced significantly. But the size of active circuit increases as more RF functions are integrated. This severely constrains the performance of the antenna as an element as well as an array. As an antenna, unwanted radiation from the rest of the circuit becomes comparable to that of the antenna. This impacts the position of the main beam and the side-lobe levels. As the size of individual element increases, the interspacing between the elements in an antenna array has to be large. With interspacing greater than one free-space wavelength, grating lobes appear [1]. These factors result in gain and power losses.

1.2 Surface Wave Excitation

The ultimate goal of AIA is to integrate the complete circuit monolithically on a semiconductor wafer. However, practical limitations, both in monolithic technology and electromagnetic characteristics of the antenna element, result in degradation of higher frequency performance [2]. The size of the wafer dictates a certain thickness for ease of handling. In general, the antenna also exhibits a greater bandwidth with a thicker substrate.

As the frequency of operation increases, the thickness of the substrate in terms of wavelength increases. With a thicker substrate, surface wave modes are easily excited. The dominant surface wave mode is TM_0, which has no lower cutoff frequency, and hence surface waves are always present to some degree. As the dielectric thickness increases, more power is transferred to the dominant surface wave mode and more higher-order modes are excited. Lower dielectric constant can also reduce such coupling but at the expense of the size of circuit. It has been shown that about 10% of the power radiated by a half-wavelength dipole is coupled to the TM_0 mode in the GaAs substrate of 40 μm thick at 60 GHz [3]. This is further increased to almost 100% when the substrate is a quarter-wavelength thick.

Figure 3.1 shows the effects of the surface wave modes on AIA systems. Such waves propagate in all directions, causing unwanted coupling among the antennas. Such coupling will affect the electrical behavior of individual elements and can reduce the output power-combining efficiency. Even a -20 to -30 dB coupling can be harmful to the antenna array, unless specifically considered from the onset of design [3]. For a finite size wafer or substrate, the abrupt change in permittivity at the edge of the wafer results in reflection of the surface wave. This sets up a standing wave pattern across the wafer, destroying the desired electrical response of each element in the array. Unwanted radiation is also excited by the surface wave at the edge of the wafer. Most designs assume that such surface-wave coupling is weak. This is justified by proper choice of

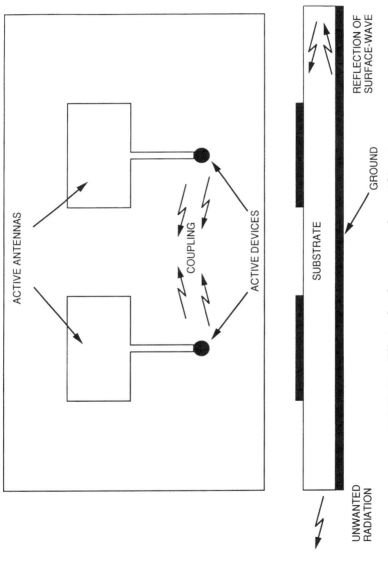

FIGURE 3.1 Effects of surface wave modes on AIA systems.

substrate thickness and dielectric constant. As such, higher-order surface modes will not be excited and power coupled to the dominant mode will be minimized. Recently, a new transmission line structure called the non-leaky coplanar waveguide (NLC) shows promising results in suppressing such modes [4]. Details of this structure and its implementation in an AIA will be discussed in one of the case studies. Another solution for millimeter-wave application is to use metal strip antenna on an ungrounded substrate [5].

1.3 Heat Sinking

One of the functions of AIAs is power generation. As such, high active device packing density is highly desirable. However, output power of active devices at millimeter-wave frequency is low with poor DC-to-RF efficiency. This causes significant electrical power to be converted into thermal energy. With such a crowded environment, heat sinking is a critical design issue. One solution is to use an antenna that has a ground plane (e.g., patch antenna), as shown in Fig. 3.2a. This allows heat to dissipate through the ground plane. Another solution is to use a material that exhibits a good conductor of heat as a dielectric layer, as shown in Fig. 3. 2b. One such dielectric material is aluminum nitride. This dielectric layer can also perturb the radiation pattern of the slot antenna to provide unidirectional radiation. A circuit using CPW, as shown in Fig. 3.2c, has also been demonstrated as a possible solution to the heat-sinking problem in monolithic integration because the real estate of the ground plane is huge [6]. Another is to use cavity-backed antenna, as shown in Fig. 3.2d, where metal contact on the ground plane allows heat sinking and unidirectional radiation [7]. Of course, the choice of active device and bias condition is important. The reader is strongly encouraged to review Chapter 13 for a detailed discussion on choice of active devices.

1.4 Free-Space Mutual Coupling

In conventional circuit design, an engineer is concerned with the coupling between circuits in close proximity. By proper circuit layout and package design, such coupling can be minimized or eliminated. However for AIA design, radiation from the circuit is encouraged, which unavoidably leads to radiative interactions between neighboring elements. This mutual coupling is one of the leading causes of discrepancies between theoretical and measured antenna patterns. This coupling can be reduced by increasing the interspacing of the elements, but at the expense of circuit real estate and possible presence of grating lobes. Another way is to deliberately introduce a stronger, more controllable coupling in the circuit that dominates the synchronization mechanism, like injection-locking [10] or mode-coupling [11]. In some cases, free-space coupling has been exploited in coupled-oscillator arrays for synchronization [9,12]. The oscillation frequency of each array elements will vary with respect to the interspacing

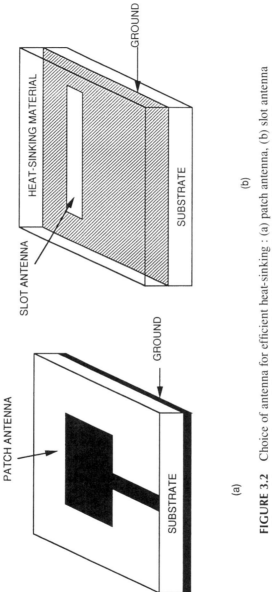

FIGURE 3.2 Choice of antenna for efficient heat-sinking: (a) patch antenna, (b) slot antenna with good heat conductor dielectric material, (c) folded slot with CPW feed, and (d) cavity-backed slot antenna. (*Continued*)

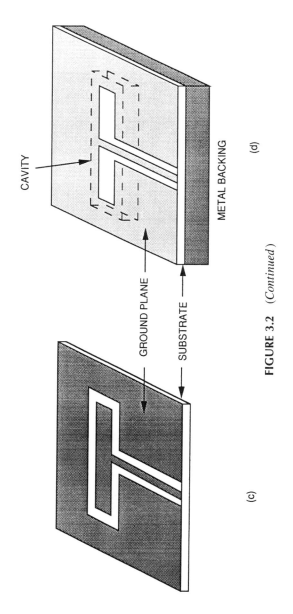

FIGURE 3.2 (*Continued*)

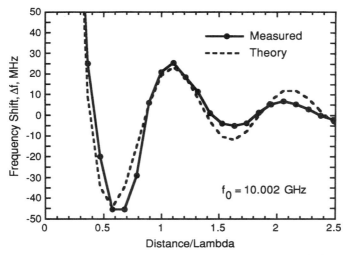

FIGURE 3.3 Measured frequency pulling due to the presence of vertical metal plate at various positions away from the oscillator. (From reference [9], © 1993 IEEE.)

between elements [8,9]. Figure 3.3 shows the variation of oscillation frequency of an oscillator with a vertical metal ground plane placed at various positions away from the oscillator to simulate the presence of an image oscillator [9]. The phase relationship among elements is also affected by such coupling. Through proper design, the oscillator elements can be synchronized to provide in-phase oscillation.

1.5 Unwanted Radiation from the RF Circuit

Besides possible unwanted radiation from the edge of the circuit board due to surface wave, radiation from the matching networks can be a problem. This is particularly true in oscillator-type active integrated antennas. The antenna functions as a resonator as well as a radiator. As such, standing waves are set up in the antenna feed and feedback networks of the oscillator. The field distribution of these standing waves then act as the sources for the unwanted radiation. They not only distort the antenna pattern but also increase the level of cross-polarized radiation.

Investigation of such unwanted radiation from matching networks of an amplifier was made in reference [13]. All of the radiating sources are modeled as dipoles with the assumption that each subcircuit is isolated electrically. The effects of horizontal or vertical microstrip lines on the antenna pattern are studied. From reference [13], it is concluded that balanced-stub matching networks result in low unwanted radiation. To minimize the complexity of the matching network, the antenna feed mechanism can be varied to provide the desired impedance match [14]. In reference [14], a cut in the radiating edge of a rectangular patch, as shown in Fig. 3.4, is made to lower the input impedance and create space for the active device. This configuration results in low unwanted radiation but maintains copolarized radiation. Another way is to isolate the circuit and the radiator through the ground plane. The signal is fed to the radiator through a via-hole wire or aperture electromagnetic coupling.

1.6 Antenna as a Resonator in Oscillator Design

For compact integration, the antenna of an AIA is commonly used as the resonator for oscillator design. Thus, the input impedance of the antenna must be precisely calculated or measured. Any discrepancy will result in a shift in oscillation frequency and a possible decrease in output power. Planar antennas are usually narrowband, which prevents wideband tuning of the oscillating frequency if the antenna is simply connected to a VCO. A solution was presented in reference [15], using a varactor-tuned patch antenna. Due to the compactness of the design and limited Q of the circuit, oscillators with antenna resonators typically suffer from excessive phase noise [16]. To reduce the complexity of

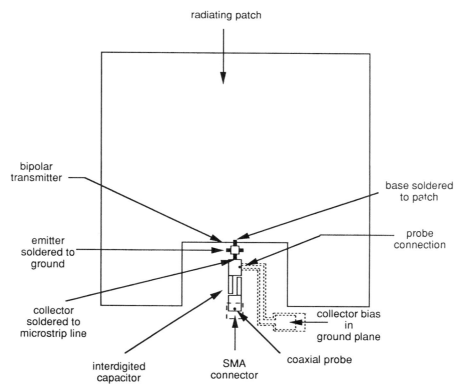

FIGURE 3.4 An quasi-optical amplifier with low cross-polarization and efficient impedance matching circuit. (From reference [14], © 1992 IEEE.)

the circuit, injection-locking is used to improve the phase noise [16], where the injected signal is assumed to possess exceptionally low phase noise. Through the injection-locking mechanism, the system will be locked to the injected signal and also inherit the phase noise characteristics of the injected signal.

1.7 Nonferrite Device Integration

As an AIA, the ideal design is to realize the whole circuit monolithically on one wafer with limited or no external interconnects. This means that all electrical functions must be implemented using semiconductor devices like MESFETs and HEMTs. These include important components like phase shifters and switches. Ferrite devices are therefore incompatible for such integration. In beam-scanning applications, the phase control can be achieved using mutual injection-locking with frequency tuning of end-elements [9] or injection-locking [17]. Examples of such phase control will be discussed later. Due to limited real estate of a wafer and high packing density of the AIA, any additional control arm that is not physically connected on the wafer is always a bonus. One such arm is the optical illumination. Through the photovoltaic and photoconductive effects, the electrical properties of the active devices can be altered [18]. In reference [19], the R_{ds} of a MESFET is varied using optical illumination for beam-switching purposes. Figure 3.5 shows the change in R_{ds} with gate bias and optical illumination.

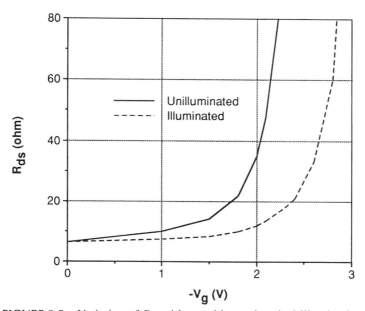

FIGURE 3.5 Variation of R_{ds} with gate bias and optical illumination.

1.8 Antenna Dynamic Load

As opposed to conventional circuit design, the termination of the circuit is no longer 50 Ω over a broadband of frequencies, but rather an antenna with limited bandwidth and multiple resonances. The separation of resonant frequencies may also be small. During simulation, all these resonances must be modeled in the load as impedance matching networks affect both in-band and out-of-band impedances directly. This potentially leads to out-of-band stability problems and possible oscillation in amplifier designs, and frequency hopping in oscillator designs.

1.9 Lack of Simulation Tool

The above-mentioned problem is one of the key reasons why there is a need for a simulation tool that can handle both the nonlinearities of the AIA and the electromagnetic properties of the antenna simultaneously. At the time of this writing, such tools are not commercially available. Most of the designs reported treat the antenna and the active circuit independently. Normally, only the in-band properties of the antenna are available using simple analysis. Effects of out-of-band termination are ignored during the initial design considerations. The design is then verified experimentally and further improved. Experience and intuitive knowledge of out-of-band antenna impedance is critical in the design process. Some of the analyses either assume weak or no electromagnetic coupling or treat it with a simple model [12]. Another key reason is that the input and/or output ports are located at free space. As such, network analysis cannot account for the losses associated with radiation, like path loss and efficiency to focus the beam.

One improved solution is to first extract the electromagnetic properties of the antenna array using full-wave analysis. The complete frequency response of the array, which includes mutual coupling, can then be absorbed into a circuit simulator as a multiport network. Once the available power to the antenna is determined by the circuit simulator, radiation characteristics can be predicted using the full-wave simulator. Figure 3.6 shows the simulation flow chart and data transfer of this method of analysis. However, this approach becomes very involved for a large network and does not allow simultaneous optimization of the circuit and antenna.

Several attempts have been made to merge these two schools of analysis. Both frequency [20] and time-domain [21,22] methods that achieve this merger have been reported. In reference [20], the passive distributed circuit is modeled as a network of lumped RLC elements. This is achieved by discretizing the domain under analysis into cells and transforming Maxwell's equations to equivalent nodal network for each cell. The currents in the network are related to the H fields, while the voltages are related to the E fields. The internodal impedance of each cell is derived from the permittivity and permeability of the medium it represents, as well as the coarseness of the grid. The analysis then uses the

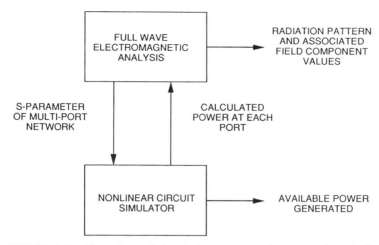

FIGURE 3.6 Flow chart of simulation process using network analysis.

harmonic balance simulator to simulate the entire AIA. Special considerations must be made to account for unwanted radiation from the rest of the circuit and mutual coupling amongst circuit elements. In the time domain, the Extended Finite Difference Time-Domain (FDTD) method is used to simulate the AIA. The nonlinear circuit is represented by lumped elements incorporated into a few FDTD cells [21] or by a SPICE-type model [22]. Of all the reported methods, FDTD shows the most promising results. As computer resources become easily available and antenna structures becomes more complicated, FDTD can become a viable and robust simulation tool. FDTD will be discussed in greater detail in a later section.

1.10 Testing in a Non-50 Ω Environment

Testing of the active antenna is an area constantly under evaluation and revision. Conventional circuits are always tested in a 50 Ω environment, and there is a vast number of equipment and measurement techniques in this testing environment. In active antenna systems the signal is transmitted into free space, and testing involves antenna measurement techniques. New measurement criteria and definitions must be established (refer to Chapter 1).

The system impedance need not necessarily be 50 Ω. An amplifier can be designed with input/output termination that matches the input impedance of the antenna [6]. The input impedance of a slot antenna is a few hundred ohms, when center-fed by a microstrip. Due to the direct integration of the active antenna, the amplifier is designed using a system impedance of several hundred ohms. As mentioned above, the antenna has a dynamic load with multiresonances with frequencies at close proximity. Such a complex load is difficult to model and substitute physically as a termination in separate component testing. This makes it a difficult task to determine the output power of the active circuit delivered to the antenna. One has to rely on the Friis transmission equation in antenna theory to extract this value. However, it is not a trivial measurement. Some reported results show combining efficiency better than 100%, indicating some discrepancy.

1.11 Others

As in passive antennas, losses in the dielectric substrate and metallization increase with frequency. Power loss becomes very high. Also, the dispersive nature of the dielectric constant makes it very difficult to achieve the design specifications in a few trials. Lower dielectric constant substrate provides better radiation efficiency but at the expense of circuit real estate. Thus, a compromise must be made between antenna radiation efficiency and compactness. As in any passive array, there is a need to be concerned with a possible blindness angle when designing beam-scanning array.

2 REVIEW OF THE FIELD

Although some form of AIA have been implemented as early as 1928 [23], quasi-optical circuits using solid-state devices at millimeter-wave frequencies was pioneered by J. W. Mink in 1986 [24]. Since then, numerous publications have appeared on this topic. An attempt is made here to review the development of single-element AIAs. These can be used as an element in an array when proper design considerations for the array have been made. Some AIA phased array will be discussed here as their main functions do not corresponds to that of power combining or beam-scanning.

The AIAs can be classified according to their applications—for example, transmitter, receiver and transceiver. All these circuits have amplifiers, differing in where the amplifiers are located in the circuit. If the input of the amplifier is connected to an antenna, the circuit functions as a receiver. If the antenna is integrated at the output of the amplifier, the circuit is a transmitter. If antennas are connected to both input and output of the amplifier, the circuit is a transceiver. This classification is ineffective in the present case since there will be a significant overlap to AIA phased arrays. Therefore, the active antenna is classified according to the RF functions of the active device used. Typical RF functions include amplification, signal generation or oscillation, and frequency conversion. Another important area that is currently under research is the integration of optical devices with active antenna. As such, these circuits will be classified into four basic groups, namely, the amplifier type, oscillator type, frequency-conversion type and optical-integrated type. The AIAs of these types are reviewed below.

2.1 Amplifier Type

This type of AIA integrates a two-port device with antennas at the input and/or output ports. When the antenna is placed at the output of the amplifier, the circuit functions as a transmitter [13,25,29]. The input is fed through a connector. When the antenna is integrated at the input of the amplifier, it is a receiver [13,14,26,32]. When both input and output of the amplifier are each terminated with an antenna, the circuit is a quasi-optical amplifier [6,27,28,30,31].

In active antenna amplifier circuits, impedance matching of the amplifier is done directly without transforming into 50 Ω [6]. Figure 3.7 shows the layout of the amplifier in [6]. The input impedance of the slot antenna is about 180 Ω. The feedback amplifier used is designed with both input and output terminated with 80 Ω. The impedance mismatch is a result of the limited range of characteristic impedances that can be synthesized in a CPW line. However, some circuits capitalized on available commercial amplifiers which are designed for 50 Ω termination [30]. Here, the antenna is matched to 50 Ω through proper antenna design.

To increase power generation, one approach is to integrate more amplifiers in one radiating element. In reference [31], a transmitting patch antenna is fed by four amplifiers. A similar configuration is used as the receiving antenna on the other side of the multilayer structure. The signal is coupled between layers through broadband slots. The advantage of this circuit is its inherent isolation between input and output ports, due to the presence of the ground plane.

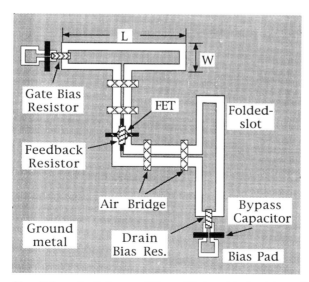

FIGURE 3.7 Circuit layout of the amplifier with input/output impedance of 80 Ω. (From reference [6], © 1994 IEEE.)

98 ACTIVATED INTEGRATED ANTENNAS

However, it suffers narrow bandwidth as patch antenna is used. Details of this circuit are presented in Chapter 6.

For quasi-optical amplifiers, the input and output antenna polarizations are normally orthogonal. This is to reduce coupling between the input and output. However, if there is a ground plane between the transmitting and receiving antenna, isolation is naturally established. In this case, polarization can be the same for both input and output signals [28].

Commonly used radiators for the active antenna amplifier circuits are patch antenna and slot. Some extend the design to traveling wave antennas, as shown in Fig. 3.8 [25]. Here, the amplifier may have zero gain but necessary phase shift for desired radiation characteristics. But with amplifier gain, the signal is amplified in the circuit, resulting in an increase in system gain.

2.2 Oscillator Type

As opposed to amplifiers, oscillator-type active integrated antenna circuits have received a lot of attention. The purpose of this type of AIA is power generation, as opposed to power amplification. As such, there is no need for a signal source.

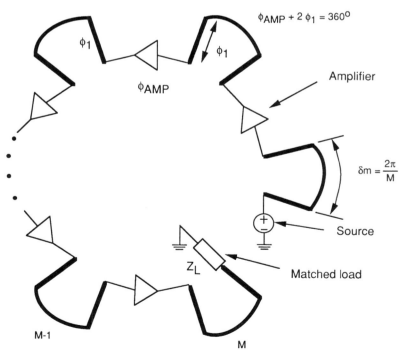

FIGURE 3.8 A traveling wave active antenna amplifier circuit. (From reference [25], © 1991 IEEE.)

To date, oscillator designs using two-terminal [33,34,36,37,51,52] and three-terminal [15,35,36,38,50,53,54,63] devices have been used. Two-terminal devices have a higher frequency of operation, but with poorer DC-to-RF efficiency. Currently, a three-terminal InP based HFET device shows promising results at the millimeter-wave frequencies [46]. As in amplifier-type circuits, high packing density design is exploited to increase power generation and suppression of harmonics. In reference [42], the oscillators are designed in the push–pull configuration. Figure 3.9 shows the circuit schematic diagram of the circuit. The patch antenna can support odd and even mode field distributions. The oscillator configuration is designed to excite the odd mode only. In reference [63], a push–push oscillator is designed as a frequency-doubling source. The balanced oscillator uses a coupled rampart line array as the antenna element and resonator.

Commonly used antennas in oscillator design are the patch antenna [33,35,36,38–40,42–44,47,49,52] and slot antenna [41,45,46,48,51]. However, for broadband applications, notch [36] and Vivaldi [50,54] antennas are

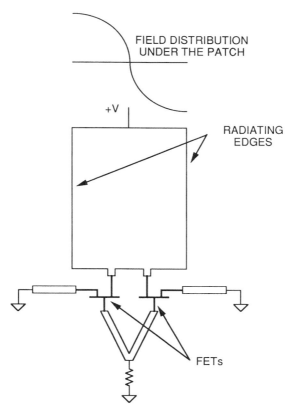

FIGURE 3.9 Schematic diagram of a push–pull quasi-optical oscillator.

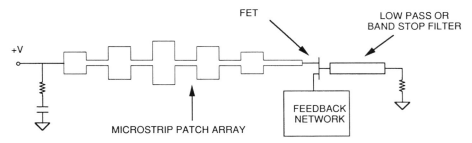

FIGURE 3.10 A leaky wave antenna quasi-optical oscillator.

used. There is also an expansion of AIA to leaky wave antennas [53,63]. As shown in Fig. 3.10, the AIA circuit uses the stopband characteristic of the antenna to satisfy the reflection condition for oscillation [63]. Normally, the antenna is used as a resonator for oscillation and as a antenna for radiation. To provide additional degrees of freedom for design, some AIAs do not rely on the antenna as a resonator, preferring to use external resonators instead [41,51].

As mentioned in the earlier section, injection locking is used to improve the phase noise characteristics of the AIA oscillator. The signal injection is made possible at circuit level [10]. However, spatial injection locking has also been demonstrated [40]. To enhance the injection- locking bandwidth, a two-port oscillator is designed [38].

Besides power generation, some of these circuits are designed for some "intelligent" functions. These functions include polarization-agile switching [43,49], beam-switching using injection-locking for radar application [84] and omnidirectional coverage [54]. Polarization-agile active antennas use two active devices to control the polarization of the transmitted signal. By switching "on" the desired active device, a particular linear polarization can be achieved. Turning both devices "on" can result in linear or circular polarizations. For proper phase reference, an injection signal is needed to lock the oscillators. In beam-switching active antenna, injection-locking is used to achieve the desired phase relationship to generate the sum or difference patterns. Both patterns in both azimuth and elevation planes can be synthesized. Detailed discussion of this circuit will be presented in the case studies section. To increase the phase tuning range, subharmonic injection-locking has been used [47]. The omnidirectional coverage Vivaldi active antenna will be discussed in Chapter 13.

2.3 Frequency-Conversion-Type Circuits

In frequency-conversion-type AIAs, integrated antennas function as a receiver, transceiver, or transponder. In these three cases, design emphasis is placed on optimizing frequency conversion loss. Some of the designs use diodes as passive mixers [55,57,59,61,65,66]. As such, a local oscillators must be integrated into the design. In reference 56, both the LO and RF signals are transmitted quasi-optically to the AIA, as shown in Fig. 3.11. Polarization diversity is used to

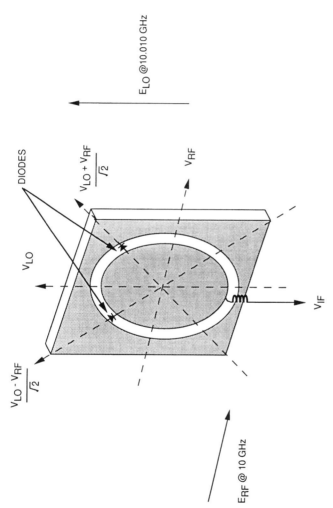

FIGURE 3.11 A quasi-optical mixer using polarization as isolation between LO and RF.

102 ACTIVATED INTEGRATED ANTENNAS

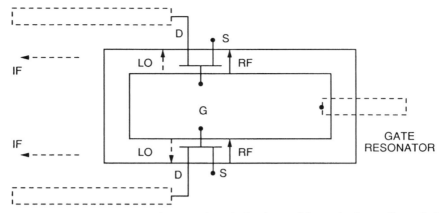

FIGURE 3.12 A self-oscillating quasi-optical mixer with mode-decoupling of the folded slot between LO and RF.

isolate both signals for good port-to-port isolation. Some circuits generate the LO internally using self-oscillating mixers [58,60,62,64,67]. Figure 3.12 shows a circuit whereby mode decoupling of the folded slot is used to prevent injection-locking of the RF to the LO signal [58]. For millimeter-wave applications and circuit simplicity, subharmonic mixing is used [57]. Here, the LO is delivered to the circuit quasi-optically. Subharmonic mixing is achieved by using the higher-order harmonics of the LO, generated by the mixer, to mix with the incoming RF signal. In reference [57], the second harmonic of the LO is used.

In transceiver AIAs, the circuit transmits a signal and processes the returned signal [62,63,65]. The frequency of the received signal is usually close to that of the transmitted signal, such as in Doppler radar applications. In reference [62], a microstrip annulus antenna with a narrow slit is integrated with two FETs as an oscillator. The FETs also function as self-oscillating mixers. From the in-phase and out-of-phase IF signals generated by mixing the oscillating signal with the returned Doppler-frequency shifted signal, the copolarized and cross-polarized returned signal can be determined.

In transponder design, the circuit first receives the signal from free space [60,61]. The signal is processed and then transmitted out. Normally the processing involves frequency transformation and signal amplification. In reference [60], the frequency conversion is performed by a self-oscillating mixer. In reference 61, subharmonic mixing is used. The details of this circuit will be discussed in a later section on case studies.

2.4 Optical-Integrated Type

Optical links or devices can be integrated to AIAs for additional degrees of freedom, improved electrical isolation, or remoting. Complete optical circuits integrated to microwave phased array have shown promising results in con-

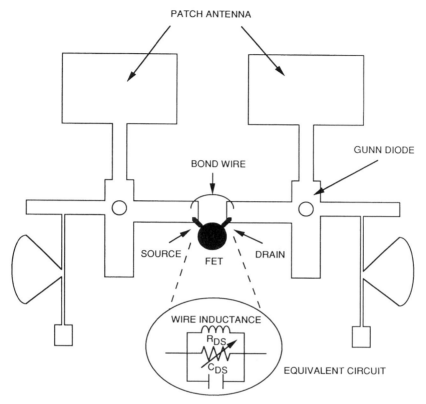

FIGURE 3.13 Schematic diagram of a two-element beam-switching array.

trolling antenna patterns [68]. In this section, emphasis is placed on integration of microwave and optical circuits to form the AIAs. There are various ways and mechanisms one can use for microwave-optical interactions. Such interactions can be broadly classified into two categories. One uses light as a simple bias control. Another uses the coherent property of the optical signal. It has been shown that the electrical properties of a semiconductor device can be controlled when illuminated optically [18]. This is achieved through photovoltaic and photoconductive effects. Thus, a FET or HEMT can be tuned optically for different oscillation frequency in an oscillator design [69,70]. Also, the R_{ds} of a FET can be controlled optically to present a high or low impedance, as shown in Fig. 3.5 [19]. This is used to control the different modes of operation in a phased array for beam-switching [19]. Figure 3.13 shows the schematic diagram of the two-element phased array. A FET, with its package removed, is placed at the center of the coupling transmission line to present either a high or low impedance termination. A bond wire is included to control the impedance level. By optical illumination, the center of the transmission line is either a

104 ACTIVATED INTEGRATED ANTENNAS

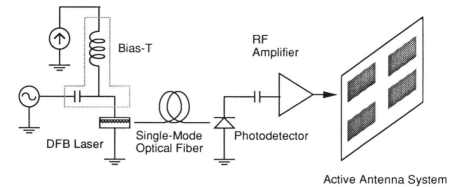

FIGURE 3.14 Schematic diagram of the optical/active antenna system.

short-circuit or inserted with some resistance. This causes the mode of the oscillators to change and the array radiates either sum or difference patterns.

When the coherent property of optical signal is used, light is used as a carrier for the RF signals. It is particularly attractive because the optical fiber is low-loss and light in weight. This makes remoting of microwave AIA systems, especially radar systems, highly viable [71]. The laser is modulated by a RF signal and light is launched into an optical fiber, as shown in Fig. 3.14. The signal is then recovered at the other end by a photodetector. In reference [71], the RF signal is the reference signal for injection-locking of the beam-scanning active antenna system. In reference [72], the optical link is not just for transmission. It also serves as an RF multiplier by configuring the laser to operate under gain-switched mode. The gain-switched laser generates a train of short pulses in time domain. In frequency domain, this corresponds to a frequency spectrum with high harmonic content. It is these generated harmonic signals that are tapped for active integrated antennas. Here, the second harmonic signal is used as a reference signal for a Doppler transceiver.

3 CHOICE OF PLANAR ANTENNAS

3.1 Introduction

Monolithic integration of active devices and antenna requires planar antennas. Even in a hybrid MIC environment, such planar antennas are advantageous as they are physically low-profile, lightweight and potentially conformal. This is particularly important for AIAs that emphasize compact integration of RF functions rather than power generation. Feed structures are easy to integrate with the antenna. By simple changes of the feed position, linear or circular polarized radiations are possible. Dual frequency antennas are also easily synthesized with planar antennas. One disadvantage of planar antennas are their limited bandwidth. They also suffer higher losses, resulting in lower antenna gain. One important disadvantage is that there is poor isolation between the antenna and the feeds. Physical size limits the power-handling capability of these planar antennas.

Here, various commonly used planar antennas are presented. Their compatibility to AIA design concept will be highlighted. Advantages and disadvantages of the antenna structures, in view of AIA design considerations, will also be included. Analysis and synthesis of such antennas will not be discussed in detail as there are extensive list of literature dedicated to such topics [73–76]. The most commonly used planar antennas in single-element AIAs are the patch antennas (including stacked patch) and the slot antennas (including folded slot). For broadband applications, commonly used antennas are the Vivaldi and notch. Some other, less commonly used antennas will be briefly presented.

3.2 Patch Antenna

A patch antenna consists of a conducting patch of arbitrary planar geometry on one side of the dielectric substrate, while the other side is the ground plane. It is a parallel resonant circuit. Common geometries are the rectangular, square, circular, ring, triangle, and ellipse. Of these, rectangular, square, and circular patch antennas are the most popularly used in AIAs. The rectangular and circular patch antennas provide linear polarization. By using the square or circular patch antennas, circularly polarized radiation can be synthesized when fed appropriately at two ports [76]. By using slant-slot in aperture-coupled patch antenna, circularly polarized radiation can also be excited. Advantages of such antennas are that they are thin and can be easily fabricated. Excitation of these antennas is easy. The disadvantage of these antennas is that they are extremely narrow band, of the order of 1–5%. For high dielectric substrate, the resonant frequency is very sensitive to dimensional deviation.

Commonly used feed structures are shown in Fig. 3.15. For a single-layer structure, the antenna is normally excited by a microstrip line connected at the center of one radiating edge, as shown in Fig. 3.15a. For multilayer substrate design, via-hole wire or aperture coupling are used, as shown in Fig. 3.15b and 3.15c,d, respectively. When aperture coupling is sought, the patch antenna is commonly referred to as *stacked patch*. The advantage of this structure is that there are two different substrate layers. The antenna layer can be of low dielectric permittivity for radiation efficiency, while the feed substrate can be of high dielectric permittivity for circuit compactness and low radiation loss. Inherent in the structure is a ground plane which acts as a natural isolator for the circuit and antenna. This prevents spurious radiation from the circuit and unwanted signal pick-up by the circuit from free space. Also, this structure exhibits a wider bandwidth than the single-layer patch antenna. The signal is coupled between the two layers through a slot cut on the ground plane. The effect of the thickness of the ground plane has also been studied [77]. Another form of aperture is that of a circular hole [78].

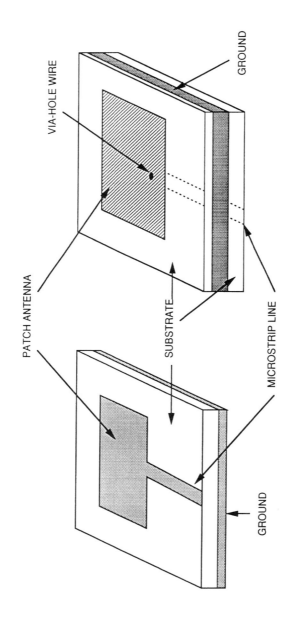

FIGURE 3.15 Various feeding mechanisms of a patch antenna: (a) Direct microstrip, (b) via-hole wire, (c) slot-aperture coupling, and (d) circular-aperture coupling.

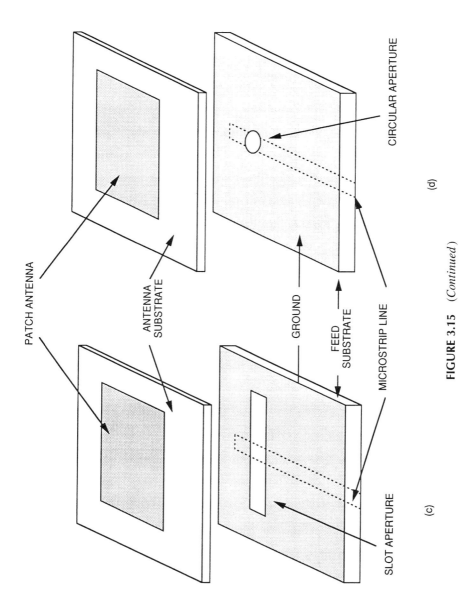

FIGURE 3.15 (*Continued*)

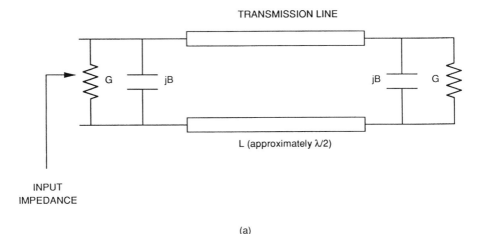

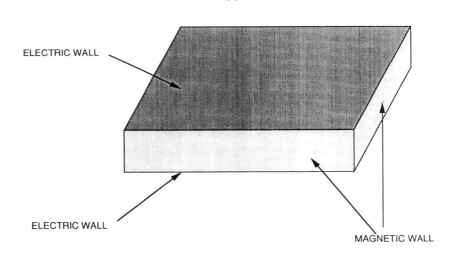

FIGURE 3.16 Various models of the patch antenna: (a) Transmission-line and (b) cavity.

Square and rectangular patch antennas can be analyzed easily using simple transmission line model. The circuit model includes two slots at the edges of the antennas connected by a microstrip transmission line, as shown in Fig. 3.16a. The length of the transmission line is of one-half guided wavelength. The antenna can also be treated as a resonant cavity with magnetic side walls, as shown in Fig. 3.16b. Being a cavity, radiation antenna efficiency is expected to be low. More complex analysis uses modal analysis or full-wave approaches. Cavity models (with Bessel functions) can also be used for circular patch antennas. Excitation is normally through via hole or direct connection at the edge.

Direct microstrip feed at the edge of the radiating edge results in higher cross-polarized and spurious copolarized radiation. This is due to the position of the feed on the edge and the impedance transforming network. By feeding off-center, cross-polarized radiation increases. Matching network is normally large in size and may have standing wave patterns on the transmission line. This causes spurious radiation. The feed structure can be modified as shown in Fig. 3.4 for proper impedance matching. By doing so, the radiation characteristic is not perturbed significantly. Due to the proper impedance termination, simplicity of the active circuit is achieved with low spurious radiation.

It is also possible to design dual frequency or frequency diversity systems with such antennas. Since the length and width of the patch antenna can be varied independently, various higher-order modes that are either harmonically or nonharmonically related can be generated. As such, the patch antenna can have resonances at the desired frequencies (normally TE_{10} and TE_{01}). These two resonances are tapped for the design of dual frequency operation. Also, the patches in different layer of a stacked patch can be of different size to create various possible dual frequency operation.

3.3 Slot Antennas

Slot antennas are exploited in active antennas due to their broader bandwidth. These antennas are slots of length either one-half or one wavelength long cut on the ground plane, as shown in Fig. 3.17. Slot antenna is a series resonant circuit. They are less sensitive to fabrication tolerance, but are more difficult to fabricate as the slot antenna is etched on the ground plane. Alignment of the antenna to the microstrip line and additional fabrication processes are needed. Another important feature is that it allows the designer to capitalize either bidirectional or unidirectional radiation. Good isolation is made between the antenna and feeds. It is not possible to create dual frequency antennas with the slot antenna. To improve bandwidth, a wider slot can be used. Simple analysis uses the Booker's relation to relate the input impedances between the slot and the well-known dipole. More complex analyses use the Method of Moments to determine the field distribution on the slot.

The slot antenna is usually fed by a microstrip line. The axis of the microstrip is perpendicular to that of the slot, as shown in Fig. 3.17a. Energy is coupled electromagnetically. The microstrip line is extended a quarter wavelength to present a short-circuit at the feed point. A via hole can be used instead, as shown in Fig. 3.17b. When excited, the slot antenna has a field distribution exhibiting a half-cosine standing wave pattern, for a one-half wavelength slot. The input impedance is high when fed at the center. To reduce the input impedance, the slot is fed off-center. It is worth noting that as the feed position changes, the resonant length of the slot also changes. A center-fed slot is longer than that of an offset-fed slot [73].

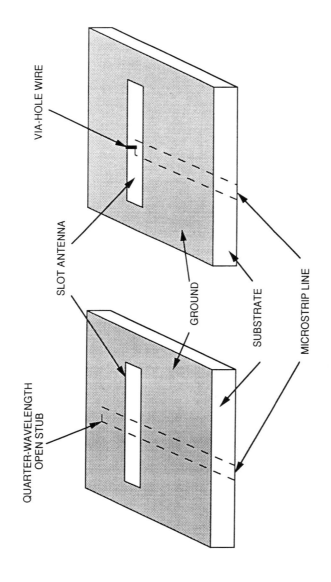

FIGURE 3.17 Various feeding mechanisms of a slot antenna: (a) Electromagnetic coupling and (b) via-hole wire.

CHOICE OF PLANAR ANTENNAS   111

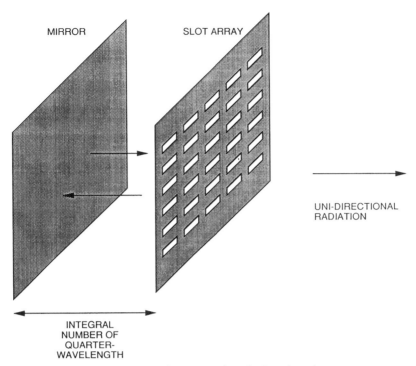

FIGURE 3.18 Mirror placed behind a quasi-optical active slot antenna to provide unidirectional radiation.

Unidirectional radiation is used commonly for quasi-optical reflection amplifier. A mirror shown in Fig. 3.18, consisting of a metal plate, is placed behind the circuit at a distance of integer number of free-space quarter wavelength. This reflected signal from the mirror then combines in-phase with the signal radiating in the other direction. Also, by placing at multiples of free-space quarter wavelength, the impedance of the mirror is transformed into an open-circuit at the circuit reference plane. However, three-quarter wavelength is commonly used as a distance of one-quarter wavelength perturbed the antenna input impedance significantly. Bidirectional radiation is used for transmission mode operation for quasi-optical amplifier design. The transmitting and receiving slots are of orthogonal polarizations. To allow signal to radiation in the desired direction, polarizers are used to screen out orthogonal polarized signal.

Besides using slotline, other forms of slot antenna uses CPW and NLC structures. CPW slot allows integration of active circuits in the same plane. To allow ease of impedance matching, a multislot antenna is used to reduced the input impedance [30].

3.4 Others

Complementary to the slot antenna is the dipole. Printed dipole antenna is used in grid-approach active antennas. Such antenna is small, hence occupying lesser space. Also, being on the same side as the active devices, integration is easy. In grid active antennas, the dipole is analyzed as an inductive stub in a waveguide cell. One disadvantage of using the printed dipole is that there is no ground plane. This will have an effect on the heat-sinking problem. For very broadband applications, bow-tie antennas are used. The disadvantage of these antennas is the requirement of a balun for proper feeding. Other structures are the notch and Vivaldi antennas. Another disadvantage of broadband antennas is the large physical size.

4 FDTD ANALYSIS AND VISUALIZATION

4.1 Introduction

As mentioned above, there is a strong need for a simulator that can provide full-wave analysis of the electromagnetic phenomena, radiation and coupling, of the active antennas and can support linear and nonlinear circuit analyses. This allows integration of the antenna and circuit at the onset of design simulation. Mutual coupling (free-space and surface wave) must be considered because it will affect the system performance, such as synchronization and phase distributions in a coupled oscillator array. To optimize the real estate of the circuit and increase the bandwidth of the antennas, multilayer structures are often sought. Such complex structures are difficult to analyze analytically. One important requirement is that the simulator must provide not only the radiation characteristic of the antenna but also any unwanted radiation from the rest of the circuit. This is critical because the system gain hinges heavily on the antenna directivity. Another key factor is that the simulation tool must be able to consider input/output ports at free space. In short, the simulation tool must be able to introduce a signal into free space, capture the signal from the AIA's effective antenna aperture, provide prediction of the linear/nonlinear mechanism in active devices, launch the signal back into free space, and receive the processed signal at free-space.

To date, such simulators are limited and preliminary. Both frequency-domain [20] and time-domain [21,22,79] approaches have been reported. In the frequency domain, a circuit-based method is used concurrently with harmonic balance method. In the time domain, the extended FDTD [21,79] and SPICE-type FDTD [22] methods are used. The complexity of the circuit and the antenna makes FDTD the favored simulation tool. The initial preparation overhead for simulation of a complex circuit is low once the algorithm is coded. Analytical formulation may not be available for certain circuit geometry. However, FDTD suffers in numerical computational efficiency and massive computer memory. These constraints are fast becoming lesser an issue as computer technology advances.

4.2 FDTD

FDTD was first formulated by Yee [80] in the 1960s. It was not popular at that time due to limited computer resources. Because there is extensive literature published in this topic, the detailed formulation of FDTD will not be discussed. However, a brief review will be presented with emphasis on the integration of linear/nonlinear circuit analysis in FDTD.

In FDTD, the region under consideration is discretized into cells, as shown in Fig. 3.19. On each side of the cell, Maxwell's equation are approximated by difference equations. The E- and H-field components are physically offset by half cell. This allows the curl relationship between the E and H fields to be discretized to second-order in accuracy. Also, the time variable is discretized into time steps. The spatial staggering of the E and H fields also requires staggering in time, which leads to a simple computational algorithm. The H-field components are first calculated using existing knowledge of the E-field component. The new E-field components are then calculated using the new H-field components. The order of which field to calculate first is arbitrary. The choice of the time step and cell size is critical to ensure numerical stability.

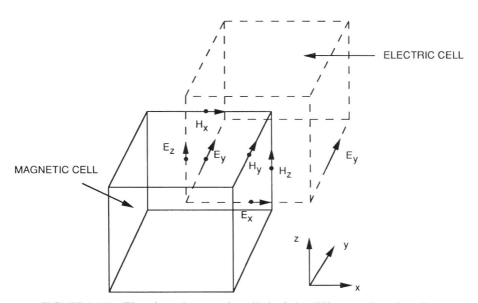

FIGURE 3.19 Electric and magnetic cells in finite difference time domain.

Associated with this method is the issue of signal excitation and boundary conditions at the boundaries of the domain under consideration. The source can be a single sinusoidal signal for single-frequency analysis or a Gaussian-modulated pulse for broadband analysis. Absorbing boundary conditions are used at the boundaries to prevent reflection of signals from the artificially created boundary of the computational volume. This reduces the domain under analysis. The analysis is considered completed once the signal reaches steady state in oscillator design, or the field strength decays to a negligible level within the domain under consideration for amplifier design. For AIA, the field distribution of a 3-D surface enclosing the circuit is constantly updated in the frequency domain. This is achieved by using discrete Fourier transform at any desired frequencies. Once the analysis converges, spatial transformation at each frequency is made to transform the near-field characteristics to the desired far-field radiation characteristics. To reduce computational time through parallel-processing, the diakoptics method can be used [81].

4.3 Integration of the Active Device in FDTD

To allow integration of active device in the analysis, the current density component in the Maxwell's equation is sought. Initial implementation incorporates the lumped circuit into the grid cell [21]. That is, the lumped component is absorbed into the coefficient of the FDTD field updating equations. In reference [21], the Gunn diode is modeled over three vertical cells and one horizontal cell. Although the modeling is distributed over a few cells, the diode characteristics cannot be calculated locally at each cell. Doing so would result in modeling of a series of diodes, instead of a single diode. This can cause DC stability problem. The voltage across the lumped component must be derived from the vertical line-integral of the E field on the cells that represent the active region. One disadvantage of this method is that a change of the lumped circuit dictates a change in the formulation of the cell characteristics. An alternative implementation is to collaborate with SPICE^{TM}-like simulator for the lumped circuit [22]. The lumped circuit is treated as a black box and is analyzed by SPICE^{TM} separately. The interaction between FDTD and SPICE^{TM} is linked through the I–V relationship. FDTD provides an input current to SPICE^{TM} through an integration of the H field at the interface. SPICE^{TM}, being a time-domain simulator, then advances the time step to generate a voltage through the circuit differential equations. This voltage is then introduced into FDTD through the update of the E field. As such, amplifying mechanisms as well as impedance mismatches are accounted for.

4.4 Simulation of Active Antennas

A two-element active antenna, as shown in Fig. 3.20, is simulated using FDTD, using both extended FDTD [21,79] and SPICE^{TM}-type [22] analyses for the lumped circuit. This antenna exhibits in-phase oscillation when a 5-Ω resistor

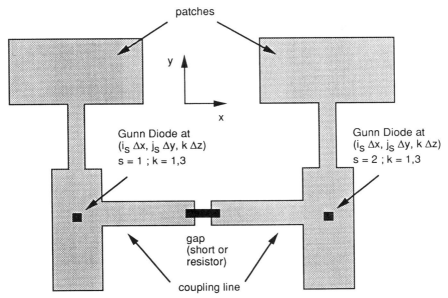

FIGURE 3.20 Two-element active antenna simulated using FDTD. (From reference [79], © 1994 IEEE.)

is placed at the axis of symmetry. An out-of-phase oscillation dominates the system when the resistor is replaced by a through line. FDTD simulation converges to the correct mode of oscillations for both cases. Figure 3.21a shows the oscillation signals at both antennas in the time domain for both cases. As expected from frequency domain analysis, FDTD shows the in-phase and out-of-phase oscillations. Figure 3.21b shows the measured and calculated antenna patterns of the array for both modes. Figure 3.21c shows the field distribution of the E field in the substrate for a particular time step for both cases. The above results are simulated using the extended FDTD analysis for the lumped circuit.

5 CASE STUDIES

As indicated above, there are several design issues that need to be addressed for a sound and workable AIA. Some of these issues will be highlighted in the following circuit examples. These selective case studies are used to provide a representative range of applications and component/antenna choices.

5.1 Gunn/Patch Oscillator [82,83]

In the first example, a rectangular patch antenna is integrated with a Gunn diode to function as a quasi-optical oscillator. The design is based on a large signal

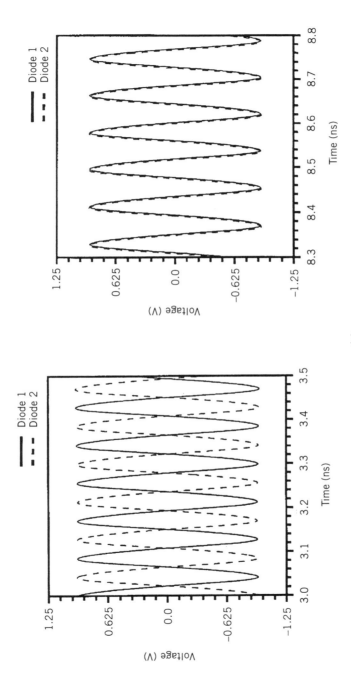

FIGURE 3.21 Measured and FDTD results of the two-element array oscillating in the in-phase and out-of-phase modes: (a) Simulated time-domain signals, (b) measured and calculated antenna patterns, and (c) simulated electric field in the substrate at a particular time step. (From reference [79]. © 1994 IEEE.)

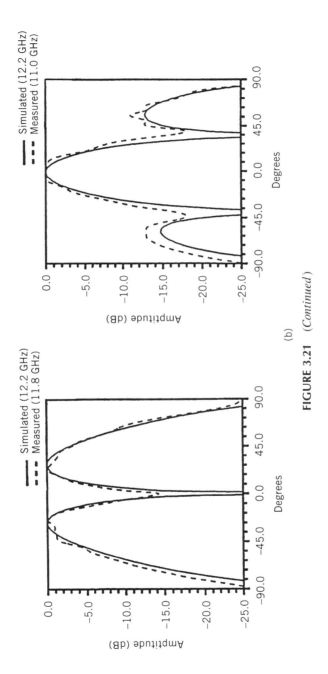

FIGURE 3.21 (*Continued*)

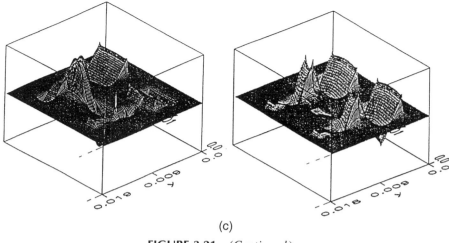

(c)

FIGURE 3.21 (*Continued*)

load-pull method. The schematic diagram is shown in Fig. 3.22. The circuit consists of a Gunn diode as an active device and a rectangular microstrip antenna as an output load. The antenna also functions as a resonator. An inductive open stub is included to cancel the capacitive part of the Gunn diode. A two-stage quarter-wavelength transformer transform the antenna input impedance to 35 Ω to satisfy oscillation condition. A two-stage transformer is used to maintain the resonator as a parallel circuit. Two half-wavelength transmission lines have been incorporated for coupling to adjacent oscillators in an array.

The antenna is first calculated using first order equations in reference [75]. Then, a rigorous analysis is made using Essof's EMSim. With a dimension of $a = 377$ mils and $b = 268$ mils, the resonant frequency of the TE_{10} mode was found to be 12.4 GHz with an input impedance of 240 Ω. The Gunn diode is first connected to an impedance tuner. The tuner is adjusted for maximum oscillation power and desired oscillation frequency. The impedance of the tuner and the embedding network is then determined using the network analyzer. With the known impedance for maximum power, impedance matching network is designed to match the negative resistance of the Gunn diode under the operating condition.

The strong-coupling theory used in the array will not be dealt with here as Chapter 6 provides a detailed discussion. The measured oscillation frequency is 12.45 GHz. The measured H-plane pattern is shown in Fig. 3.23.

5.2 Active Slot Antenna [48]

Here, a quasi-optical oscillator using an HEMT and a slot antenna will be discussed. As mentioned in the first section, conductor-backed coplanar wave-

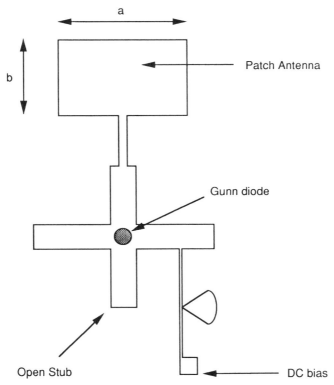

FIGURE 3.22 Schematic diagram of a single-element quasi-optical Gunn oscillator.

guide (CBCPW) and slotline (CBSL) are used in MMIC for their mechanical strength and heat-sinking properties. However, the presence of a conductor backing causes leaky modes to exist. One solution is to use a non-leaky coplanar (NLC) waveguide developed at UCLA [4]. This waveguide has an additional lower dielectric layer inserted between the original substrate and ground plane, as shown in the inset of Fig. 3.24. The oscillator is designed using the NLC structure.

First, the substrate thickness and dielectric constant of each layer is calculated using spectral domain method to ensure that there is no leakage. Figure 3.24 shows that for the frequency band of interest, there is no presence of leaky modes. The thickness and dielectric constant of each layer are shown in the caption of Fig. 3.24. Once, the substrate characteristics are determined, the one-wavelength slot antenna was designed. The slot is 240 mils long and 24 mils wide. Spectral domain method is used to design the slot antenna at 20 GHz. FDTD is also used to determine the input impedance of the slot. Both SDA and FDTD show good agreement in the calculation of input impedance. At 20 GHz, the input impedance is $Z = 13.2 - j4.1 \, \Omega$.

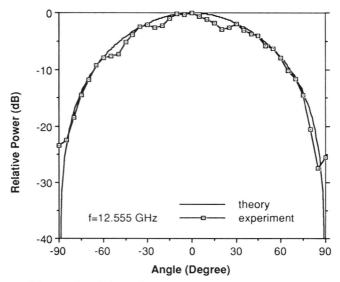

FIGURE 3.23 Measured and theoretical antenna patterns of the single-element quasi-optical Gunn oscillator.

Figure 3.25 shows the layout of the circuit. A GaAs HEMT is used as the active device. Short-circuited stubs are used at the source of the HEMT for feedback and DC ground. An open-circuited stub is connected at the gate to satisfy the oscillation condition. Narrow slits are cut in the ground plane to provide DC isolation between the drain and gate ports.

Measurements show that the oscillation frequency is at 19.82 GHz when the HEMT is biased at $I_{ds} = 10$ mA and $V_{ds} = 2.0$ V. The measured EIRP is 5.2 dBm with an isotropic conversion efficiency of 16%. Measured copolarized and cross-polarized antenna patterns are shown in Fig. 3.26.

5.3 Noncontact ID Transponder [61]

In this example, the design of a transponder is discussed. A transponder receives a signal from free space and processes it (e.g. frequency conversion). Instead of residing in the system, the processed signal of a transponder is transmitted back into free space. There is a wide range of applications for such systems—for example, surveillance, repeater, and identification. Here, the transponder under discussion is used for noncontact identification. This can be used in computerized systems for security and control tasks. The attractive feature of this design is that microwave signals are used, which have lower signal noise and interference levels than conventional fundamental or harmonic receivers.

The schematic diagram of the transponder circuit is shown in Fig. 3.27. All the circuit functions are integrated on a substrate of size comparable to that of a credit card. An interrogating system transmits a signal of frequency f_0 (6 GHz is used). This signal is received by the transponder on a card through a planar

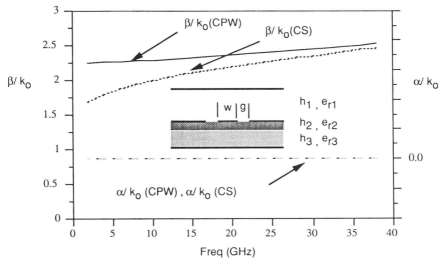

FIGURE 3.24 Calculated normalized phase and attenuation constants vs. frequency of a non-leaky coplanar waveguide ($\epsilon_{r1} = 1$, $\epsilon_{r2} = 10.8$, $\epsilon_{r3} = 2.2$, $h_1 = 30.0$ mm, $h_2 = 0.635$ mm, $h_3 = 3.175$ mm, $w = 0.813$ mm, $g = 0.406$ mm). (From reference [48], © 1995 IEEE.)

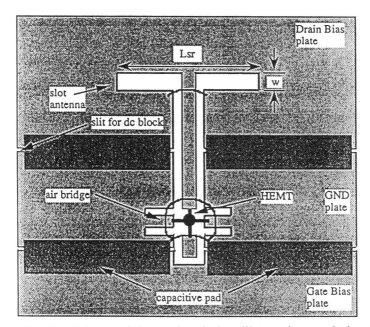

FIGURE 3.25 Circuit layout of the quasi-optical oscillator using non-leaky coplanar waveguide. (From reference [48], © 1995 IEEE.)

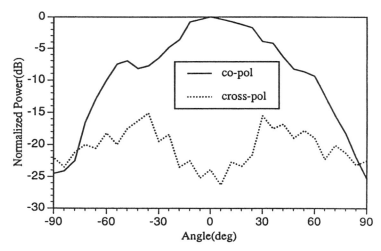

FIGURE 3.26 Measured antenna pattern of the non-leaky coplanar waveguide quasi-optical oscillator. (From reference [48], © 1995 IEEE.)

bowtie antenna. An antiparallel diode pair is placed at the terminals of the antenna. This design is based on a quasi-optical mixer structure first proposed by Stephan and Itoh [57]. A bow-tie antenna is chosen because its broad bandwidth allows reception of the interrogating signal and simultaneous transmission of a processed signal an octave apart in frequency. The antiparallel diode pair is used as a subharmonic mixer and generates only odd order mixing products. All even order products circulate in the diode loop. The dominant mixing term is $2f_0 \pm f_d$, where f_d is the locally generated frequency in the card (10 MHz is used). This allows the processed signal to be not harmonically related to the transmitted signal, hence eliminating interference. The 10-MHz signal is generated using discrete transistor and a crystal. Low-power CMOS digital circuit is used to generate the digital word for ASK modulation of the 10-MHz signal. All digital circuitry is supported by a 1.5-V silver oxide battery. To conserve DC power, a one-wavelength slot antenna in the ground plane with an impedance-matched Schottky diode is used to detect the presence of an interrogating signal, thus triggering the modulation. The conversion loss of the diode-pair with respect to isotropic transmitted power is shown in Fig. 3.28. The interrogating signal suffers $1/r^2$ path loss, where r is the distance between the interrogator and transponder. Since subharmonic mixing is used, the resulting signal shows a dependence of $1/r^4$. As the signal is transmitted back to the interrogator, the returned signal shows a resultant relationship of $1/r^6$ path loss.

5.4 Monopulse Switch [84]

Here, an active antenna monopulse system is designed to replace the complex waveguide feed structure of the phased array. This feed provides the necessary

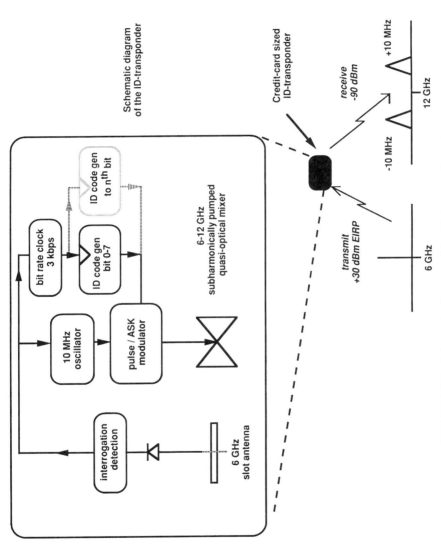

FIGURE 3.27 Schematic diagram of the ID transponder system.

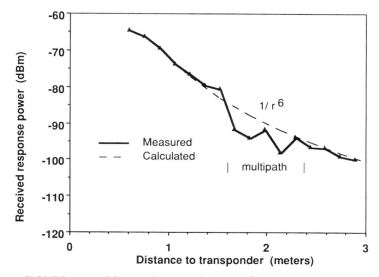

FIGURE 3.28 Measured conversion loss of the ID transponder.

phase and power relationships for the synthesis of the sum and difference patterns. Using different combinations of phase relationship results in various patterns in both planes. However, such feed is bulky and adds weight to the antenna platform. The active antenna system synthesizes these phase relationships using injection-locking.

When a reference signal, whose frequency is close to that of the free-running oscillator, is injected into the oscillator, injection-locking phenomenon occurs. The oscillating signal is locked to the reference frequency, but exhibits a phase difference with respect to that of the reference signal. Based on Kurokawa's theory [85], this phase difference, ϕ, is related by the following equation:

$$\phi = \sin^{-1}\left(\frac{\omega_f - \omega_0}{\omega_m}\right) \tag{3.1}$$

where ω_f is the free-running angular frequency, ω_0 is the injected angular frequency, and $2\omega_m$ is the locking bandwidth. From (3.1), ϕ can be tuned to a maximum of $\pm 90°$ by varying either ω_f or ω_0. ω_m increases with higher injected power. Here, the injected reference signal is fixed with an angular frequency, ω_0. From (3.1), all the oscillating elements will be in-phase if the free-running angular frequency of each element is tuned to that of the injection frequency. By tuning to the band edge, the phase difference between the injected and oscillating signals is either $+90°$ or $-90°$. The tuning can be continuous but is discretized in this design. The phase states of interest are $\pm 90°$ and $0°$. These states are realized by tuning the free-running frequency as dictated by (3.1).

The schematic diagram of the 2 × 2 array is shown in Fig. 3.29. Each element consists of a rectangular patch antenna and an FET. The antenna is

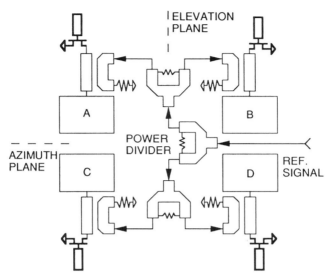

FIGURE 3.29 Schematic diagram of the 2 × 2 beam-switching array.

edge-fed with a microstrip line. The input impedance of the patch is measured using TRL calibration technique on HP8510B. A coupler is integrated to cater for injection-locking. The oscillating frequency is about 6.62 GHz and can be varied independently by at least 40 MHz via their respective drain bias. The upper quadrants and the lower quadrants are fed on the opposite sides of their respective patches, resulting in 180° excitation. This is due to space constraint. Also, the design places emphasis on the difference pattern for a well-defined null. By providing 180° excitation, instability due to injection-locking is avoided [86]. A comparison is made with the theoretical patterns, calculated from a simple model [1].

The measured sum and difference patterns in the azimuth plane is shown in Fig. 3.30a. The measured ERP of the sum pattern is 30 dBm. The null at broadside of the difference pattern is better than −30 dB. There is asymmetry in the difference pattern. This is due to the difference in output power level of the oscillator for different free-running frequency. In the elevation plane, the sum and difference patterns are shown in Fig. 3.30b. The null in the difference pattern is better than −25 dB. However, these plots in Fig. 3.30b do not match the theoretical results very well. This may be due to radiation from the rest of the circuit. Nevertheless, the patterns agree well in general.

5.5 Doppler Transceiver [87]

In this example, an active antenna phased array radar is discussed. A Doppler transceiver using a two-element array will be presented. To incorporate more functions into the circuit, direction tracking in the azimuth plane is implemented

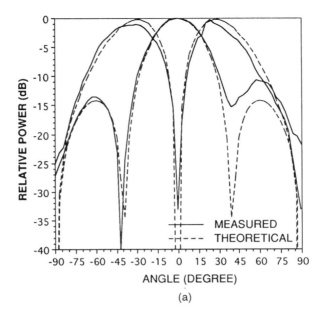

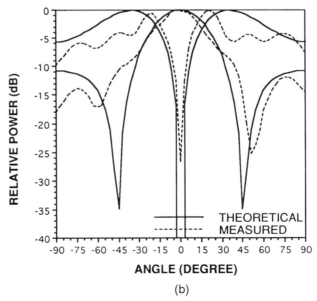

FIGURE 3.30 (a) Measured sum and difference patterns in the azimuth plane. (b) Measured sum and difference patterns in the elevation plane.

at the IF. In the design of this active antenna radar, three important analog functions are considered : (1) quasi-optical power combination in the transmitter mode, (2) frequency conversion in receiver mode, and (3) direction tracking.

The design also hinges on injection-locking to provide necessary phase relationship. In-phase synchronization is needed for broadside radiation. Upon reflection by a moving target, the returned signal suffers a Doppler frequency shift. This frequency-shifted signal is then received by the system for signal processing. Self-oscillating mixer is used for the recovery of the Doppler frequency signal. This is particularly favorable as the transmitting signal served also as the local oscillator signal, reducing component count. In-phase oscillation of each element also allows the phase relationship amongst the RF signals of the array to be maintained in the IF. To provide direction tracking, the monopulse concept is adopted at IF. This is particularly useful because the desired antenna beamwidth need not be large.

The system was implemented using a 2 × 1 active antenna array. The schematic diagram of the system is shown in Fig. 3.31. Similar to the previous example, an FET is used to provide oscillation with the rectangular antenna as the resonator and radiator. Bias condition and placement of the antenna at the drain port of the FET are chosen for power transmission. A transformer is used at the drain bias circuit to tap the IF signal. To create the sum and difference channels, a 0°/180° planar drop-in power divider is used. Signals for both channels are then amplified to compensate for mixer loss.

The measured oscillation frequency is 6.52 GHz. The EIRP of the transmitter is about 22 dBm. The sum and difference channels are then measured with

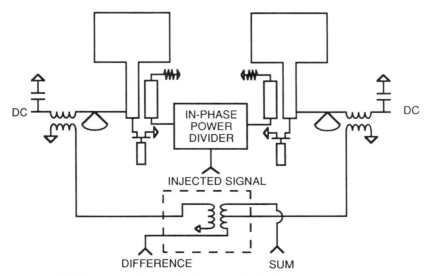

FIGURE 3.31 Schematic diagram of the Doppler transceiver.

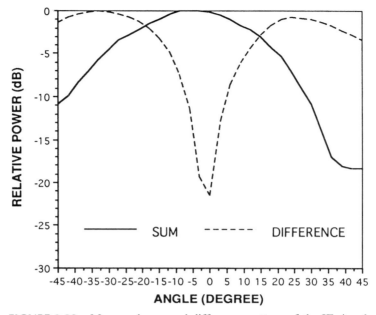

FIGURE 3.32 Measured sum and difference pattern of the IF signal.

a signal radiating at broadside from a standard gain horn to simulate the Doppler frequency shift. The measured receiver patterns of the sum and difference channels are shown in Fig. 3.32. There is asymmetry in the all the measured patterns. This is due to poor mixer performance as a pair and phase imbalance in the injection-locking and the 0°/180° power divider.

REFERENCES

[1] C. A. Balanis, *Antenna Theory: Analysis and Design*, Wiley, New York, 1982.
[2] D. W. Griffin, "Monolithic active array limitations due to substrate modes," *IEEE AP-S Int. Symp. Dig.*, vol. 2, Newport Beach, CA, 18–23 June 1995, pp. 1300–1303.
[3] D. M. Pozar, "Considerations for millimeter wave printed antennas," *IEEE Trans. Antennas Propagat.*, vol. AP-31, pp. 740–747, Sept. 1983.
[4] Y. Liu, K. Cha, and T. Itoh, "Non-leaky coplanar (NLC) waveguides with conductor backing," *IEEE Trans. Microwave Theory Tech.*, vol. MTT-43, pp. 1067–1072, May 1995.
[5] A. J. Parfitt, D. W. Griffin, and P. H. Cole, "Analysis of infinite arrays of substrate-supported metal strip antennas," *IEEE Trans. Antennas Propagat.*, vol. AP-41, pp. 191–199, Feb. 1993.

[6] H. S. Tsai, M. J. W. Rodwell, and R. A. York, "Planar amplifier array with improved bandwidth using folded-slots," *IEEE Microwave Guided Wave Lett.*, vol. 4, pp. 112–114, Apr. 1994.

[7] H. P. Moyer and R. A. York, "Active cavity-backed slot antenna using MESFET's," *IEEE Microwave Guided Wave Lett.*, vol. 3, pp. 95–97, Apr. 1993.

[8] S. Sancheti and V. F. Fusco, "Modeling of active antenna array coupling effects—a load variation method," *IEEE Trans. Microwave Theory Tech.*, vol. MTT-43, pp. 1805–1808, Aug. 1995.

[9] P. Liao and R. A. York, "A new phase-shifterless beam-scanning technique using arrays of coupled oscillators," *IEEE Trans. Microwave Theory Tech.*, vol. MTT-41, pp. 1810–1815, Oct. 1993.

[10] J. Birkeland and T. Itoh, "A 16 element quasi-optical FET oscillator power combining array with external injection locking," *IEEE Trans. Microwave Theory Tech.*, vol. MTT-40, pp. 475–481, Mar. 1992.

[11] J. Lin and T. Itoh, "Two-dimensional quasi-optical power-combining arrays using strongly coupled oscillators," *IEEE Trans. Microwave Theory Tech.*, vol. MTT-42, pp. 734–741, Apr. 1994.

[12] R. A. York, "Nonlinear analysis of phase relationships in quasi-optical oscillator arrays," *IEEE Trans. Microwave Theory Tech.*, vol. MTT-41, pp. 1799–1809, Oct. 1993.

[13] P. S. Hall, "Analysis of radiation from active microstrip antennas," *Electron. Lett.*, vol. 29, pp. 127–129, Jan. 1993.

[14] B. Robert, T. Razban, and A. Papiernik, "Compact amplifier integration in square patch antenna," *Electron. Lett.*, vol. 28, pp. 1808–1810, Sept. 1992.

[15] P. Liao and R. A. York, "A varactor-tuned patch oscillator for active arrays," *IEEE Microwave Guided Wave Lett.*, vol. 4, pp. 335–337, Oct. 1994.

[16] X. Cao and R. A. York, "Phase noise reduction in scanning oscillator arrays," *IEEE MTT-S Int. Microwave Symp. Dig.*, vol. 2, Orlando, FL, 16–20 May 1995, pp. 769–772.

[17] J. Lin, S. T. Chew, and T. Itoh, "A unilateral injection-locking type active phased array for beam scanning," *IEEE MTT-S Int. Microwave Symp. Dig.*, vol. 2, San Diego, CA, 23–27 May 1994, pp. 1231–1234.

[18] A. A. A. De Salles, "Optical control of GaAs MESFET's," *IEEE Trans. Microwave Theory Tech.*, vol. MTT-31, pp. 812-820, Oct. 1983.

[19] M. Minegishi *et al.*, "Control of mode-switching in an active antenna using MESFET," *IEEE Trans. Microwave Theory Tech.*, vol. MTT-43, pp. 1869–1874, Aug. 1995.

[20] D. S. McDowall and V. F. Fusco, "Concurrent large signal simulation of an active microstrip antenna," *Int. Journal of Numerical Modelling : Electronic Networks, Devices and Fields*, vol. 8, pp. 3–12, Jan.–Feb. 1995.

[21] B. Toland, J. Lin, B. Houshmand, and T. Itoh, "FDTD analysis of an active antenna," *IEEE Microwave Guided Wave Lett.*, vol. 3, pp. 423–425, Nov. 1993.

[22] V. A. Thomas *et al.*, "FDTD analysis of an active antenna," *IEEE Microwave Guided Wave Lett.*, vol. 4, pp. 296–298, Sept. 1994.

[23] H. A. Wheeler, "Small antennas," *IEEE Trans. Antennas Propagat.*, vol. AP-23, pp. 462–469, July 1975.

[24] J. W. Mink, "Quasi-optical power combining of solid-state millimeter-wave sources," *IEEE Trans. Microwave Theory Tech.*, vol. MTT-34, pp. 273–279, Feb. 1986.

[25] D. J. Roscoe, A. Ittipiboon, and L. Shafai, "The development of an active integrated microstrip antenna," *IEEE AP-S Int. Symp. Dig.*, vol. 1, London, Ont., Canada, pp. 48–51, 24-28 June 1991.

[26] R. Gillard et al., "Rigorous modelling of receiving active microstrip antenna," *Electron. Lett.*, vol. 27, pp. 2357–2359, Dec. 1991.

[27] C-Y. Chi, and G. M. Rebeiz, "A quasi-optical amplifier," *IEEE Microwave Guided Wave Lett.*, vol. 3, pp. 164–166, June 1993.

[28] T. P. Budka et al., "A 75 GHz to 115 GHz quasi-optical amplifier," *IEEE AP-S Int. Symp. Dig.*, vol. 2, Ann Arbor, MI, 28 June–2 July 1993, pp. 576–579.

[29] M. G. Keller et al., "Active millimetre-wave aperture-coupled microstrip patch antenna array," *Electron. Lett.*, vol. 31, pp. 2–4, Jan. 1995.

[30] H. S. Tsai and R. A. York, "Multi-slot 50-W antennas for quasi-optical circuits," *IEEE Microwave Guided Wave Lett.*, vol. 5, pp. 180–182, June 1995.

[31] T. Ivanov and A. Mortazawi, "A double layer microstrip spatial amplifier with increased active device density," *25th Eur. Microwave Conf. Proc.*, vol. 1, Bologna, Italy, 4–7 Sept. 1995, pp. 320–323.

[32] M. G. Keller et al., "Active aperture-coupled rectangular dielectric resonator antenna," *IEEE Microwave Guided Wave Lett.*, vol. 5, pp. 376–378, Nov. 1995.

[33] H. J. Thomas, D. L. Fudge, and G. Morris, "Gunn source integrated with microstrip patch," *Microwave RF*, vol. 24, pp. 87–91, Feb. 1985.

[34] N. Camilleri and B. Bayraktaroglu, "Monolithic millimeter-wave IMPATT oscillator and active antenna," *IEEE Trans. Microwave Theory Tech.*, vol. MTT-36, pp. 1670–1676, Dec. 1988.

[35] R. A. York, R. D. Martinez, and R. C. Compton, "Active patch antenna element for array applications," *Electron Lett.*, vol. 26, pp. 494–495, Mar. 1990.

[36] J. A. Navarro, K. A. Hummer, and K. Chang, "Active integrated antenna elements," *Microwave J.*, vol. 34, pp. 115, 117–119, 121–122, 124, 126, Jan. 1991.

[37] R. E. Miller and K. Chang, "Integrated active antenna using annular ring microstrip antenna and Gunn diode," *Microwave Optical Techno. Lett*, vol. 4, pp. 72–75, Jan. 1991.

[38] J. Birkeland and T. Itoh, "Two-port FET oscillators with applications to active arrays," *IEEE Microwave Guided Wave Lett.*, vol. 1, pp. 112–113, May 1991.

[39] J. Birkeland and T. Itoh, "A circularly polarized FET oscillator active radiating element," *IEEE MTT-S Int. Microwave Symp. Dig.*, vol. 3, Boston, MA, 10–14 June 1991, pp. 1265–1268.

[40] J. Birkeland and T. Itoh, "An FET oscillator element for Spatially Injection Locked Arrays," *IEEE MTT-S Int. Microwave Symp. Dig.*, vol. 3, Albuquerque, NM, 1–5 June 1992, pp. 1535–1538.

[41] M. J. Vaughan and R. C. Compton, "Resonant-tee CPW oscillator and the application of the design to a monolithic array of MESFETs," *Electron. Lett.*, vol. 29, pp. 1477–1479, Aug. 1993.

[42] J. Birkeland and T. Itoh, "Spatial power combining using push–pull FET oscillators with microstrip patch resonators," *IEEE MTT-S Int. Microwave Symp. Dig.*, vol. 3, Dallas, TX, 8–10 May 1990, pp. 1217–1220.

[43] P. M. Haskins, P. S. Hall, and J. S. Dahele, "Polarization-agile active patch antenna," *Electron. Lett.*, vol. 30, pp. 98–99, Jan. 1994.

[44] T. Razban, H. Frances, B. Robert, and A. Papiernik, "A compact oscillator in a microstrip patch antenna," *Microwave J.*, vol . 37, pp. 110, 112–115, Feb. 1994.

[45] B. K. Kormanyos, W. Harokopus, Jr., L. P. B. Katehi and G. M. Rebeiz, "CPW-fed active slot Antennas," *IEEE Trans. Microwave Theory Tech.*, vol. MTT-42, pp. 541–545, Apr. 1994.

[46] B. K. Kormanyos et al., "Monolithic 155 GHz and 215 GHz quasi-optical slot oscillators," *IEEE MTT-S Int. Microwave Symp. Dig.*, vol. 2, San Diego, CA, 23–27 May 1994, pp. 835–838.

[47] A. Zarroug, P. S. Hall, and M. Cryan, "Active antenna phase control using subharmonic locking," *Electron. Lett.*, vol. 31, pp. 842–843, May 1995.

[48] K. Cha, Y. Liu, C.-Y. Lee, and T. Itoh, "Non-leaky Coplanar Waveguide Active Antenna," *IEEE MTT-S Int. Microwave Symp. Dig.*, vol. 2, Orlando FL, 16–20 May 1995, pp. 765–767.

[49] P. M. Haskins and J. S. Dahele, "Compact active polarization-agile antenna using square patch," *Electron. Lett.*, vol. 31, pp. 1305–1306, Aug. 1995.

[50] M. J. Vaughan and R. C. Compton, "28 GHz oscillator for endfire quasioptical power combining Arrays," *Electron. Lett.*, vol. 31, pp. 1453–1455, Aug. 1995.

[51] Z. Ding, L. Fan, and K. Chang, "A new type of active antenna for coupled Gunn oscillator driven spatial power combining arrays," *IEEE Microwave Guided Wave Lett.*, vol. 5, pp. 264–266, Aug. 1995.

[52] J. Bartolic, J. Sanford, Z. Sipus, and D. Bonefacic, "Voltage controlled self-oscillating microstrip patch antenna," *25th European Microwave Conf. Proc.*, vol. 1, Bologna, Italy, 4–7 Sept. 1995, pp. 316–319.

[53] C.-K. C. Tzuang, and G.-J. Chou, "An active microstrip leaky-wave antenna employing uniplanar oscillator," *25th European Microwave Conf. Proc.*, vol. 1, Bologna, Italy, 4–7 Sept. 1995, pp. 308–311.

[54] M. J. Vaughan and R. C. Compton, "28 GHz omni-directional quasi-optical transmitter array," *IEEE Trans. Microwave Theory Tech.*, vol. MTT-43, pp. 2507–2509, Oct. 1995.

[55] A. R. Kerr, P. H. Siegel, and R. J. Mattauch, "A simple quasi-optical mixer for 100–120 GHz," *IEEE MTT-S Int. Microwave Symp. Dig.*, San Diego, CA, 21–23 June 1977, pp. 96–98.

[56] K. D. Stephen, N. Camilleri, and T. Itoh, "A quasi-optical polarization-duplexed balanced mixer for millimeter-wave applications," *IEEE Trans. Microwave Theory Tech.*, vol. MTT-31, pp. 164–170, Feb. 1983.

[57] K. D. Stephen and T. Itoh, "A planar quasi-optical subharmonically pumped mixer characterized by isotropic conversion loss," *IEEE Trans. Microwave Theory Tech.*, vol. MTT- 32, pp. 97–102, Jan. 1984.

[58] V. D. Hwang and T. Itoh, "Quasi-optical HEMT and MESFET self-oscillating mixers," *IEEE Trans. Microwave Theory Tech.*, vol. MTT-36, pp. 1701–1705, Dec. 1988.

[59] S. V. Robertson et al., "A folded slot antenna for planar quasi-optical mixer applications," *IEEE AP-S Int. Symp. Dig.*, vol. 2, Ann Arbor, MI, 28 June– 2 July 1993, pp. 600–603.

[60] K. Cha, S. Kawasaki, and T. Itoh, "Transponder using self-oscillating mixer and

active antenna," *IEEE MTT-S Int. Microwave Symp. Dig.,* San Diego, CA, vol. 1, 23–27 May 1994, pp. 425–428.

[61] C. W. Pobanz and T. Itoh, "A microwave noncontact identification transponder using subharmonic interrogation," *IEEE Trans. Microwave Theory Tech.,* vol. MTT-43, pp. 1673–1679, July 1995.

[62] J. Birkeland and T. Itoh, "A microstrip based active antenna Doppler transceiver module," *19th Eur. Microwave Conf. Proc.,* London, United Kingdom, 4–7 Sept. 1989, pp. 172-175.

[63] J. Birkeland and T. Itoh, "FET-based planar circuits for quasi-optical sources and transceivers," *IEEE Trans. Microwave Theory Tech.,* vol. 37, pp. 1452–1459, Sept. 1989.

[64] R. D. Martinez and R. C. Compton, "A quasi-optical oscillator/modulator for wireless transmission," *IEEE MTT-S Int. Microwave Symp. Dig.,* San Diego, CA, vol. 2, 23–27 May 1994, pp. 839–842.

[65] R. Flynt *et al.,* "Low cost and compact active integrated antenna transceiver for system applications," *IEEE MTT-S Int. Microwave Symp. Dig.,* vol. 2, Orlando, FL, 16–20 May 1995, pp. 953–956.

[66] A. Nesic, S. Jovanovic, and I. Radnovic, "Integrated uniplanar oscillator-transmitter and mixer with active antennas," *25th Eur. Microwave Conf. Proc.,* vol. 1, Bologna, Italy, 4–7 Sept. 1995, pp. 324–328.

[67] G. Soffici, G. Avitabile, and G. B. Gentili, "A Versatile Active Antenna for low-cost microwave link applications," *25th Eur. Microwave Conf. Proc.,* vol. 1, Bologna, Italy, 4–7 Sept. 1995, pp. 312–315.

[68] W. Ng *et al.,* "The first demonstration of an optically steered microwave phased array antenna using True-Time-Delay," *J. Lightwave Technol.,* vol. 9, pp. 1124–1131, Sept. 1991.

[69] S. Kawasaki and T. Itoh, "Optical control of active integrated antenna," *22nd European Microwave Conf. Proc.,* vol. 1, Espoo, Finland, 24–27 Aug. 1992, pp. 697–701.

[70] S. Kawasaki, and T. Itoh, "Optical control on 2-element CPW active integrated antenna array with strong coupling," *IEEE AP-S Int. Symp. Dig.,* vol. 3, Ann Arbor, MI, 28 June–2 July 1993, pp. 1616–1619.

[71] S. T. Chew, T. K. Tong, M. C. Wu, and T. Itoh, "An active phased array with optical input and beam-scanning capability" *IEEE Microwave Guided Wave Lett.,* vol. 4, pp. 347–349, Oct. 1994.

[72] S. T. Chew, D. T. K. Tong, M. C. Wu, and T. Itoh, "Use of direct-modulated/gain-switched optical links in monopulse-type active phased array systems" *IEEE Trans. Microwave Theory Tech.,* vol. MTT-44, pp. 326–330, Feb. 1996.

[73] I. J. Bahl and P. Bhartia, *Microstrip Antennas,* Artech House, Boston, 1980.

[74] P. Bhartia, K. V. S. Rao, and R. S. Tomar, *Millimeter-Wave Microstrip and Printed Circuit Antennas,* Artech House, Boston, 1991.

[75] J. R. James, P. S. Hall, and C. Wood, *Microstrip Antenna Theory and Design,* P. Peregrinus (on behalf of the Institution of Electrical Engineers), New York, 1981.

[76] R. C. Johnson and H. Jasik, *Antenna Engineering Handbook,* 2nd ed, McGraw-Hill, New York, 1984.

[77] K. Takeuchi, I. Chiba, and Y. Karasawa, "A slot coupled microstrip antenna with a multi-layer thick ground plane," *IEICE Trans. Electron.*, vol. E78-C, pp. 988–994, Aug. 1995.

[78] J. Lin and T. Itoh, "A 4 × 4 spatial power-combining array with strongly coupled oscillators in multi-layer structure," *IEEE MTT-S Int. Microwave Symp. Dig.*, vol. 2, Atlanta, GA, 14–18 June 1993, pp. 607–610.

[79] B. Toland, J. Lin, B. Houshmand, and T. Itoh, "Electromagnetic Simulation of mode control of a two-element active antenna," *IEEE MTT-S Int. Microwave Symp. Dig.*, vol. 2, San Diego, CA, 23–27 May 1994, pp. 883–886.

[80] K. S. Yee, "Numerical solution of initial boundary value problems involving Maxwell's equations in isotropic media," *IEEE Trans. Antennas Propagat.*, vol. AP-14, pp. 302–307, May 1966.

[81] T.-W. Huang, B. Houshmand, and T. Itoh, "Fast sequential FDTD Diakoptics method using the system identification technique," *IEEE Microwave Guided Wave Lett.*, vol. 3, pp. 378–380, Oct. 1993.

[82] J. Lin, "Strongly coupled active antenna array: analysis and application in quasi-optical power generation," Ph.D dissertation, University of California, Los Angeles, 1994.

[83] J. Lin, private communication.

[84] S. T. Chew and T. Itoh, "A 2 × 2 Beam-switching active antenna array," *IEEE MTT-S Int. Microwave Symp. Dig.*, vol. 2, Orlando, FL, 16–20 May 1995, pp. 925–928.

[85] K. Kurokawa, "Injection locking of microwave solid-state oscillators," *Proc. IEEE*, vol. 61, pp. 1386–1410, Oct. 1973.

[86] P. S. Hall *et al.*, "Phase control in injection locked microstrip active antennas," *IEEE MTT-S Int. Microwave Symp. Dig.*, vol. 2, San Diego, CA, 23–27 May 1994, pp. 1227–1230.

[87] S. T. Chew and T. Itoh, "An active antenna phased array doppler radar with tracking capability," *IEEE AP-S Int. Symp. Dig.*, vol. 3, Newport Beach, CA, 18–23 June 1995, pp. 1368–1371.

CHAPTER FOUR

Coupled-Oscillator Arrays and Scanning Techniques

JONATHAN J. LYNCH
Hughes Research Laboratories, Inc., California

HENG-CHIA CHANG
ROBERT A. YORK
University of California, Santa Barbara

Spatial or quasi-optical combining techniques can be used to create high-power microwave and millimeter-wave sources from an array of low-power solid-state sources, provided that the array elements are mutually coherent. Bilateral and/or unilateral injection-locking techniques are one possible method for achieving synchronous operation of the array elements. An attractive feature of injection-locking techniques is the ability to manipulate the phase distribution without additional phase-shifting circuitry, suggesting a potential for low-cost beam scanning systems. Both power-combining and beam-scanning applications require a sound understanding of the nonlinear dynamics of interacting oscillators to ascertain regions of mutual coherence and to establish conditions for achieving prescribed phase distributions. This chapter develops analytical methods for coupled or injection-locked oscillator arrays, suitable for also predicting mode stability, transient response, and noise properties. Several beam-scanning possibilities are explored, and experimental microwave arrays using

Active and Quasi-Optical Arrays for Solid-State Power Combining, Edited by Robert A. York and Zoya B. Popović.
ISBN 0-471-14614-5 © 1997 John Wiley & Sons, Inc.

both bilateral and unilateral injection-locking are described. It is shown that certain coupling topologies lead to a phase noise reduction proportional to the number of oscillators, an important result for most millimeter-wave applications.

1 INTRODUCTION

The modern concept of a quasi-optical power-combining system was significantly influenced by Mink [1], who described a fundamental-mode oscillating system composed of an active array in an open cavity resonator. This combiner system is related to older waveguide-based multiple device oscillators [2], but can also be viewed as the electrical analogue of a laser (see Chapter 1). Since Mink's proposal, other types of oscillator arrays have been described, which may or may not employ a quasi-optical cavity. These are summarized in Fig 4.1. Two basic approaches have been identified: (1) a distributed periodic negative resistance structure, which includes the grid oscillators of Chapter 8 and the extended-resonance method of Chapters 6 and 2) a coupled set of distinct, injection-locked oscillators with integrated antennas, which are described in this chapter. In either case, the operation is highly dependent on interactions between the nonlinear active devices, which serve to maintain mutual coherence at a single frequency with a prescribed phase relationship. The interaction medium could be a quasi-optical cavity (as in Mink's original proposal) or circuit coupling networks, or some combination of both.

Though the many types of oscillator arrays share common features, the analysis and conceptual understanding of each system is somewhat different, reflecting the design of the unit cell and the coupling network. The coupled-oscillator systems described in this chapter are treated on the assumption that the individual oscillators are single-mode systems that have been perturbed by energy injected from neighboring oscillators. The physical basis for synchro-

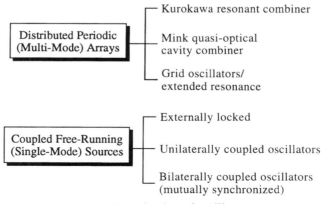

FIGURE 4.1 Classification of oscillator arrays.

nization in this case is the injection-locking phenomenon [3,4]. This effect has a long history dating back to observations of synchronized mechanical pendulums by Christian Huygens [5] and has since been observed in a wide variety of self-sustained (nonlinear) biological and physical oscillators. The analysis of coupled-oscillator arrays involves relatively minor extensions of elementary injection-locking theory, but yields significantly more complex and interesting dynamical phenomena.

Application of mutual injection-locking concepts to synchronization of more than one oscillator for spatial power-combining was reported contemporaneously with Mink's proposal by Stephan [7,8], who described a theoretical analysis and low frequency prototypes, and Dinger *et al.* [6], who presented results using a small X-band array. Many suitable active antenna designs have been reported for such arrays (see Chapter 3). Unusual dynamics have been identified for beam scanning [7,9], mode-locked pulse generation [10], and possibly controlled chaos [11]. It is interesting to note that many of the synchronization phenomena to appear later in this chapter have analogues in a diversity of biological and physical systems, and in fact the mathematical models are in some cases virtually identical. Examples include modeling of neural activity [17], swarms of synchronously flashing fireflies [18], the coordinated firing of cardiac pacemaker cells [12], rhythmic spinal locomotion in vertebrates [13,14], the synchronized activity of nerve cells in response to external stimuli [18], and synchronized menstrual cycles in groups of women [15]. In the physical sciences, examples include oscillations in certain nonlinear chemical reactions [16], the collective behavior of Josephson junction arrays [19,20], laser diode arrays [21], and possibly quantum dots [22].

The analysis presented here is only one of many different approaches to the theory of nonlinear oscillations. It was first utilized by Kurokawa [23] to describe the operation and noise characteristics of microwave oscillators. The method relies on a frequency-domain impedance, admittance, or scattering parameter description and is therefore particularly well suited to microwave oscillators. As we will show, Kurokawa's method is derived from a rigorous expansion of the time-domain response through the inverse Fourier transform. Other analysis methods include those of Van der Pol in his classic paper [24], and later refined by Bogoliubov and Mitropolsky [25] in which a method of averaging is applied to the system of differential equations that describe relevant system variables in the time domain. The method of averaging yields a new (hopefully tractable) system of differential equations that describe only the slowly varying amplitude and phase of the oscillator, suppressing the time variations of the carrier. This averaging method is derived from a rigorous asymptotic expansion in powers of a nonlinearity parameter and is valid for weak nonlinearities giving rise to "nearly sinusodal" oscillations. Interestingly, the methods of Kurokawa and Van der Pol give nearly identical results in many cases of practical interest. Since Kurokawa relies heavily on network concepts, his approach gives considerable engineering insight into the behavior of oscillators and has been adopted in the present work.

The averaged potential theory developed by Kuramitsu and Takase [26] also relies on the method of averaging to describe the slowly varying amplitude and phase of an oscillator, but the differential equations are derived from a single potential function. The formulation is derived using the nonlinear network theory developed by Brayton and Moser [27]. One advantage of the averaged potential method is that the stable oscillating states are given by minima of the potential function. In the important case of the Van der Pol oscillator (a particular class of oscillator—not related to Van der Pol's analysis method) the potential takes on a quadratic form and gives global stability information. When applied to systems of coupled oscillators, this method is essentially a modal analysis and can be very useful with strongly coupled (multimode) oscillator systems [28]. In addition to the methods mentioned here, many other techniques exist for the study of nonlinear oscillations. However, much of the literature is concerned with various mathematical details, such as behavior near a synchronization boundary or other bifurcation points, which are regions of operation that would probably be avoided in practical systems.

2 OSCILLATOR MODELING

Following Kurokawa [4], our approach involves writing appropriate circuit equations in the frequency domain, and transforming these to the time domain under simplifying assumptions to generate dynamic equations. A general model for a single oscillator circuit is shown in Fig. 4.2a. The oscillator admittance, Y_{osc}, describes the nonlinear active element and embedding circuit, which is coupled to a load admittance Y_L. We assume the nonlinear active element is described by an instantaneous current–voltage relationship $I(V)$ in the time domain, so the nonlinear part of the device admittance is frequency independent and will depend only on the amplitude of the output voltage $|V|$ and not the phase. Any *linear* reactive part of the device admittance can be considered part of the embedding circuit. The possibility of externally injected signals are included via the independent current source I_{inj}. From Kirchhoff's Current Law (KCL) we have the relationship

$$\tilde{I}_{inj}(\omega) = \tilde{V}(\omega) Y_t(\omega, |V|) \tag{4.1}$$

where for later notation convenience we have defined a "total admittance" across the output terminals as

$$Y_t(\omega, |V|) = [Y_{osc}(\omega, |V|) + Y_L(\omega)] \tag{4.2}$$

The tilde (~) denotes a frequency-domain (phasor) quantity. The next critical assumption in the analysis is that the oscillator is a single-mode system, designed to produce nearly sinusoidal oscillations around a nominal center frequency ω_0; this is the "free-running" or unperturbed oscillation frequency. Most practical oscillators can be designed to satisfy this criteria by (1) using an embedding

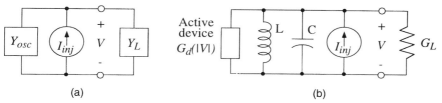

FIGURE 4.2 (a) General parallel model for oscillators, suitable for negative conductance devices, (b) A simple circuit for sinusoidal oscillators, suitable for analysis of array dynamics.

circuit with a well-defined and isolated resonance at ω_0 with a sufficiently high Q factor ($Q > 10$) and (2) terminating the device so as to provide narrowband gain around this resonance. The time-dependent output voltage can then be written (using complex notation) in the following useful forms:

$$\begin{aligned} V(t) &= A(t)\, e^{j[\omega_r t + \phi(t)]} \\ &= A(t)\, e^{j\theta(t)} \\ &= V'(t)\, e^{j\omega_r t} \end{aligned} \quad (4.3)$$

where A and ϕ are dynamic amplitude and phase variables, $V' = A \exp(j\phi)$ is the output "phasor" voltage, $\phi(t) = \omega_r t + \phi(t)$ is the instantaneous phase, and ω_r is a "reference" frequency that is presumably close in magnitude to ω_0, but otherwise somewhat arbitrary and chosen for convenience for a particular problem. The reason for defining $V(t)$ in terms of this reference frequency will become clear later, but in the meantime can be considered as establishing a variable harmonic reference for the time variation of ϕ, which proves convenient when the oscillator system is perturbed from its free-running state. Naturally the *true* time variation is obtained by taking the real part of (4.3).

Applying the inverse Fourier transform to (4.1) and exploiting the slowly varying amplitude and phase assumption, it can be shown (Appendix A) that (4.1) transforms to

$$I_{inj}(t) = V(t)\left[Y_t(\omega_r, A) + \frac{\partial Y_t(\omega_r, A)}{\partial \omega}\left(\frac{d\phi}{dt} - j\frac{1}{A}\frac{dA}{dt}\right)\right] \quad (4.4)$$

This result is equivalent to using "Kurokawa substitution" for the frequency in (4.1), and it can be solved for the amplitude and phase variations by separating real and imaginary parts to give

$$\frac{dA}{dt} = A\,\mathrm{Re}\{F(A, \phi)\} \quad (4.5)$$

$$\frac{d\phi}{dt} = \mathrm{Im}\{F(A, \phi)\} \quad (4.6)$$

where

$$F(A, \phi) = \frac{I_{\text{inj}}/V - Y_t(\omega_r, A)}{\frac{\partial Y_t(\omega_r, A)}{\partial (j\omega)}} \quad (4.7)$$

Note that (4.5) and (4.6) can be concisely written in complex form using (4.3) as

$$\frac{dV'}{dt} = F(V')V' \quad (4.8)$$

which is a standard form for nonlinear analysis (note $F(V')$ stands for $F(A, \phi)$). As described in the Appendix, (4.4) is an approximation, valid when higher-order time derivatives of amplitude and phase, as well as higher-order frequency derivatives of the total admittance, are negligible compared with the first two terms of the expansion (the two terms in the square brackets in (4.4)). This is generally satisfied for oscillation around an isolated resonance, as well as for amplitude and phase fluctuations which are slow compared to the carrier (an excellent assumption in practice).

To proceed further we need to specify an embedding and load circuit, such as shown in Fig. 4.2b. For this case, the total admittance near resonance is

$$Y_t(\omega, A) = G_d(A) + G_L + 2jC(\omega - \omega_0) \quad (4.9)$$

where $\omega_0 = 1/\sqrt{LC}$ is the resonant frequency. It is convenient to express the dynamic equations at microwave frequencies in terms of easily measurable quantities like resonant frequency and Q factor. For a parallel resonator, $Q = \omega_0 C/G_L$. Assuming a free-running oscillator (no injection-source, $I_{\text{inj}} = 0$), we have

$$F(A, \phi) = -\frac{\omega_0}{2Q}\left[\frac{G_d(A)}{G_L} + 1\right] + j(\omega_0 - \omega_r) \quad (4.10)$$

which gives the dynamic equations

$$\frac{dA}{dt} = -\frac{\omega_0}{2Q}A\left[\frac{G_d(A)}{G_L} + 1\right] \quad (4.11)$$

$$\frac{d\phi}{dt} = \omega_0 - \omega_r \quad (4.12)$$

The phase equation (4.12) can be easily integrated to give $\phi(t) = (\omega_0 - \omega_r)t + \phi_0$, where ϕ_0 is an arbitrary constant. Substitution into (4.3) shown that the free-running frequency is always at the resonant frequency ω_0, irrespective of the choice of ω_r. From the perspective of solving for the phase dynamics, the clear choice for ω_r in this case is $\omega_r = \omega_0$, which eliminates any time dependence of ϕ. This approach, though trivial in the present case, will prove especially useful later for injection-locked oscillators or multiple oscillator systems. The important point here is that regardless of the choice of ω_r, a unique result will always be obtained for the actual time variation of the output signal, but certain choices of reference frequency can simplify the description of the oscillator phase.

A steady-state solution for the free-running amplitude α is determined from (4.11) by setting $dA/dt = 0$, which gives the oscillation condition

$$G_d(\alpha) + G_L = 0 \tag{4.13}$$

The device conductance can be represented by a polynomial in powers of A. Following Van der Pol [24] we write

$$G_d(A) = -G_0 + G_2 A^2 \tag{4.14}$$

where G_0 is the magnitude of the small-signal negative conductance. The oscillation condition (4.13) can be combined with (4.14) to write

$$\left[\frac{G_d(A)}{G_L} + 1\right] = -\mu(1 - A^2/\alpha^2) \tag{4.15}$$

where the dimensionless parameter μ is defined as $\mu = (G_0 - G_L)/G_L = G_2\alpha^2/G_L$. This is a convenient form for modeling since the device is now specified in terms of the easily measurable free-running amplitude, α, and a dimensionless "nonlinearity parameter" μ that can be determined from the oscillator transient response. The dynamic equations for a free-running sinusoidal Van der Pol oscillator can be rewritten as

$$\frac{dA}{dt} = \mu \frac{\omega_0}{2Q} A(1 - A^2/\alpha^2) \tag{4.16}$$

$$\frac{d\phi}{dt} = 0 \tag{4.17}$$

The steady-state solution $A = \alpha$ and $\phi = \phi_0$ must then be checked for stability. This is done by a perturbation analysis. Writing $A = \alpha + \delta A$, (4.16) becomes

$$\frac{d(\delta A)}{dt} = -\frac{\mu\omega_0}{Q}\delta A \tag{4.18}$$

The perturbation δA will decay with time as long as $\mu > 0$, which in turn requires that $G_0 > G_L$ and $G_2 > 0$. The former is also the "start-up" condition for a parallel oscillator; the second is equivalent to Kurokawa's stability condition $dG_d/dA > 0$.

2.1 Injection-Locking

When an externally injected signal is present, (4.9) and (4.7) give

$$F(A, \phi) = \frac{\omega_0}{2Q}\left[\mu S(A) + \frac{I_{inj}}{G_L V}\right] + j(\omega_0 - \omega_r) \quad (4.19)$$

where $S(A) \equiv (1 - A^2/\alpha^2)$. Writing the injected signal as

$$I_{inj} = \rho G_L e^{j\omega_{inj}t + \psi(t)} \quad (4.20)$$

and using (4.3) gives the oscillator amplitude and phase dynamics

$$\frac{dA}{dt} = \mu\frac{\omega_0}{2Q}AS(A) + \rho\frac{\omega_0}{2Q}\cos[(\omega_{inj} - \omega_r)t + (\psi - \phi)] \quad (4.21)$$

$$\frac{d\phi}{dt} = \omega_0 - \omega_r + \frac{\rho}{A}\frac{\omega_0}{2Q}\sin[(\omega_{inj} - \omega_r)t + (\psi - \phi)] \quad (4.22)$$

In this form ρ, which describes the strength of the injected signal has the same units as A. Here it is convenient to choose $\omega_r = \omega_{inj}$ to eliminate the explicit time dependence on the right-hand side. In the general case those coupled differential equations can not be solved analytically, but if the injected signal is significantly smaller than the free-running amplitude ($\rho \ll A$), then the steady-state amplitude should remain close to its free-running value, $A \approx \alpha$, and (4.22) is decoupled from (4.21). The oscillator is then described by Adler's equation

$$\frac{d\phi}{dt} = \omega_0 - \omega_{inj} + \underbrace{\frac{\rho}{\alpha}\frac{\omega_0}{2Q}}_{\Delta\omega_{inj}}\sin(\psi - \phi) \quad (4.23)$$

This is the key equation for injection-locking. If a steady-state solution can be found for the phase such that $d\phi/dt = 0$, this indicates that the oscillator is synchronized to the injected signal. Solving for the steady-state phase difference between the oscillator and the injected signal ($\Delta\phi = \psi - \phi_0$) gives

$$\Delta\phi = \sin^{-1}\left(\frac{\omega_{inj} - \omega_0}{\Delta\omega_m}\right) \quad (4.24)$$

which indicates that an injection-locked solution is possible only when the injected signal frequency lies within the "locking range" of the oscillator, $\omega_0 \pm \Delta\omega_m$ (note that $\Delta\omega_m$ is often referred to as the locking range, but physically represents *half* the locking range). The inverse sine function gives two possible solutions for the phase difference in this range; the correct result is found from a stability analysis. Following the previous section, the phase is perturbed from its free-running state (ϕ_0) by writing $\phi = \phi_0 + \delta\phi$, which reduces (4.23) to

$$\frac{d\delta\phi}{dt} = -\delta\phi \, \Delta\omega_m \cos \Delta\phi \qquad (4.25)$$

The perturbation will decay in time provided that $\cos \Delta\phi > 0$, which restricts the phase difference to the range

$$-\frac{\pi}{2} \leq \Delta\phi \leq \frac{\pi}{2} \qquad (4.26)$$

As the injected signal frequency is tuned over the locking range, the phase difference between the oscillator and the injected signal will vary between $\pm\pi/2$ according to (4.24); this is illustrated in Fig. 4.3. This induced phase shift suggests possible schemes for phased arrays using injection-locking, which are described later. Note that the locking range is proportional to the injected signal strength and is inversely related to the Q factor. To the extent that a large locking range is desired in a practical system, low-Q oscillators with large injected signal strengths are required.

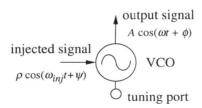

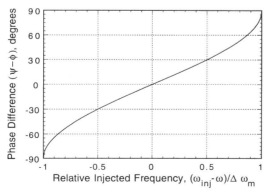

FIGURE 4.3 Relative phase shift between the oscillator output and injected signal versus the normalized injected frequency.

144 COUPLED-OSCILLATOR ARRAYS AND SCANNING TECHNIQUES

In the general case when the injected signal strength is sufficient to cause a significant perturbation of the free-running amplitude and hence a coupling of the dynamic equations (4.21)–(4.22), then both equations must be solved simultaneously. The math is straightforward but sufficiently messy that only the results will be mentioned here. The resulting locking range is smaller than that suggested by $\Delta\omega_m$, and the range of induced phase shifts is reduced. This can be important in practical systems where a large locking range is desired.

It should also be noted that while the simple circuit model of Fig. 4.2b predicts many of the qualitative features of injection locking, it may not accurately represent the dynamics in an absolute sense; in particular, real oscillators exhibit (1) significant asymmetric power variation as the signal is tuned across the locking range and (2) asymmetric locking characteristics around the free-running frequency. These effects have been found to be modeled well by the following empirical admittance model for the oscillator [29]:

$$Y_{\text{osc}} = Y_0 + Y_\omega(\omega - \omega_0) + Y_V |V_0|^2 \quad (4.27)$$

where Y_0, Y_ω, and Y_V are all complex (i.e., $Y_x = G_x + jB_x$) and are empirically determined quantities. By comparison with the simple parallel-resonant oscillator admittance in (4.9) we see that the model (4.27) includes three new terms which apparently have the following influence [29]:

$B_0 \Rightarrow$ shift resonance away from ω_0

$G_\omega \Rightarrow$ asymmetrical amplitude variation

$B_V \Rightarrow$ bend equifrequency loci on a Reicke diagram

The model has accurately described real oscillators. For computation, the new model (4.27) is just as easy to use as the simple model (4.9), but for analysis aimed at physical insight it is an unnecessary complication.

2.2 Oscillator Noise

Oscillator noise can be treated in a simple fashion using the injection-locking equations, (4.21) and (4.22), with the external current source used to represent an internal noise generator. As Kurokawa showed [30], the two injection terms on the right-hand side of (4.21)–(4.22) can be considered as the in-phase and quadrature noise terms, respectively, which we write as

$$\frac{dA}{dt} = \mu \frac{\omega_0}{2Q} AS(A) - \frac{\omega_0}{2Q} AG_n(t) \quad (4.28)$$

$$\frac{d\phi}{dt} = -\frac{\omega_0}{2Q} B_n(t) \quad (4.29)$$

Alternatively, these same equations can be derived by adding a normalized "noise admittance," $Y_n/G_L = G_n + jB_n$, into the oscillator embedding circuit, as described by Okabe and Okamura [31,32] and Makino et al. [33]. Assuming that the amplitude and phase fluctuations due to these noise sources are small and given by δA and $\delta\phi$, respectively, we linearize (4.28) and (4.29) around the "noise-free" (i.e., free-running) solution by writing $A = \alpha + \delta A$ and $\phi = \phi_0 + \delta\phi$ and solving for the fluctuations. Fourier transforming the resulting equations gives

$$\left(\frac{j\omega}{\omega_{3\,dB}} + 2\mu\right)\delta A = -\alpha\tilde{G}_n \qquad (4.30)$$

and

$$\frac{j\omega}{\omega_{3\,dB}}\widetilde{\delta\phi} = -\tilde{B}_n \qquad (4.31)$$

where $\omega_{3\,dB} \equiv \omega_0/2Q$, half the 3-dB bandwidth of the oscillator tank circuit. The power spectral density of the amplitude and phase fluctuations (i.e., the amplitude and phase noise) are found by finding the magnitude-square of $\widetilde{\delta A}$ and $\widetilde{\delta\phi}$. Therefore, the AM noise for single oscillator is

$$|\widetilde{\delta A}|^2_{\text{uncoupled}} = \frac{\alpha^2|\tilde{G}_n|^2}{(\omega/\omega_{3\,dB})^2 + 4\mu^2} \qquad (4.32)$$

and the PM noise of single oscillator is

$$|\widetilde{\delta\phi}|^2_{\text{uncoupled}} = \frac{|\tilde{B}_n|^2}{(\omega/\omega_{3\,dB})^2} \qquad (4.33)$$

These results are equivalent to those derived by Kurokawa [30] and Schlosser [34] and others. Note that for most oscillators, noise close to the carrier ($\omega \ll \omega_{3\,dB}$) is dominated by phase noise. The result (4.33) features prominently in the later derivation of the noise properties of N coupled oscillators, where output noise relative to a single free-running oscillator is examined.

3 SYSTEMS OF COUPLED OSCILLATORS

The results of the previous section can be extended to a multiple-oscillator system. An oscillator array can be interconnected or externally injection-locked in a number of possible configurations; a few examples from the literature are shown in Fig. 4.4, such as a corporate fed externally locked array [35], unilateral nearest-neighbor injection-locking [36], and "inter-injection-locked" [7,8] or

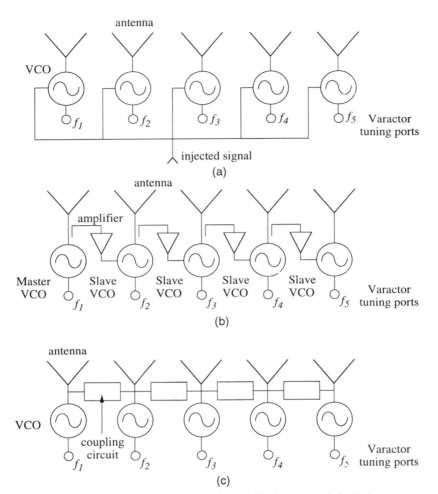

FIGURE 4.4 A few common array topologies. (a) Each element is locked to a common reference signal, which is distributed by an RF feed network. (b) Unilateral injection-locking where each oscillator is slaved to the preceding element. (c) Bilateral nearest-neighbor coupling.

"mutually synchronized" [37–39] bilateral nearest-neighbor injection-locking. A general model which includes all these examples as a special case is the oscillator system shown in Fig. 4.5. Each oscillator circuit is coupled to an N-port network, which will be described in terms of Y parameters for compatibility with the parallel oscillator model. The N-port includes both coupling circuits and the load. In addition, there is an independent source at each port to model external injection-locking. The externally locked array of Fig. 4.4a would be modeled in this case by (1) a coupling network with only diagonal entries ($Y_{ij} = 0$ for $i \neq j$) representing the independent loads (antennas in the limit of

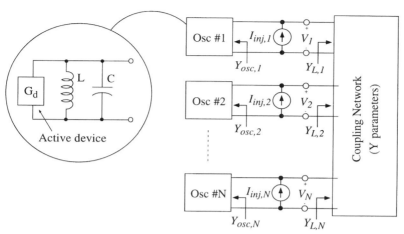

FIGURE 4.5 General model for oscillator array analysis, which includes the topologies of Fig. 4.4 as special cases.

negligible mutual coupling) and (2) a synchronous set of injection sources with an amplitude and phase distribution that mimics the corporate feed network. The unilaterally coupled array of Fig. 4.4b would require a nonreciprocal coupling network ($Y_{ij} \neq Y_{ji}$), with all of the independent injection sources set to zero (with the possible exception of the source at port 1). The bilaterally coupled array of Fig. 4.4c would similarly have all of the independent sources set to zero, with a reciprocal coupling network that might represent an antenna array with radiative mutual coupling [37] and possibly additional coupling circuits [39].

3.1 Derivation of the Dynamic Equations

At each oscillator port we can write dynamic equations of the form (4.5)–(4.6). For an N-oscillator system, we will have $2N$ coupling first-order differential equations. It is assumed that all of the oscillators are constructed similarly, having (to first order) the same Q-factors and amplitude saturation factors (μ), but possibly slightly different free-running frequencies and free-running amplitudes. Using the parallel oscillator model of Fig. 4.2b, the ith oscillator admittance (near resonance) is

$$Y_{\text{osc},i} \approx G_d(A_i) + 2jC(\omega - \omega_i) \qquad (4.34)$$

where ω_i is the free-running frequencies of the ith oscillator. Assuming that each oscillator is designed to feed a nominal load admittance G_L and follows the Van der Pol nonlinearity given by (4.15), (4.34) can be written as

$$Y_{\text{osc},i}(\omega, A_i) = G_L\left[-S_\mu(A_i) + j\frac{2Q}{\omega_i}(\omega - \omega_i)\right] \qquad (4.35)$$

where

$$S_\mu(A_i) \equiv [1 + \mu(1 - A_i^2/\alpha_i^2)] \quad (4.36)$$

and α_i are the free-running amplitudes. Each oscillator sees a "load" admittance given by (at port i)

$$Y_{L,i}(\omega) = \sum_{j=1}^{N} Y_{ij}(\omega)\frac{V_j}{V_i} \quad (4.37)$$

where V_i is the terminal voltage at port i, again represented in the time domain as in (4.3):

$$V_i(t) = A_i e^{j(\omega_r t + \phi_i)} = A_i e^{j\theta_i} = V_i' e^{j\omega_r t} \quad (4.38)$$

Defining the "normalized coupling matrix," $\overline{\overline{\kappa}}$, as

$$\kappa_{ij} \equiv \frac{Y_{ij}}{G_L} \quad (4.39)$$

then total admittance at the ith port can be written as

$$Y_{t,i}(\omega, A_i) = G_L \frac{2Q}{\omega_i}\left[-\frac{\omega_i}{2Q} S_\mu(A_i) + j(\omega - \omega_i) + \frac{\omega_i}{2Q}\sum_{j=1}^{N} \kappa_{ij}\frac{V_j}{V_i}\right] \quad (4.40)$$

Substituting this into (4.7) gives a set of N differential equations like (4.8),

$$\frac{dV_i'}{dt} = F_i(\overline{V}) V_i', \quad i = 1, \ldots, N \quad (4.41)$$

where

$$F_i(\overline{V}) = \frac{\frac{\omega_i}{2Q}\left[\frac{I_{\text{inj},i}}{G_L V_i} + S_\mu(A_i) - \sum_{j=1}^{N} \kappa_{ij}(\omega_r)\frac{V_j}{V_i}\right] + j(\omega_i - \omega_r)}{1 + \frac{\omega_i}{2Q}\sum_{j=1}^{N} \frac{\partial \kappa_{ij}(\omega_r)}{\partial(j\omega)}\frac{V_j}{V_i}} \quad (4.42)$$

and where the independent sources representing externally injected signals are assumed to be mutually coherent at a common frequency ω_{inj} and are written in the form of (4.20):

$$I_{\text{inj},i} = \rho_i G_L e^{j\omega_{\text{inj}} t + \psi_i(t)} \quad (4.43)$$

We are most interested in steady-state solutions to (4.41) where all oscillators are synchronized to a common frequency, ω, which occurs when

$$\frac{dA_i}{dt} = 0 \quad \text{and} \quad \frac{d\phi_i}{dt} = \omega - \omega_r, \quad i = 1, \ldots, N \quad (4.44)$$

In this case the oscillator phases will be bounded in time if the reference frequency is chosen to be $\omega_r = \omega$, in which case the steady-state solutions are determined by the set of nonlinear algebraic equations

$$F_i(\overline{A}, \overline{\phi}) = 0, \quad i = 1, \ldots, N \quad (4.45)$$

If externally injected signals at frequency ω_{inj} are present, the synchronized frequency (and reference frequency) can be taken as $\omega = \omega_{inj} = \omega_r$. However, in the absence of externally injected signals ($I_{inj,i} = 0$ for all i), the steady-state synchronized frequency is not known *a priori*, but must be determined from (4.45). The real and imaginary parts of F_i must separately equate to zero, so (4.45) represents a set of $2N$ equations. Since the amplitude and phase of each oscillator are $2N$ unknowns that must be solved for, it would appear that when ω is also unknown there would be $2N + 1$ unknowns, but in this case one of the phases is arbitrary and can be set to zero (only the relative phases are important physically). For externally locked arrays, all of the oscillator phases are unknown, since the injected signals establish a phase reference.

3.2 Stability of Solutions

Let $(\hat{A}, \hat{\phi})$ be a solution vector to (4.45). The stability of these solutions can be investigated by linearizing (4.41) around them:

$$\frac{d\hat{A}_i}{dt} + \frac{d\delta A_i}{dt} = (\hat{A}_i + \delta A_i) \, \text{Re}\{F_i(\hat{\overline{A}} + \overline{\delta A}, \hat{\overline{\phi}} + \overline{\delta\phi})\} \quad (4.46)$$

$$\frac{d\hat{\phi}_i}{dt} + \frac{d\delta\phi_i}{dt} = \text{Im}\{F_i(\hat{\overline{A}} + \overline{\delta A}, \hat{\overline{\phi}} + \overline{\delta\phi})\} \quad (4.47)$$

where δA and $\delta\phi$ are small perturbations about the amplitude and phase solutions, respectively. Using the first two terms of a Taylor series expansion we can write

$$F_i(\hat{\overline{A}} + \overline{\delta A}, \hat{\overline{\phi}} + \overline{\delta\phi}) \approx F_i(\hat{\overline{A}}, \hat{\overline{\phi}}) + \sum_k \left(\frac{\partial F_i}{\partial A_k} \delta A_k + \frac{\partial F_i}{\partial \phi_k} \delta\phi_k \right) \quad (4.48)$$

where the derivatives are evaluated at the point $(\hat{A}, \hat{\phi})$. Substituting (4.48) into (4.47) and using (4.45) gives linearized equations for the perturbations:

$$\frac{d\delta A_i}{dt} = \hat{A}_i \sum_k \text{Re}\left\{\frac{\partial F_i}{\partial A_k}\delta A_k + \frac{\partial F_i}{\partial \phi_k}\delta\phi_k\right\} \tag{4.49}$$

$$\frac{d\delta\phi_i}{dt} = \sum_k \text{Im}\left\{\frac{\partial F_i}{\partial A_k}\delta A_k + \frac{\partial F_i}{\partial \phi_k}\delta\phi_k\right\} \quad i = 1, \ldots, N \tag{4.50}$$

This can be written as a matrix equation

$$\frac{d}{dt}\begin{bmatrix}\overline{\delta A}\\\overline{\delta\phi}\end{bmatrix} = -\overline{\overline{M}}\begin{bmatrix}\overline{\delta A}\\\overline{\delta\phi}\end{bmatrix} \tag{4.51}$$

A stable solution requires the perturbations to decay away in time, which occurs if the eigenvalues of the stability matrix $\overline{\overline{M}}$ have *positive* real parts. In general the eigenvalues must be evaluated numerically, but certain simple coupling networks result in a stability matrix with mostly zero entries and symmetry properties that can be exploited for analytical evaluation of the eigenvalues. For example, if the matrix is positive definite, all eigenvalues will be positive. A diagonally dominant matrix will be positive definite if the diagonal elements are positive. Note that for the case when there are no injected signals ($I_{\text{inj},i} = 0$ for all i), one of the eigenvalues will always be zero and should be ignored, reflecting the arbitrariness of the phase reference for that case.

3.3 Broadband Coupling Networks

The two terms in the denominator in (4.42) are proportional to the first-order frequency derivatives of the oscillator admittance ($\partial Y_{\text{osc},i}/\partial(j\omega)$), and coupling/load network port admittance ($\partial Y_{L,i}/\partial(j\omega)$), respectively. It should be remembered that these terms result from a Taylor series approximation of the respective transfer functions (Appendix A), where higher-order derivatives were assumed negligible. Our assumption of a simple parallel resonator with a linear transfer function near resonance makes the Taylor approximation extremely accurate, even for large Q factors. Similarly, if the coupling network is operated near an isolated parallel resonance, the approximation is quite good, but in practice it would be extremely difficult to align the resonances of all the oscillators with that of the coupling network.

Furthermore, employing a resonant coupling network together with resonant oscillator circuits results in a multimode system that can have more than one stable oscillation frequency, and steps must then be taken to suppress undesired modes of the system. We avoid this situation entirely by assuming that the coupling circuit is relatively broadband compared to the oscillator tank circuits. This would be the case if the oscillators were resistively coupled or if the coupling network has a significantly lower Q factor than that of the oscillators.

Under this assumption the second term in the denominator of (4.42) is generally negligible compare to the first; that is,

$$\frac{\omega_i}{2Q} \sum_{j=1}^{N} \frac{\partial \kappa_{ij}(\omega_r)}{\partial(j\omega)} \frac{V_j}{V_i} \ll 1 \qquad \text{for all } i \qquad (4.52)$$

The amplitude and phase dynamics in (4.41) then simplify to

$$\frac{dA_i}{dt} = \frac{\omega_i}{2Q}\left[A_i S_\mu(A_i) - \sum_{j=1}^{N} A_j \operatorname{Re}\{\kappa_{ij}e^{j(\phi_j-\phi_i)}\} + \rho_i \cos(\psi_i - \phi_i)\right] \qquad (4.53)$$

$$\frac{d\phi_i}{dt} = (\omega_i - \omega) + \frac{\omega_i}{2Q}\left[-\sum_{j=1}^{N} \frac{A_j}{A_i} \operatorname{Im}\{\kappa_{ij}e^{j(\phi_j-\phi_i)}\} + \frac{\rho_i}{A_i} \sin(\psi_i - \phi_i)\right] \qquad (4.54)$$

for $i = 1, \ldots, N$. It is understood that $\omega = \omega_{\text{inj}}$ when externally injected signals are present, or ω is an unknown to be solved for otherwise. Although formidable, considerable insight into the behavior of injection-locked oscillator systems can be obtained from (4.53)–(4.54) using simple analytical techniques, as will be done in the remainder of this chapter, following some further simplifications. Alternatively, these equations can be numerically integrated to find the amplitude and phase distributions for a given set of oscillator parameters.

3.4 Analysis, Synthesis, and Simplifications

The form of the coupling matrix, $\overline{\overline{\kappa}}$, will have a profound influence on the dynamics of the system. Note also the dependence on the free-running frequencies, ω_i. In a practical array the coupling network is usually fixed, but the free-running frequency distribution can be dynamically manipulated by employing voltage-controlled oscillators (VCOs) with a varactor tuning element. Therefore, for a given coupling network we have the following analysis and synthesis problems: (1) Given a set of free-running frequencies, solve for the resulting steady-state phase; and (2) given a desired phase distribution, solve for the free-running frequencies that will produce this phase relationship. The former is extremely difficult in the general case and must be carried out numerically. However, the linearity of (4.53)–(4.54) with respect to the free-running frequencies makes the synthesis procedure very straightforward.

In the steady state, (4.54) can be solved for a desired phase relationship, denoted by a circumflex ($\hat{\phi}_i$), giving

$$\omega_i = \omega\left[1 - \frac{1}{2Q}\sum_{j=1}^{N} \frac{\hat{A}_j}{\hat{A}_i}\operatorname{Im}\{\kappa_{ij}e^{j(\hat{\phi}_j-\hat{\phi}_i)}\} + \frac{1}{2Q}\frac{\rho_i}{\hat{A}_i}\sin(\psi_i - \hat{\phi}_i)\right]^{-1} \qquad (4.55)$$

Notice that (4.53), which does not depend on ω_i in the steady-state, requires that the amplitudes and phases have a relationship that is fixed by the coupling

152 COUPLED-OSCILLATOR ARRAYS AND SCANNING TECHNIQUES

network and the device nonlinearity, and hence we cannot arbitrarily specify both amplitude and phase simultaneously in the general case. Given an assumed phase distribution, the steady-state amplitudes, $\hat{A}_i$, corresponding to this phase distribution can be determined from (4.53). These are then inserted into (4.55) to determine the required oscillator tunings.

This discussion seems to imply that we can synthesize any arbitrary phase relationship by suitably adjusting the free-running frequencies, but recall that any realizable phase distribution must also satisfy a stability constraint, which requires that the matrix $\overline{\overline{M}}$ in (4.51) have eigenvalues with positive real parts. This condition must be checked after solving (4.55).

In practice it is helpful to sacrifice some accuracy in the modeling to allow for more convenient analysis and hopefully better insight into the operation of the system. Unless the coupling is very strong, the amplitudes will generally remain close to their free-running values. The small perturbation in amplitude introduces only a second-order correction to the phase. Therefore, most of the interesting dynamics are contained in (4.54) with $A_i \approx \alpha_i$, and the amplitude dynamics (4.53) can be neglected. This approximation is strictly valid only in the limit of weak coupling, but gives much insight into the operation of the arrays, even when the coupling is relatively strong. Since it also results in a significant reduction in mathematical complexity, this approximation will be used throughout the remainder of the chapter. In addition, we introduce the following notation for the coupling parameters:

$$\kappa_{ij} \equiv \epsilon_{ij} e^{j(\pi - \Phi_{ij})} \qquad (4.56)$$

where ϵ_{ij} and Φ_{ij} are the "coupling strength" and "coupling phase," respectively, from oscillator j into oscillator i. The reason for defining the coupling phase in this manner will become clearer when the theory is applied to a particular circuit, but it can be justified as follows: The terms involving κ_{ij} for $i \neq j$ represent the current injected into oscillator i from other oscillators in the system, but the standard definition for current flow in N-port network theory is *into* the network, which corresponds to current flow *out of* the oscillator. This accounts for the π in the exponent of (4.56). The negative sign in front of Φ is used because in a physical system the coupling involves a time *delay*. For $i = j$ this definition of Φ_{ij} is confusing, but for the cases we will consider κ_{ii} is real and therefore the definition is unimportant.

Using this notation and the above assumptions, (4.53)–(4.54) reduce to

$$\frac{d\phi_i}{dt} = (\omega_i - \omega) + \frac{\omega_i}{2Q}\left[-\sum_{j=1}^{N}\frac{\alpha_j}{\alpha_i}\epsilon_{ij}\sin(\Phi_{ij} + \phi_i - \phi_j) + \frac{\rho_i}{\alpha_i}\sin(\psi_i - \phi_i)\right] \qquad (4.57)$$

To proceed further we must specify a coupling network, and/or a feeding arrangement if externally injected signals are present. There are many possible choices, and no comprehensive comparison of the merits of different coupling schemes has been carried out. The specific requirements of the coupling scheme are application-specific. For spatial power combining, a baseline configuration might be an array of identical oscillators (identical free-running parameters) that are coupled in such a way as to encourage rapid and robust locking, with all the oscillators in phase. Intuitively this would require relatively strong coupling to provide a large locking range, and hence tolerance to small variations in the resonant frequencies (free-running frequencies) of the individual oscillators. It also seems logical to expect that the in-phase mode would be established for coupling angles at or near multiples of 2π.

The coupling schemes shown in Fig. 4.4 have been explored experimentally, along with radiative coupling [37] and spatially fed injection-locked systems [40]. In practice, multiple oscillator system transmission-line coupling circuits provide the strongest coupling, and they are easiest to construct in a nearest-neighbor coupling configuration, which could be unilateral (nonreciprocal) or bilateral (reciprocal), as shown in Fig 4.4b and 4.4c. Many of the experimental results described later were obtained with bilateral nearest-neighbor coupling. Nearest-neighbor coupling is also convenient from an analytical standpoint, and it will be considered next for the bilateral case.

3.5 Linear Arrays with Nearest-Neighbor Bilateral Coupling

Using the form (4.56), nearest-neighbor bilateral coupling in a linear chain of oscillators can be described by the coupling parameters

$$\kappa_{ij} = \begin{cases} 1, & i = j \\ \epsilon e^{j(\pi - \Phi)}, & i - j = 1 \\ 0, & \text{otherwise} \end{cases} \quad (4.58)$$

Assuming identical free-running amplitudes ($\alpha_i = \alpha_j$) gives

$$\frac{d\phi_i}{dt} = (\omega_i - \omega) - \frac{\epsilon \omega_i}{2Q} \sum_{\substack{j=i-1 \\ j \neq i}}^{i+1} \sin(\Phi + \phi_i - \phi_j) + \frac{\omega_i}{2Q} \frac{\rho_i}{\alpha_i} \sin(\psi_i - \phi_i) \quad (4.59)$$

for $i = 1, \ldots, N$, where it is implicitly understood that any terms containing subscripts 0 or $N + 1$ are ignored. Our first goal will be to show that $\Phi = 0$ is a desirable characteristic for coupling networks, and for convenience we will assume there are no externally injected signals ($\rho_i = 0$). Since we are primarily interested in the case when the oscillators can mutually lock, a useful simplification can be made by defining

$$\Delta \omega_m \equiv \frac{\epsilon \omega_i}{2Q} \quad (4.60)$$

By comparison with (4.23), $\Delta\omega_m$ can be interpreted as the locking range of the ith oscillator due to injection by a neighboring oscillator when all the amplitudes are identical. Since the free-running frequencies ω_i must be similar (within $\Delta\omega_m$ or each other) for locking to occur, then we can take $\Delta\omega_m$ to be the same for all oscillators. We then cast (4.59) into a form containing only relative phases by defining

$$\Delta\phi_i \equiv \phi_{i+1} - \phi_i, \quad i = 1, \ldots, N-1$$
$$\Delta\beta_i \equiv \omega_{i+1} - \omega_i$$

giving

$$\frac{d\Delta\phi_i}{dt} = \Delta\beta_i - \Delta\omega_m[-\sin(\Phi + \Delta\phi_{i-1}) + 2\cos\Phi\sin\Delta\phi_i + \sin(\Phi - \Delta\phi_{i+1})] \quad (4.61)$$

for $i = 1, \ldots, N-1$. This eliminates the problem of having one arbitrary phase and reduces the order of the system by one. It also eliminates the unknown frequency ω, but after solving for the $\Delta\phi_n$ we can find ω from any one of equations (4.59). The equations can be written in matrix form as

$$\frac{d}{dt}\overline{\Delta\phi} = \overline{\Delta\beta} - \Delta\omega_m[\cos(\Phi)\overline{\overline{A}}\,\overline{s} + \sin(\Phi)\overline{\overline{B}}\,\overline{c}] \quad (4.62)$$

where $\overline{\Delta\phi}$ and $\overline{\Delta\beta}$ are vectors with elements $\Delta\phi_i$ and $\Delta\beta_i$, and

$$\overline{\overline{A}} = \begin{pmatrix} 2 & -1 & & 0 \\ -1 & 2 & -1 & \\ & & \ddots & \\ 0 & & -1 & 2 \end{pmatrix}, \quad \overline{s} = \begin{pmatrix} \sin\Delta\phi_1 \\ \sin\Delta\phi_2 \\ \vdots \\ \sin\Delta\phi_{N-1} \end{pmatrix}$$

$$\overline{\overline{B}} = \begin{pmatrix} 0 & 1 & & 0 \\ -1 & 0 & 1 & \\ & & \ddots & \\ 0 & & -1 & 0 \end{pmatrix}, \quad \overline{c} = \begin{pmatrix} \cos\Delta\phi_1 \\ \cos\Delta\phi_2 \\ \vdots \\ \cos\Delta\phi_{N-1} \end{pmatrix}$$

Stability is examined by linearizing (4.62) around a steady-state solution. Denoting the perturbation to $\overline{\Delta\phi}$ as $\overline{\delta}$ gives

$$\frac{d}{dt}\overline{\delta} = -\overline{\overline{M}}\,\overline{\delta} \quad (4.63)$$

where the $(N-1) \times (N-1)$ stability matrix $\overline{\overline{M}}$ is

$$\overline{\overline{M}} = \Delta\omega_m[\cos(\Phi)\overline{\overline{A}}\,\overline{\overline{C}} + \sin(\Phi)\overline{\overline{B}}\,\overline{\overline{S}}] \tag{4.64}$$

and we have defined the $(N-1) \times (N-1)$ diagonal sine and cosine matrices as

$$\overline{\overline{S}} = \begin{pmatrix} \sin\Delta\phi_1 & & & 0 \\ & \sin\Delta\phi_2 & & \\ & & \ddots & \\ 0 & & & \sin\Delta\phi_{N-1} \end{pmatrix}$$

$$\overline{\overline{C}} = \begin{pmatrix} \cos\Delta\phi_1 & & & 0 \\ & \cos\Delta\phi_2 & & \\ & & \ddots & \\ 0 & & & \cos\Delta\phi_{N-1} \end{pmatrix}$$

As mentioned earlier, a baseline configuration for power-combining can be taken as an array of identical oscillators operating simultaneously in-phase, which would produce a broadside beam in a spatially combined array. For $\Delta\beta_i = 0$ (identical free-running frequencies) and $\Delta\phi_i = 0$ (identical phases), (4.62) requires that $\Phi = n\pi$, where n is an integer. The stability matrix then becomes $\overline{\overline{M}} = \Delta\omega_m \cos(\Phi)\overline{\overline{A}}$. A stable mode requires that all the eigenvalues of $\overline{\overline{M}}$ have positive real parts. Since $\overline{\overline{A}}$ is symmetric and positive definite, it will have real, positive eigenvalues. Therefore stability requires that $\cos(\Phi) > 0$, which gives the final result for a stable in-phase mode:

$$\Phi = 2n\pi, \qquad n = 0, 1, 2, \ldots \tag{4.65}$$

At the other extreme, examining (4.62) for the case of $\Phi = \pi/2$ we find that the matrix $\overline{\overline{B}}$ is singular for even values of N. This implies that the equations are not linearly independent and we cannot determine the phase from a given set of oscillator tunings. For odd N the matrix $\overline{\overline{B}}$ is nonsingular, so an inverse exists, but the stability matrix $\overline{\overline{M}} = -\Delta\omega_m \overline{\overline{B}}\,\overline{\overline{S}}$ always has zero trace. Since the trace of a matrix is the sum of its eigenvalues, the real parts of the eigenvalues cannot all be positive and hence no stable state exists.

When Φ is some other value the stability matrix is a weighted combination of $\overline{\overline{A}}\,\overline{\overline{C}}$ and $\overline{\overline{B}}\,\overline{\overline{S}}$, and we can probably infer that the stability region is maximized for $\Phi = 0$, but we have not proved this. It can be shown that the tendency for synchronization is maximized at $\Phi = 0$ for the case of $N = 2$, but this is difficult to prove for larger N. The condition (4.65) will be subsequently assumed in all analyses, and is an important design constraint for practical coupling networks.

Analysis with Zero Coupling Phase Inserting $\Phi = 0$ into (4.62) gives

$$\frac{d}{dt}\overline{\Delta\phi} = \overline{\Delta\beta} - \Delta\omega_m \overline{\overline{A}}\,\overline{s} \qquad (4.66)$$

Setting the time derivative equal to zero gives an algebraic equation for the steady-state phase differences in terms of the oscillator free running frequencies:

$$\overline{s} = \frac{1}{\Delta\omega_m}\overline{\overline{A}}^{-1}\overline{\Delta\beta} \qquad (4.67)$$

which can be solved by inverting the matrix $\overline{\overline{A}}$ and solving for the phase using the inverse sine function. The matrix $\overline{\overline{A}}$ has a simple inverse

$$A_{ij}^{-1} = \frac{j(i-N)}{N} \quad i \geq j, \qquad A_{ij}^{-1} = A_{ji}^{-1} \qquad (4.68)$$

Clearly there are no possible solutions of (4.67) if any element of the column vector $\overline{\overline{A}}^{-1}\overline{\Delta\beta}$ has a magnitude greater than $\Delta\omega_m$. When there is a valid solution for the sine vector, this will correspond to 2^{N-1} different solutions for the phase differences, since the inverse sine function is multivalued. The correct solution is found by stability analysis. The stability matrix for this case is

$$\overline{\overline{M}} = \Delta\omega_m \overline{\overline{A}}\,\overline{\overline{C}} \qquad (4.69)$$

The matrix $\overline{\overline{A}}$ is positive definite, and the matrix $\overline{\overline{C}}$ is also positive definite when each of the phases lies in the range

$$-\frac{\pi}{2} < \Delta\phi_i < \frac{\pi}{2}, \qquad i = 1,\ldots,N-1 \qquad (4.70)$$

Since the product of two positive definite matrices is also positive definite, the eigenvalues of the stability matrix are all real and positive when the phases lie in the above range. This range of phases is sufficient to cause the vector $\overline{s}$ in (4.67) to span all of its possible values, which proves that the stability region fills the entire existence region. Furthermore, over this range of phases the sine functions in $\overline{s}$ are one-to-one. Thus for each set of tunings within the stability region there is a unique phase vector which implies that a unique stable synchronized state exists for a given tuning.

As mentioned in the previous section, once the phases or relative tunings are determined, the steady-state synchronized frequency is found by inserting back into (4.59). Assuming both (4.60) and (4.65) hold, we find the interesting result for the steady-state frequency:

$$\omega = \frac{1}{N}\sum_{i=1}^{N}\omega_i \qquad (4.71)$$

In other words, the synchronized frequency is just the average of the free-running frequencies.

Phase Sensitivity A convenient measure of phase sensitivity is the change in length of the phase vector for a given change in the length of the frequency tuning vector. Using (4.66) in the steady state, a differential change in phase is related to a change in tuning through the stability matrix:

$$d\overline{\Delta\beta} = \Delta\omega_m \overline{\overline{A}}\,\overline{\overline{C}}\, d\overline{\Delta\phi} = \overline{\overline{M}}\, d\overline{\Delta\phi} \qquad (4.72)$$

The phase sensitivity as defined above is then given by

$$S_{\Delta\beta}^{\Delta\phi} \equiv \left(\frac{d\overline{\Delta\phi}^T\, d\overline{\Delta\phi}}{d\overline{\Delta\beta}^T\, d\overline{\Delta\beta}}\right)^{1/2} = \left(\frac{d\overline{\Delta\beta}^T((\overline{\overline{M}}^{-1})^T\, \overline{\overline{M}}^{-1})d\overline{\Delta\beta}}{d\overline{\Delta\beta}^T\, d\overline{\Delta\beta}}\right)^{1/2} \qquad (4.73)$$

If the change in tuning is an eigenvector of the matrix $\overline{\overline{M}}^{-1}$ with eigenvalue λ^{-1}, then the sensitivity is simply $S_{\Delta\beta}^{\Delta\phi} = |\lambda^{-1}|$. Note that the eigenvectors of $\overline{\overline{M}}^{-1}$ and $\overline{\overline{M}}$ are indentical and the eigenvalues are reciprocals of one another. One can show that the eigenvectors of $\overline{\overline{M}}$ are the directions for which the sensitivity is either a local maximum or minimum. Thus a tuning change in the direction of an eigenvector of the stability matrix $\overline{\overline{M}}$, which we call a "characteristic tuning" for the array, causes an extremum of the phase sensitivity, which is given by the reciprocal of the corresponding eigen value of $\overline{\overline{M}}$. This result, together with the results of the previous section, shows that the size of the locking region is directly related to the phase sensitivity. Indeed, we expect to lose synchronization more quickly if we tune in the direction of high phase sensitivity. The characteristic tunings also play an important role in the transient phase response of the array, as will be discussed later.

3.6 Experimental Results

A useful practical example of a nearest-neighbor bilateral coupling network is shown in Fig. 4.6 for a linear array [30]. The oscillators are connected in parallel

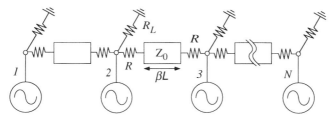

FIGURE 4.6 Simple coupling and load network, suitable for parallel oscillator circuits, which gives nearest-neighbor coupling parameters [39].

with sections of resistively loaded transmission lines. The resistors are used to manipulate the coupling strength and also to satisfy the broadband condition (4.52). This circuit has desirable propeties for the special case of $R = Z_0$, for which the Y-parameters become

$$Y_{ij} = \begin{cases} \dfrac{\eta_i}{2Z_0} + G_L, & i = j \\ \dfrac{-e^{-j\beta L}}{2Z_0}, & |i - j| = 1 \\ 0, & \text{otherwise} \end{cases} \quad \dfrac{\partial Y_{ij}}{\partial(j\omega)} = \begin{cases} 0, & i = j \\ \dfrac{\tau_g e^{-j\beta L}}{2Z_0}, & |i - j| = 1 \\ 0, & \text{otherwise} \end{cases} \quad (4.74)$$

where βL is the electrical length of the transmission-line, $\eta_i = (2 - \delta_{i1} - \delta_{iN})$, δ_{ij} is the Kronecker delta function, and τ_g is the group delay through the transmission line. For TEM or quasi-TEM lines, $\tau_g = \beta L/\omega$; and when $\beta L \leq 2\pi$, then condition (4.52) is satisfied if $\pi R_L/QZ_0 \ll 1 (R_L = 1/G_L)$. Assuming that this constraint holds, and letting $\epsilon = R_L/2Z_0$ and $\Phi = \beta L$, then (4.57) reduces to exactly the same form as our ideal nearest-neighbor coupling equation, (4.59).

This structure was experimentally investigated using the six-element patch array shown in Fig. 4.7 [39]. The coupling network is connected at a radiating edge of the patch. At patch resonance, the load impedance at a radiating edge is a parallel combination of the two edge radiation resistances. The oscillator and patch circuits were designed on 31-mil (0.787-mm) Rogers Duroid 6810 with $\epsilon_r = 10.8$, for operation at 4.0 GHz with an array period of $d = 20$ mm. The

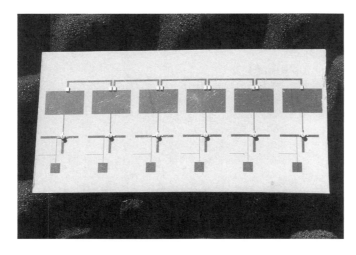

FIGURE 4.7 Photograph of a six-element MESFET microstrip oscillator array using patch antennas and a transmission-line coupling network based on that of Fig. 4.6. The array operated at 4 GHz [39,38].

patch dimensions gave a radiation resistance of approximately 800Ω at each edge. A one-wavelength, 50Ω coupling line with 50Ω chip resistors were used, which should provide a coupling angle of $\Phi = 0°$ and a coupling strength of $\epsilon \approx 4.0$

The oscillators were a common-gate configuration with the output taken from the source lead, which typically results in a parallel oscillator model. A low-noise packaged GaAs MESFET, NEC 32184A, was used. No tuning element was employed, but some variation in output frequency was achieved through bias tuning (this also affects the amplitude of oscillation, which is a drawback for comparison with simple dynamic array models). When the oscillators were all powered on and mutually locked, the measured synchronization frequency was centered in the range of 4.3 GHz, or 7% above the design frequency. As a result, the coupling phase was not exactly 0°, and with all the oscillators set to a common free-running frequency, the measured radiation pattern was skewed away from the broadside position (approximately $-5°$) as shown in Fig. 4.8a. The nonzero coupling phase, the invariable presence of slight randomness in the free-running frequencies, and measurable differences in the free-running amplitudes of the oscillators all conspire to produce a small average phase progression along the array. Combining a 6-element array factor (calculated using a phase shift of 8° along the array) with the H-plane patch antenna element pattern (essentially given by cos θ) gives the theoretical pattern shown in Fig. 4.8a for comparison with the measurements. Good agreement is observed with the measured pattern. The scanned pattern in Fig. 4.8b is achieved by antisymmetrically detuning the end elements and will be described later.

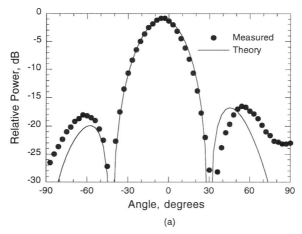

(a)

FIGURE 4.8 Measured H-plane radiation patterns for the array of Fig. 4.7 [48]. (a) A nearly broadside pattern obtained with all the oscillators set to approximately the same free-running frequency. (b) A scanned pattern obtained by antisymmetrically detuning the end elements. (*Continued*)

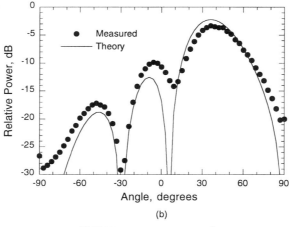

(b)

FIGURE 4.8 (*Continued*)

For power-combining purposes, our best result using the coupled-oscillator concept has been a 6 × 2 array operating at 11 GHz [41]. This array used a coupling network similar to the one shown in Fig. 4.7, with two linear arrays of size oscillators interconnected at the edges to form a 12-element loop. The oscillators were designed around packaged GaAs power FETs (NEC 9008-11) in a feedback oscillator topology. Since this was a power design, care was taken to provide adequate heatsinking for the device. Consequently a "tray" approach was adopted as illustrated in Fig. 4.9, where each oscillator circuit was con-

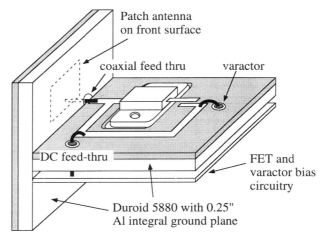

FIGURE 4.9 Hardware configuration for a two-dimensional power oscillator array [41]. The feedback circuits were isolated from the output radiation using a tray approach, which the oscillators feeding a patch antenna via a coaxial feed-through. Thick ground planes were required for heat-sinking and mechanical stability. Each oscillator produced 1 W CW at 11 GHz and 15% DC-to-RF efficiency.

structed on a Duroid substrate with a thick (quarter inch) aluminum ground plane. Each oscillator fed a patch antenna on a separate board via short coaxial feed-throughs. The feed point on the probe-fed patches was selected to provide an optimal impedance match for maximum output power, which was determined empirically using a mechanical tuner and network analyzer. Each oscillator produced approximately 1 W in this configuration, under typical bias conditions of 8.5 V @ 900 mA.

The array was built from two linear array oscillator "palettes," one of which is shown in Fig. 4.10. Patterns for the 6×2 arrays are shown in Fig. 4.11. The measured EIRP for this array was 933 W. The E-plane patterns were distorted by the presence of additional parasitic patches on one side of the array (the array was actually a subarray of a larger 6×6 version [41]). Since the E-plane

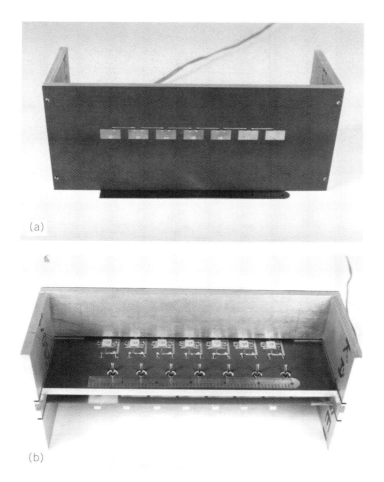

FIGURE 4.10 Photographs showing (a) front and (b) back views of a prototype oscillator "pallette" used to construct a modular two-dimensional array.

pattern for this array does not closely match the theoretical pattern for a 12-element patch array, the total power could not be reliably estimated by finding the theoretical directivity for the array as described in Chapter 1. An empirical estimate of the directivity was made by constructing an approximate three-dimensional pattern from the measured principal pattern cuts and integrating according to the standard definition of directivity (see Chapter 1). This gave a directivity of $D_0 = 81$ (19 dB), which leads to a total radiated power of 11.7 W

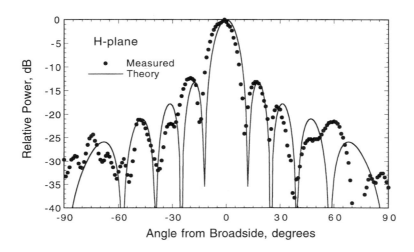

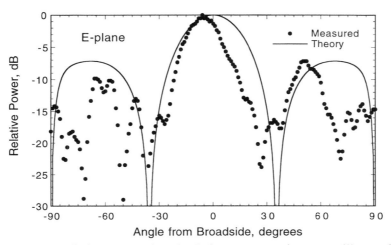

FIGURE 4.11 Radiation patterns for a 6 × 2 element array using two oscillator pallettes similar to those shown in Fig 4.10. The E-plane patterns were distorted by the presence of inactive patch antennas surrounding the two active pallettes.

and an estimated power-combining efficiency of 96% based on the direct power measurements of the individual oscillators. The total arrays drew 9 A of current at 8.5 V, giving a wall-plug efficiency of 15%.

3.7 Transient Response to Tuning Variations

Up to this point we have ignored the dynamic problems associated with the modulation of the carrier frequency, such as modulation bandwidth and array settling time. However, these are important concerns for the system designer. For beam steering systems (described later) there is a settling time required for the oscillator phases to achieve a new steady state after the beam has suffered a step change. Similarly, for communication systems relying on phase modulation this settling time implies a finite modulation bandwidth. The settling time not only depends on the coupling strength, oscillator bandwidth, and the number of elements, but also depends on the manner that the array is modulated. The following analysis shows that the complex phase behavior due to an arbitrary change in the array element tunings can be decomposed into a combination of responses to characteristic tunings, each giving rise to a characteristic phase perturbation that possesses a unique exponential decay time. The analysis assumes small tuning changes near the center of the synchronization region, but the computed settling times give approximate results for large tuning variations, and the physical insight developed is useful in understanding the dynamic phase behavior.

Although a general analysis of the effects of arbitrary time-dependent tuning changes is possible, in the following we will consider only step changes of element tunings. This type of tuning might occur in beam steering systems where a step change in the transmitting beam direction is desired. We assume that we can control the oscillator free running frequencies, or tunings, with infinite rapidity. In practice, we will not have such control, but the following results will hold approximately if the element tunings respond faster than the fastest time constant of the array. If this is not the case the time constants of the tunings must be included in the analysis.

For small phase changes (keeping the tunings constant for the moment) the phase response is governed by the system (4.63), which has a solution [42]

$$\Delta\delta(t) = e^{-\bar{\bar{M}}t}\Delta\delta(0) \tag{4.75}$$

where the matrix exponential is defined through the usual power series expansion

$$e^{-\bar{\bar{M}}t} = I - \bar{\bar{M}}t + \frac{1}{2}(\bar{\bar{M}}t)^2 + \cdots \tag{4.76}$$

Thus the behavior of the phases for a stable frequency-locked state always involves exponential decay, at least for small phase changes. Large phase changes may be brought about by beam switching across large angles, or by operation near the edge of the frequency locking region. Such phase changes may not initially behave according to equation (4.75), but as the phase approach the steady state the exponential decay of equation will prevail.

A general result of linear system theory tells us that the eigenvalues of the stability matrix in equation (4.63) are the reciprocals of the time constants that govern the exponential decay of the phases into the steady state for particular phase distributions, the eigenvectors. These characteristic phase perturbations can be caused by particular tuning changes that we will call characteristic tunings. As mentioned previously, the eigenvalues and eigenvectors of the matrix $\overline{\overline{M}}$ are directly related to the size, shape, and orientation of the locking region.

The change in steady-state phase due to an arbitrary, but infinitesimal, tuning change is given by (4.72). This shows that if the phase change is an eigenvector of the stability matrix, then the corresponding tuning change is proportional to the same eigenvector and the constant of proportionality is the eigenvalue; that is

$$\overline{\overline{M}}\, d\overline{\Delta\phi} = \lambda d\overline{\Delta\phi} = d\overline{\Delta\beta} \tag{4.77}$$

Thus, the characteristic tuning changes are found by computing the eigenvectors of the stability matrix $\overline{\overline{M}}$.

Practical systems will be designed to operate near the center of the locking region to avoid the excessive phase sensitivity that occurs near the region edge. Near the region center we can make the approximation in equation (4.66) that $\overline{s} \cong \overline{\Delta\phi}$, which gives a linear relation between the phases and oscillator tunings:

$$\overline{\Delta\beta} \cong \Delta\omega_m \overline{\overline{A}}\overline{\Delta\phi} \tag{4.78}$$

We now consider the dynamic phase response to an arbitrary step change (at $t = 0$) of the oscillator tunings near the center of the locking region. The tuning difference vector prior to the step change is denoted as $\overline{\Delta\beta}^{\text{init}}$, with a corresponding phase difference vector $\overline{\Delta\phi}^{\text{init}}$. After the change we have $\overline{\Delta\beta}^{\text{final}}$ and the steady-state phase vector is $\overline{\Delta\phi}^{\text{final}}$. Just after the step change the phase vector is time-dependent, $\overline{\Delta\phi}(t)$, and the difference, or perturbation, $\overline{\Delta\delta}(t) = \overline{\Delta\phi}(t) - \overline{\Delta\phi}^{\text{final}}$ satisfies the linear system (4.63) with the initial value $\overline{\Delta\phi}^{\text{init}} - \overline{\Delta\phi}^{\text{final}}$. The initial and final phase vectors are directly related to the oscillator tunings through (4.67). Therefore, the system dynamics will evolve according to (4.75):

$$\overline{\Delta\delta}(t) = \frac{1}{\Delta\omega_m}[\overline{\overline{A}}^{-1}(\overline{\Delta\beta}^{\text{init}} - \overline{\Delta\beta}^{\text{final}})]e^{-\Delta\omega_m \overline{\overline{A}} t} \tag{4.79}$$

Among other things, this expression illustrates the importance of having a large locking range. The general behavior is quite complicated and probably best understood by decomposing the response into a superposition of characteristic responses. The eigenvalues λ_k and eigenvectors $\hat{e}^{(k)}$ are defined by the relation $\overline{\overline{A}}\hat{e}^{(k)} = \lambda_k \hat{e}^{(k)}$ and are easily determined using difference equation methods by noting the similarity to the derivative operator d^2/dx^2, giving

$$\lambda_k = 4 \sin^2\left(\frac{k\pi}{2N}\right), \quad k = 1, \ldots, N-1 \qquad (4.80)$$

$$[\hat{e}^{(k)}]_n = \sqrt{\frac{2}{N}} \sin\left(k\frac{n\pi}{N}\right), \quad n = 1, \ldots, N-1 \qquad (4.81)$$

If the tuning change is proportional to one of the eigenvectors—that is, if $\overline{\Delta\beta}^{\text{init}} - \overline{\Delta\beta}^{\text{final}} = \Omega\hat{e}^{(k)}$ where Ω sets the step size—then the phase perturbations will decay exponentially with a unique time constant $\tau_k = 1/\Delta\omega_m\lambda_k$ according to

$$\overline{\Delta\delta}(t) = \Omega\tau_k e^{-t/\tau_k} \hat{e}^{(k)} \qquad (4.82)$$

This shows that the effect of the tuning change on the phases increases as the eigenvalue decreases. This is consistent with the statement earlier concerning the sensitivity of phase with respect to frequency variations and is directly related to the size of the locking region in this particular tuning direction. The longest time constant, given by the smallest eigenvalue, represents the array settling time (assuming this mode is excited). The eigenvectors form a complete set (an orthonormal set for the matrix $\overline{\overline{A}}$) and can be used to represent an arbitrary tuning change. Once the change is decomposed into a linear combination of the characteristic tuning changes the phase dynamics can be analyzed by superposing the individual responses, each of which has a characteristic decay time directly related to its eigenvalue. Certain tuning variations may excite some modes to such a small degree that the phase perturbations can be neglected.

Example: Settling Time for Beam Steering Using the example of electronic beam steering discussed later, we will treat the case of a step change away from broadside for a five-element array. Initially, broadside radiation implies that the elements are identically tuned so that $\overline{\Delta\beta}^{\text{init}} = 0$. To steer a transmitting beam the end elements must be tuned in equal but opposite direction; thus the final vector of tuning *differences* is $\overline{\Delta\beta}^{\text{final}} = (\Omega, 0, \ldots, 0, \Omega)^T$, where Ω is the amount of the tuning change. Expressing the applied tuning as a linear combination of characteristic tunings, we have

$$\overline{\Delta\beta} = \sum_{k=1}^{N-1} a_k \hat{e}^{(k)} \qquad (4.83)$$

and taking the dot product of both sides with $\hat{e}^{(l)}$ and using the eigenvector orthogonality relation $\hat{e}^{(l)} \cdot \hat{e}^{(k)} = \delta_{kl}$ gives the weighting factors

$$a_l = \overline{\Delta \beta} \cdot \hat{e}^{(l)} \tag{4.84}$$

For our example we find that the weighting factors are

$$a = (0.743 \quad 0 \quad 1.203 \quad 0)^T \tag{4.85}$$

with corresponding time constants

$$\tau = \frac{1}{\Delta \omega_m}(2.618 \quad 0.724 \quad 0.382 \quad 0.276)^T \tag{4.86}$$

The zero entries are due to orthogonal symmetries between the tuning variation and two of the characteristic tunings. We can see that the applied tuning couples quite efficiently into the dominant time constant, which is the first element of each vector. However, for the type of tuning considered here, where only the end elements are varied, the lowest-order mode is coupled less efficiently as the number of oscillator elements increases. In other words, it takes longer for the array to settle to the final steady-state phase distribution for a given step change as the size of the array increases.

3.8 Phase Noise Analysis

Oscillator noise is a complicated subject even for a single oscillator. For simplicity we will consider the behavior of noisy coupled oscillators by restricting attention to phase fluctuations, ignoring the possibility of conversion of amplitude noise to phase noise. Following the procedure described earlier, a system of noisy coupled oscillators is modeled by replacing the ith external injection term in (4.57) by a quadrature noise component $-B_{ni}(t)$, which is equivalent to adding a noise susceptance in the oscillator equivalent circuit [31–33,43]. This gives

$$\frac{d\phi_i}{dt} = (\omega_i - \omega) - \frac{\omega_i}{2Q}\left[\sum_{j=1}^{N} \frac{\alpha_j}{\alpha_i} \epsilon_{ij} \sin(\Phi_{ij} + \phi_i - \phi_j) + B_{ni}(t)\right] \tag{4.87}$$

Assuming the oscilllator phase fluctuations due to the presence of these noise sources are small, we can linearize (4.87) around a "noise-free" solution, $\hat{\phi}_i$, which is a stable solution to (4.87) when $B_{ni} = 0$. Writing $\phi_i = \hat{\phi}_i + \delta\phi_i$ gives

$$\frac{d\delta\phi_i}{dt} = -\omega_{3\,dB}\left[\sum_{j=1}^{N} \epsilon_{ij}(\delta\theta_i - \delta\theta_j)\frac{\hat{\alpha}_j}{\hat{\alpha}_i}\cos(\hat{\phi}_i - \hat{\phi}_j + \Phi_{ij}) + B_{ni}(t)\right] \tag{4.88}$$

where $\omega_{3\,dB} \equiv \omega_i/2Q$, half the 3-dB bandwidth of the oscillator tank circuits (assumed identical to first order). In the following we once again invoke the coupling condition (4.65), so that for identical oscillators and free-running frequencies the stable locked state has all oscillators in-phase. We will further assume that all the steady-state amplitudes are identical. The spectral characteristics of the noise fluctuations are found by Fourier transforming (4.88) to give

$$j\left(\frac{\omega}{\omega_{3\,dB}}\right)\widetilde{\delta\phi_i} = -\sum_{j=1}^{N} \epsilon_{ij}(\widetilde{\delta\phi_i} - \widetilde{\delta\phi_j}) \cos(\hat{\phi}_i - \hat{\phi}_j) - \tilde{B}_{ni} \quad (4.89)$$

for $i = 1, \ldots, N$, where the tilde ($\tilde{\ }$) denotes a transformed or spectral variable, and ω is the noise frequency measured relative to the carrier. The matrix form of (4.89) is

$$\overline{\overline{N}}\,\widetilde{\overline{\delta\phi}} = \tilde{\overline{B}}_n \quad (4.90)$$

where

$$\widetilde{\overline{\delta\phi}} = \begin{pmatrix} \widetilde{\delta\phi_1} \\ \widetilde{\delta\phi_2} \\ \vdots \\ \widetilde{\delta\phi_N} \end{pmatrix}, \quad \hat{\overline{B}}_n = \begin{pmatrix} \tilde{B}_{n1} \\ \tilde{B}_{n2} \\ \vdots \\ \tilde{B}_{nN} \end{pmatrix}$$

The quadrature noise sources $\tilde{B}_{n1}, \tilde{B}_{n2}, \ldots, \tilde{B}_{nN}$ are assumed to be uncorrelated and have the same power spectral density:

$$|\tilde{B}_{n1}|^2 = |\tilde{B}_{n2}|^2 = \cdots = |\tilde{B}_{nN}|^2 = |\tilde{B}_n|^2 \quad (4.91)$$

The matrix $\overline{\overline{N}}$ will reflect the coupling topology of the N-element coupled oscillator array. The phase fluctuations of the individual oscillator are then determined by the matrix equation

$$\widetilde{\overline{\delta\phi}} = \overline{\overline{P}}\tilde{\overline{B}}_n \quad (4.92)$$

where $\overline{\overline{P}} = \overline{\overline{N}}^{-1}$. Since many of the coupling topologies possess some intrinsic symmetry which leads to common solutions for all of the phase fluctuations, it is useful to simplify (4.92) by writing

$$\widetilde{\delta\phi_i} = \sum_{j=1}^{N} p_{ij}\tilde{B}_{nj} \quad (4.93)$$

where p_{ij} is an element of the matrix $\bar{\bar{P}}$. Using (4.91) we can write the power spectral density of the ith-oscillator phase fluctuation (i.e., the phase noise) as

$$|\widetilde{\delta\phi_i}|^2 = |\tilde{B}_n|^2 \sum_{j=1}^{N} |p_{ij}|^2 \tag{4.94}$$

which indicates that the ith element phase noise is found by summing the magnitude square of the elements in the ith row of the matrix $\bar{\bar{P}}$.

The combined output of all the array elements is the most important quantity of interest in coupled-oscillator array applications. Assuming that the outputs are combined efficiently, the combined output signal is given by

$$V(t) = A \sum_{j=1}^{N} \cos(\omega_0 t + \delta\theta_j) \tag{4.95}$$

where the oscillators are locked to a common frequency ω_0. Using the small fluctuation assumption allows (4.95) to be written as

$$V(t) = NA \cos(\omega_0 t + \delta\theta_{\text{total}}) \tag{4.96}$$

where

$$\delta\theta_{\text{total}} = \frac{1}{N} \sum_{j=1}^{N} \delta\theta_j \tag{4.97}$$

Using (4.93) we can write (4.97) as

$$\widetilde{\delta\phi}_{\text{total}} = \frac{1}{N} \sum_{j=1}^{N} \left(\sum_{i=1}^{N} p_{ij} \right) \tilde{B}_{nj} \tag{4.98}$$

Again using (4.91) we can write the total phase noise as

$$|\widetilde{\delta\theta}_{\text{total}}|^2 = \frac{|\tilde{B}_n|^2}{N^2} \sum_{j=1}^{N} \left| \sum_{i=1}^{N} p_{ij} \right|^2 \tag{4.99}$$

Many of the commonly encountered coupling matrices have properties that make carrying out the indictated sums in both (4.99) and (4.94) very easy.

Example: Nearest-Neighbor Bilateral Coupling For the case of nearest-neighbor bilateral coupling we have

$$\epsilon_{ij} = \begin{cases} \epsilon, & |i - j| = 1 \\ 0, & \text{otherwise} \end{cases} \tag{4.100}$$

where ϵ is a constant that can be related to the circuit design [39]. To keep the math tractable we will also assume a constant phase progression along the array so that $\hat{\phi}_i - \hat{\phi}_{i+1} = \Delta\hat{\phi}$. As described later, this phase progression can be established by varying the free running frequencies of the both end oscillators, while keeping the central elements at a common free-running frequency. For this configuration the matrix $\bar{\bar{N}}$ in (4.90) becomes

$$\bar{\bar{N}} = \epsilon \cos \Delta\hat{\phi} \begin{pmatrix} -1-jx & 1 & & & & 0 \\ 1 & -2-jx & 1 & & & \\ & 1 & -2-jx & 1 & & \\ & & & \ddots & & \\ 0 & & & & 1 & -1-jx \end{pmatrix}$$

(4.101)

where $x = \omega/(\Delta\omega_m \cos \Delta\hat{\phi})$. The inverse of $\bar{\bar{N}}$ is not easily expressed for the general case. However, note that from the relation $\bar{\bar{N}}\bar{\bar{P}} = \bar{\bar{I}}$ we can write

$$\sum_{j=1}^{N} p_{ij} n_{jk} = \delta_{ik}$$

and consequently

$$\sum_{k=1}^{N} \sum_{j=1}^{N} p_{ij} n_{jk} = \sum_{j=1}^{N} p_{ij} \left(\sum_{k=1}^{N} n_{jk} \right) = 1 \quad (4.102)$$

By inspection of (4.101) we can easily see that the term in parentheses in (4.102), which is the sum of the jth row, the simply $-j\omega/\omega_{3\,\text{dB}}$, for all j. Therefore

$$\sum_{j=1}^{N} p_{ij} = \frac{-1}{j\omega/\omega_{3\,\text{dB}}} \quad (4.103)$$

so, from (4.97), the total output phase noise is

$$|\widetilde{\delta\phi}_{\text{total}}|^2 = \frac{1}{N} \frac{|\tilde{B}_n|^2}{(\omega/\omega_{3\,\text{dB}})^2} \quad (4.104)$$

Comparing (4.104) with the single-oscillator noise result in (4.33) we find

$$|\widetilde{\delta\phi}_{\text{total}}|^2 = \frac{1}{N} |\widetilde{\delta\phi}_i|^2_{\text{uncoupled}} \quad (4.105)$$

The total phase noise is reduced by a factor of $1/N$, independent of the phase difference $\Delta\hat{\phi}$. Evidently the mutual synchronization does not lead to any significant correlation of the oscillator phases (to first order).

170 COUPLED-OSCILLATOR ARRAYS AND SCANNING TECHNIQUES

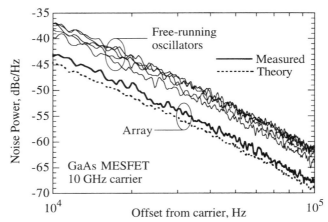

FIGURE 4.12 Measured phase noise for a five-element X-band MESFET array for 10 kHz to 100 kHz offset from the carrier. Noise powers of the individual free-running oscillators are compared with noise power of the synchronized array, showing a noise reduction in general agreement with the simple models described in the text.

Experimental Verification A five-element linear coupled oscillator chain was built for experimental verification of the theory for a nearest-neighbor bilaterally coupled array. The array was composed of five varactor-tuned MESFET VCOs with a nominal tuning range of 8.0–9.0 GHz. These VCOs use NE32184A packaged MESFETs and MA-COM 46600 varactor diodes. The VCOs are coupled together by one-wavelength (at about 8.5 GHz) microstrip transmission lines and are resistively loaded with two 75Ω chip resistors, using the coupling topology of Fig. 4.6. As described earlier, this technique provides coupling parameters $\epsilon \approx 0.5$ and $\Phi = 0°$. Each oscillator is designed to deliver power to a 50Ω load. The oscillators were "connectorized" using SMA-to-microstrip transitions, which allowed for simple testing and later connection to an external five-element patch antenna array.

When all elements are set to a common free-running frequency, the elements are nominally in phase and a broadside beam is expected. This was verified by the measurements (see reference [43]). It was found that the array can remain locked within a maximum end-element detuning of approximately ±125 MHz, which gives a useful estimate of the locking range.

Because of VCO's inherent poor phase noise behavior and comparatively large thermal drift, a frequency discriminator technique was used for the phase noise measurement (see reference 43 for details). This apparatus was used to characterize both the total output array noise and the individual oscillator fluctuations in a variety of conditions. For total array noise, the oscillators were connected to a patch antenna array and the output signal was measured with a detector in the far field. Isolators were placed between the oscillator output and the antenna feed to maintain a 50Ω load impedance. For the individual oscillator measurements, the oscillators were connected directly to the measurement system using SMA cables.

Figure 4.12 shows (1) the phase noise of the individual array elements when free-running and (2) the total array output under synchronized conditions (all oscillators set to a common free-running frequency). The total output phase noise is clearly reduced as compared to those of the free-running oscillators. The theoretical result using (4.105) is shown for comparison, and it shows close agreement to the measurements. The small difference between the measurement and theoretical values could be due to a number of influences that are neglected in our analysis, including the assumption of the nearest-neighbor coupling, the approximation for total output phase noise (4.99), and the neglect of transformation of amplitude noise to phase noise.

4 SCANNING OSCILLATOR ARRAYS

The analytical techniques developed above have demonstrated an ability to synthesize certain phase relationships in an oscillator array through the use of a suitable coupling scheme and by manipulating the free-running frequencies in the array. It is natural to explore the possibility of manipulating the phase for beam control, in which case a constant phase progression is typically desired. Beam-scanning arrays using coupled oscillators have been developed for several coupling schemes, as will be described below. The chief advantage of this concept is the total elimination of phase shifter circuitry, and in most cases an elimination of the RF feed network. The disadvantages of this approach are a relatively small bandwidth, limited scanning range, and difficulties in constructing oscillators with highly reproducible characteristics with high yield. However, such problems can be alleviated through advances in oscillator design, the use of phase-locked-loop techniques, and clever circuit augmentations [44].

4.1 Unilateral Injection-Locking

The most straightforward application of injection-locking to phased arrays is shown in Fig. 4.4a. Each array element is a self-contained, voltage-controlled oscillator that delivers its energy to an antenna. The oscillators are all slaved to a common signal (the desired output signal) which is distributed using a standard feed structure like a corporate feed network. The phase of each oscillator can be changed relative to the reference signal (and hence the other oscillators) by adjusting the oscillator tuning voltage (the free-running frequency) according to Fig. 4.3.

A 4×4 array using the topology of Fig. 4.4a has been reported [35], although the emphasis was on the power-combining aspects of the array and not the possible scanning feature. All the techniques shown in Fig. 4.4 will be limited in bandwidth by the locking range of the oscillators, which in turn is

172 COUPLED-OSCILLATOR ARRAYS AND SCANNING TECHNIQUES

related to the Q factor and injected signal strength. This array topology should degrade gracefully. It may also be possible to use self-oscillating mixers as the array elements in order to combine transmit and receive functions. The external locking approach has been used in conjunction with optical signal distribution for large aperture phased arrays by Daryoush [45] and others. This scheme is similar to that of Fig. 4.4a except that the RF signal is distributed to each array element via an optical carrier. A photodetector at each array element converts the optical signal to RF which then directly injection-locks a microwave oscillator.

A variation of this idea that eliminates the corporate feed structure involves an injection-locked cascade whereby each array element is slaved to the preceding element in the array. This was illustrated conceptually in Fig. 4.4b and has been tested experimentally for scanning applications [36]. To ensure that the injection-locking is unilateral, amplifiers are used between the oscillators, resulting in a nonreciprocal active coupling matrix $\bar{\bar{\kappa}}$. As a result, the dynamics of each oscillator are essentially governed by Adler's equation (4.23), so the relative phasing is established by adjusting the free-running frequencies of each array elements according to Fig. 4.3. The amplifiers increase the circuit complexity, but also help boost the signal strength for a better locking range and hence bandwidth. The amplifiers also are used so that only a small amount of power needs to be coupled out of each oscillator, so the coupler does not strongly affect the oscillator performance and most of the oscillator power is delivered to the antenna. The extension to two-dimensional arrays is straightforward.

A 4×1 active antenna array was designed and tested to demonstrated this concept, as shown in Fig. 4.13 (a similar array with optical signal distribution was also reported [46]). The array is fabricated on Duroid substrate with dielec-

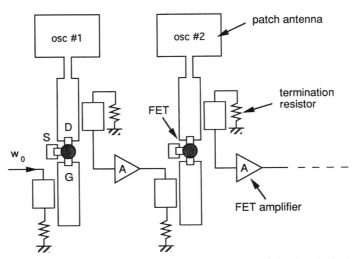

FIGURE 4.13 Illustration of an experimental unilaterally injection-locked oscillator array, using microstrip MESFET oscillators and patch antennas [36].

tric constant 2.33 and thickness 31 mil. Rectangular patch antennas were used in this design. Microstrip couples are incorporated at the gate and drain ports of the FET for injection and tapping of signals, respectively. The active device (NEC 72084) is self-biased at $V_{ds} = 3$ V and $I_{ds} = 30$ mA. The free-running frequency of the oscillators can be tuned independently through the drain bias, but this also affects the output power. The amplifier used in the circuit is a MGA 64135 from Hewlett-Packard. It has a gain of 9 dB with an isolation of 26 dB. The oscillators operate at 5.8 GHz with a tuning range of approximately 30 MHz. The antenna elements are placed 0.89 λ_0 apart, where λ_0 is the free-space wavelength at 5.8 GHz. The measured pattern is shown in Fig. 4.14. With this spacing the theoretical scan angle is 33°; the measured scan angle is about 24°. The scanned pattern is asymmetrical with varied peak transmitted power. These discrepancies are partly due to the difficulty in achieving band-edge injection locking and uncompensated coupling phase. However, the main cause of the difference in peak power is due to the difference in drain voltages of the oscillators. This is a compromise for circuit simplicity.

One feature of all injection-locked arrays is that the oscillators will roughly assume the noise properties of the master oscillator of reference signal, even if the oscillators themselves are quite noisy. The coupling network will have some influence, however, on the specific noise reduction. The array in figure 4.4b will collectively have a low phase noise as long as the injected signal is derived from a "quiet" source. However, an analysis shows a slight noise degradation as the number of oscillators increases [43].

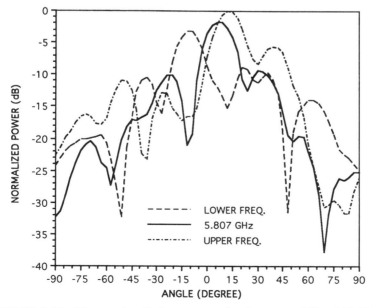

FIGURE 4.14 Measured radiation patterns for the array of Fig. 4.13 [36].

4.2 Stephan's Scanning Approach

One of the first oscillator systems for beam scanning was proposed by Stephan, using the configuration shown in Fig. 4.15 [7,8]. Stephan's array is a nearest-neighbor bilaterally coupled or "inter-injection-locked" array, with external signals injected at opposite ends of the array. The two injected signals are coherent with a variable relative phase, which is implemented by splitting the signal from a common master oscillator and delaying one of the channels. Stephan found the interesting property that under certain conditions the phase difference between the two injected signals is divided uniformly along the array to produce a constant phase progression. This can be explained using (4.59) as a starting point. Assuming $\Phi = 0$, (4.59) reduces to

$$\frac{d\phi_i}{dt} = (\omega_i - \omega_{\text{inj}}) - \Delta\omega_m[\sin(\phi_i - \phi_{i-1}) + \sin(\phi_i - \phi_{i+1})]$$
$$+ \frac{\omega_i}{2Q}\frac{\rho_i}{\alpha_i}\sin(\psi_i - \phi_i) \tag{4.106}$$

for $i = 1, \ldots, N$, where it is implicitly understood that any terms containing subscripts 0 to $N + 1$ are ignored. For the configuration of Fig. 4.15, we have only two injected signals, so

$$\rho_i = \begin{cases} \rho, & i = 1 \text{ or } i = N \\ 0, & \text{otherwise} \end{cases} \tag{4.107}$$

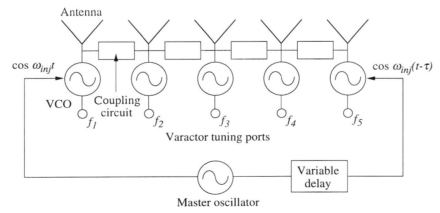

FIGURE 4.15 Scanning array configuration proposed by Stephan [7], where a nearest-neighbor bilaterally coupled array is injection-locked at either end by external signals with a prescribed phase difference ($\psi_N = -\omega_{\text{inj}}\tau$). The adjustable phase difference between the ends is found to be divided equally along the chain.

We will assume that the strength of the injected signal is the same as the strength of the coupling between oscillators, $\rho = \epsilon$. Therefore, a constant phase progression, given by $\phi_i - \phi_{i-1} = \Delta\phi$, is a valid solution to (4.106) provided that

$$\omega_{inj} = \begin{cases} \omega_1 + \Delta\omega_m[\sin\Delta\phi + \sin(\psi_1 - \phi_1)], & i = 1 \\ \omega_i & 2 \leq i \leq N - 1 \\ \omega_N + \Delta\omega_m[-\sin\Delta\phi + \sin(\psi_N - \phi_N)], & i = 1 \end{cases} \quad (4.108)$$

Assuming all of the free-running frequencies are the same as the injected signal frequency, $\omega_i = \omega_{inj}$, we must have $\phi_1 - \psi_1 = \Delta\phi$ and $\psi_N - \phi_N = \Delta\phi$. We can assume that one of the injected signals establishes a common phase references, which we will take as $\psi_1 = 0$, which means that $\phi_i = i\Delta\phi$, and therefore

$$\psi_N = (N + 1)\Delta\phi \quad (4.109)$$

In words, a constant phase progression is a solution to (4.106) provided that the injected signals have a relative phase difference given by (4.109), which implies that the phase difference between the injected signals is divided uniformly along the array.

This result must be checked for stability. Writing the phase perturbation as $\overline{\delta\phi}$, we find from (4.106)

$$\frac{d}{dt}\overline{\delta\phi} = -\Delta\omega_m \cos\Delta\phi \overline{\overline{A}}\,\overline{\delta\phi} \quad (4.110)$$

where $\overline{\overline{A}}$ is the same matrix as discussed in conjunction with (4.62). Therefore, the constant phase solution is stable provided that $\cos\Delta\phi > 0$, which again implies the constraint (4.70). As with all simple injection-locked systems, the maximum phase shift between adjacent oscillators is $\pm\pi/2$, which for an array with $\lambda_0/2$ spacing gives a maximum scan angle of $\pm 30°$ off broadside.

For sinusoidal signals the largest unambiguous relative phasing between the two injected signals is $\psi_N = \pi$, which implies an interelement phase shift of $\Delta\phi = \pi/(N + 1)$. However, for a modulated source using a true time delay (variable delay line) in the path of one injected signal, this constraint on the maximum phase shift is relaxed provided that the modulation period is long compared to the maximum time delay. In other words, the bandwidth is linearly related to the array size (assuming it does not exceed the locking range).

Morgan and Stephan [47] reported a four-element X-band prototype using the configuration of Fig. 4.15, with microstrip Gunn diode oscillators and tapered-slot antennas. The measured output phase versus injected control phase difference ($\psi_N - \psi_1$) is shown in Fig. 4.16 and is in qualitative agreement with the simple theory described above. Departures from a perfect uniform phase progression and a linear dependence on the injected phase difference can be

176 COUPLED-OSCILLATOR ARRAYS AND SCANNING TECHNIQUES

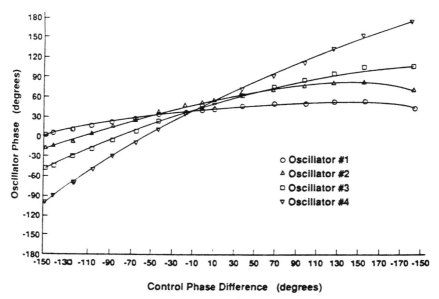

FIGURE 4.16 Measured results using Stephan's approach for a four-element Gunn-oscillator array. (From reference [47], © 1988 IEEE.)

modeled by including many of the effects neglected in the above analysis such as nonuniform oscillator free-running parameters, amplitude dynamics, and nonuniform loading and coupling of the array elements.

4.3 Bilateral Coupling with Symmetric End Tuning

As mentioned in Section 3.5, a constant phase progression can also be established in a nearest-neighbor bilaterally coupled array *without* externally injected signals, simply by manipulating the free-running frequencies, or oscillator "tunings." From (4.66), we can establish the required conditions for a uniform phase progression $\Delta\phi$ by inserting the $(N - 1)$ element sine vector

$$\bar{s} = \sin \Delta\phi (1, 1, \ldots, 1)^T \tag{4.111}$$

which gives the result

$$\Delta\beta_i = \begin{cases} \Delta\omega_m \sin \Delta\phi, & i = 1 \text{ and } i = N - 1 \\ 0, & \text{otherwise} \end{cases} \tag{4.112}$$

This implies that a uniform phase shift is induced simply by detuning the end elements of the array (relative to the central elements) by equal amounts and in opposite directions, with the amount of detuning establishing the amount of the induced phase shift. Inserting the result back into (4.71) we find that the steady-state synchronized frequency is the same as the free-running frequencies of the central elements, independent of the end-element tuning since the ends are tuned in opposite directions. To summarize, the required frequency distribution is

$$\omega_i = \begin{cases} \omega[1 - \Delta\omega_m \sin \Delta\phi]^{-1} & \text{if } i = 1 \\ \omega & \text{if } 1 < i < N \\ \omega[1 + \Delta\omega_m \sin \Delta\phi]^{-1} & \text{if } i = N \end{cases} \quad (4.113)$$

The range of phase shifts that can be synthesized has already been determined from a stability analysis resulting in (4.70).

This simple and useful result has the disadvantage of being difficult to explain in a physically appealing manner. Fortunately, we have found that equivalent continuum models provide a striking analogy between linear oscillator chains and a parallel-plate capacitor. The basic idea is to pass to the limit of a continuous chain of oscillators (rather than a discrete chain) and find the partial differential equation that governs the system [14]. Applying this concept to our nearest-neighbor oscillator equations (4.59), we write

$$\phi_i - \phi_{i-1} \Rightarrow \left.\frac{\partial\phi}{\partial i}\right|_{i-1/2} \quad \text{and} \quad \phi_i - \phi_{i+1} \Rightarrow -\left.\frac{\partial\phi}{\partial i}\right|_{i+1/2}$$

Then (4.59) becomes

$$\frac{\partial\phi}{\partial t} = \omega(i) - \omega + \Delta\omega_m \frac{\partial}{\partial i} \sin\left(\frac{\partial\phi}{\partial i}\right) \quad (4.114)$$

The phase and free-running frequency variables are now continuous functions of the position i. The entire set of coupled differential equations (4.59) is therefore just a discretized version of (4.114). In the steady synchronized state and by assuming that the phase is a slowly varying function of position (small phase change between neighboring oscillators), then (4.114) is approximated by

$$\frac{\partial^2\phi}{\partial i^2} = \frac{\omega - \omega(i)}{\Delta\omega_m} \quad (4.115)$$

This is just the one-dimensional Poisson equation. In electromagnetic terms, the phase plays the role of the electrostatic potential, while the relative oscillator tunings (which can be viewed as proportional to the DC voltage distribution for the varactor tuning elements in a VCO array) play the role of the charge distribution.

Using this result we can understand the beam-scanning method easily by analogy to electrostatics. The frequency distribution required to scan the beam is uniform except at the boundaries of the array. This would be analogous to a parallel-plate capacitor where the charge is zero between the plates and is nonzero on the plates, with opposite signs on each plate. As we know, the potential varies linearly between the plates of a capacitor, with a slope proportional to the charge; therefore in the oscillator problem we expect a linear phase distribution with the phase shift proportional to the frequency tunings at the end of the array.

This is a very powerful result. All of our intuition from electrostatics can be used to understand and predict the behavior of coupled-oscillator systems. In particular, we now have a formal solution (for linear chains) for the steady-state phase distribution in terms of the frequency tunings,

$$\theta(i) = c_1 + \int di \sin^{-1}\left[\int \rho(i)di + c_2\right] \quad (4.116)$$

where c_1 and c_2 are integration constants that are found by satisfying the appropriate boundary conditions. For the two-dimensional array case, Green's function or modal expansion techniques can be used advantageously. Naturally (4.116) will not represent the exact steady-state phase distribution in a real (discrete) array, but it provides a good approximation and a simple method for prediction.

The beam-scanning technique has been verified experimentally. The six element array shown in Fig. 4.7 was first used, with representative patterns shown in Fig. 4.8 [48]. Due to a small antenna spacing of $d = 0.3\lambda_0$, the maximum scan range for this array was $\pm 60°$ off broadside. The beam angle was scanned experimentally within $\pm 40°$ off broadside, suggesting an interelement phase shift was achieved in the range of $|\Delta\phi| \leq 70°$. A second, larger array was constructed using a similar coupling topology as shown in Fig. 4.17 [49]. The array was designed to operate at 8.4 GHz, and featured an element separation of one half wavelength. The maximum beam-scan range for this array was therefore $\pm 30°$ off broadside. The array was comprised of eight VCOs, each of which used an NEC 900276 packaged GaAs MESFET (two parasitic elements are shown in the figure). The oscillator was designed in a feedback topology using a single-device amplifier with a tunable microstrip patch antenna in the feedback path. This varactor tuned patch served as a resonant load for the amplifier [50]. By adjusting the varactor bias, the free-running frequency of each oscillator was variable over a range of 150 MHz.

The effective radiated power (ERP) of the array was measured to be 8.5 W at 8.43 GHz. Figure 4.18a illustrates a typical broadside pattern which was obtained when the free running frequencies of the oscillators in the array were set to 8.45 GHz. Varying the end-element tunings in the manner predicted by (4.113), it was possible to scan the array from $-15°$ to $+30°$ off broadside; a representative scanned pattern in shown in Fig. 4.18b. This scanning range indicates that progressive phase shifts were achieved in the range $-47° \leq \Delta\theta \leq +90°$. A maximum phase difference of 630° was established between the first and last elements in the array. That the scan range is centered away from broadside is consistent with a nonzero coupling phase angle. This could be a

FIGURE 4.17 Photograph of an 8.4-GHz oscillator array using feedback oscillators with a varactor-tuned patch antenna [49]. The transmission-line coupling network is based on the configuration of Fig. 4.6. The two end elements were inactive and served to equalize the loading along the remaining oscillators. The red bias switches were used to determine the individual (free-running) oscillator tuning characteristics.

result of a propagation delay through the chip resistors in the coupling network, which was not accounted for in this first-iteration design. Since the patterns closely resemble the theoretical patterns for a patch array, the directivity could be accurately estimated from theory, yielding a total radiated power of over 1 W for this array.

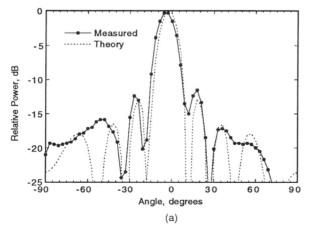

(a)

FIGURE 4.18 Representative measured patterns for the array of Fig. 4.17. (a) A nearly broadside pattern was obtained with all the elements set to approximately the same free-running frequency. (b) A second pattern in the range of $-15°$ to $+30°$ off broadside was obtained by symmetrically detuning the end oscillators. (*Continued*)

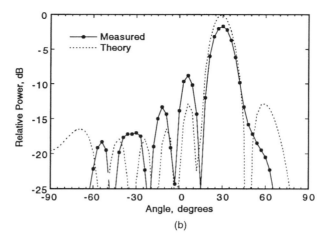

FIGURE 4.18 (*Continued*)

4.4 Variations

One apparent limitation of the injection-locked or coupled-oscillator topologies (for some applications) is the limited range of phase shifts that can be synthesized. This could be improved by introducing a frequency doubler circuit after each oscillator, which effectively doubles the interelement phase shift. Despite additional circuit complexity, this technique has some additional benefits: The oscillators can be designed at a lower frequency (half the desired output frequency), which is useful because osciallators are sensitive to parasitic reactances, and also because oscillator design is simpler when the device has high gain, which is more easily achieved at lower frequencies. The range of oscillator tuning required to achieve a given scan range is also significantly reduced, which is advantageous since operation of the array near the locking band edge is undesirable due to increased phase noise, reduced modulation range, and increased sensitivity to environmental distrubances. FET-based doublers also provide a useful measure of isolation between the oscillator and load, so a stable broadband load impedance is presented to the oscillators, with the possibility of conversion gain. A prototype array using PET doublers has been demonstrated at X-band [51], achieving a total phase variation of $\approx 260°$.

Phase noise is also a potential problem. As shown earlier, coupled-oscillator systems possess an interesting noise reduction property, but for small arrays this effect is not sufficient to overcome the poor noise performance of the individual oscillators arising from the low-Q circuits. We have found that low noise can be achieved by injecting a stable reference signal (which could also contain modulation) into the *center* element of the array, without disrupting the scanning properties described above [52].

The scanning oscillator configurations can also be used for receiving applications [53]. This is accomplished by using the scanning osciallator array as the local oscillator for a set of mixers. Using one of our early array prototypes and some commercial packaged mixers, this concept was tested by first measuring the scanning properties in transmit mode (osciallators coupled directly to antennas), followed by the receive mode. Identical scan ranges and patterns were observed in each case, as expected [53]. It may be possible to merge the transmit and receive functions, especially for FMCW imaging arrays, by making each array element a self-contained FMCW transmitter and receiver, with each array element coupled to its neighbors. Alternatively, each array element could be a self-osciallating mixer. These concepts have not yet been tested.

A further refinement of the coupled oscillator technique is the use of phase-locked-loop (PLL) techniques to phase-lock the array elements [54,55], instead of the nonlinear injection-locking process. The phase dynamics of simple phase-locked loops are almost identical to that predicted by Adler's equation; hence all of the previously developed scanning techniques would be preserved. The benefit of this technique is more robust operation; the array elements are forced to lock to each other, and the locking or capture range can be much larger than that achieved with injection-locking techniques. In addition, the PLL technique does not suffer adversely from dynamic amplitude variations when strongly coupled, which appears to be limiting factor in the coupled-oscillator technique. The disadvantage is, of course, the added complexity of the PLL circuitry. However, it is conceivable that a suitable PLL MMIC chip could be developed for use in such arrays, which would be cost effective in comparison to stand-alone VCO chips. Microwave prototypes of both a mutually synchronized PLL array and a "mode-locked" (pulsed) array have been demonstrated [56].

APPENDIX: KUROKAWA'S SUBSTITUTION

Consider a linear, time-invariant network described by the voltage transfer function $H(\omega)$:

$$\tilde{V}_{out}(\omega) = H(\omega)\tilde{V}_{in}(\omega) \tag{A.1}$$

This equation can be expressed in the time domain using the Fourier transform pair

$$\tilde{V}(\omega) = \int_{-\infty}^{\infty} v(t)e^{-j\omega t} \, dt \quad \longleftrightarrow \quad v(t) = \frac{1}{2\pi} \int_{-\infty}^{\infty} \tilde{V}(\omega)e^{j\omega t} \, d\omega \tag{A.2}$$

If the input signal V_{in} is a modulated sinusoid at a carrier frequency ω_0, we can represent the time variation in the complex form

$$v_{in}(t) = A(t)e^{j[\omega_0 t + \phi(t)]} = v'_{in}(t)e^{j\omega_0 t} \tag{A.3}$$

182 COUPLED-OSCILLATOR ARRAYS AND SCANNING TECHNIQUES

where it is understood that the actual time variation is found by taking the real part of (A.3). Note that the spectrum of the modulation alone,

$$\tilde{V}'_{in}(\omega) = \int_{-\infty}^{\infty} v'_{in}(t) c^{j\phi(t)} \epsilon^{-j\omega t} \, dt \tag{A.4}$$

is related to the input signal spectrum by the well-known shifting property

$$\tilde{V}'_{in}(\omega) = \tilde{V}_{in}(\omega + \omega_0) \tag{A.5}$$

Taking the inverse Fourier transform of (A.1) gives

$$v_{out}(t) = \frac{1}{2\pi} \int_{-\infty}^{\infty} H(\omega) \tilde{V}_{in}(\omega) e^{j\omega t} \, d\omega \tag{A.6}$$

$$= \frac{1}{2\pi} \int_{-\infty}^{\infty} H(\omega' + \omega_0) \tilde{V}'_{in}(\omega') e^{j(\omega' + \omega_0)t} \, d\omega' \tag{A.7}$$

The transfer function can be expressed as a Taylor series about the carrier frequency ω_0,

$$H(\omega' + \omega_0) = \sum_{n=0}^{\infty} \frac{1}{n!} \frac{d^n H(\omega_0)}{d\omega^n} (\omega')^n \tag{A.8}$$

Substituting (A.8) into (A.7) and interchanging the order of integration and summation gives

$$v_{out}(t) = \sum_{n=0}^{\infty} \frac{1}{n!} \frac{d^n H(\omega_0)}{d(j\omega)^n} \left[\frac{1}{2\pi} \int_{-\infty}^{\infty} (j\omega')^n \tilde{V}'_{in}(\omega') e^{j\omega' t} \, dt \right] e^{j\omega_0 t} \tag{A.9}$$

The integral in square brackets is recognized as the nth derivative of the modulation, so from (A.3) we can write

$$v_{out}(t) = \sum_{n=0}^{\infty} \frac{1}{n!} \frac{d^n H(\omega_0)}{d(j\omega)^n} \frac{d^n (Ae^{j\phi})}{dt^n} e^{j\omega_0 t} \tag{A.10}$$

The actual output is found by taking the real part of (A.10). In the case of a slowly varying modulation and/or a broadband network, the higher-order derivatives in (A.10) diminish rapidly and the output voltage is well described by the first two terms of the expansion. Taking the first two terms and using (A.3), (A.10) reduces to

$$v_{out}(t) \approx v_{in}(t) \left[H(\omega_0) + \frac{dH(\omega_0)}{d\omega} \left(\frac{d\phi}{dt} - j\frac{1}{A}\frac{dA}{dt} \right) \right] \tag{A.11}$$

Kurokawa [23] appears to be the first to recognize that this result can be obtained from the state equation (A.1) by substituting

$$\omega \to \omega_0 + \frac{d\phi}{dt} - j\frac{1}{A}\frac{dA}{dt} \quad (A.12)$$

in the transfer function, and invoking a "slowly varying amplitude and phase" assumption,

$$\frac{d\phi}{dt} \ll \omega_0, \quad \frac{1}{A}\frac{dA}{dt} \ll \omega_0 \quad (A.13)$$

to expand the transfer function around the frequency ω_0. This is referred to as the "Kurokawa substitution" in the text.

Note also that we did not have to choose the carrier frequency ω_0 as the reference point for the Taylor expansion in (A.8). It is a natural choice in this example, but in the case of a multiple oscillator system there are several "natural" frequencies to choose from. We can express the input signal (A.3) in terms of any arbitrary reference frequency by suitably redefining the phase variable to include a linearly increasing part,

$$v_{in}(t) = A(t)e^{j[\omega_r t + \phi'(t)]} \quad (A.14)$$

where $\phi'(t) = \phi(t) + (\omega_0 - \omega_r)t$. The resulting expression for $v_{out}(t)$ is then exactly the same as (A.10) if ω_r replaces ω_0 and if ϕ' replaces ϕ. One must be careful, however, when truncating the expansion under the assumption of slowly varying amplitude and phase in this case, since ϕ' can vary rapidly in time compared with ϕ if the new reference frequency ω_r differs significantly from the "true" carrier frequency, which was assumed to be ω_0 in the example.

ACKNOWLEDGMENTS

The authors thank Siou Teck Chew and Prof. Tatsuo Itoh (Chapter 3) for contributing material relating to their work on unilateral injection-locked arrays. This work was supported by the U.S. Army Research Office under contract DAAH04-93-G-0210 and by the Hughes Research Laboratories (Malibu, CA).

REFERENCES

[1] J. W. Mink, "Quasi-optical power combining of solid-state millimeter-wave sources," *IEEE Trans. Microwave Theory Tech.*, vol. MTT-34, pp. 273–279, Feb. 1986.

[2] K. Kurokawa, "The single-cavity multiple-device oscillator," *IEEE Trans. Microwave Theory Tech.*, vol. MTT-19, No. 10, pp. 793–801, Oct. 1971.

[3] R. Adler, "A study of locking phenomena in oscillators," *Proc. IRE,* vol. 34, pp. 351–357, June 1946; also reprinted in *Proc. IEEE,* vol. 61, pp. 1380–1385, Oct. 1973.

[4] K. Kurokawa, "Injection-locking of solid-state microwave oscillators," *Proc. IEEE,* vol. 61, pp. 1386–1409, Oct. 1973.

[5] A. E. Siegman, *Lasers,* University Science Books, San Jose, CA, 1986.

[6] R. J. Dinger, D. J. White, and D. R. Bowling, "10 GHz space power-combiner with parasitic injection-locking," *Electronics Lett.,* vol. 23, pp. 397–398, 9 April 1987.

[7] K. D. Stephan, "Inter-injection-locked oscillators for power combining and phased arrays," *IEEE Trans. Microwave Theory Tech.,* vol. MTT-34, pp. 1017–1025, Oct. 1986.

[8] K. D. Stephan and W. A. Morgan, "Analysis of inter-injection-locked oscillators for integrated phased arrays," *IEEE Trans. Antennas Propagat.,* vol. AP-35, pp. 771–781, July 1987.

[9] P. Liao, and R. A. York, "A new phase-shifterless beam scanning technique using arrays of coupled oscillators," *IEEE Trans. Microwave Theory Tech.,* pp. 1810–1815, Oct. 1993.

[10] R. A. York and R. C. Compton, "Mode-locked oscillator arrays," *IEEE Microwave Guided-Wave Lett.,* vol. 1, pp. 215–218, Aug. 1991; R. A. York and R. C. Compton, "Experimental observation and simulation of mode-locking in coupled-oscillator arrays," *J. Appl. Phys,* vol. 71, no. 6 pp. 2959–2965, 15 March 1992.

[11] R. J. Ram, R. Sporer, H.-R. Blank, P. Maccarini, H.-C. Chang, and R. A. York, "Chaos in microwave antenna arrays" (invited paper), presented at *1996 IEEE MIT-S International Microwave Symposium* (San Francisco).

[12] P. C. Matthews and S. H. Strogatz, "Phase diagram for the collective behavior of limit-cycle oscillators," *Phys. Rev. Lett.,* vol. 65, pp. 1701–1704, 1990.

[13] A. H. Cohen, P. J. Holmes, and R. H. Rand, "The nature of the coupling between segmental oscillators of the lamprey spinal generator for locomotion: a mathematical model," *J. Math. Biol,* vol. 13, pp. 345–369, 1982.

[14] N. Koppel, "Toward a theory of modelling central pattern generators," *Neural Control of Rhythmic Movements in Vertebrates,* A. H. Cohen, ed., Wiley, New York, 1988, Chapter 10.

[15] J. Buck, *Q. Rev. Biol,* vol. 63, p. 265, 1988: T. J. Walker, *Science,* vol. 166, p. 891, 1969; M. K. McClintock, *Nature (London),* vol. 229, p. 244, 1971.

[16] Y. Yamaguchi, K. Komentani, and H. Shimizu, "Self-synchronization of nonlinear oscillations in the presence of fluctuations," *J. Stat. Phys.,* vol. 26, pp. 719–743, 1981.

[17] Hoppensteadt, *Introduction to the Mathematics of Neurons,* 1991.

[18] I. Peterson, "Step in time: exploring the mathematics of synchronously flashing fireflies," *Sci. News,* vol. 140, pp. 136–137, Aug. 1991.

[19] P. C. Matthews, R. E. Mirollo, and S. H. Strogatz, "Dynamics of a large system of coupled nonlinear oscillators," *Physica D,* vol. 52, p. 293, 1991.

[20] K. Y. Tsang, R. E. Mirollo, S. H. Strogatz, and K. Wiesenfeld, "Dynamics of a globally coupled oscillator array," *Physica D,* vol. 48, pp. 102–112, 1991.

[21] S. S. Wang and H. G. Winful, "Dynamics of phaso-locked semiconductor laser arrays," *Appl. Phys. Lett.,* vol. 52, pp. 1774–1776, 1988.

[22] Private communication with Atac Imamoglu, University of California at Santa Barbara, NSF Center for Quantized Electronic Structures (QUEST).

[23] K. Kurokawa, "Some basic characteristics of broadband negative resistance oscillator circuits," *Bell Syst. Tech. J.,* Aug. 1969.

[24] B. Van der Pol, "A theory of the amplitude of free and forced triode librations," *Radio Review,* vol. 1, pp. 701–754, 1920; also B. Van der Pol, "The nonlinear theory of electric oscillations," *Proc. IRE,* vol. 22, pp. 1051–1085, Sept. 1934.

[25] N. N. Bogoliubov and Y. A. Mitropolsky, *Asymptotic Methods in the Theory of Nonlinear Oscillations,* Hindustan Pub. Corp., 1961.

[26] M. Kuramitsu and F. Takase, "An analytical method for multimode oscillators using the averaged potential," *Trans. IECEJ,* vol. J66-A, pp. 336–343, April 1983 (in Japanese).

[27] R. K. Brayton, J. K. Moser, "A theory of nonlinear networks," *Q. Appl. Math.,* vol. XXII, no. 1, April 1964.

[28] S. Nogi, J. Lin, and T. Itoh, "Mode analysis and stabilization of a spatial power-combining array with strongly coupled oscillators," *IEEE Trans. Microwave Theory Tech.,* Special Issue on Quasi-Optical Techniques, Oct. 1993.

[29] K. Fukumoto, M. Nakajima, and J.-I. Ikenoue, "Mathematical representation of microwave oscillator Characteristics by use of the Rieke Diagram," *IEEE Trans. Microwave Theory Tech.,* vol. MTT-31, pp. 954–959, Nov. 1983. K. Fukumoto, M. Nakajima, and J. -I. Ikenoue, "Mathematical expression of the loading characteristics of microwave oscillators and injection-locking characteristics," *IEEE Trans. Microwave Theory Tech.,* vol. MTT-33 pp. 319–323 April, 1985.

[30] K. Kurokawa, "Noise in synchronized oscillators," *IEEE Trans. Microwave Theory Tech.,* pp. 234–240, April 1968.

[31] Y. Okabe and S. Okamura, "Analysis of the stability and noise of oscillators in free, synchronized, and parallel running modes," *Electron. Communi. Japan,* vol. 52-B, no. 12, pp.102–110, 1969.

[32] Y. Okabe and S. Okamura, "Stability and noise of many oscillators in parallel running," *Electron. Communi. Japan,* vol. 53-B, no. 12, pp. 94–103, 1970.

[33] T. Makino, M. Nakajima, and J. Ikenoue, "Noise reduction mechanism of a power combining oscillator system," *Electronics Communi. Japan,* vol. 62-B, no. 4, pp. 37–44, 1979.

[34] W. O. Schlosser, "Noise in mutually synchronized oscillators," *IEEE Trans. Microwave Theory Tech.,* pp. 732–737, Sept. 1968.

[35] J. Birkeland and T. Itoh, "A 16 element quasi-optical FET oscillator power combining array with external injection locking," *IEEE Trans. Microwave Theory Tech.,* vol. MTT–40, pp. 475–481, March 1992.

[36] J. Lin, S. T. Chew, and T. Itoh, "A unilateral injection-locking type active phased array for beam scanning," *IEE MTS-S Int. Microwave Symp. Dig.* (San Diego), pp. 1231–1234, June 1994.

[37] R. A. York and R. C. Compton, "Quasi-optical power-combining using mutually synchronized oscillator arrays," *IEEE Trans. Microwave Theory Tech.,* vol. MTT-39, pp. 1000–1009, June 1991.

[38] R. A. York, "Nonlinear analysis of phase relationships in quasi-optical oscillator arrays," *IEEE Trans. Microwave Theory Tech.,* vol. MTT-41, pp. 1799–1809, Oct. 1993.

[39] R. A. York, P. Liao, and J. J. Lynch, "Oscillator array dynamics with broadband N-port coupling networks," *IEEE Trans. Microwave Theory Tech.*, vol. MTT-42, pp. 2040–2045, Nov. 1994.
[40] M. J. Vaughan, W. Wright, and R. C. Compton, "Active antenna elements for millimeter-wave cellular communications," *1995 URSI International Symposium on Signals, Systems, and Electronics* (San Francisco, CA), Oct. 1995.
[41] P. Liao and R. A. York, "A high power two-dimensional coupled oscillator array at X-band," *1995 IEEE MTT-S International Microwave Symposium* (Orlando), pp. 909–912.
[42] G. Strang, *Linear Algebra and its Applications*, Academic Press, New York, 1980.
[43] H. C. Chang, X, Cao, U. K. Mishra, and R. A. York, "Phase noise in coupled oscillators: theory and experiment," *IEEE Trans. Microwave Theory Tech.*, in press.
[44] R. A. York, "Novel beam scanning techniques for low cost commercial applications" (invited paper), *SPIE Int. Conf. on Millimeter Waves and Applications* (San Diego), July 1995.
[45] A. S. Daryoush, "Optical synchronization of millimeter-wave oscillators for distributed architectures," *IEEE Trans. Microwave Theory Tech.*, vol. MTT-38, pp. 467–476, May 1990.
[46] S. T. Chew *et al.*, "An active phased array with optical input and beam-scanning capability," *IEEE Microwave Guided Wave Lett.*, vol. 4, no. 10, pp. 347–349, Oct. 1994.
[47] W. A. Morgan and K. D. Stephan, "An X-band experimental model of a millimeter-wave interinjection-locked phased array system," *IEEE Trans. Antennas Propagat.*, vol. AP-36, pp. 1641–1645, Nov. 1988.
[48] P. Liao and R. A. York, "A six-element scanning oscillator array," *IEEE Microwave Guided Wave Lett.*, vol. 4, no. 1, pp. 20–22, Jan. 1994.
[49] P. Liao and R. A. York, "A 1 watt X-band power-combining array using coupled VCOs," *IEEE MTT-S Int. Microwave Symp. Dig.* (San Diego), pp. 1235–1238, June 1994.
[50] P. Liao and R. A. York, "A varactor-tuned patch oscillator for active arrays," *IEEE Microwave Guided Wave Lett.*, vol. 4, no. 10, pp. 335–337, Oct. 1994.
[51] A. Alexanian, H. C. Chang and R. A. York, "Enhanced scanning range in coupled oscillator arrays utilizing frequency multipliers," *1995 IEEE Antennas Propaga. Soc. Symp. Dig.* (Newport Beach, CA), pp. 1308–1310.
[52] X. Cao and R. A. York, "Phase noise reduction in scanning oscillator arrays," *1995 IEEE MTT-S Int. Microwave Symp.* (Orlando), pp. 769–772.
[53] X. Cao and R. A. York, "Coupled oscillator scanning technique for receiver applications," *1995 IEEE Antennas Propag. Soc. Symp.* (Newport Beach, CA), pp. 1311–1314.
[54] R. D. Martinez and R. C. Compton, "Electronic beamsteering of active arrays with phase locked loops," *IEEE Microwave Guided Lett.*, vol. 4, no. 6, pp. 166–168, June 1994.
[55] J. J. Lynch and R. A. York, "Mode-locked arrays of coupled phase-locked loops," *IEEE Microwave Guided Wave Lett.*, vol. 5, no. 7, pp. 213–215, July 1995.
[56] J. J. Lynch, "Analysis and Design of Systems of Coupled Microwave Oscillators," Ph. D. dissertation, University of California at Santa Barbara, Jan. 1996.

CHAPTER FIVE

Quasi-Optical Antenna-Array Amplifiers

ZOYA B. POPOVIĆ
University of Colorado, Boulder

ROBERT A. YORK
University of California, Santa Barbara

EMILIO A. SOVERO
Rockwell International Science Center, California

JON SCHOENBERG
Phillips Laboratory, Kirtland Air Force Base, New Mexico

Antenna-array amplifiers are based on classical antenna-array topologies, where each array element contains an input antenna, an amplifier, and an output antenna. An attractive feature of this approach is the compatibility with a wide variety of existing antenna geometries and circuit designs, building on decades of research in active phased arrays and conformal planar antennas. Flexibility in design allows the arrays to be used in many systems configurations, from enclosed waveguides to quasi-optical (Gaussian-beam) systems. Important issues relating to design, fabrication, measurements, and packaging of these arrays are described in this chapter, by way of several case studies using a variety of topologies in the frequency range from 5 GHz to 60 GHz. Further discussions of active amplifier arrays can also be found in Chapter 6, with an emphasis on multilayer geometries.

Active and Quasi-Optical Arrays for Solid-State Power Combining, Edited by Robert A. York and Zoya B. Popović.
ISBN 0-471-14614-5 © 1997 John Wiley & Sons, Inc.

1 INTRODUCTION

As described in Chapter 1, modern active antenna systems have a long history. However, significant progress in quasi-optical amplifier arrays did not take place until the first report of a grid amplifier [1]. This grid amplifier presented a clever technique for achieving input/output isolation in a single-layer array, through the use of orthogonally polarized input and output dipoles coupled by a differential amplifier circuit. Subsequently, larger amplifier grids have been developed for more output power, as well as for higher frequency (millimeter-wave) operation; these results are described in detail in Chapter 9.

Following the grid work, quasi-optical amplifiers employing more conventional active antenna elements have been developed. Like the grid amplifier, each array element in this "antenna-array amplifier" scheme uses an input and output antenna interconnected by an amplifier, illustrated in Fig. 5.1 [2–6]. However, in this scheme the array elements are not tightly coupled, and a wide variety of antenna geometries and amplifer circuits are possible. By careful design (choice of antenna or array topology), isolation may be achieved within the array structure, thus eliminating the need for external polarizers and allowing for polarization flexibility between input and output [2,7].

Conventional antennas are typically standing-wave (resonant) structures and are consequently larger than a typical grid element. Amplifier matching circuitry and resistive feedback stabilization networks consume much more array area than grid-embedded devices. Therefore, active array amplifiers have a lower density of amplifiers than grids (see Chapter 1 for a discussion of maximum power density). The arrays are potentially easier to analyse, at least in an approximate sense, since the elements in the array are spaced far enough from

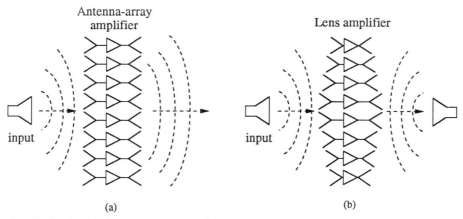

FIGURE 5.1 (a) Antenna array amplifier, fed by a horn antenna. (b) Lens amplifier, where path delays are adjusted to compensate for spherical phase fronts on the input and focusing on the output.

each other to reduce the influence of mutual coupling. The impedances presented to the inputs and outputs of the amplifiers can be designed for given specifications (high gain, low noise, high efficiency, etc.). In addition, a single cell of the array can be tested independently, so the power-combining efficiency is well defined and easy to quantify.

Still, there are several challenges. In addition to the input/output isolation problem mentioned above, the planar radiating structures must have a high radiation efficiency and must be easily integrated with devices on a semiconducting substrate; this has proved difficult due to the large dielectric constant of the common semiconductors. Another challenge is the limited bandwidth of common planar antenna structures. Furthermore, unlike conventional circuit design in a broadband 50-Ω environment, the load impedance of antennas can vary wildly outside of the operating bandwidth, which can easily lead to instabilities. For this reason, mutual coupling between antennas, even though comparatively weak, must be examined over the entire band in which the amplifier has gain in order to determine possible conditions for oscillation. This is especially important at low frequencies where three-terminal devices typically have high gain. These observations suggest that an ideal amplifier circuit for active arrays should be unconditionally stable, and have gain only in the desired bandwidth of operation.

The array power combiner, with the discrete input and output antenna arrays, lends itself to lens focussing techniques to increase the efficiency of the amplifier and overcome the losses due to diffraction, as demonstrated in [8–10]. These active arrays may be fed from a focal point very close to their input side, reducing the losses due to diffraction. The array uses delay lines so that the received spherical wave is converted into a collimated beam at the output. The active lens power combiner arrays have the added features of beamforming and beamsteering without the use of phase shifters.

It is interesting to compare the power combining efficiency at the chip level and in free space. In reference [12], two class-E amplifiers are presented: One uses the Fujitsu FLK052WG MESFET, and the other uses the Fujitsu FLK202MH-14 MESFET. The intrinsic semiconductor chips inside these two MESFETs are identical, except that the FLK202MH-14 has four times the gate periphery of the FLK052WG. In operation, the FLK052WG class-E amplifier delivered 0.61 W of output power into 50 Ω, while the FLK202MH-14 delivered a maximum of 1.8 W using the same circuit configuration. Considering that the FLK202MH-14 is physically four times as large as the FLK052WG, the output power might be expected to be four times as large as well, or 2.44 W. This suggests a device-level combining efficiency of 74%. Using the quasi-optical techniques demonstrated in this chapter, a higher power-combining efficiency of 84% has been demonstrated. This advantage is expected to become more significant when larger numbers of elements are combined, and it suggests the use of smaller devices with higher internal efficiency (and hence power density), as pointed out in Chapter 1.

1.1 Quasi-Optical Amplifier Gain

The gain of the amplifier arrays was described in Chapter 1; for convenience, some of the material is repeated here with a slightly different notation. In the most common experimental configuration, a large absorbing screen is placed between two antennas. A window corresponding to the physical array aperture is cut in the screen. The input and output feed and collector antennas (usually linearly polarized horns) are polarization-matched to the array input and output elements and are placed in the far field of the array. We define the gain of an individual input array as G_{in}, the gain of the output array as G_{out}, and the array factor is A_{arr}. The active gain contributed by the amplifiers in the array is G_a. Two power measurements are made, one with the array in the window, giving P_{meas}, and another without the array present, giving P_{cal}. Using the Friis transmission formula, the ratio between these two measured powers, which is effectively an "insertion loss" for the amplifier array, is

$$\frac{P_{meas}}{P_{cal}} = \underbrace{G_{in} G_a G_{out}}_{\text{EIPG}} \frac{(r_1 + r_2)^2}{r_1^2 r_2^2} \left(\frac{\lambda}{4\pi}\right)^2 \tag{5.1}$$

where r_1 and r_2 are the distances from the horn antennas to the device under test. The formula simplifies if r_1 and r_2 are chosen to be equal. The product of the power gain and array directive gains is analogous to the EIRP for transmitters and has been called the *effective isotropic power gain* (EIPG) (see Chapter 1).

Another method of calibrating a free-space amplifier is presented in reference [10]. A passive array identical to the active one is made with through-lines replacing the amplifiers between the input and output antenna elements. In this method, the gain contributed by the amplifier is separated from the input and output antenna insertion loss, array feed efficiency, and other loss mechanisms.

1.2 Quasi-Optical Amplifier Array Feed

Amplifier arrays can be fed in several different ways, illustrated in Fig. 5.2. The first quasi-optical grid amplifier employed a far-field feed with polarizers for isolation of input and output waves [1]. The 5×5 grid of differential-pair MESFET amplifiers covered a 64-cm^2 area. The grid amplifier was fed at a range of 50 cm, and the investigators assumed that the effective array area equals the physical area, A_p. Substituting $G_{in} = G_{out} = (4\pi/\lambda_0^2)A_p$ into (5.1), setting $r_1 = r_2 = R$ and rearranging results in an equation for the active gain contributed by the amplifiers:

$$G_a = \frac{P_{meas}}{P_{cal}} \left(\frac{\lambda R}{A_p}\right)^2 = \frac{\text{EIPG}}{A_p^2} \left(\frac{\lambda^2}{4\pi}\right)^2 \tag{5.2}$$

INTRODUCTION   191

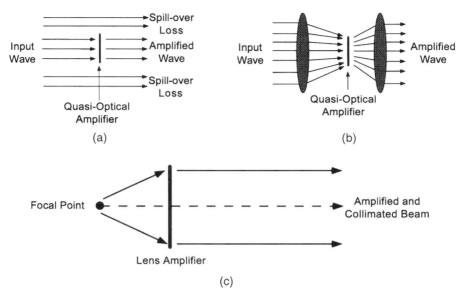

FIGURE 5.2 Feed methods for quasi-optical amplifiers: (a) Plane wave feed from a source in the far-field; (b) feed at the beam waist of a dielectric lens pair; (c) focal point feed of a lens amplifier.

Using (5.2), a peak gain G_a of 11 dB at 3.3 GHz was reported in reference [13]. However, the grid amplifier exhibited an effective insertion loss $P_{\text{meas}}/P_{\text{cal}} = 0.1$ dB. The purpose of the work done in subsequent power combiner amplifier arrays was to overcome the insertion loss problem and develop quasi-optical power combiners with absolute power gain.

To capture more of the feed power, dielectric lenses with an aperture much greater than that of the amplifier array may be used, as shown in Fig 5.4b. The distance between the lenses, assuming they are identical, is approximately twice the focal length. The active array is placed at the beam waist, where the wavefront is planar. However, the amplitude distribution is Gaussian in the transverse direction, leading to an amplitude taper across the array. Dielectric lenses add weight, size, and complexity to the quasi-optical system. If lenses with a shorter focal length to diameter ratio (f/D) are used to reduce system size, the system is more difficult to align and prone to aberrations.

Table 5.1 presents a comparison of feed mechanisms used in several quasi-optical amplifiers demonstrated to date. Rutledge *et al.* used the dielectric lens feed in a second grid amplifier utilizing HBTs[12]. This amplifier consists of a 10×10 grid of differential-pair amplifiers covering a 64-cm^2 area. Assuming that no power is lost in the dielectric lens to waveguide horn transitions, the resulting insertion gain was 10 dB at 10 GHz with a 3-dB bandwidth of 1 GHz. However, a 16% loss can be calculated for two polyethylene lenses used in the

TABLE 5.1 Comparative of Active Gain and Absolute Power Gain for Different Feeds of Quasi-Optical Amplifiers Demonstrated to Date

Ref.	Feed Type	f (GHz)	BW (%)	G_{act} (dB)	G_{abs} (dB)	Loss (dB)	Other
[2]	Far-field	10.0	3	7.1	−2.9	10.0	
[1]	Far-field	3.3	3	11.0	−0.1	10.9	
[12]	Far-field	10.0	10	10.0	3.5	6.5	NF = 3.5 dB
[12]	Dielectric lens	10.0	10	10.0	10.0(9.25)	0.75	
[30]	Far-field	11.0	5	8.0			
[10]	Focal point	9.75	3	9.5	8.1	1.4	NF = 3.4 dB
[10]	Focal point	10.0	11	18.5	13.0	5.5	NF = 1.9 dB
[46]	Hard horn	10.0	1	18.0	16.2	1.8	

measurement, which results in 9.25 dB of asbsolute power gain. A comparison to the far-field feed was performed at a range $r_1 = r_2 = 90$ cm, resulting in a peak insertion gain of 3.5 dB at 10 GHz. Employing the dielectric lens feed improved the system insertion loss of the grid amplifier by 6.5 dB.

Lens amplifiers, described in Section 4, demonstrate *absolute* power gain without the requirement for external polarizers and lenses. What is meant by *absolute* is an increase in the effective radiated power (ERP) of a source due to using the quasi-optical amplifier in a free-space power measurement.

From Table 5.1, it can be seen that the amplifier grids and arrays that are fed from the far-field suffer from feed loss since the absolute gain of the array is much lower than the active gain. The dielectric lens and the focal point feed significantly reduce the feed loss, thus greatly improving the absolute power gain. Table 5.2 summarizes the output power and power-added efficiency of several reported quasi-optical amplifiers.

TABLE 5.2 Comparison of Quasi-Optical Amplifiers with Respect to Active Gain and Efficiency

Ref.	N	f (GHz)	BW (%)	G_{act} (dB)	P_{out} (mW)	P/elem (mW)	PAE (%)
[11]	4	5.2	6.0	8.8	2400	600	64
[33]	6	29.0	1.5	6.0	1000	167	5
[47]	36	40.0	4.5	5.0	670	18.6	4
[34]	56	42.0	6.0	3.0	260	4.6	2.5
[46]	9	10.0	1.0	18.0	90	10	18

2 ANTENNA ELEMENTS FOR AMPLIFIER ARRAYS

In amplifier arrays, the antenna element choice is a very important one. Unlike in standard amplifier design, where the input and output ports are usually 50 Ω over the entire frequency range of interest, in an active amplifier array the input and output ports for each amplifier unit cell are frequency dependent. The amplifiers need to be designed with this in mind, and a good choice of antenna element and its characterization is important for successful operation of the entire power-combining array. For example, patch antennas are norrowband and their impedance varies significantly over on octave, in some common designs presenting a short-circuit to the source at a frequency which is roughly 40% lower than the resonance. At this frequency, the amplifier has a good chance of becoming unstable and the design needs to account for this. Another important consideration is the guiding medium chosen for the circuit. Microstrip amplifiers can feed patch antennas or slots in the ground plane, but typically require via-hole processing and other backside-alignment steps. Single-layer processing favors coplanar waveguide (CPW) circuits, which are compatible with slot antennas. Broadband (traveling-wave) designs require some other configuration, such as Vivaldi slot arrays. This section discusses the antenna elements used in active amplifier arrays demonstrated to date.

2.1 Patch Antennas

The microstrip patch antenna was selected for several quasi-optical amplifiers because of its compatibility with microstrip circuitry and ground-plane isolation technique. The extensive information and availability of modeling software made the element design relatively straightforward [13,14]. In a non radiating-edge (NRE) microstrip-fed design, a wide variety of impedances can be chosen. For example, in reference [9], a resonant input impedance of 100 Ω was selected so that the microstrip delay lines connecting the patch with its amplifier circuitry had a narrow width and minimal dispersion. The distance from the feed line center to the closer radiating edge is the critical dimension determining the resonant input impedance. The width of the radiating edge was set to be narrower than its resonant length for a compact design, and sets the fundamental orthogonal (the radiating-edge-fed) mode at a higher frequency where the amplifier has much lower gain. As will be shown later, the bias and matching network are an integrated part of the antenna design. For example, the patch has a natural low-impedance point along the center of a nonradiating edge where there is an (approximate) field null, which is exploited for biasing.

2.2 Slot Antennas

The uniplanar nature of CPW circuits is attractive for monolithic array fabrication. All devices, circuitry, transmission lines and antenna elements of the array are contained on one side of the dielectric, thus removing the need for vias.

194 QUASI-OPTICAL ANTENNA-ARRAY AMPLIFIERS

Active devices and lumped components are easily connected in series or shunt in CPW, which also provides flexibility in choosing the ground-to-ground spacing for a desired impedance. Slot antennas may use CPW as a compatible feed line a uniplanar circuit. CPW-fed slot antennas have been demonstrated in a quasi-optical mixer [15] and amplifier arrays [6,16], where it was recognized that slots can significantly increase the bandwidth of quasi-optical components.

Analyses of slot antennas defined in a metallic sheet on thin dielectric substrates have been performed with FDTD [17] or with a space-domain integral equation solved using the method of moments [15]. These methods are numerically intensive and may take considerable time to generate an s-parameter model over a sizeable frequency range for simulation with an amplifier circuit, as will be discussed later in this section. A moment-method-based solution with entire-domain basis functions was used for the design of some of the slot antennas used in quasi-optical arrays [18,19]. The theory was developed for slot antennas on a thin dielectric using the concept of complementary electromagnetic structures [20].

A CPW-fed resonant slot is the dual structure of a comparable resonant dipole antenna. Booker's relation gives the input impedance of an infinitely thin half-wave resonant slot in an inifinite conducting sheet in free space as $Z_{slot} = \eta_0^2/(4Z_{dipole})$, where Z_{dipole} is the impedance of a center-fed half-wave resonant dipole in free space. Since the impedance of the dipole at first resonance is $Z_{dipole} \approx 73\,\Omega$, the slot impedance is $Z_{slot} \approx 500\,\Omega$, which is a difficult impedance to match over a broadband in passive or active CPW circuits. Two

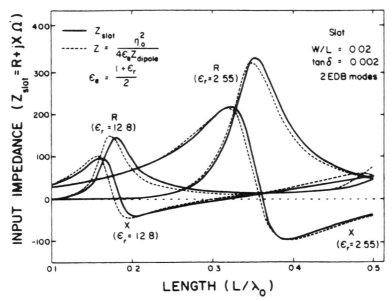

FIGURE 5.3 Input impedance of slot antennas on semi-infinite substrates. (After reference [21].)

interesting variations, the antiresonant slot and folded or multiple slots, are described below which not only lower the bandwidth but also have inherently lower Q factors. Finally, arrays of broadband Vivaldi antennas have been used in traveling-wave quasi-optical amplifiers.

Antiresonant Slot Antennas A slot antenna operated near the first resonance (where the length is approximately one-half wavelength in the surrounding medium) is relatively narrowband (high Q), with a parallel equivalent circuit model and a resonant conductance of $\approx 1/500 \ \Omega$ in free space. Figure 5.3 shows the computed input impedance of a center-fed slot on a semi-infinite substrate

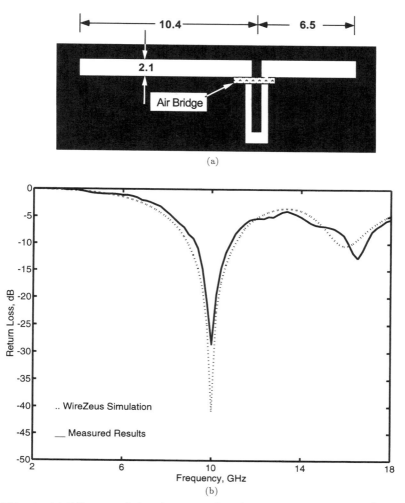

FIGURE 5.4 (a) Off-center-fed antiresonant slot. (b) Return loss of the 100 Ω antiresonant slot used in a low-noise amplifier array. Measured results (solid line) and simulated data (dashed line) show excellent agreement. The 2:1 VSWR bandwidth is 18%.

[21] for two representative dielectric constants. Note that the second resonance (which has a series model) by contrast, is relatively broadband. Unfortunately the radiation resistance is very small. This can be addressed by feeding the slot assymetrically.

An example of an off-center-fed slot is illustrated in Fig. 5.4a with the dimensions designed using a full-wave theory [20]. This was used in a lens amplifier described later. Its length is $0.57\lambda_0$ and its width is $0.067\lambda_0$. The off-

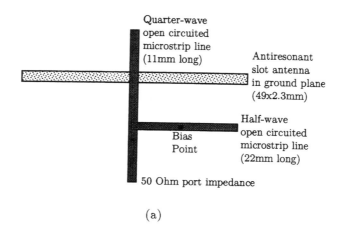

(a)

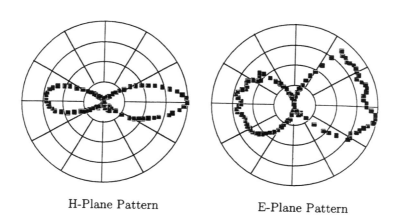

(b)

FIGURE 5.5 (a) A 50-Ω antiresonant slot used for a 5-GHz high-efficiency amplifier array. All microstrip lines are 50 Ω, and the substrate is a 0.508-mm-thick Duroid with $\epsilon_r = 2.2$. (b) Measured rediation patterns in the E and H planes of the slot show about 1.5 dB more power radiated on the dielectric side.

center feed position is one variable used to tailor the input impedance of the slot to the desired value of 100 Ω, with the other variable being the slot width for a given substrate dielectric and metallization. Figure 5.4b shows comparison between measured and simulated slot reflection coefficient. The measured 2:1 VSWR bandwidth is 18% with a center frequency of 10.0 GHz. The measured resonant impedance is 10 Ω, a deviation of +7% from design, which may be attributed to the measured slot width being slightly wider than design specification.

The antiresonant slot shown in Fig. 5.5a was used in a high-efficiency amplifier array [19]. The slot is fed with an open-circuited 50-Ω microstrip line which extends quarter of a wavelength beyond the center of the slot. For this antenna, a 2:1 VSWR bandwidth of more than 20% and a crosspolarization ratio of 23 dB were measured at 5 GHz. A half-wavelength-long open microstrip line is used to provide bias to the amplifiers by connecting a thin high-impedance line to the voltage null. The measured radiation patterns are shown in Fig. 5.5b. The antenna radiates approximately 1.5 dB more power into the dielectric/air side.

Folded and Multiple-Slot Antennas The folded-slot and multiple slot antennas are extensions of ideas used for years on television sets for manipulating input impedance and improving bandwidth of dipole antennas. The concept is illustrated in Fig. 5.6. The basic idea for the folded dipole is the use of additional

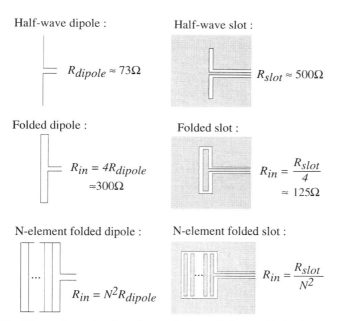

FIGURE 5.6 Illustration showing the evolution of the multiple-slot concept from analogous consideration of resonant radiation resistance for folded dipoles [23].

resonators coupled to the driven element. For an N-element dipole, each branch carries approximately the same current as a single-arm dipole for a given applied voltage, which increases the radiated power by a factor of N^2. This in turn means the radiation resistance increases by a factor of N^2 compared to a single-arm dipole. The multiple slot is the dual of the multiple dipole, which from Babinet's principle [22] means that the input impedance *decreases* by a factor of N^2 (or equivalently, the radiation *conductance* increases by N^2). Note that this not a Yagi–Uda design: The parasitic elements are the same size, and all are tightly coupled and in close proximity with the driven element. As a result, there is little change in the directive characteristics, but a strong effect on the circuit characteristics.

From a circuit perspective, the addition of parasitic resonators is analogous to adding extra reactive elements to increase bandwidth in a broadband filter or impedance-matching network. The scaling laws were varified in the microwave range (X-band) for 2, 3, 4, and 5 slots on an alumina substrate [23]. The results are summarized in Fig. 5.7, along with theoretical impedance curves computed using finite-difference time-domain (FDTD) codes developed at UC Santa Barbara [23,25]. Although it is not clear from the figure, the resonant frequencies of the antennas (10.5 GHz) do not change appreciably with the addition of the extra slots. This is because the slots are of the same physical length. However, the additional slots do provide reactive compensation which increases the bandwidth.

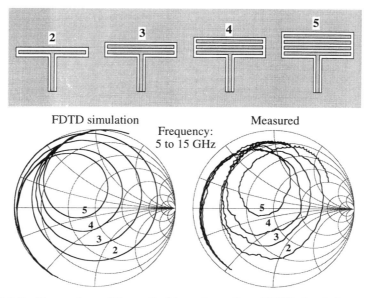

FIGURE 5.7 Comparison of theoretical impedance from FDTD simulations and measurements for CPW-fed 2-8-1-and 5-slot antennas on 25 mil alumina in the 5 to 15 GHz range [24,25]. Resonant frequencies are all approximately 10.5 GHz.

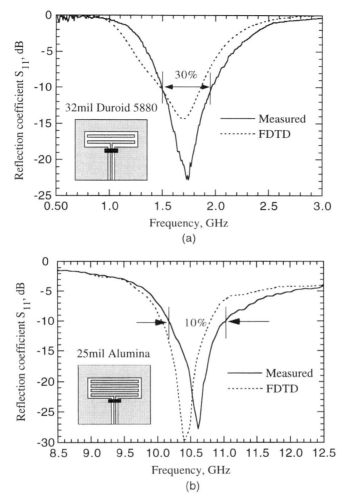

FIGURE 5.8 Two examples of using impedance scaling to create a 50-Ω input impedance. (a) 32-mil Duroid 5880 with $\epsilon_r = 2.2$ and (b) 25-mil alumina with $\epsilon_r = 9.8$. Dimensions are given in the text. These antennas were not optimized for bandwidth.

A single resonant slot has a very narrow bandwidth and large resonant impedance (around 500 Ω in air, increasing with substrate dielectric constant for thin substrates). Multiple slots can be used to scale the impedance down to the optimum load for power (determined by a load-line or nonlinear analysis), and the design can be optimized for much wider bandwidth than conventional planar antennas, on the order of 25–30% fractional bandwidth [23].

The impedance scaling property can be used to create broadband planar antennas with 50-Ω resonant impedance for compatibility with conventional MMIC circuits [24]. Example results for Duroid 5880 and alumina are shown in Fig. 5.8. The number of slots required for a good match to 50 Ω depends on the substrate parameters. For Duroid 5880 with a thickness of 0.813 mm and $\epsilon_r = 2.2$, a single slot has a measured radiation resistance of around 500 Ω at 2 GHz, which suggests the use of three slots for a good match to 50 Ω. The result for three-slots shown in Fig. 5.8a used a slot length of $D = 78$ mm, a slot width of $W = 2$ mm, and a slot separation of $S = 2$ mm. The measured input impedance at resonance is approximately 60 Ω at 1.66 GHz, which is close to what is expected. The bandwidth at -10 dB return loss is close to 30%.

For an alumina substrate with a thickness of 0.635 mm and $\epsilon_r = 9.8$, a single slot has a radiation resistance of approximately 1200 Ω at 10 GHz, which suggests using a 5-slot antenna for a good match to 50 Ω. The measured result for five slots shown in Fig. 5.8b used a slot length $D = 7.2$ mm, slot width $W = 0.4$ mm, and slot separation $S = 0.2$ mm. The bandwidth at -10 dB return loss is 10%. In both

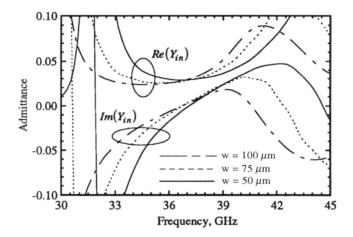

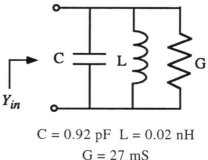

FIGURE 5.9 Influence of slot width W on input admittance for a Ka-band 5-slot antenna on 500 μm GaAs, along with an approximate parallel equivalent circuit valid from 34 to 38 GHz.

these examples, relatively wide bandwidths were observed, although these antennas were not specifically designed to maximize bandwidth, and even higher fractional bandwidths have been obtained on simulations [23]. It can be shown that the slot gap width is an important parameter in determining bandwidth. The influence of slot gap width on impedance is shown in Fig. 5.9, where it is clear that near resonance the slope of the reactance curve is changing, while the real part (radiation conductance) and resonant frequency remain relatively unchanged.

Loaded Folded-Slot Antennas A loaded center-fed folded slot was used as the input element of a lens amplifier array described in Section 4. The center-fed folded slot is loaded with a series-connected 51-Ω resistor and 10-pF capacitor at each of its ends (Fig. 5.10a). The series parasitic inductance of the resistor and capacitor are additive, so a total series inductance of 1.2 nH is added in the simulation model [20]. The folded slot is designed to have a 20-Ω impedance at the second resonance. This impedance was chosen so that a single series-line match could provide an optimum noise match to the first-stage PHEMT amplifier used in the lens, which is $\Gamma_{opt} = 0.66e^{j148°}$. The folded slot layout is very compact, with a length of $0.50\lambda_0$ and a width of $0.077\lambda_0$. The folded slot input return loss simultation is compared to the measured data in Fig 5.10b. The simulation data include the 15°, 100-Ω series-line match required to provide an optimum noise match to the first-stage PHEMT amplifier in the final array. Without this series line, the simulation indicates a 2:1 VSWR bandwidth of 15% where $Z_0 = 20\ \Omega$.

2.3 Broadband Tapered-Slot Antennas

For very wideband operation, large-traveling-wave or self-complementary antennas must be used. An attractive choice for arrays is the end-first tapered slot structure [26], shown in Fig 5.11, which lends itself to the "tray" array architecture (Chapter 2). The input impedance of these structures is essentially that of the feed transmission line; the structure is a broadband impedance transformer. These antennas are highly compatible with traveling-wave or distributed circuits [27], such as nonlinear transmission lines and traveling-wave amplifiers. At millimeter wavelengths, both these technologies would benefit from the increased power-handling capability afforded by quasi-optical arrays, and in turn the quasi-optical array concept benefits from the large bandwidth and reduced sensitivity to device or processing variations afforded by distributed circuits.

A wideband passive array, or "quasi-optical transition" based on this concept, is illustrated in Fig. 5.12. This array can be analyzed using the equivalent waveguide approach (see Chapter 1). Gradual transitions of this sort exhibit large bandwidths, and some literature exists for such transitions in closed metal rectangular waveguides; for example, linear, exponential or circular arc tapers have been examined in a finline topology [28,29]. Due to the different boundary conditions in our problem, the prior work cannot be used directly, but is useful as a guide. Each substrate carries a number of slotlines which are ideal for

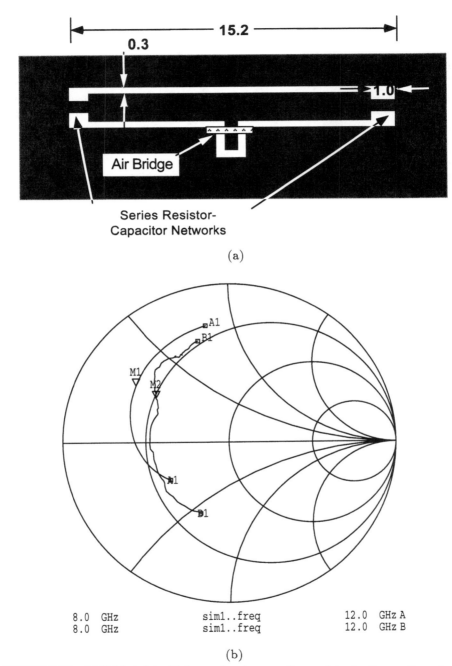

FIGURE 5.10 (a) Folded slot with lumped series capacitor-resistor loading on its ends; (b) folded slot return loss comparison between simulation (trace A) and measurement (trace B). Both traces include a single series-line match for optimum noise match to the first stage of a low-noise amplifier.

ANTENNA ELEMENTS FOR AMPLIFIER ARRAYS 203

Broadband Antennas
Tapered slot
Planar end-fire structure

Distributed Circuits
Broadband & Insensitive to device parameter variations

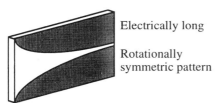
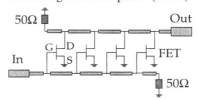

FIGURE 5.11 End-fire tapered slot structure and traveling-wave amplifier, which can be combined for ultra-wideband quasi-optical arrays.

mounting two-terminal devices (diodes). For three-terminal devices a slot-to-CPW transition or a finline-to-microstrip transition can be used. The extra available substrate surface area is extremely important for heat removal purposes, which has been an important limitation in the monolithic arrays published to date.

Input/output isolation is achieved naturally with traveling-wave arrays because of the directional nature of the radiators. It has been found that in an array environment the aperture opening can be reduced dramatically, leading to higher packing density. Usually, tapered slots require an aperture opening of at least $\lambda/2$ at the lowest desired frequency of operation, in order to provide a good impedance match to the feed line. In an array this is not required since the impedance of the equivalent waveguide into which the antenna is radiating is not necessarily 377 Ω and depends on the aspect ratio of the cross section. Adjusting the wafer spacing provides a mechanism for independently adjusting the antenna characteristics.

Some experimental work has been done to test this concept. A Ka-band array was constructed on Duroid using a 16 × 19 array of tapered-slot antennas. The

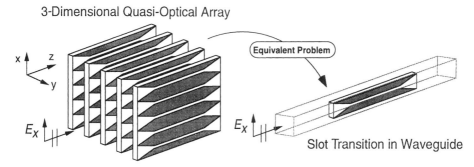

FIGURE 5.12 Broadband quasi-optical passive array of tapered slot and equivalent waveguide for analysis.

measured and theoretical results are shown in Fig. 5.13, along with the slot dimensions. Nearly perfect transmission through the structure was measured (within experimental uncertainty) in a lens-focused Gaussian beam setup. Also shown for comparison are the results of a simulation using FDTD grid codes developed at the University of California at Santa Barbara [25], which analyze the unit cell using the equivalent waveguide approach with Floquet boundary conditions. Although measurements could only be made in the 26 to 40-GHz range, the simulations show less than 1-dB insertion loss from 20 to 80 GHz. Note that an isolated tapered slot would require an aperture opening roughly 5 times bigger for similar performance in this range [26].

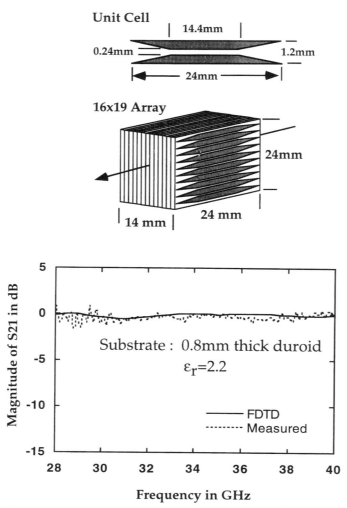

FIGURE 5.13 Experimental passive array prototype demonstrating low insertion loss and wide bandwidth, with a high packing density.

3 PLANE-WAVE-FED AMPLIFIER ARRAYS

Several case stuides of plane-wave-fed quasi-optical amplifier arrays are presented in this section. The first five arrays operate at microwave frequencies, and the others operate at millimeter wavelengths (30, 42 and 60 GHz). Most are fabricated using hybrid technology, and the two highest-frequency arrays are monolithic. In one case, the amplifiers in the array are designed to operate at very high efficiency in order to demonstrate that these two-dimensional arrays can be adequately heat-sunk.

3.1 Polarization-Preserving Class-A 24-Element MESFET Patch Array

One of the first demonstrated transmission-mode quasi-optical amplifiers was a MESFET patch antenna amplifier array in which the input and output wave polarizations may be arbitrarily selected [2]. This amplifier has an array of rectangular patch antennas on input and output sides of the substrate (Fig. 5.14). Each amplifier is coupled and matched to its input and output patch elements with microstrip lines. A substrate via connects the output of each amplifier to its output patch printed on the opposite side of the substrate. Alternating ground planes effectively isolate input and output sections of the amplifying structure. The ground planes are connected by vias periodically spaced at quarter-wavelength intervals.

Amplifier Unit Cell An individual amplifier unit cell using a low-cost packaged Avantek ATF-13484 GaAs MESFET was built on a teflon substrate with

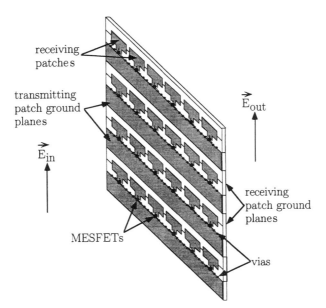

FIGURE 5.14 A 24-element patch-patch quasi-optical transmission wave amplifier.

relative permittivity $\epsilon_r = 2.2$ and dielectric thickness $h = 0.508$ mm. To save space, minimize parasitics, and maximize bandwidth, a single microstrip transmission line matches the radiating-edge impedance of the patch to the MESFET. The patches are fed at the center of their radiating edges with an input impedance of 140 Ω. The use of a frequency-dependent one-port equivalent circuit for the patch during the design process was essential for analyzing the stability of the amplifier. A bilateral design at 10 GHz was performed with a 100-Ω stabilizing resistor between the gate and source of the MESFET. The DC and RF voltages across the gate-to-source terminals of the MESFET are smaller in magnitude than across the drain-to-source terminals, so degradation of efficiency and heat dissipation due to the stabilization resistor is minimized. The design is stable and the maximum gain contributed by the MESFET amplifier is 8.4 dB at 10 GHz. The final design of the amplifier unit cell is shown in Fig. 5.15a. The polarization of input and output waves in this unit cell are linear and copolarized.

The ground-plane isolation structure of the amplifier unit cell provides the flexibility to arbitrarily select the polarities of the input and output waves without affecting stability. An amplifier unit cell with a linearly polarized input and a circularly polarized output patch was designed as illustrated in Fig. 5.15b. The output square patch is fed on two adjacent edges with a 300-Ω edge impedance and a 90° separation in phase between the two feeds. The drain bias of the MESFET is provided through a via to the output patch at its voltage null.

In the free-space measurement setup, a known value for the patch gain was used, along with the theoretical array directivity, to determine the gain contributed by the amplifiers. When the MESFET is biased on, a maximum gain G_a of 7.1 dB is measured at 10 GHz. The frequency-dependent gain of the patch antennas, calculated from a multiport network model [13] of the patch and experimentally verified at 10 GHz, is used for G_{in} and G_{out} in (5.1). The insertion gain (or loss) of the amplifier is -23.9 dB. Using the same measurement methodology, the gain of the rectangular-to-circularly polarized amplifier cell was evaluated. An amplifier added gain G_a of 6.8 dB and an output axial ratio of 0.3 dB were measured at 10 GHz.

Active Array Performance The 24 elements of this array are arranged in a 4×6 periodic rectangular lattice. The horizontal spacing between elements is $2\lambda_0/3$ in the horizontal direction and the vertical spacing between elements is $4\lambda_0/3$. The large vertical spacing is required to accommodate the ground-plane isolation, and it could be reduced by using a substrate with higher dielectric permittivity. The measured gain of the array is 21 dB higher than that of the amplifier unit cell. Therefore, the insertion gain of the array is -23.9 dB (unit cell) + 21 dB = -2.9 dB, and thus the array exhibits insertion loss. The single biggest factor of the insertion loss is diffraction loss in the amplifier feed. Illuminating the array at the design frequency in the broadside direction with a standard gain horn at the far-field range of 2.1 meters, the array captures only 1.1% of the source power. This represents a feed loss of -19.6 dB due to the diffraction from the far-field source.

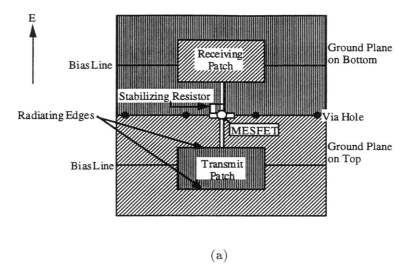

(a)

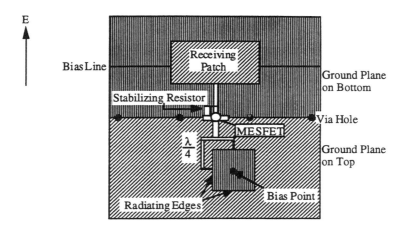

(b)

FIGURE 5.15 Planar transmission-wave amplifier unit cells for (a) linear polarization for input and output and (b) linear polarization input and circular polarization output.

The polarization-preserving 24-MESFET patch array was designed for maximum gain at a bias point of 20 mA and 2 V per device. The s parameters at a higher bias point (40 mA and 4 V per device) are practically identical, except for an increase in $|s_{21}|$. Since, to a first approximation only, this parameter changes (decreases) when a MESFET is saturated, the design for a saturated Class-A amplifier at the high bias point should be approximately the same as the low-bias point linear design.

First a single cell was measured between two polarization-aligned 16-dB gain horn antennas, placed in the far field of the active antenna. The horns were connected to a HP71500A Transition Analyzer, used to measure the power saturation curves of the unit cell at several bias points. In Fig. 5.16, these saturation curves are plotted for a unit cell. Then the array was tested in the setup, which required increasing the range from the horns to the array to 2.1 meters. A 10-W traveling-wave tube (TWT) amplifier was connected between the frequency sweeper and the feed horn to drive the entire array into saturation from this range. The TWT amplifier provided a total input power to the array of 64 mW (2.7mW/device) at 10 GHz. The entire array results are superimposed on the curves. The average power per element in the array is slightly lower than that of a single element tested alone, which means that the array combining is lower than that in an ideal, uniformly illuminated array.

The on/off ratio of this quasi-optical array is 10 dB. With a constant input power and total array drain current of 1 A, the output power was measured for several drain-bias voltage points. A maximum saturated Class-A drain efficiency of 24% with an associated power-added efficiency (PAE) of 22% with 0.6 W of output power was found assuming a gain of 5.5 dB of the patch.

3.2 High-Efficiency Class-E MESFET Slot Array

A potential problem with quasi-optical transmission-mode amplifiers is that the dissipated heat needs to be removed laterally from a two-dimensional structure with heat sinks on the perimeter. While both the output power and the dissipated power grow as the area of the 2D array, the heat sink grows as only the perimeter. Thus the center elements of a large array would operate at possibly much

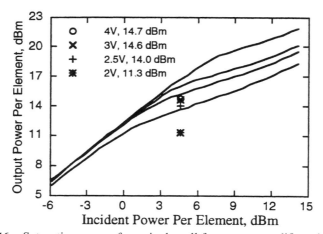

FIGURE 5.16 Saturation curves for a single-cell free-space amplifier with a constant drain current and several drain bias voltages. The superimposed symbols represent measured array power levels, divided by the number of elements ($N = 24$).

higher temperatures than those at the edges. For example, a Ka-band array presented in Section 4 was designed to operate as a high-power amplifier in saturated Class-A mode with only 11% PAE. Despite a carefully designed liquid-cooled heatsink, after 30 minutes of continuous operation, the array cleaved due to excess heat. One solution to the problem is to design high-efficiency amplifiers. If the efficiency of an amplifier is increased from 50% to 90%, for the same output power, the dissipated heat decreases by a factor of nine, reducing the required size of the heat sink. In this section, a 5-GHz high-efficiency quassi-optical amplifier is presented. The microstrip amplifier circuits in the array operate in Class-E mode [30]. In this mode, the active device operates as a switch and the antenna serves as part of the output tuned circuit. The output capacitance of the device, which limits the switching speed, is used as part of the tuned circuit. A transistor can operate in this high-efficiency mode up to about a third of its cutoff frequency, but the degradation in efficiency is graceful with increasing frequency [11].

The 4-element quasi-optical Class-E power amplifier in shown in Fig. 5.17. A modular approach was taken in designing the active array. The FLK052WG 5-GHz Class-E amplifier, the antiresonant 50-Ω slot antenna described in Section 2, and the bias network were all integrated into the quasi-optical structure. The Class-E amplifier sees 50 Ω at the fundamental frequency of operation at both its input and output ports. At the second harmonic frequency, the input impedance of the antiresonant slot antenna is unimportant, since the

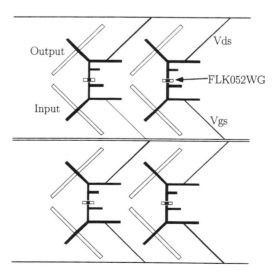

FIGURE 5.17 The quasi-optical Class-E power amplifier. This 4-element power-combining structure demonstrated 2.4 W of output power at 5.05 GHz, with 8.8 dB compressed gain, 74% drain efficiency, and 64% power-added efficiency. Antiresonant 50-Ω orthogonal slot antennas couple the power between the amplifier and free space, and half-wave microstrip lines provide bias to the transistor gate and drain terminals.

open-circuit condition at the output port of the FLK052WG is preserved independent of the load impedance at the second harmonic [11]. The input and output antennas are orthogonally polarized.

First a single cell of the quasi-optical Class-E amplifier was measured in free space, and it demonstrated 0.67 W of output power at 5 GHz, with a drain efficiency of 80%, a PAE of 71%, and a compressed gain of 9.3 dB.

To calibrate the free-space amplifier array measurement, a passive array with 50-Ω lines replacing the Class-E amplifiers was fabricated. First the passive structure was placed between two horn antennas, and the power received in the second horn relative to the power delivered to the first horn was measured. Approximately half of the power loss was assumed to be due to the input side of the array. A correction was made in the calculations to compensate for the 1.5-dB difference in power radiated from the two sides of the slot antennas.

Then the quasi-optical Class-E power amplifier was placed in the system, and enough power was delivered to the first horn to drive the input of the amplifier into compression: 10 W of power was delivered to the first horn, and 31 mW of power was received by the amplifier. The final amplifier was not experimentally adjusted between fabrication and measurement, and the best performance was observed at 5.05 GHz, where an output power of 2.4 W was measured, with a corresponding drain efficiency of 74%, a PAE of 64%, and a compressed gain of 8.8 dB. These numbers represent the total power delivered to the output antennas in the amplifier, and do not take antenna losses into account. This method of calibration resulted in measurements of output power and efficiency which were repeatable to within a few percent; and it is probably more accurate than using the Friis formula for calibration, but requires fabrication of a passive array.

When we divide the output power per element in the 2 × 2 array by the output power of the single cell, a power-combining efficiency of 84% is found. A power sweep was performed on the single cell quasi-optical Class-E amplifier, and it is shown in Fig. 5.18a. The results of the 2 × 2 quasi-optical Class-E power amplifier are superimposed on the graph; the output power in this case is plotted *per element* for direct comparison with the single cell results.

The 2 × 2 quasi-optical Class-E amplifier was then calibrated over a 10% frequency range. The output power, drain efficiency, and PAE frequency sweep is shown in Fig. 5.18b. The efficiency is above 50% over a 3.5% bandwidth, and it is above 30% over a 6% bandwidth. The best performance was observed at 5.05 GHz (1% higher than the design frequency), where an output power of 2.4 W was measured, with a corresponding drain efficiency of 74% and a PAE of 64%. The compressed gain at 5.05 GHz was 8.8 dB.

3.3 C-Band Folded-Slot Array

An early test of the active array amplifier concept was a demonstration of a 4 × 4 array operating at 4 GHz using broadband folded-slot antennas [31]. This array used resistive-feedback amplifier circuits with packaged 0.25-μm GaAs

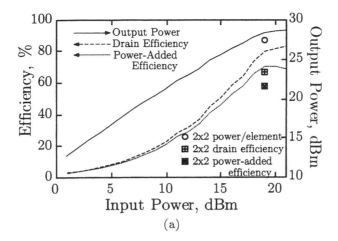

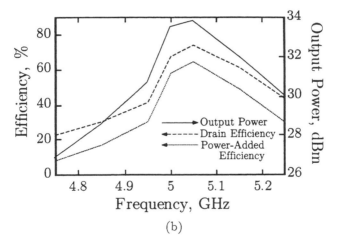

FIGURE 5.18 (a) Power sweep curve for the single cell quasi-optical Class-E power amplifier element. The results of the 2 × 2 quasi-optical Class-E power amplifier are superimposed on the graph; the output power in this case is plotted *per element*. (b) Frequency sweep curve for the 2 × 2 quasi-optical Class-E power amplifier array. The input power level is +25 dBm at 5 GHz.

MESFETs. The unit cell for this amplifier has already been shown in Chapter 3, and it consists of two orthogonally polarized folded slots coupled by the active amplifier circuit. The unit cell was fabricated on Rogers Duroid 6010($\epsilon_r = 10.8$) with a thickness of 0.635 mm. For this substrate the measured folded-slot impedance was 180 Ω. The input/output impedance of the resistive feedback amplifier was 125 Ω. The amplifier and antennas were connected with 80-Ω

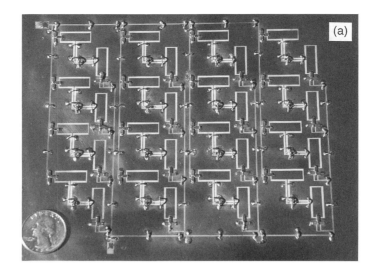

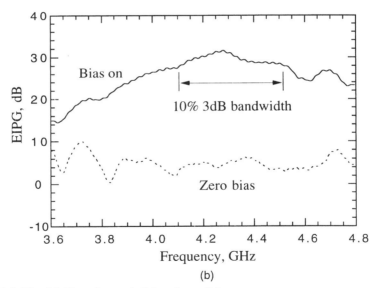

FIGURE 5.19 (a) Experimental C-band amplifier array prototype using folded-slot antennas and packaged MESFET amplifiers. (b) Measured response in a far-field (plane-wave) measurement setup.

coplanar waveguides. The gain of the entire structure was nominally 6 dB according to simulations on EESof's LIBRA, which was confirmed by measurements of an isolated unit cell in free space [31].

A photograph of the array, and corresponding measurements, are shown in Fig. 5.19. For this measurement no external polarizers were used. The maximum EIPG was 32 dB at 4.2 GHz, with a 10% 3-dB bandwidth. Using array theory and a simple model for the folded-slot pattern [31], the directivity of the structure was estimated to be 13.5 dB, which gives a real power gain of $32 - 2(13.5) = 5$ dB, which is consistent with the expected gain (6 dB) based on simulations and measurements of the isolated unit cell. The discrepancy could be due to an inaccurate estimation of the array directivity, or due to circuit nonuniformities over the array. When biased for Class-A operation, no oscillation was observed, but parasitic oscillation did occur for a small range of gate bias below the Class-A point, which indicates either excessive coupling between neighboring elements (which could be occurring through substrate modes) or an improper matching network to a conditionally stable device.

3.4 X-Band Multiple-Slot Array with Commercial MMICs

The multiple-slot antenna structure can be used to create 50-Ω antennas, which in turn suggests direct integration of commercial MMIC chips that are designed for a 50-Ω input/output impedance. To explore this idea, a quasi-optical amplifier cell at X-band was constructed by bonding a commercial MMIC amplifier block between two of the five-slot antennas on alumina (Fig. 5.8b), without any external matching networks [24,32]. The MMIC amplifier was a Rockwell HBT gain stage. This device was first characterized using HB 8510B with a Cascade wafer probing setup and had a 9-dB gain at 10 GHz, with a unity-gain cutoff of approximately 18 GHz. The amplifier cell was then measured in a oversized square waveguide using a HP 8720 network analyzer. This measurement setup provided a more direct measurement of cell gain, unlike many quasi-optical amplifier measurements which rely on the Friis transmission equation. Two waveguide transitions were machined to make a smooth taper from standard X-band waveguide to a one inch square aperture. The one-inch square middle section allowed for the propagation of signals with both vertical and horizontal polarizations. A polarizer was used in front of the amplifier cell in order to direct the output radiation in the forward direction; this also provided some tuning for better output match. Measurements confirmed a gain of 8 dB at 11 GHz, with a 3-dB bandwidth of $\approx$5%. The narrower bandwidth of this circuit compared to that of the single antenna measurement is expected, due to the use of two antennas and the frequency-dependent gain of the amplifier in this frequency range. Larger bandwidth could be obtained with additional matching circuitry if needed, and also by optimizing the antenna.

Using this unit cell, a 4 $\times$ 4 (16-element) array was constructed for operation at 11 GHz [32]. A photograph of this array is shown in Fig. 5.20a, along with measured results in Fig. 5.20b (the inset of Fig. 5.20b shows the unit cell for the array). This array was tested in a far-field (plane-wave) transmission system, without input/output polarizers. The measurements yielded as effective isotropic power gain (EIPG) of 35 dB at 11 GHz, with a 5% bandwidth. Using an estimated directivity of 13.5 dB gives a power gain of 8 dB, which agrees with measurements on the unit cell in a closed waveguide.

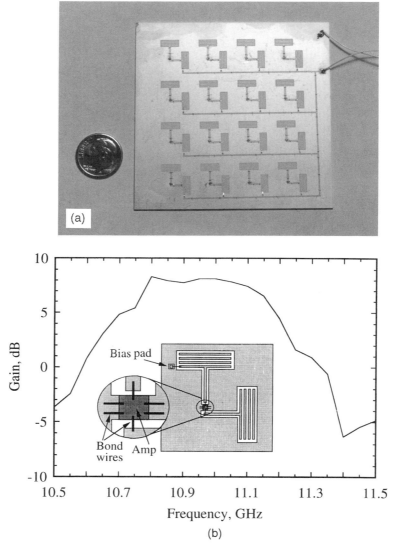

FIGURE 5.20 (a) Experimental X-band amplifier array prototype using commercial HBT gain-block chips bonded between two 50 Ω multiple-slot antennas on alumina. (b) Unit cell and measured response in a far-field (plane-wave) measurement setup.

3.5 X-Band Tapered-Slot Array in Waveguide

An illustration of a high-power combiner using broadband tapered-slot transitions is shown in Fig. 5.21 [33]. This approach was originally intended as a waveguide simulator for a quasi-optical traveling-wave amplifier array [34], but has since been suggested for us "as is" at lower frequencies. In this approach, two or more circuit-combined power amplifier circuits are integrated with tapered

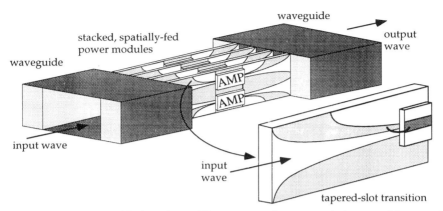

FIGURE 5.21 Waveguide-based amplifier array using tapered-slot transitions and dielectric side-wall loading for uniform excitation.

slots and stacked vertically to form a power "card." These cards are stacked together and placed in a metallic waveguide enclosure, which confines the energy and also serves as a heatsink. Since the power-combining takes place in the waveguide interior (free space, as opposed to transmission-line circuits), the combining efficiency is extremely high. A secondary advantage of this approach is that the amplifier circuit design is no longer constrained by the conventional 50-Ω environment; the driving-point impedances of the antennas can be chosen freely. This could be an advantage in matching for optimum power. Furthermore, the spatial combining structure takes full advantage of available volume, unlike conventional two-dimensional (planar) combining techniques.

Some preliminary tests of the concept have been made. A 4×2 array of tapered slot structures. Four cards using 10 mil aluminium nitride were constructed, each with two tapered slots and two Texas Instruments broadband power amplifier MMICs (#TGA8014-SCC), bonded in using a simple slot-to-microstrip transition, as illustrated in Fig. 5.21. These amplifier MMICs are unconditionally stable with a nominal small-signal power gain of 11 dB from 6 to 18 GHz, and an output power of 27 dB at 1-dB gain compression, with a 16% power-added efficiency. The tapered-slots were designed semiempirically with a cosine taper from 50-Ω slotline to a 5-mm aperture opening, over a distance of 3 cm. The power cards were epoxy-mounted to small metal holders, which held the cards rigidly in place while also providing additional heatsinking. These were then inserted in standard X-band waveguide. Preliminary results over the full waveguide band are shown in Fig. 5.22.

This approach is highly modular, since the concept is roughly independent of the number of amplifier cards that are stacked inside the waveguide. This would have long-term advantages for later manufacturing. In fact, the waveguide cross section can be widened to accommodate additional cards, but this will require careful modeling to ensure that the higher-order modes are not excited. The extra available substrate space will be extremely important for heat removal purposes, which has been an important limitation in the monolithic arrays

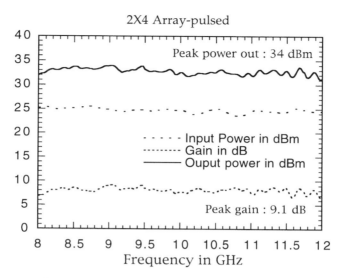

FIGURE 5.22 Preliminary measurements of an X-band combiner using a 4 × 2 array in waveguide, as shown in Fig. 5.21. This protoype combines the output of eight Texas Instruments TGA8014 medium-power GaAs MMIC amps.

published to date. In future work using several power cards, dielectric side-wall loading will be required to ensure uniform excitation to all of the amplifiers and hence optimum efficiency.

3.6 Ka-Band Quasi-Monolithic 2.4-W Array

The Lockheed-Martin Corporation (Orlando) and the University of Colorado jointly worked on the design of a high-power hybrid Ka-band quasi-optical amplifier [35]. The amplifier unit cell consists of a driver amplifier chip followed by a two-stage high-power amplifier chip. Input as well as output antennas are antiresonant slots fabricated on a GaAs substrate. A 6 × 6 array was designed for an output power in excess of 8 W. The array allows a variety of amplifier chips to populate any number of cells up to the full array size. A 3 × 3 subarray achieved a small-signal absolute power gain of 6 dB at 29.0 GHz with 10% PAE and a saturated output power of 2.4 W. A liquid-cooled test fixture was designed to remove excess heat from the amplifiers. This was the first working quasi-optical millimeter-wave amplifier [35], and it is the highest-power Ka-Band quasi-optical amplifier that has been reported to date.

Single Cell Design   The amplifier consists of two MMIC chips per element mounted on a GaAs monolithic antenna array. GaAs is chosen as the antenna array substrate to address issues related to high-power amplifier design of a monolithic high-power amplifier which will be fabricated in the future. The hybrid amplifier is a uniformly spaced 6 × 6 square array. Each of the unit cells is designed for an output power level of 300 mW, small signal gain of 16 dB, and 10% PAE.

The unit cell contains a first-stage driver amplifier MMIC followed by a two-stage high-power amplifier MMIC, as illustrated in Fig. 5.23a. The driver consists of a self-biased single-stage power MESFET amplifier that produces 100-mW output power with 6-dB small-signal gain, 4.5-dB large-signal gain and 11.8% PAE at a bias point of $V_{DS} = 5V$ and 140-mA drain current. (The high-power MMIC amplifier chip consists of four combined 400-μm power MESFETs which produce 400-mW output power with 12-dB small-signal gain, 6-dB large-signal gain, and 11.8% PAE at a bias point of $V_{DS} = 5$ V and 500-mA total drain current.) The 1-dB bandwidth of each MMIC amplifier in the cascade is 5 GHz, centered at 32.5 GHz. The chips were individually probed before being mounted onto the hybrid antenna array, shown in Fig. 5.23b.

Each unit cell has a pair of center-fed antiresonant slot antennas. The input and output slots are orthogonally polarized, which facilitates free-space measurements and is necessary for amplifier stability. The antiresonant slots operate in the second resonant mode and offers broader bandwidth and a convenient input impedance compared to resonant slots. The slot dimensions (4220 × 471 μm) are designed to provide a 25-Ω impedance on the 100-μm-thick GaAs substrate at 33 GHz. The slot length is shorter than a half-wave free-space wavelength, which is convenient for array design. The slot antennas are matched to the 50-Ω amplifier port impedances with quarter-wave transformers. Low-loss CPW-to-microstrip transitions are used at the interface between the CPW-antenna feeds and the microstrip MMICs. Additional hybrid decoupling capacitors are added along the bias lines for stability.

Array Performance The hybrid array's substrate is a 3.8-cm square GaAs chip. The substrate is thinned to 100 μm which is substantially less than 90 electrical degrees at 30 GHz and therefore is unlikely to support substrate modes. The thin substrate also allows better heat transfer from the amplifier chips to the backside of the substrate, which may be cooled. Input and output polarizers are used to reduce input feed loss and increase transmitted output power. In the array, the gate and drain bias lines are laid out in a square mesh with air bridges at the intersections.

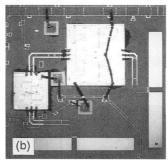

FIGURE 5.23 (a) Photograph of the amplifier unit cell for the K_a-band amplifier. Hybrid fabrication with MMIC driver and power amplifier chips is used. (b) Populated quasi-monolithic array. (From reference [9], © 1994 IEEE.)

A test fixture capable of removing up to 33 W of power dissipated from the array amplifier was designed. Fluorinert FC-43 was chosen as the coolant liquid since it has a low loss tangent of 0.0036 and a reasonable dielectric constant of 1.90 (measured at 8.5 GHz). The liquid directly cools the backside of the array and flows through a plastic channel which covers the entire backside of the array and is 540 electrical degrees thick at 33 GHz. The plastic and the liquid coolant have the same permittivity of 1.9 and represent a second dielectric layer to the transmitted amplified wave. Since the heat generated by the transistors has to propagate through 200 μm of GaAs before it reaches the cooled backside of the substrate, the transistors are 64°C higher in temperature than the coolant. The material of the GaAs chip carrier is Thermcon, which has the same thermal coefficient of expansion as GaAs.

Free-space array measurements were made with a scalar network analyzer with round corrugated waveguide horns. The horns produce a Gaussian transverse-field profile that focuses 63% of the power on a 2.5-cm spot size midway between the horns (the power density at the spot edge is $1/e^2$ lower than at the center). Passive array measurements were conducted before the MMIC amplifiers were mounted with no coolant present. The passive array had a peak in its transmission frequency response at 34.5 GHz with a 10% bandwidth. The insertion loss of the passive array is 10 dB. Simulated CPW line loss within the unit cell contributes 1 dB of the loss, indicating that there is 4.5 dB of loss on both the input and output sides of the array.

A 3 × 3 subarray was populated with the MMIC amplifiers, and the substrate was mounded on the cooling structure. Small-signal measurements give a transmission frequency response of the subarray shown in Fig. 5.24a. The peak absolute power gain was 6 dB, measured at 29.0 GHz, and the 3-dB bandwidth is 440 MHz, or 1.5% of center frequency. The measured frequency is shifted toward lower frequencies as compared to the passive array because the antenna frequency response is changed when it is dielectrically loaded by the cooling liquid. The measured 6 dB agrees well with the expected gain when the afore-

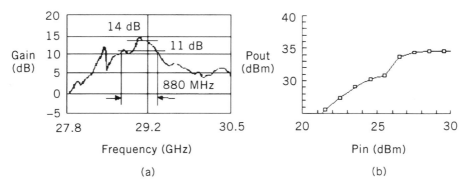

FIGURE 5.24 (a) Measured frequency sweep of the array power gain. (From reference [35], © 1995 IEEE.) (b) Measured saturation curve.

mentioned losses are taken into account and the expected MMIC amplifier small-signal gain is 16 dB.

Large signal measurements were performed with a TWT amplifier connected to the feed horn and a power meter connected at the output horn. An output power output of 2.5 W was measured with −2 dB absolute power gain. The drop in power gain is due to reduced MMIC gain under saturation. The saturation curve of the array is shown in Fig. 5.24b. The center elements of the array saturate first with increased feed power since they receive more of the Gaussian beam power. Therefore, the saturation curve shows two distinct gain slopes as the array feed power is increased. One slope indicates that the center elements saturate at the lower feed power, and the second slope indicates that the edge elements saturate at the higher feed power.

3.7 Monolithic 42-GHz Quasi-Optic Amplifiers

A guided wave system in which the input and output fields are constrained in an oversized waveguide is a solution alternative to using lenses in order to reduce diffraction losses. The challenge in such a system is to control the waveguide modes in order to have a uniform field distribution, which translates into more efficient use of the quasi-optical amplifier cross section. Such a 42-GHz monolithic quasi-optic amplifier is presented in this section. The guided beam amplifier was demonstrated using PHEMT MMIC amplifiers, slot-input antennas, and patch-output antennas [36]. The design of the 42-GHz monolithic GaAs PHEMT amplifier separately considers the antenna elements, the array, amplifier circuit, the bias distribution network, thermal management, unwanted waveguide and surface modes, electromagnetic field uniformity in the oversized waveguide, and finally the amplifier package.

Antenna Design The choice of the antenna elements is dictated by some fabrication parameters. A typical MMIC process uses microstrip transmission lines on a 75-μm GaAs substrate (dielectric constant 12.8), where thru-substrate vias connect the ground plane to the front circuitry. Two or three metallization steps allow the implementation of additional circuit components such as transmission lines, MIM capacitors, air bridges, and so on. To be compatible with this process, the choice of antenna elements is limited to printed circuit antennas. Resonant patch and slot antennas were selected in this design mainly for their preferential directionality and their compatibility with monolithic microstrip transmission lines.

In a transmission amplifier, it is desirable that the input and output antennas show gain in opposite directions. This property, combined with orthogonal polarization, minimizes the risk of oscillations due to feedback and allows for easy cascading. Patch antennas radiate only in the hemisphere above their ground plane, while slot antennas can be designed to have a 10 dB or better directionality in one hemisphere over the other by using dielectric loading. In the amplifier presented here, the slot array is used at the input while the patch array is used at the output. Maxwell Eminence (TM by Ansoft Corp.) was used for the design, where an infinite array approximation was used in order to reduce the problem to a manageable size.

The input slot antennas are dielectrically loaded (a high dielectric material is placed on one side of the slot). The ratio of the fields in the forward (high dielectric side) to reverse directions is proportional to the ratio of two dielectric constants of the materials to the power of $3/2$ [2]. For example, aluminum nitride (AlN) and air give a ratio of $(7.6/1.0)^{1.5} = 14$ dB. The dielectric serves as a mechanical support structure to the 75-μm-thick GaAs chip, a heatsink and chip carrier for bias distribution. The unidirectional property of the slot can be further improved by using a polarizer.

A microstrip patch is chosen as the output antenna. A patch has a narrow bandwidth and poor efficiency on a 75-μm-thick GaAs substrate at millimeter wavelengths. For example, the calculated radiation efficiency for a patch at 44 GHz is below 50% for any substrate thickness, assuming a tan $\delta = 0.001$ for GaAs. A parasitic patch was added to improve both bandwidth and efficiency. To increase the active device density and output, both input and output antennas were driven at two feed points in push–pull configuration, as shown in Fig. 5.25.

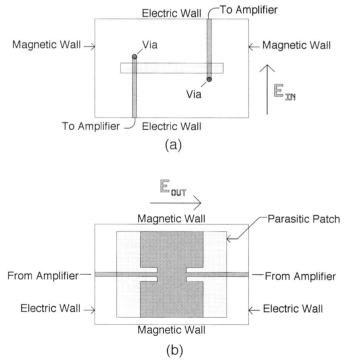

FIGURE 5.25 Input (a) and output (b) antenna elements used in the 42-GHz array.

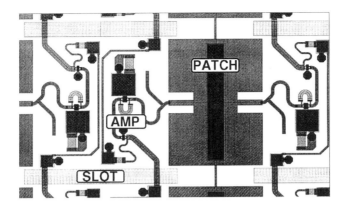

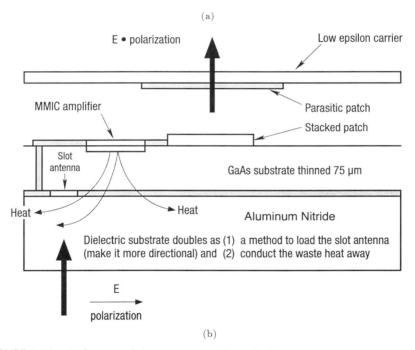

FIGURE 5.26 (a) Layout of the power amplifier cell. The output antenna is driven in push–pull by outputs from two PHEMT amplifiers, each of which receives its input from a slot antenna in the microstrip ground plane. Amplifier inputs are in antiphase. The patch and slot dimensions are 1200 by 860 μm and 1400 by 140 μm, respectively. (b) Cross section of a unit cell of the transmission QO amplifier. The input signal is coupled by a slot antenna to the microstrip. MMIC which drives a patch antenna from its output port. The output patch antenna is tuned by a parasitic patch, and the input slot antenna is loaded by a slab of AlN.

Unit Cell Design The antenna impedances calculated by the EM simulation program are used to design the two-stage PHEMT MMIC amplifier with standard circuit design techniques [37] and under unit cell dimension constraints. A self-biased two-stage PHEMT design was used (Fig. 5.26a) which shows that each amplifier is one of a pair that drives a patch antenna in push–pull mode. The PHEMT amplifier has a predicted gain of at least 7 dB at 44 GHz over a 10% bandwidth. The devices have sub-quarter-micron gate lengths and gate widths of 80 and 160 μm, respectively. Extra slot cells were added at the edges in order to preserve the symmetry of the array resulting in a 5 × 7 input array connected to a 4 × 8 output array with a 8 × 7 amplifier array.

A cross section of the amplifier cell is shown in Fig. 5.26b. The GaAs chip is placed on an aluminum nitride chip carrier that provides a heat removal path and also electrically loads the slot antenna to increase gain in the direction of the incident wave. The input signal is received by the slot antenna in the ground plane of the microstrip. The signal is delivered to the amplifier through a microstrip-via hole loop which is positioned along the slot for a good impedance match.

Array Performance The quasi-optical amplifier was tested in a waveguide test fixture shown in Fig. 5.27. The fixture consists of two pyramidal horn transitions from U-band waveguide (WR19) to a straight 15-mm long section. The straight section transverse dimension is the size of the amplifier (about 10 × 10 mm), which is placed between the two transitions. The calibration is performed by removing the amplifier carrier and measuring the insertion loss (less than 0.5 dB) for copolarized input and output. When the quasi-optical

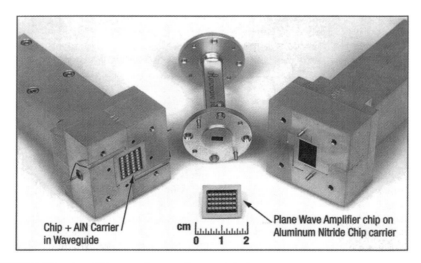

FIGURE 5.27 Amplifier test fixture disassembled to show the amplifier IC mounted. For comparison, a standard WR19 flange is also shown. Both sides of the fixture taper down to a WR19 flange from the amplifier section shown.

amplifier is measured, the transitions are connected in the cross-polarized mode and a polarizer is placed in the output waveguide. The polarizer reflects any fraction of the input wave that is transmitted through the amplifier array. The position of the polarized is adjustable and provides a small amount of tuning.

The measured gain of one of the prototype amplifiers is shown in Fig. 5.28. Positive gain is seen from 41 to 43 GHz with a maximum gain of 3 dB at 42 GHz. The maximum input return loss is 15 dB at 43 GHz. The gain peaks at a lower frequency than designed due to the shift in patch antenna resonant frequency. The maximum output power was just over 0.26 W, or 5 mW per amplifier, corresponding to a 1-dB compression point. The DC input power was 10 W at a 5-V drain bias. The measured output power is roughly an order of magnitude lower than expected. This is probably due in part to the losses in the antennas which are higher than expected, and in part to the nonuniform illumination of the array. Namely, the waveguide was not dielectrically loaded [36, 38] to make the incident power density uniform over the entire amplifier array. As a result, the power is very low at the edges and maximum in the middle, so that at 1-dB compression only the cells in the middle of the array give predicted output power.

3.8 60-GHz Monolithic Patch/Slot Amplifier Array

Another example of a quasi-optical amplifier with patches and slots is shown in Fig. 5.29. This V-band amplifier is designed for 5.8 dB of gain at 60 GHz with a 10% 3-dB bandwidth into 75-Ω input and output impedances. The amplifier

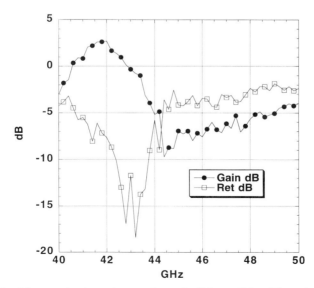

FIGURE 5.28 Measured gain and return loss of a QO amplifier. The gain is 3 dB at the point of maximum response at 42 GHz.

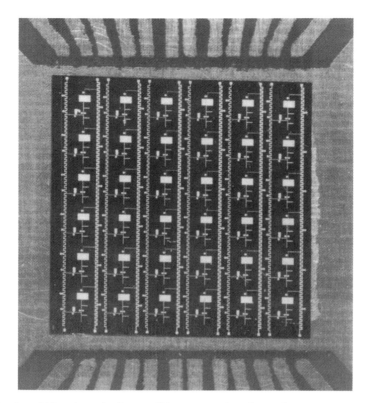

FIGURE 5.29 V-band patch-slot amplifier array. The microstrip patch is the receiving antenna, and the microstrip coupled resonant slot is the transmitting antenna. Meandered bias lines with capacitors prevent low-frequency oscillations.

is unconditionally stable above 100 MHz. Each amplifier in the 36-element array is coupled into free space by a nonradiating edge patch antenna on the circuit side of the array chip, along with a microstrip-coupled resonant slot antenna on the back side of the chip. The input and output antennas are orthogonally polarized. Th impedance of each antenna at resonance is 75 Ω at 60 GHz with a bandwidth of 1%. The square unit cells are half of a free-space wavelength on the side.

Martin Marietta Research Labs (Baltimore) fabricated several monolithic arrays on 100-μm GaAs substrates with gate-periphery enhancement-mode PHEMTs. A 36-element array measuring 1.5 cm on the side with a DC yield of approximately 80% was tested as a free-space amplifier in both transmission and reflection, as shown in Fig. 5.30. In the transmission case, the calibration was performed by measuring the frequency response of an open aperture the size of the array, and with the horns copolarized. Then the horns were orthogonally polarized and the array was placed in the beam waist. The amplifier array was stable and had an on/off ratio of 5 dB at 60 GHz, and this ratio was positive

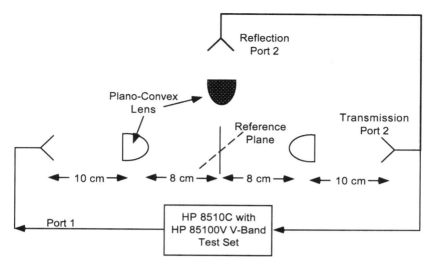

FIGURE 5.30 Free-space V-band measurement system for evaluating the array at the reference plane (solid line) for return loss and transmission, or as a reflection amplifier (dashed reference plane and shaded lens).

between 58.0 and 62.1 GHz. The 3-dB bandwidth was 1.4%. The measurement showed large etalon effects, which can be reduced in a reflection measurement since in that case the experimental setup no longer resembles a cavity, and reflected power is not returned to the feed horn.

The reflection-amplifier mode of operation is possible since the slot antennas radiate preferentially on one side of the array. The array is illuminated at a 45° angle, which is not ideal in terms of array gain, but was the only angle practical with the given measurement apparatus. The calibration was made with polarization-matched horns with a metal plate placed at the plane of reference. When the array was inserted and the output horn rotated for polarization match, a power measurement shown in Fig. 5.31 resulted. The array amplifier produced an on/off ratio of 4 dB, and this ratio was positive between 58.1 GHz and 60.8 GHz. While this is a lower ratio and a smaller bandwidth than the transmission-through mode, the measured absolute power gain in the reflection mode is 2 dB higher. Therefore, the reflection amplifier mode is a viable means of producing power in a quasi-optical amplifier array.

This 60-GHz amplifier is the first fully monolithic quasi-optical amplifier array operational at V-band. The array is stable and shows an on/off ratio, but does not exhibit real power gain (its peak response is at -13.5 dB with respect to the through calibration). There are several reasons for this poor performance. Unfortunately, the gate and drain terminals for the PHEMTs were inadvertently reversed in the fabrication, resulting in bias resistor connection to the output circuit and a nonideal match. This reduced the simulated gain from 7.1 dB to 5.8 dB, but unconditional stability was maintained. As a result of reversing the

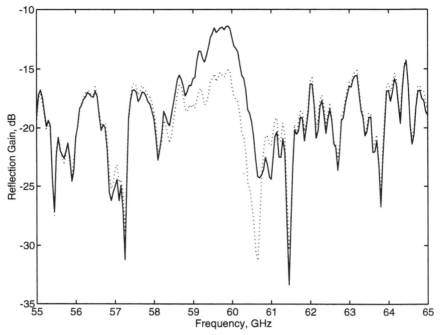

FIGURE 5.31 Measured reflection-mode frequency response for the V-band amplifier array in the biased-on (solid line) and biased-off (dotted line) states.

input and output, the output slots are not well matched to free space, and their output radiation is preferentially directed into the substrate. The slot and patch frequency responses are both narrow banded, and it is possible that the frequency range over which they are simultaneously matched to the amplifier is narrow or nonexistent.

4 LENS AMPLIFIERS

In the previous sections, quasi-optical amplifiers were shown to exhibit poor insertion gain (or have overall insertion loss) unless the energy from the source is efficiently coupled into the amplifier. Dielectric lenses were shown to greatly improve the feed efficiency for QOPC amplifiers, but the increased system complexity, size, and weight may make such an arrangement undesirable. To preserve the proper phase relationship among elements in an active array that is fed from the near field, geometric optics principles may be employed. The nonuniform phase front input into the array will require the designer to vary certain parameters of each element pair of the array, such as the elements' positions or delay-line length connecting the element pair. The theory of constrained lenses, demonstrated for passive lens arrays, may be applied to active arrays. In this section, we discuss transmission-wave amplifier arrays fed from a single focal point or multiple focal points in the near field of the input side of the array. The constrained lens array receives a spherical wave, processes the signal at each element, and re-radiates an amplified collimated beam.

In addition to serving as a free-space power combiner, the constrained lens amplifier can also perform beamsteering, beamforming, and beamswitching over a wide scan angle range with a relatively small cluster of feed horns [39]. The feed may be provided by any one horn in the cluster to produce a single beam in a particular direction, or the horns may simultaneously feed the array for multiple beam forming [9]. In addition, reflect arrays, polarization rotations, and amplitude or phase taper across the array may be performed. On the receiving end, a lens LNA can focus the received power onto a mixer [40]. In this case, since each unit cell amplifier handles only a fraction of the entire signal power at the array input surface, the dynamic range of a receiver is improved by using quasi-optical power combining in a receiver array. Also, several mixers can be placed along the focal surface to receive signals from different incidence directions, resulting in a very simple quasi-optical receiver with angle diversity. These properties of lens amplifiers are described in more detail in Chapter 12 in the context of systems applications.

4.1 Historical Development of Constrained Lenses

The earliest work in microwave antenna design using geometrical optics was done during World War II for enhancing the detection and tracking functions of early radar systems. This work concentrated on the use of artificial dielectrics and metal parallel-plate configurations [41]. The metal plate lens, made up of a stack of parallel-plate waveguides, imitates a dielectric with a curved surface for collimation. First described as "metal plate dielectrics" [42], it was not until later that the device was perceived as an array structure. Since the phase velocity of parallel–plate waveguide is faster than that of free space, a stack of waveguides open at both ends appears electrically as a dielectric with an index of refraction less than unity. The open ends of the waveguides allow a TEM wave to be coupled between free space and the waveguide. A wave traveling through the guide is *constrained* to that waveguide path length between the input and output. For a stack of thin metal plates spaced distance a apart, the index of refraction inside the guide is [43] $n = \sqrt{1 - (\lambda_0/2a)^2}$ for a wave polarized parallel to the plates. The metal parallel-plate structure imitates the focusing properties of the homogeneous plano-convex dielectric lens. To obtain dual-plane scan capability, rows of vertical parallel plates were combined with a stack of horizontal plates, resulting in a lattice of rectangular waveguides. Since the phase velocity of rectangular waveguide is always faster than free space, a lens using rectangular waveguides is thinnest in the center and grows progressively thicker toward its edge. This structure, known as the "egg crate" lens, focuses in both transverse planes down to one perfect focal point and may be scanned to wide angles along two perpendicular focal lines [43].

With Gent's introduction of the "bootlace aerial" in 1957 [44], the perception of microwave lenses changed from to that of an antenna array. His experimental lens was made of pairs of folded dipole radiators separated by a ground screen and interconnected by a parallel wire transmission line, or "twin lead." Since the transmission lines were longest near the lens center and progressively shorter toward its edge, an incident wave from a feed suspended behind the lens would be delayed near the center relative to the edges, or be collimated. Possibly the most importat capability of the new lens was for very broadbanded operation since it could use nondispersive tranmission lines, like coax or twin lead, for the delay lines. Although Gent alluded to the possibility of multiple-focal point lens design by incorporating the lens variables listed above, it was Rotman [45] who was the first to show a solution to the path-length equality conditions necessary to produce three perfect focal points. His solution results in expressions for the back face element locations and transmission line lengths for the two-dimensional geometry shown in Fig. 5.32. Using Fermat's principle for the path-length equality condition, it may be shown that three focal points may be simultaneously satisfied with the given lens geometry.

As bandwidth requirements increase, necessitating higher frequency of operation, either hybrid or monolithic techniques may be used for lens fabrication. These fabrication techniques are inherently planar, so a lens design with two interconnected planar printed circuit arrays is most applicable. McGrath [46] proposed using two back-to-back arrays of printed microstrip patch antennas. The central ground plane provides isolation to both input and output elements. A through-substrate via or a coupling slot in the ground plane connects the elements. Delay lines are made with microstrip transmission lines and their meandered pattern is mirrored on both lens surfaces. With the constraints that both back side and front side surfaces are planar, the only design degrees of

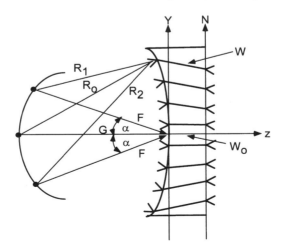

FIGURE 5.32 Geometry of the Rotman (trifocal) lens used in the solution. The Rotman lens scans in one plane only.

freedom available for this lens are (1) the relative positive of the feed side element with respect to its aperture side element and (2) the transmission line length connecting them.

Linear 7-Element Patch Lens Array The constrained lens concept allows integration of amplifiers in each unit cell of the lens. Such a quasi-optical component can be called a *lens amplifier*. The first demonstrated lens amplifier was a 7-element linear array designed for 10 GHz using a 0.787-mm-thick Arlon substrate with a relative permittivity of 2.17. The lens amplifier, as shown in Fig. 5.33a, has an array of receive and transmit patch antennas on opposite sides of the substrate. Each common-source PHEMT amplifier is coupled and matched to its input and output patch elements with microstrip lines. A substrate via connects each amplifier's output to its transmit-side patch. Alternating ground planes effectively isolate input and output sides of the amplifier structure, allowing input and output wave polarization to be arbitrarily selected [2], and are chosen to be orthogonal in this case for ease of measurement.

The lens amplifier element is shown in Fig. 5.33b. An input nonradiating edge impedance of the patch of 100 Ω is used. The patches were designed as described in Section 2. Avantek ATF-35576 PHEMTs are impedance-matched for gain to 100-Ω microstrip delay lines using single-stub matching sections. The gate bias network is similar to that described in [49], and provides unconditional stability for the PHEMTs. The gate bias network consists of a 90° section of 150-Ω microstrip line, terminated by a 70-Ω, 90° open stub in parallel with a

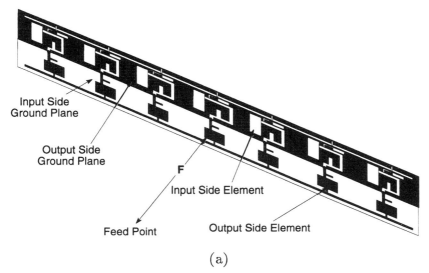

(a)

FIGURE 5.33 (a) The linear 7-element lens amplifier array with a focal-point feed. A horn illuminates the input side of the lens from the focal length $F = 2D$, where D is the lens length. (From reference [9], © 1994 IEEE.) (b) Topology of an amplifier unit cell used in the patch-patch linear and two-dimensional lens array. (*Continued*)

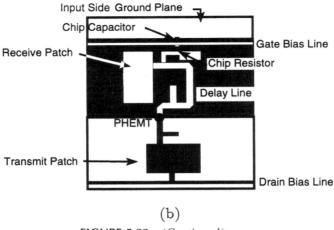

(b)

FIGURE 5.33 (*Continued*)

100-Ω chip resistor connected to the gate bias line. The 1-pF chip capacitor couples the resistive stability network to the feed-side ground plane. Drain bias is provided through the RF null of the transmit patch. Each microstrip delay line has the same number of discontinuities (mitred bends) for reproducible control of phase delay.

The lens' focal length-to-diameter ratio was chosen to be $f/D = 2.0$ in order to match the radiation pattern of the feed horn. The interelement spacing is $0.75\lambda_0 = 22.5$ cm on the output side, and the design focal length F is 27 cm [18]. The lens was horn-fed at the focal point and first tested with the five center elements biased, and then with all seven elements biased. The resulting output power is 3.1 dB higher for 7 elements than for 5 elements biased. This is 0.2 dB above the theoretical increase of 2.9 dB ($20 \log_{10}(7/5)$) expected from a uniformly driven array factor and indicates that the output of the planar lens has a uniform phase and that its effective area increases linearly with the number of elements. The maximum output power of the array was measured at 9.70 GHz, which is 3% below the design frequency due to the lower-than-expected resonant frequency of the microstrip patch antennas.

The isolation, defined as the biased-on to biased-off transmission power ratio, is 29 dB at center frequency. The theoretical beamwidth for a 7-element uniformly illuminated array is 14.6°, and its side-lobe level is -12.6 dB. The measured side-lobe level at a feed distance of 26 cm (the design focal length was 27 cm) was -12 dB. The cross-polarization ratio for the lens is 14.0 dB at the experimentally determined focal length of 26 cm. The lens preserves the incident phase of the input wave, so it may be fed with a progressive phase angle for beamsteering. Beamsteering of lens amplifiers is described in Chapter 12.

An HP Momentum simulation was performed to calculate mutual coupling effects between the 7 antenna elements. A port was defined at the end of the feedline of each patch, and each port is driven with the same amplitude and phase. The simulated magnitude of the s-parameters of the coupled ports are

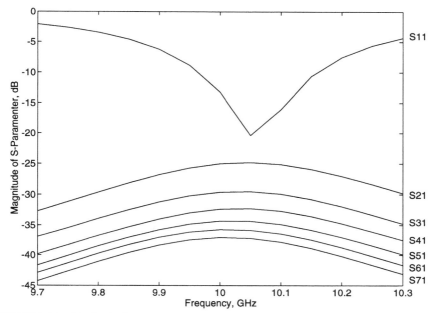

FIGURE 5.34 Method of moments simulation of a 7-element array of patches showing the magnitude of mutual coupling between elements.

shown in Fig. 5.34 defining port 1 (an edge element) as the excitation port. The s_{11} curve shows the same resonant behavior as that shown in the simulation of an uncoupled patch above. The s_{m1} curves show the magnitude of the coupling between ports, with each patch most strongly coupled at resonance. The strongest coupling is with the neighboring patch, with a maximum coupling of -25 dB, and the minimum coupling is between edge elements at -45 dB. The entire s_{mn}-parameter array was generated and examined and was found to be symmetrical with all $s_{mn} = s_{nm}$. Since the magnitude of the coupling is small between the elements, we feel that simulating the array patterns without mutual coupling effects is a good approximation.

4.2 Two-Dimensional Patch Lens Amplifier Array

For practical applications, two-dimensional array surfaces are required, which greatly reduces diffraction loss. A 24-element two-dimensional array, shown in Fig. 5.35, was designed. The lens diameter is $4\lambda_0$ the focal length is 25 cm, and the ratio f/D equals 2. The output side has uniform interelement spacing of $0.75\lambda_0$ in the horizontal direction and 0.90λ in the vertical direction. The relatively large spacing in the vertical direction is required to accommodate the via holes connecting the feed-side and output-side ground planes crucial to the isolation and stability of the lens amplifier. A triangular interelement spacing was selected for improved array scan performance with the large interelement spacing in the vertical direction. The microstrip patch antennas are fed on the

232 QUASI-OPTICAL ANTENNA-ARRAY AMPLIFIERS

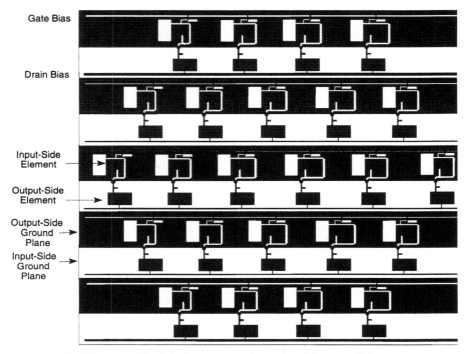

FIGURE 5.35 The 24-element two-dimensional lens amplifier array.

nonradiating edge with 100-Ω input impedance at resonance. The beam-steering and beamforming properties of this array are discussed in Chapter 12.

The design of the single-stage PHEMT amplifier was performed using HP MDS to account for all discontinuities in the amplifier layout. A microstrip amplifier without antennas as loads was tested on an HP 8510B network analyzer, with a measured peak gain of 13.12 dB at 9.92 GHz and a 3-dB bandwidth of 1.4 GHz. With a single element of the lens biased, a free-space gain measurement yields an amplifier gain of 12.95 dB, indicating good agreement with the network analyzer measurements.

Several symmetrical configurations of lens elements were biased to study how the lens insertion gain (or loss) is affected by the number of elements. The results, plotted in Fig. 5.36, show a dramatic improvement in absolute power gain as the number of elements increases. With all 24 elements biased, the lens delivered 5.7 dB of absolute power gain. The measured response showed a bandwidth of 320 MHz over which the isolation was at least 17 dB. A peak isolation value of 21 dB occurred at 10.25 GHz. The frequency response of the 24-element lens was broader than that of the 7-element lens and has two peaks. One peak was at 10.25 GHz and the other peak, 2.1 dB lower than the first, was at 9.7 GHz. The reason for this was the inadvertent variation in the resonant frequency of the patches, since for some input-side elements the radiating edges

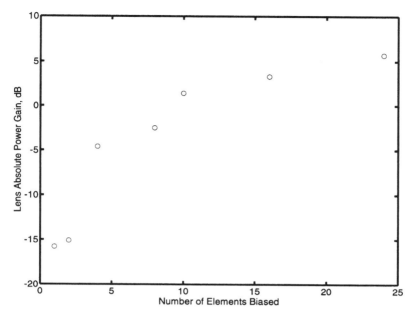

FIGURE 5.36 Absolute power gain (or loss) of the lens amplifier as a function of the number of elements biased. The one- and two-element cases were fed in the far field, while the other data points were obtained with a focal-point feed.

were closer to their neighboring ground planes. This was confirmed with HP Momentum simulations [18]. When parts of these ground planes were removed, the measured absolute power gain increased, as shown in Fig. 5.37. Before the modification, the amplifier shows two frequency peaks: The one at 9.70 GHz is where the input elements were resonant, and the other at 10.25 GHz is where the output elements radiated best. After ground plane metal was removed from around the radiating edges of the input patches, the absolute power gain of the array improved by 2.3 dB to a peak gain of 8.0 dB at 9.75 GHz.

As the input power to the lens amplifier is increased, the center elements saturate first. This makes the saturation curve of the array more gradual than that of a single element. However, as the lens becomes nonuniformly saturated, its radiation pattern changes: The main lobe broadens and the side lobes are reduced [9].

An estimate of the feed efficiency of the lens was performed by measuring the peak power reflected by the input side of the biased-on lens, and comparing that power to the peak power reflected off of a metal plate of the same dimensions as the lens. For both measurements, the horn feed was placed on the focal arc at an angle of 30° to avoid blockage by the horn [45]. The peak reflected power was measured at $-30°$ as expected, and 4 dB more power was reflected by the metal plate than by the lens. Therefore, we estimate that about 39% of the power incident on the lens is reflected and not coupled into the amplifiers. Increasing

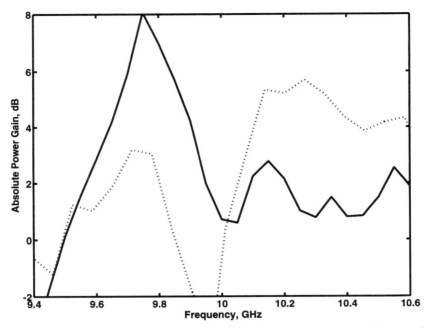

FIGURE 5.37 Lens amplifier gain and change in frequency response after top-side ground plane modification (solid line) compared to before (dotted line).

the element density, matching the resonant frequency of all elements, and reducing the reflection due to the isolating ground planes should reduce the reflected power.

4.3 Low-Noise CPW Slot Lens Amplifier

In the previous section, a lens amplifier designed as a transmitting array was described. A lens amplifier can also be designed as a receiver. In this case, an amplifier matched for noise is designed. Higher gain for each amplifier cell is necessary to improve the receiver performance and increase the overall insertion gain of the lens amplifier. For a given signal-to-noise (SNR) criterion, a 1-dB reduction in the amplifier noise figure allows a 20% reduction in the array size. Before discussing the specific design and results, a method for measuring the noise figure is outlined and the experimental procedure for measuring the noise figure in free space is described. A single amplifier cell is measured and compared to its theoretical added noise. Finally, the results of the free-space noise figure and associated gain of a 24-element, two-stage-per-element low-noise lens amplifier array are presented.

Noise Sources in a Quasi-Optical System Important issues in measuring noise of a lens amplifier include identifying and modeling the sources of noise in a free-space system and determining the reference planes at which the input and output "ports" of the quasi-optical lens array amplifier are placed. The sources of noise which are accounted for in the measurements presented in this section are: (1) far-field sources entering the input of the amplifier array, such as background noise and the noise standards used to determine the noise figure; (2) background noise bypassing the amplifier array and going directly to the collection horn; (3) equivalent noise temperature due to losses in the feed; and (4) noise generated by the amplifier array itself, which is the quantity of interest. A detailed description of these noise sources is given in [18].

The method used to evaluate the added noise T_e of the amplifier is the hot–cold or Y-factor test, where the noise standard temperature T_S is periodically switched from a high temperature T_{hot} to a lower temperature T_{cold}. The noise figure measurement determines the ratio $Y = T_{hot} + T_e/T_{cold} + T_e$. The susceptibility of the measured effective noise temperature to error may be expressed as

$$\Delta T_e = \frac{1}{Y-1} \Delta T_{hot} - \frac{Y}{Y-1} \Delta T_{cold} - \frac{T_{cold} + T_e}{Y-1} \Delta Y$$

assuming that the noise standard impedance does not change between hot and cold states, which is usually the case in a free-space measurement. The accuracy of the measurement is completely dominated by the uncertainty in the nonambient standard temperature T_{hot}. This uncertainty is reduced if $Y \gg 1$, so T_{hot} needs to be significantly larger than that of T_{cold} and T_e. The free-space measurement compounds the difficulty to minimize uncertainty because of the path loss between the noise standard and the input to the amplifier, which must be in its far-field. In the case presented in this section, a ratio of $T_{hot}/T_0 \gg 18.5$ dB is required. The effective noise ratio (ENR) is used to specify T_{hot} in relation to T_{cold} through the relation ENR (dB) = $10 \log_{10} (T_{hot}/T_0 - 1)$.

The noise power contribution at the collection horn due to the noise at the array input, modeled by the antenna temperature T_A, is $N_A = \kappa B T_A G_{AMP} L_F$, where κ is Boltzmann's constant, B is the spectrum over which the noise power is analyzed, G_{AMP} is the relationship between the total RF power output to the total power input at the output and input array surfaces, respectively, and L_F is the free-space path loss over the focal length F between the output of the array and the collection horn, given by Friis formula. The background noise arriving directly at the collection horn, bypassing the array, along with noise generated from losses in the collection horn, may be modeled like the noise from an attenuator with free-space loss L_F and temperature which is some mean value of the physical temperatures of the horn, the output array, and the background. If we assume that all of these temperatures are equal to the ambient temperature of the anechoic chamber T_0, then the noise power due to the background noise bypassing the array and arriving at the collection horn is $N_F = \kappa B T (1 - L_F)$.

The noise-equivalent temperature of the entire amplifier array at its output, T_{AMP}, or that for each amplifier T_i, is the desired quantity to be determined. The value of T_i includes the noise generated by the active device and any losses in the amplifier circuitry. The noise power delivered at the output of each amplifier circuit is $N_i = \kappa B T_i G_{\text{active}_i}$, where G_{active_i} is the overall gain of the ith amplifier circuit. An overall noise temperature of the entire amplifier array at its output, T_{AMP}, may be defined from the total noise power at the grid amplifier output that is collected by the collection horn. Noise contribution to the collection horn from each individual amplifier will add as independent sources, since the noise generated in one transistor amplifier is uncorrelated from that in any other amplifier. The total noise power contributed by the amplifier array and detected at the collection horn is

$$N_{\text{AMP}} = \kappa B T_{\text{AMP}} G_{\text{AMP}} L_F = \kappa B \sum_{i=1}^{N} T_i G_{\text{active}_i} \frac{\lambda^2 G_{Ei}(\theta_i, \phi_i) G_R(\theta_i, \phi_i)}{[4\pi f(\theta_i, \phi_i)]^2}$$

where $G_{Ei}(\theta_i, \phi_i)$ and $G_R(\theta_i, \phi_i)$ are the gains of the ith output element and collection horn, respectively, at the angle (θ_i, ϕ_i), and $f(\theta_i, \phi_i)$ is the range between them. For the relatively long f/D ratio chosen for the lens array, the angles (θ_i, ϕ_i) are limited to a range where the angular change in the element and horn gains is negligible. If the individual transistor amplifiers have the same gain and same noise temperature, then the overall noise temperature of the array will be the same, assuming $G_{\text{AMP}} = N \cdot G_a$. The expression for the total noise temperature is

$$T_R = T_{\text{AMP}} G_{\text{AMP}} L_F + T_S G_{\text{AMP}} L_R L_F + T_0 \{1 - L_F[1 - G_{\text{AMP}}(1 - L_R)]\}$$

This equation is a variation of the equation presented in reference 13 to determine the noise figure of an HBT grid amplifier in free space.

Measuring Noise Figure of a Lens Amplifier Array   In the free space measurement, a Noise Com 3208A noise source with an ENR of 29 dB, an HP8593E Spectrum Analyzer, and a 38-dB preamplifier are used. In the calibration, each horn is placed a distance $r_1 = r_2$ away from a passive unit cell. The passive unit cell contains the input and output elements used in the active unit cell, but with a matched CPW line replacing the active circuitry. Illuminating and collection horns are identical and are polarization-matched to their respective passive-cell elements. The noise meter determines the entire path loss present between the two horns, with the resulting total equally attributed to the input-side and output-side losses in this symmetrical system. The loss of this symmetrical system is equally divided between input and output sides of the passive cell and is used to calibrate out free-space losses L_R and L_R, reflective losses from the input and output surfaces, and background sources for the HP8593E spectrum analyzer. This calibration places the reference plane of the noise diode at the transverse plane of the passive cell. Factoring in the losses effectively transfers the standard noise source temperature T_S from its feed horn to the unit cell's input surface. Likewise, the plane of reference for the collection horn is transferred up to the output surface of the unit cell. With this calibration complete, the passive unit cell is replaced by the active cell.

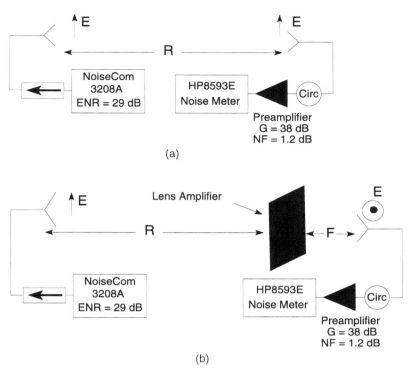

FIGURE 5.38 Measurement scheme for determining the noise figure of the lens amplifier array. In (a), the path losses L_R and L_F are each determined to define the reference plane up for the array. In (b), the feed horn is rotated for polarization match to the array.

The calibration and measurement scheme for the entire lens array is illustrated in Fig. 5.38. First, the path losses L_F and L_R are each determined using the free-space calibration measurement between the illumination and collection horns as shown in Fig. 5.38a. The loss factor at the output is then adjusted by the known value of the absolute power gain of the lens amplifier, which is performed through a free-space measurement using the microwave transition analyzer as reported below. The calibration sets the reference plane of the noise standard to the transverse plane containing the input surface of the array. The active array is then placed at the reference plane, and the feed horn is rotated for polarization match to the array input, as illustrated in Fig. 5.38b.

Slot-Slot Low-Noise Lens Amplifier A two-stage CPW low-noise amplifier was designed to receive a plane wave and focus it down to a focal point at which a mixer can be placed. The topology of a single cell and its schematic is shown in Fig. 5.30. This two-stage amplifier is biased with one drain bias and one gate bias line to simplify its application to an array. Before making a free-space unit cell, the amplifier design was evaluated in a CPW test fixture on an HP 8510B network analyzer. It has a peak gain of 20 dB and a 1 GHz bandwidth centered at 10 GHz. The HP Avantek ATF-35376 PHEMT is selected for the low-noise design because of its good noise performance (NF$_{opt}$ = 0.83 dB, R_n = 7Ω at 10 GHz) and low cost. The manufacturer's optimum noise match is transformed to a 100-Ω system, resulting in Γ_{opt} = 0.66 $e^{j \cdot 148°}$. The center-fed folded-slot antenna described in Section 2 is designed to provide an impedance of 20 Ω at its second resonance at 10 GHz. This folded slot is CPW-fed and loaded with a series-connected 10-pF capacitor and 51-Ω resistor at each end of the folded slot. The lumped load enables the folded slots to act as part of the stabilization and biasing network. The folded slot has a 15% 2:1 VSWR bandwidth. The output antenna is an antiresonant off-center fed slot with an input impedance of 100 Ω and a 25% bandwidth at the 10-GHz second resonance. The amplifier is fabricated on an ϵ_r = 2.17, 0.79-mm-thick teflon substrate.

The free-space measurements described in the previous section show that the unit-cell two-stage amplifier has a noise figure of 2 dB at 10 GHz with an associated gain of 18.5 dB (Fig. 5.39c). The minimum noise figure is 1.7 dB at 10.2 GHz with an associated gain of 15.1 dB. The noise and gain were simulated using HP's MDS with the aforementioned noise model data. The frequency-dependent impedances presented by the folded slot and the antiresonant slot are used as the input and output port impedances, respectively, in the MDS noise modeling. The noise figure simulation shows excellent agreement with the measurement. The measured gain is similar to that observed in a microstrip amplifier, built for comparison purposes.

Figure 5.40 shows a 24-element two-dimensional CPW lens array using the 2-stage low-noise amplifier cell. The lens diameter is 6 λ_0, the focal length is 20 cm, and f/D = 1.2. The input array of folded slots has a uniform interelement spacing of 0.9 λ_0. The output antiresonant slots are connected to delay lines and placed in such a way that the array focuses down to a point and the power is collected by a horn antenna. Interelement spacing would be significantly reduced using a higher permittivity substrate and monolithic fabrication.

Figure 5.41a shows the absolute power gain and isolation frequency response of the lens. The array demonstrates a peak absolute power gain of 13 dB at 10 GHz and a 3-dB bandwidth of 1.1 GHz over which the absolute power gain is above 10 dB. The on/off ratio is over 30 dB over the bandwidth. The noise figure and associated gain are shown in Fig. 5.41b, showing a minimum noise

---

FIGURE 5.39 A single cell of an X-band slot- slot CPW-based low-noise amplifier array with two-stage low noise amplifier. The topology of the cell is shown in (a), and the circuit schematic is shown in (b). (From reference [10], © 1995 IEEE.) (c) Measured noise figure and associated gain (symbols) of a single cell of the low-noise amplifier array. Simulated noise and gain performance (lines) using the antenna impedances at input and output ports show excellent prediction of the noise figure and predicts the trend in the gain.

LENS AMPLIFIERS   239

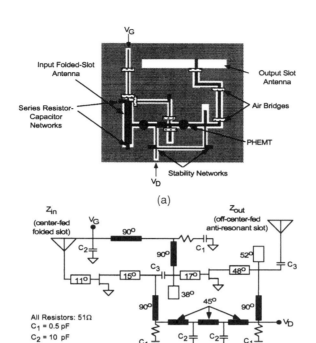

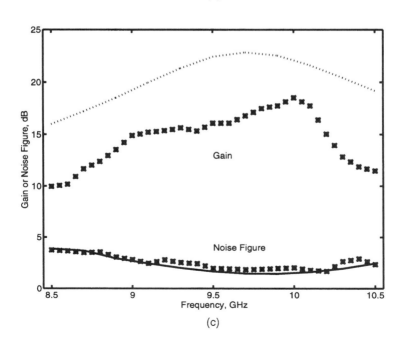

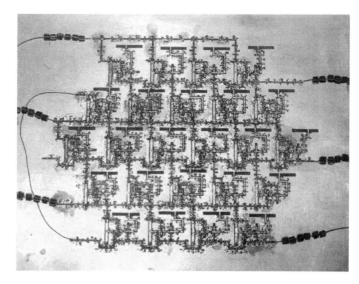

FIGURE 5.40 A 24-element two-dimensional CPW low-noise amplifier lens array. The lens diameter is $6\lambda_0$, the focal length is 20 cm, and $f/D = 1.2$.

figure of 1.7 dB at 9.80 GHz and an average noise figure of 2.3 dB across the 11% 3-dB bandwidth. The individual amplifier cells making up the array are producing approximately the same gain and noise temperature, and the lens is focused at the design focal distance.

5 CONCLUSIONS

In this chapter, we have made an attempt to present the potential advantages of active quasi-optical amplifier arrays for microwave and millimeter wave solid-state power generation. The results presented here have all been obtained in the past few years by a relatively small group of researchers. We feel that with some additional development work, quasi-optical amplifier arrays can solve the problem of millimeter-wave power generation for some applications. With achievements such as 2.4 W at Ka band, 65% PAE with 85% combining efficiency and 2.5 W at 5 GHz, demonstrated phase-shifterless beamforming, 1.9-dB noise figure with 13 dB gain at X band, and broadband operation over an entire waveguide X band, these components seem to be an attractive lower-cost alternative to more conventional approaches. The problems that remain to be solved include: efficient feeding for absolute end-to-end power gain; good PAE and low-power consumption at millimeter-wave frequencies; good linearity for communication applications; proven high reliability and increased dynamic range; and packaging including adequate heat sinking.

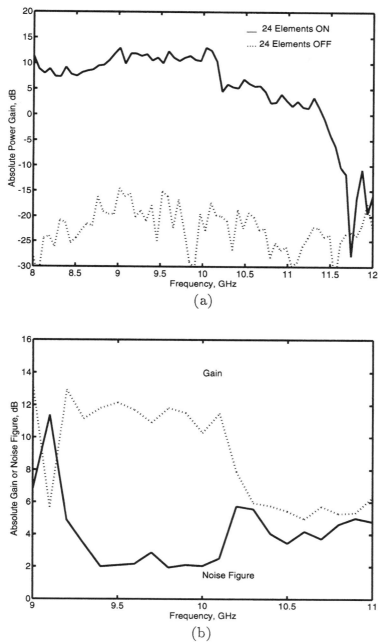

FIGURE 5.41 (a) Measured absolute power gain as a function of frequency for the low-noise amplifier lens array with all 24 elements biased on (solid line) and biased off (dotted line). (b) Measured gain and noise figure of the 24-element LNA lens array.

REFERENCES

[1] R. M. Weikle II, M. Kim, J. B. Hacker, M. P. DeLisio, and D. B. Rutledge, "Planar MESFET grid oscillator using gate feedback," *IEEE Trans. Microwave Theory Techn,* vol. 40, no.11, pp, 1997–2003, Nov. 1992.

[2] T. Mader, J. Schoenberg, L. Harmon, and Z. B. Popović, "Planar MESFET transmission amplifier," *IEE Electron. Lett.,* vol. 29, no. 19, pp. 1699–1701, Sept. 1993.

[3] N. J. Kolias and R. C. Compton, "A microstrip-based unit cell for quasi-optical amplifier arraya," *IEEE Microwave and Guided Wave Lett.,* vol. 3, pp. 330–332, Sept. 1993.

[4] N. Sheth, T. Ivanov, A. Balasubramaniyan, and A. Mortazawi, "A nine HEMT Spatial Amplifier," *1994 IEEE Int. Microwave Theory Tech. Symp. Dig.,* pp. 1239–1242, May 1994.

[5] H. S. Tsai, and R. A. York, "Polarization-rotating quasi-optical reflection amplifier cell," *IEE Electron. Lett,* vol. 29, pp. 2125–2127, Nov. 1993.

[6] H. S. Tsai, M. J. W. Rodwell, R. A. York, "Planar amplifier array with improved bandwidth using folded slots, " *IEEE Microwave Guided Wave Lett.,* vol. no. 4, pp. 112–114, April 1994.

[7] T. Ivanov, and A. Mortazawi, "Two-stage double layer microstrip spatial amplifiers," *1995 IEEE Int. Microwave Theory Tech. Symp. Digest,* pp. 589–592, May 1995.

[8] J. S. H. Schoenberg, and Z. B. Popović. "Planar lens amplifier," *IEEE 1994 Int. Microwave Theory. Tech. Symp. Dig.,* pp. 429–432.

[9] J. S. H. Schoenberg, S. C. Bundy, and Z. B. Popović, "Two-level power combining using a lens amplifier," *IEEE Trans. Microwave Theory Tech.,* vol. MTT-42, no. 2, pp. 2480–2485, Dec. 1994.

[10] J. Schoenberg, T. Mader, B. Shaw, and Z. B. Popović, "Quasi-optical antenna array amplifiers," *IEEE 1995 Int. Microwave Theory Tech. Symp. Dig.,* pp. 605–609, May 1995.

[11] T. B. Mader, "Quasi-optical Class-E Power Amplifiers," Ph.D. dissertation. University of Colorado, Boulder, 1995.

[12] M. Kim, E. A. Sovero, J. B. Hacker, M. P. DeLisio, J. C. Chiao S. J. Li, D. R. Gagnon, J. J. Rosenberg, and D. B. Rutledge, "A 100-Element HBT grid amplifier," *IEEE Trans. Microwave Theory Tech.,* vol. MTT-41, no. 11, pp. 1762–1771, Nov. 1993.

[13] K. C. Gupta, "CAD of active microstrip antennas and microstrip arrays," *1995 Asia-Pacific Microwave Conf. Proc.,* pp. 89–895, Dec. 1994.

[14] K. C. Gupta, "Computer-aided design of microstrip antennas," ECEN 5004-3, *Special Topics on Computer Aided Design of Microstrip Antennas Class Notes,* University of Colorado, 1993.

[15] S. V. Robertson, N. I. Dib, G. Yang, and L. P. B. Katehi, "A folded slot antenna for planar quasi-optical mixer applications," *IEEE Antennas Propag. Soc. Int. Symp. Proc.* 1993, pp. 600–603, May 1993.

[16] H. S. Tsai and R. A. York, "Quasi-optical amplifier array using direct integration of MMICs and 50 ohm multi-blot antennas," 1995 *IEEE Int. Microwave Theory Tech. Symp. Dig.,* pp. 593–596, May 1995.

[18] J. S. H. Schoenberg, *Quasi-optical constrained lens amplifiers,* Ph.D. dissertation, University of Colorado, Boulder, 1995.

[19] T. B. Maoder, *Quasi-optical Class-E power amplifiers,* Ph.D. dissertation, University of Colorado, Boulder, 1995.

[20] B. D. Popović, *CAD of Wire Antennas and Related Radiating Structures,* Wiley, Inc., New York, 1991.

[21] M. Kominami, D. M. Pozar, and D. H. Shaubert, "Dipole and slot elements and arrays on semi-infinite substrates," *IEEE Trans. Antennas Propag.,* vol. AP-33, pp. 600–607, June 1985.

[22] D. B. Rutledge, D. P. Neikirk, and D. P. Kasilingam, "Integrated circuit antennas," in *Infrared and Millimeter Waves,* vol. 10, K. J. Button, ed., Academic Press, New York, 1983, Chapter 1, pp. 1–90.

[23] H. S. Tsai and R. A. York, "FDTD analysis of folded-slot and multiple slot antennas on thin substrates," *IEEE Trans. Antennas Propag.,* vol. AP-44, pp. 217–226, Feb. 1996.

[24] H. S. Tasi and R. A. York, "Multi-slot 50 Ω antennas for quasi-optical circuits," *IEEE Microwave Guided Wave Lett.,* vol. 5, no. 6, pp. 180–182, June 1995.

[25] A. Alexanian, N. J. Kolias, R. C. Compton, and R. A. York, "FDTD analysis of quasi-optical arrays using cyclic boundary conditions and berenger's PML," *IEEE Microwave Guided Wave Lett.,* pp. 138–140, March 1996.

[26] R. Janaswamy and D. H. Schaubert, "Analysis of the tapered slot antenna," *IEEE Trans. Antennas Propag,* vol. 35, pp. 1058–1065, Sept. 1987.

[27] M. Rodwell *et. al.,* "Active and nonlinear wave propagation devices in ultrafast electronics and optoelectronics," *Proc. IEEE,* vol. 82, pp. 1037–1058, July 1994.

[28] A. Boyer and I. Wolff, "Finline tapper design made easy," *IEEE MTT-S Dig.,* S-14, pp. 493–496, 1985.

[29] C. J. Verver and W. J. R. Hoefer, "Quarter-wave matching of waveguide-to-finline transitions," *IEEE Trans. Microwave Theory Tech.,* vol. 32, pp. 1645–1648, Dec. 1984.

[30] T. B. Mader and Z. B. Popović, " The transmission-line high-efficiency. Class-E amplifier," *IEEE MTT Microwave Guided Wave Lett,* vol. 6, no. 10, pp. 290–293, Oct. 1995.

[31] H. S. Tsai, M. D. W. Rodwell, and R. A. York, "Planar amplifier array with improved bandwidth using folded slots," *IEEE Microwave Guided Wave Lett.,* vol. 4, pp. 112–114, April 1994.

[32] H. S. Tsai and R. A. York, "Quasi-optical amplifier array using direct integration of MMICs and 50 Ω multi-slot antenna," *1995 IEEE MTT-S Int. Microwave Symp.* (Orlando), pp. 593–596.

[33] Unpublished work done at UC Santa Barbara; patent pending.

[34] A. Alexanian, H. S. Tsai, and R. A. York, "Quasi optical travelling wave amplifiers," *1996 IEEE MTT-S Int. Microwave Symp.* (San Francisco).

[35] J. Hubert, J. Schoenberg, and Z. B. Popović, "High-power hybrid quasi optical Ka-band amplifier design," *1995 IEEE Int. Microwave Theory Tech. Symp. Dig,* pp. 585–588, May 1995.

[36] E. A. Sovero, Y. Kwon, D. S. Deakin, A. L. Sailer, and J. A. Higgins, "A PHEMTB based monolithic plane wave amplifier for 42 GHz," *1996 IEEE MTT-S Int. Microwave Symp. Dig.,* June 1996.

[37] G. D. Vendelin, A. M. Paviod, and U. L. Rohde, *Microwave Circuit Design Using Linear and Nonlinear Techniques,* Wiley, (New York,) 1990.

[38] J. A. Higgins, E. A. Sovero, and W. J. Ho, "44 GHz monolithic plane wave amplifiers," *IEEE Microwave Guided Wave Lett.,* vol. 5, no. 10, Oct. 1995.

[39] P. J. B. Clarricoats, A. D. Monk, and Z. Hai, "An array-fed reconfigurable reflector for flexible coverage," *23rd Eur. Microwave Conf. Proc.,* pp. 194–197, Sept. 1993.

[40] W. A. Shiroma, E. Bryerton, S. Holliung, and Z. B. Popović, "A quasi-optical receiver with angle diversity," *1996 IEEE MTT-S Int. Symp. Dig.* (San Francisco), June 1996.

[41] W. E. Koch, "Metal lens antennas," *Proc. IRE,* vol. 34. pp. 828–836, Nov. 1946.

[42] J. Brown, "Microwave lenses," *Methuen's Monographs on Physical Subjects,* Methuen, London 1953.

[43] J. Ruze, "Wide-angle metal plate optics," *Proc. IRE,* vol. 38, pp. 53–59, Jan. 1950.

[44] H. Gent, "The bootlace aerial," *Royal Rader Establishment J.,* pp. 47–57, Oct. 1957.

[45] W. Rotman and R. Turner, "Wide angle microwave lens for line source applications," *IEEE Trans. Antennas Propag.,* vol. AP-11, no. 6, pp. 623–632, Nov. 1953.

[46] D. T. McGrath, "Planar three-dimensional constrained lenses," *IEEE Trans Antennas Propag.,* vol. AP-34, no 1, pp. 46–50, Jan. 1986.

[47] R. T. Webster, A. J. Slobodnik, and G. A. Roberts, "Monolithic InP HEMT V-band low noise amplifier, " *IEEE Microwave Guided Wave Lett.* vol. 2, no. 6, pp. 236–238, June 1992.

[48] T. Ivanov and A. Mortazawi, "A two stage spatial amplifier with hard horn feeds," to appear in the *IEEE Microwave Guided Wave Lett,* 1996.

[49] C-M. Liu, E. Sovero, W. J. IIo, J. A. Higgins, M. P. DeLisio, and D. B. Rutledge, "Monolithic 40 GHz 670 mW grid amplifier," *IEEE MTT-S Int. Symp., June 1996.*

CHAPTER SIX

Multilayer and Distributed Arrays

AMIR MORTAZAWI
University of Central Florida, Orlando

CARL L. BROCKMAN
JOHN F. HUBERT
Lockheed/Martin Electronics and Missiles, Florida

1 INTRODUCTION

Over the last decade, significant progress has been made in the extension of solid-state technology to the microwave and millimeter—wave region of the frequency spectrum. Indeed, the MIMIC and MAFET programs sponsored by the U.S. Government have resulted in the development of monolithic gallium arsenide devices and chips capable of performing all of the conventional functions of RF systems in radars, radios, communications links, and passive radiometry sensors. The devices are now being combined in large multifunction chips capable of performing these functions in cross-sectional areas and volumes that are orders of magnitude less than the technology of the 1970s and 1980s. These developments are spawning wireless and vehicular technology applications of RF sensors and inexpensive military weapons applications that just were not possible for an affordable price with hybrid and discrete component RF systems.

Active and Quasi-Optical Arrays for Solid-State Power Combining, Edited by Robert A. York and Zoya B. Popović.
ISBN 0-471-14614-5 © 1997 John Wiley & Sons, Inc.

TABLE 6.1 Generic Military Applications Versus the Requirements for Frequency and Power

Market Segment	Band of Interest	Power (Watts)	Phased Arrays
Ground vehicle FCR digital battlefield	Ka or W	10–100	Yes
Rotary wing strike	Ka or W	20–50	Yes
Fixed wing strike RECCE	Ku	2000	Yes
Smart weapons upgrades 20	Ka and W Improbable	5–	
Aircraft communications	Ka and Q	10–100	Yes
Hit-to-kill seekers	Ka or W	100–1000	Probable

However, despite these advances, the system designers of the 1990s are limited in the application of the solid-state technology in two ways. First, the technology provides only modest quantities of solid-state RF power. Second, the application of MMICs to active array technology has seen cost limitations, which ultimately may be insolvable by conventional techniques given the nature of market forces on desired cost goals.

Table 6.1 lists some generic military applications versus the requirements for frequency, power, and phase arrays. Table 6.2 provides a similar list for commercial applications. Power levels from fractions of a watt to kilowatts are forecast for these system applications. Phased arrays are preferred for many of these applications as well. As discussed in Chapter 1, due to the limitations of current technology, the available RF power from conventional power-combined solid-state sources cannot meet many of the emerging requirements shown in Tables 6.1 and 6.2. However, by spatially combining a two-dimensional amplifier array configuration containing many unit cells (with 3-dB total combining loss), the power available from solid-state sources would vary with frequency as shown in Fig. 6.1. Three curves are shown in Fig. 6.1: One is shown for the assumed level of a single cell amplifier versus frequency based on reported power for various transistor devices (such as HBTs and PHEMTs),

TABLE 6.2 Generic Commercial Applications Versus the Requirements for Frequency and Power

Market Segment	Band of Interest	Power (Watts)	Phased Arrays
Global communications	Ka to V	10–400	Yes
Terrestrial or uplink	Ka to V	10–100	Yes
Transportation radar applications	Ka to W	10–100	Maybe
Personal communications	Ka to Q	1–5	No
Commercial lab equipment	Ka to W	5–10	No

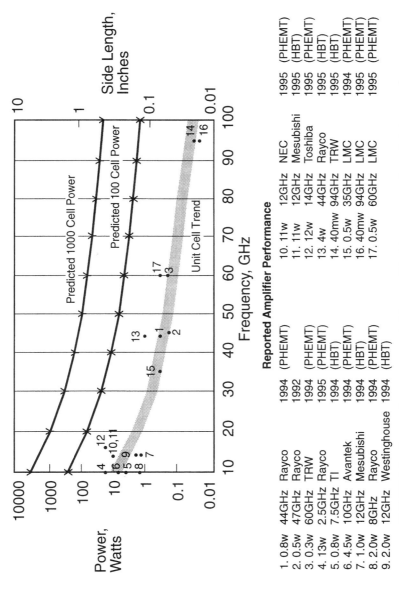

FIGURE 6.1 Reported power generated from solid-state sources and predicted power for a 100- and a 1000-cell spatial power combiner versus frequency.

while the other two are shown for predicted performance of a 100-cell and 1000-cell spatial (quasi-optical) power combiner. We have assumed that the cells are spaced at quarter wavelengths (or two devices per cell with half wave length spacings) since device size is roughly proportional to wavelength in each dimension. Furthermore, existing cell designs comply with this limitation, and this cell size is convenient in the future as the technology is extended to phased arrays.

This leads to an interesting conclusion about power availability from spatial (quasi-optical) arrays. As discussed earlier, available power varies as $1/f^2$ for applicable device types, as shown in the lefthand scale of Fig. 6.1. However, as just discussed, power is roughly proportional to cell size, and cell size varies as $1/f^2$ as well. Then the power versus frequency and the cell size versus frequency factors effectively cancel each other for an array of a given area so that the absolute value of the power available from a spatial combined array of this type would be dependent on the side length at any frequency, as shown on the right-hand scale of Fig. 6.1. Moreover, the sizes of arrays under these assumptions are both practical to build and useful from a packaging perspective. It must be mentioned that Fig. 6.1 is intended to show trends and is not included for design purposes.

Based on the above discussion, the amplifier results in Fig. 6.1 require multilayered array construction in reality because of the practical limitations in gain associated with each array and its associated cell design.

2 A MULTILAYER AMPLIFIER ARRAY

The general concept of a multilayer MMIC architecture for the construction of millimeter-wave active arrays was introduced in reference [1] in connection with work on phased arrays. The idea was to group radiating elements of a large array into subarrays on a common semiconductor substrate, where each subarray contains amplifiers, power dividers, and phase shifters. The active devices are placed on one layer and are coupled through interconnects to the next layer which contains the radiating elements. The challenges that are involved in implementing this type of architecture are choices of interconnects, heat removal, thermal expansion, and board alignment.

In this chapter the construction of multilayer and distributed arrays is discussed. Figure 6.2 shows the general concept of a multilayer spatial amplifier array. The input signal is incident upon the front side of the first amplifier layer via a horn antenna. The amplified signal is then coupled through the first layer to the other side of that layer. The amplified signal is consequently coupled to the second amplifier layer for further amplification, and so on for N layers. Finally the output signal from the last layer is captured by the collecting horn antenna. Each of the amplifying layers shown in Fig. 6.2 is comprised of two microstrip receiving and transmitting layers coupled together forming a "double-layer

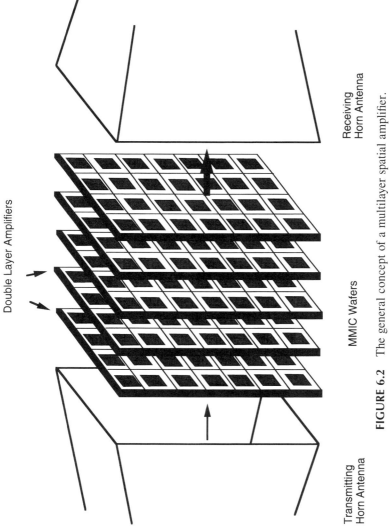

FIGURE 6.2 The general concept of a multilayer spatial amplifier.

array." In this chapter the term "double-layer array" will be used to describe the two-dimensional array shown in Fig. 6.3, comprised of two microstrip circuits placed back to back with a shared ground plane. Both vias and microstrip to slot to microstrip transitions are used as a means of interconnecting between the front and back of two microstrip circuit layers. The term "multilayer array" will be used to describe the three-dimensional structure formed by cascading several "double-layer arrays" as was shown in Fig. 6.2. Broadside microstrip couplers are proposed to couple signals from one double-layer array to the next in order to form a multilayer structure. Some of the challenges in the design and construction of multilayer arrays are: the method excitation and coupling the input signal to the array, isolation between the input and the amplified signals, coupling between the layers, and heatsinking.

In a double-layer array, the active devices can be grown on one or both layers and hence use GaAs wafers. The coupling between the two layers is accomplished either through via interconnects or microstrip to slot to microstrip transitions. The advantage of such circuits are as follows:

(a) The incident and transmitted waves are isolated from each other by a continuous ground plane, as opposed to the use of polarizers to isolate the input and output which can increase the system size.
(b) A thick ground plane can be used for heatsinking. Provisions can also be made for coolant flow in the thick ground plane to improve its heatsinking capability.

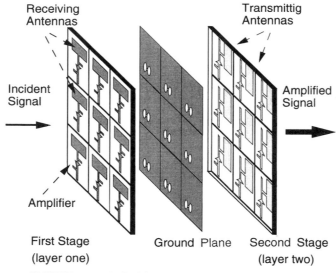

FIGURE 6.3 A double-layer spatial amplifier array.

(c) Such circuits are readily cascadable. This allows the construction of multiple-stage cascadable units to obtain the appropriate gain and power required for a variety of applications.

(d) Active devices can be placed on both sides of the double-layer board, thereby reducing the heat density on each side for a given power level.

The excitation of spatial amplifiers arrays is another important issue. For efficient power combining, incident waves must be focused entirely on the surface of amplifier array. Furthermore, in order to saturate all the devices across the surface of the spatial amplifier simultaneously (assuming that all the devices have the same periphery), the incident power distribution across the plane of the amplifier array must be uniform. The technique that will be described here takes advantage of electromagnetic "hard" surfaces to design horn antennas for excitation of spatial power-combining amplifiers [2]. These "hard horns" [3] have near-uniform aperture field distributions which allows the simultaneous saturation of active devices across the spatial amplifier's surface.

A multilayer spatial amplifier can readily be constructed by cascading several of the double-layer spatial amplifiers. This is because there is no need for polarizers to isolate the stages. In order to accomplish the cascaded structure there is a need for a coupling mechanism between the stages. A technique to couple energy from one double-layer amplifier stage to another based on nearfield coupling between the layers using broadside microstrip couplers will be discussed. Electromagnetic simulation of these couplers at millimeter wave frequencies predicts the feasibility of such an approach. This type of coupling is advantageous due to the fact that it results in compact and efficient multilayer power amplifiers at millimeter-wave frequencies.

2.1 Double-Layer Amplifier Architecture

The double-layer amplifier array which is the building block for the multilayer amplifiers is constructed by placing two GaAs wafers back to back. The basic double-layer spatial amplifier array structure was shown is Fig. 6.3. The amplifier array consists of two microstrip layers with a shared ground plane. Each layer is comprised of many unit cells, each containing active devices and a receiving microstrip patch antenna (on the receiving layer) or a transmit patch antenna (on the transmitting layer). The incident signal is received by the patch antenna on the receiving layer and amplified. The output of each amplifier is coupled to the transmitting layer by a microstrip–slot–microstrip transition (through wafer coupler). The transmitting layer can contain another stage of amplification. Finally the amplified output signal on the transmit layer is fed to a patch antenna which re-radiates it back into space. The ground plane provides an effective isolation between the received and transmitted signals. It also used to conduct the heat generated by power devices into a heatsink. This can be more

252 MULTILAYER AND DISTRIBUTED ARRAYS

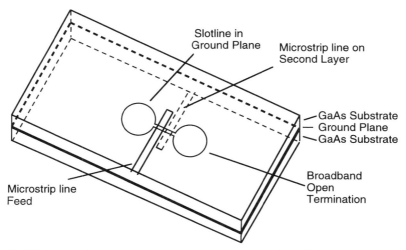

FIGURE 6.4 A microstrip–slot–microstrip (through wafer) interconnect.

easily accomplished by using a thick ground plane. An electromagnetic simulation of the through wafer coupler consisting of microstrip–slot–microstrip transition will be discussed next.

2.2 Through Wafer Coupling Mechanism

Recently there has been an increased interest in design and analysis of three-dimensional high-frequency circuits. This is due to increased demand for high density MMIC circuits. By using a shared ground plane between two microstrip layers, two circuits can be formed which utilize the same area. Two different techniques has been adopted for construction of double-layer amplifier arrays. One approach to couple signals from receiving to transmitting layer of a double-layer array is to use vias to physically connect two microstrip layers through an opening in the ground plane [4]. An alternate approach is to use microstrip–slot–microstrip transitions for this purpose [5]. An advantage of the latter technique is that it does not physically connect the two layers. This simplifies the construction of double-layer arrays and provides DC isolation between the two layers. This is important when both layers contain active devices. A detailed view of a microstrip–slot–microstrip transition is shown in Fig. 6.4. Two microstrip lines fabricated on a GaAs substrate are coupled to each other through a slot on the ground plane. This type of transition can provide a wideband response with a small insertion loss. Figures 6.5a and 6.5b show simulated performance of such a transition at Ka band obtained using the Hewlett Packard High-Frequency Structure Simulator (HFSS™) software. All the conductor and dielectric losses were taken into consideration. As can be seen, the insertion loss of approximately 0.5 dB over a wide frequency range is predicted. To

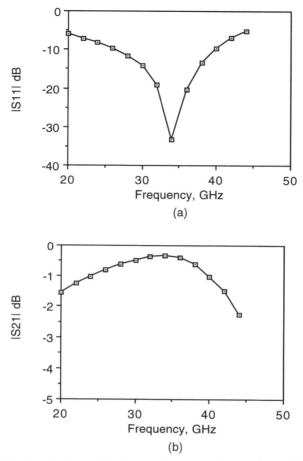

FIGURE 6.5 Simulated S_{11} and S_{21} for a through wafer coupler at Ka band (GaAs thickness = 4 mils; microstrip line width = 2.5 mils; slotline width = 6 mils).

enhance the ground plane's utility as a heatsink, its thickness can be increased (Fig. 6.6). For a very thick ground plane the slot length approaches one-half wavelength in free space and behaves like an ordinary waveguide. However, for the ground plane thicknesses comparable to a fraction of a wavelength, the resonant length of the thick slot is smaller than a half wavelength in free space. Figures 6.7a and 6.7b show the simulation results for a transition at Ka band using a 20-mil-thick ground plane. The predicted insertion loss for the microstrip–slot–microstrip transition is approximately 0.5 dB. The cylindrical cuts in the ground plane are used to increase the bandwidth. In order to reduce the size of the slot transition when thick metal planes are used, one can fill the void in the ground plane with high dielectric-constant material.

254 MULTILAYER AND DISTRIBUTED ARRAYS

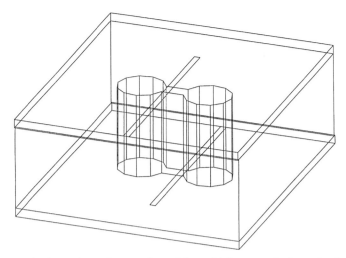

FIGURE 6.6 A through wafer coupler with a thick ground plane for heatsinking.

2.3 The Amplifier's Unit Cells

The basic design of a spatial power-combining amplifier array involves the design of a unit cell. An example of the unit cell for the double-layer spatial amplifier array discussed above is shown in Fig. 6.8. This unit cell contains two amplification stages. It employs microstrip patch antennas to receive the input signal and to radiate the amplified signal into free space. The open-circuited

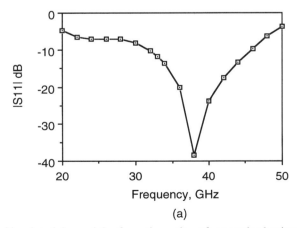

(a)

FIGURE 6.7 Simulated S_{11} and S_{21} for a through wafer coupler having a 20-mil-thick ground plane.

A MULTILAYER AMPLIFIER ARRAY   255

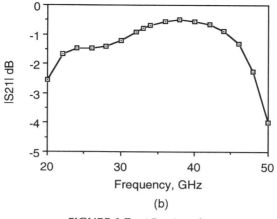

FIGURE 6.7 (*Continued*)

stubs in both the gate and the drain circuits provide the simultaneous conjugate match susceptances to the active devices. Therefore what is left is purely real and can be matched to the microstrip patch radiation resistance via quarter-wave transformers. The input and output patches are cross-polarized. This was done to simplify the amplifier measurements in the far field and is not necessary when near-field excitation via horn antennas is employed. The isolation in between the incoming and outgoing waves is provided via the continuous shared ground plane in between the two layers. The circuit was fabricated on RT-Duroid substrate with $\epsilon_r = 2.33$ and thickness of 31 mils [2].

2.4 A Unit Cell with High Active Device Density

Due to the large physical dimensions of patch antennas compared to the active devices used, most of the chip real state is occupied by antennas. In order to use the chip's real state efficiently and to be able to construct small circuits containing many devices, one can increase the number of devices that are connected to

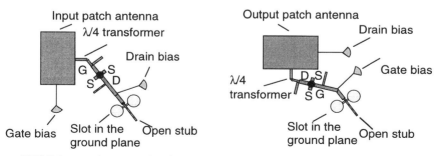

FIGURE 6.8 An example of a unit cell for the double-layer spatial amplifier.

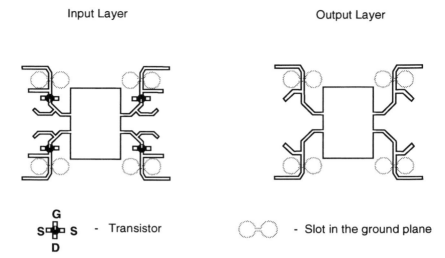

FIGURE 6.9 A unit cell with an increased active device density.

each patch antenna. Figure 6.9 shows the diagram of a unit cell for the double-layer spatial amplifier with a high active device density [6]. Each patch antenna is connected to four devices. Devices are connected to the nonradiating edges of microstrip patch antennas. This type of configuration uses microstrip patch antennas to form a simple-circuit-level power combiner (divider). As before, input and output antennas are isolated by the ground plane. This unit cell forms a single-stage amplifier where the transmitting layer does not contain any active devices. Of course it is obvious that one can add another stage of amplification on the transmitting layer. The gain of this one-stage spatial amplifier was measured in the far field to be 10 dB at 10.1 GHz.

2.5 Near-Field Excitation of Spatial Amplifiers

One of the aspects of the design of spatial amplifiers that requires more research is an efficient method of excitation. The excitation method must efficiently distribute the input power among many unit cells uniformly and collect the amplified power without introducing much loss or having a major impact on the system bandwidth. Furthermore, it should not increase the amplifiers size and weight significantly. In the first demonstration of spatial amplifiers [7,8] a far-field measurement technique was used to determine the gain of such amplifiers. Figures 6.10a and 6.10b show a typical far-field measurement setup to determine the spatial amplifiers' gain. The gain of the spatial amplifier can be calculated using the following equation derived from Friis's transmission formula:

$$G = \frac{P_r}{P_c} \frac{r_1^2 r_2^2}{(r_1 + r_2)^2} \left(\frac{4\pi}{\lambda}\right)^2 G_{\text{in}}^{-1} G_{\text{out}}^{-1} \qquad (6.1)$$

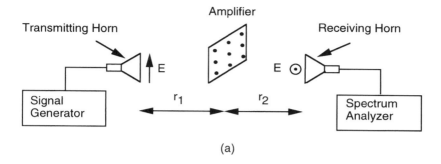

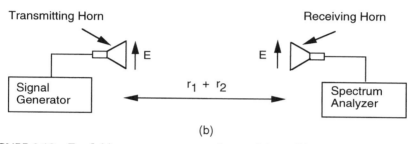

FIGURE 6.10 Far-field measurement setup for spatial amplifier gain determination.

where r_1 and r_2 are the distances from the circuit to the transmitting and the receiving horn antennas, respectively, λ is the free-space wavelength, and G_{in} and G_{out} are the receiving and transmitting microstrip antenna gains as a function of frequency. P_r is the power received by the receiving horn when the two horns are cross-polarized with the circuit placed in the path between the two horns (Fig. 6.10a). P_c is the power received when the two horns are copolarized with the circuit removed from the path (Fig. 6.10b). Even though the far-field measurement technique is capable of predicting the amplifier's gain, due to the spillover losses, it does not represent a practical way of excitation of spatial amplifiers for power generation purposes.

From the system application point of view it is important to have a spatial amplifier with net system gain including any actual internal $1/r^2$ losses from the transmitting horn to the plane of the spatial amplifier and from the spatial amplifier to the receiving horn. In other words, rather than correcting for the spillover losses as was done in the far-field measurement technique, one must be able to excite the spatial amplifier from two clearly defined ports and the gain measurement should represent all the losses that occur between the two ports. Spatial amplifiers should be tested in the same manner as conventional amplifiers with results based on two port measurements. A review of different techniques for excitation of spatial amplifiers is given in reference [9]. One way of doing so is by focusing the millimeter-wave energy in and out of the spatial amplifier through lenses. By using this technique one can dramatically reduce

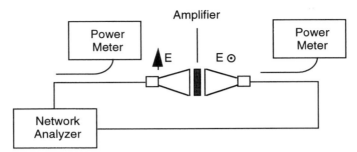

FIGURE 6.11 Near-field measurement setup to form a constrained package amplifier.

the spillover losses. However, one disadvantage of this technique is that it increases the size and weight of the spatial amplifier and complicates the packaging of the entire structure. Another disadvantage of this technique is that it does not excite all the devices across the surface of the spatial amplifier uniformly. In reference 4 it was demonstrated that reactive field coupling (the spatial amplifier was placed near the aperture of horn antennas) can be used to couple energy into and out of a spatial amplifier circuit. The setup shown in Fig. 6.11 was used to eliminate the $1/r^2$ losses associated with the far-field excitation of spatial amplifiers. This method has an inherent advantage over techniques using lenses for focusing since the latter requires a finite separation between the receiving and transmitting horns, the focusing lenses, and the spatial amplifier. Thus the overall system size and weight, using the near-field coupling to horns, are smaller.

The use of regular horn antennas for the excitation of spatial amplifiers introduces a difficulty. This is due to the nonuniformity of the aperture field distribution across the horn antennas. Figure 6.12 shows the power distribution across the aperture of an X-band horn antenna measured using an X-band waveguide (with its flange removed) as it was swept across the horn antenna's aperture. As can be seen, the power drops very rapidly close to the E-walls of the horn antenna. Hence the power received by various unit cells across the surface of the spatial amplifier varies depending on their positions. The unit cells close to the center of the horn antenna receive most of the power while the antennas placed at the edges which are close to the E-walls of the horn would receive very small amounts of power. This causes different cells in the spatial amplifier array to be excited differently, and the cells close to the center of the spatial amplifier circuit saturate faster than those cells close to the edges. This prevents efficient power combining from taking place. In order to alleviate such a problem it is necessary to use horn antennas with a uniform field distribution across their apertures for the excitation of spatial amplifiers.

2.6 Excitation Using Hard Horn Feeds

Recent advances in the theory of hard electromagnetic surfaces have led to the development of feeds with enhanced aperture efficiency [3]. An important

A MULTILAYER AMPLIFIER ARRAY 259

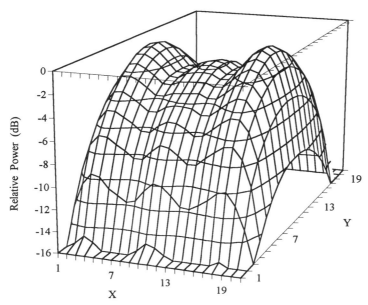

FIGURE 6.12 Measured power distribution across the aperture of an X-band horn antenna.

characteristic of a horn antenna utilizing hard electromagnetic surfaces "hard horn" is that it can support a plane wave. This means that the field distribution over the aperture of the hard horn is uniform, resulting in the maximum available directivity for that aperture. The near-uniform field distribution across the aperture of the hard feed horns allows us to saturate individual amplifiers across the surface of the spatial amplifier simultaneously, thus obtaining an efficient power combination.

Longitudinal metallic corrugations with an appropriate depth filled with dielectric material (Fig. 6.13) approximate the hard boundary conditions [10]. The direction of propagation is denoted as $\hat{k}$. By definition the hard surface

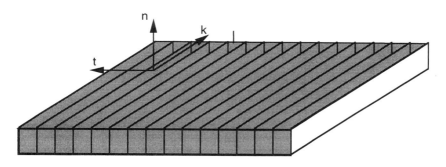

Dielectric Filled Metallic Grooves
FIGURE 6.13 A hard electromagnetic surface.

impedance is $Z_t = \infty$ and $Z_k = 0$. Corrugations act as shorted transmission lines for the transverse polarization, and their depth is chosen such that $Z_t = \infty$ at the surface of corrugations. To obtain $Z_t = \infty$, we must satisfy $k_n d = \pi/2$. The depth of corrugations is given by $d = \lambda/4\sqrt{\epsilon_r - 1}$, where d is the depth of the corrugation and ϵ_r is dielectric constant of the filling material. Several corrugations per unit wavelength are needed. One should note that the bandwidth of the hard surface is proportional to $1/\sqrt{\epsilon_r}$. Therefore in order to obtain a large bandwidth the dielectric constant of the filling material should be small. For example, if the $\epsilon_r = 2$, then the corrguations depth is equal to quarter of the wavelength.

The construction of rectangular hard horn antennas for excitation of spatial amplifiers was reported in reference [11]. Two dielectric slabs with $\epsilon_r = 2.2$ placed on the E-walls of the X-band horn antennas were used to construct the hard horns (Fig. 6.14). Figure 6.15 shows the measured aperture field distribution for an X-band hard horn antenna. The maximum field variation over the 75% of the horn aperture which does not contain dielectric material is ±1 dB. The comparison of the results obtained from regular and hard horn excitation of spatial amplifiers introduced will be given next.

2.7 Spatial Amplifier Measurements

Results obtained from near-field excitation of a nine-cell HEMT spatial amplifier will be presented here. Near-field measurements are performed by employing regular as well as hard horn feeds. It will be shown that the hard horn feeds dramatically improve the power added efficiency for the nine-HEMT spatial amplifier [2].

As mentioned earlier, the far-field results are only a good test of the amplifier performance in the laboratory and do not represent a practical method to excite the spatial amplifiers for system applications. The measurement setup when regular rectangular horn antennas were used to couple the energy to and out of the spatial amplifier circuit was shown in Fig. 6.11. Figure 6.16 shows the measured amplifier's gain versus frequency for this case. The calibration was performed by connecting the flanges of the coax to waveguide couplers together

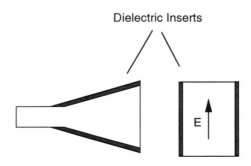

FIGURE 6.14 Construction of a hard horn antenna.

A MULTILAYER AMPLIFIER ARRAY   261

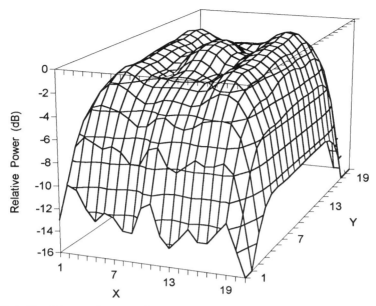

FIGURE 6.15 Measured power distribution across the aperture of X-band hard horn antenna.

(therefore any losses introduced by the feed horns are considered to be a part of the amplifier losses). The maximum small signal gain was 17 dB.

The amplifier's gain compression curve for nearfield measurement is shown in Fig. 6.17. At 2-dB compression the circuits provided 30 mW of output power, which is less than one-third of the power expected from the spatial amplifier. This is due to the nonuniform power distribution across the aperture

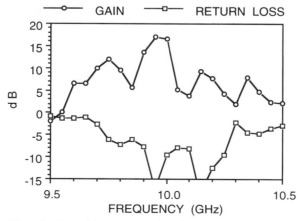

FIGURE 6.16 Plot of gain and input return loss for a nine-cell HEMT spatial amplifier excited with a regular horn feed.

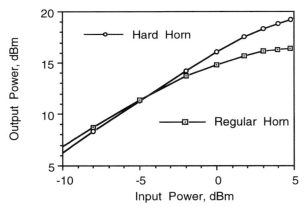

FIGURE 6.17 Plot of gain compression for regular and hard horn excitation of a nine-HEMT spatial amplifier.

of the horn antenna. Only the center column of active devices can be saturated. Furthermore, the collecting horn antenna does not receive much signal from the devices near the E-walls. In order to increase the power combining efficiency it is necessary to saturate all devices simultaneously. Hence, a feed with uniform aperture fields must be used. Next two pyramidal horns employing electromagnetic hard surfaces were designed and constructed. The spatial amplifier performance was measured while placed in this hard horn system. The gain and the return loss of the circuit versus frequency are shown in Fig 6.18. A maximum small signal gain of 16.2 dB at 9.98 GHz was obtained. The gain compression characteristics of the amplifier with regular and hard horn feeds are compared

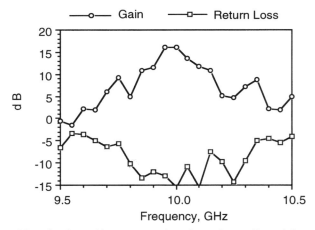

FIGURE 6.18 Plot of gain and input return loss for a nine-cell spatial amplifier excited with a hard horn feed.

TABLE 6.3 Summary of the Results Obtained from a Two-Stage Constrained Package Spatial Amplifier at X-Band

	Parameter	Value
Hard horn feed	Frequency	X-band
	Array size	9 elements
	Power out	90 mW
	Total gain	14.2 dB
	PAE	10.5%
	Drive power	5 dBm
	Horn pair loss including the passive circuit	2.2 dB
Regular horn feed	Power out	30 mW
	Total gain	14.7 dB
	PAE	5.4%
	Drive power	0 dBm
	Horn pair loss including the passive circuit	1.4 dB

in Fig. 6.17. For the hard horn excitation, the output power at 2 dB compression is 90 mW. This indicates almost uniform excitation of the active devices across the spatial amplifier surface. The output power obtained when the circuit was excited with the hard horns was 4.8 dB larger than the output power obtained using regular horns. The insertion loss when no bias was applied to the spatial amplifier was greater than 25 dB for regular as well as hard horn excitations. Table 6.3 summarizes the results obtained from the two-stage constrained package amplifier at X-band. Power-added efficiencies obtained are for the complete amplifier array. It should be noted that the first stage was not saturated. The estimated power-added efficiency for the second stage alone was approximately 25%. It is expected that with improvements in hard horn design and better unit cell configurations the power-added efficiency will approach 40%.

3 MULTILAYER SPATIAL AMPLIFIER ARRAYS

In order to be able to increase the gain and power obtainable from spatial amplifiers, one can construct multilayer spatial amplifiers where many amplifier layers are cascaded. Figure 6.2 showed the general concept for a multilayer spatial amplifier array. Double-layer spatial amplifiers discussed so far are easily amenable to cascading and are used in construction of a three-dimensional array. In reference [4] the cascading of two double-layer amplifier arrays employing near-field coupling between them was demonstrated. One method of cascading the spatial amplifiers is to use the broadside coupled inverted microstrip line couplers to couple one layer to the next. Microstrip patch antennas

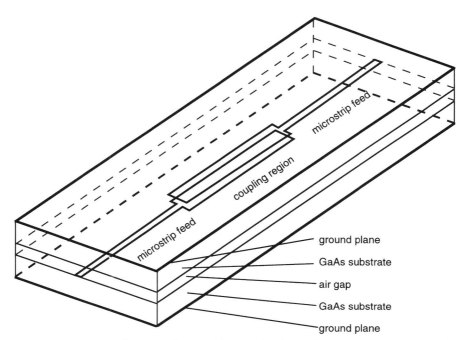

FIGURE 6.19 A wafer to wafer coupling mechanism using a broadside microstrip coupler.

on the transmitting side of the first stage and the receiving patch antennas on the second stage are replaced by broadside inverted microstrip couplers. These couplers have low insertion loss and broadband response at millimeter-wave frequencies which make them attractive for this task.

A detailed diagram of a broadside inverted microstrip coupler is shown in Fig 6.19. Figures 6.20a and 20b show S_{11} and S_{21} for a broadside coupler at millimeter-wave frequencies. These results were obtained from the HP High-Frequency Structure Simulator™ and take into account both the conductor and the dielectric losses. The dielectric material is GaAs and the air gap between the two wafers is 2 mils. The design and construction of multiwafer spatial amplifiers using broadside coupled microstrip lines is currently underway at the University of Central Florida and Lockheed Martin Corporation.

3.1 Monolithic Design of Double-Layer Arrays

Using a monolithic approach for array design has several clear advantages. A monolithic array overcomes unit point failures, demonstrates lower assembly costs and occupies less space than alternative approaches. A hybrid array consisting of MMICs mounted on a monolithic antenna array is another alternative. A hybrid array overcomes unit point failures and also occupies less space;

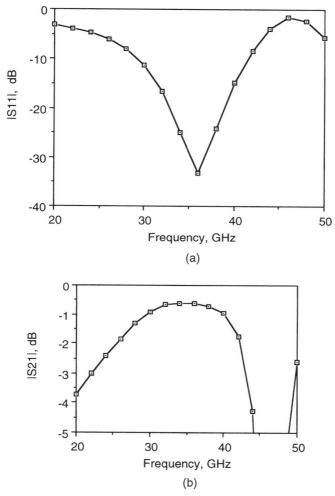

FIGURE 6.20 Simulated performance of a broadside coupler at Ka band (GaAs thickness = 4 mils; coupler width = 5 mils; coupler length = 60 mils; air gap = 2 mils).

however, assembly costs would be higher than a high-yield monolithic array. The hybrid array would, however, be a good choice to use during monolithic array development or for low-production-rate jobs.

Some considerations needed to design a monolithic high power amplifier array include, power, gain, efficiency, stability, yield, size, and heat removal. Throughout the design cycle, trade-offs between these design considerations will need to be made in order to design the array to meet the goals of a given performance specification. Each of these design considerations, and some of the trade-offs between them, will be discussed herein.

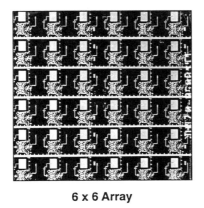
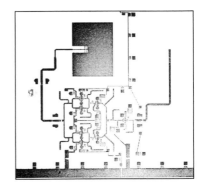

6 x 6 Array

5.8 x 5.8 x 0.1 mm
(0.25 μm PMESFET)

Potential Po$_{6 \times 6}$ = 15 Watts

FIGURE 6.21 A Ka-band monolithic multilayer spatial amplifier array.

An example of a monolithic high-power amplifier array and unit cell are shown in Fig. 6.21. In this example an input patch antenna, multistage power amplifier, and through-wafer coupler are cascaded together to form the unit cell. The unit cells are combined to form the rows and columns of the monolithic array.

High power is achieved by utilizing amplifiers with as much output FET periphery as possible. The FET periphery may be either distributed evenly or concentrated in areas across the array. In the example shown in Fig. 6.21, FET periphery is concentrated in areas across the array. One can address on how much FET periphery to include in the unit cell by considering the trade-off between effective use of GaAs real-estate and processing yield. The cell radiator is usually the driving factor in determining the size of the unit cell. A good approach would be to fill the usable areas unoccupied by the radiator within the unit cell with the highest output-power amplifier possible.

Gain performance goals can be met by cascading multiple amplifier stages within the unit cell. Gain should be increased beyond performance goals to compensate for losses internal to the array, losses external to the array, and losses due to processing variations. Internal array losses include circuit losses and radiator losses. External array losses include losses due to excitation of the array. There are trade-offs between high gain versus stability and high gain versus yield.

Several factors add to the stability of a spatial amplifier including, orthogonally located antennas on the receiving and transmitting layers, orthogonal biasing, via holes surrounding elements within the unit cell, and the use of a thin substrate. Orthogonally locating the antennas helps to reduce mutual coupling [12].

Efficiency is enhanced by optimizing power amplifier design and reducing losses wherever possible. It is critical to reduce losses following the last power amplifier stage. Critical loss areas include transmission line to the antenna and the antenna coupling efficiency to collecting hard horn.

Yield is kept high by using uniform unit cells, removable decoupling networks, multiple make-or-break contact points, and conservative foundry design rules. To achieve array stability it is often necessary to use multiple RF decoupling networks on the DC bias comprised of large, low-yield capacitors. Arrays with faulty (shorted) decoupling networks may be salvaged by using removable air bridges to disconnect the decoupling network from the array. Make-or-break networks can be used to effectively remove faulty areas of the array. Using conservative design rules allows the processing foundry to fabricate the arrays with the least number of defects.

Size is reduced by selecting small, closely spaced radiators and locating power amplifiers in available space: Gate biases may meander to accommodate size. Drain bias should use a wide transmission line following the shortest path possible to minimize voltage drops.

Heat can be removed with via holes, a thinner substrate, distributed FET periphery, a second thermally conductive substrate, and liquid cooling. The method used for heat removal should be decided upon early in the design cycle since the layout of the array can be impacted drastically.

4 SPATIAL POWER-COMBINING OSCILLATORS BASED ON AN EXTENDED RESONANCE TECHNIQUE

Spatial power-combining oscillators have been around longer than spatial amplifiers. Again the idea is to power combine many solid-state oscillators though injection locking. Extensive work has been conducted in this area for the last several years [13–20]. The typical method of combining microwave and millimeter-wave power from many active devices utilizes locally resonant individual active devices to form many individual oscillators. In order to obtain coherent power combining, these individual oscillators are phase-locked together through some strong or weak coupling mechanism. A power-combining oscillator based on an extended resonance was introduced in reference [21]. In reference 22 this technique was expanded to design double-layer spatial power-combining oscillators. In reference [23] a single-layer spatial power-combining oscillator based on an extended resonance technique was described.

An equivalent circuit for a power-combining oscillator employing two-terminal devices is shown in Fig. 6.22. Here many active devices are connected through a network of transmission lines of length L. Active devices are modeled as admittances $-G + jB$, and G_r represents the radiation conductance of antenna connected to each device port. An important feature of this circuit is that the individual devices are not resonant by themselves. The design is such that the

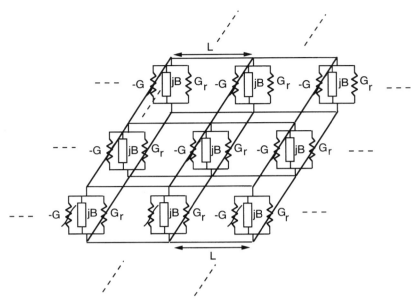

FIGURE 6.22 A two-dimensional spatial power-combining oscillator employing two-terminal devices.

active devices resonate one another through their interconnection via transmission lines. The resonance is achieved by choosing the length of the line L such that one quarter of each device's susceptance is transformed to a susceptance with the same magnitude but the opposite sign ($-jB/4$). The edge devices' extra susceptances can be compensated with open or short circuited stubs of appropriate length.

There are several configurations by which one can incorporate three-terminal devices into extended resonance circuits. One possible circuit topology for a spatial power-combining oscillator using three-terminal devices is shown in Fig. 6.23. Here the gate circuit susceptances are incorporated into the extended resonance structure. A series inductor is used in the source circuit as feedback in order to make the device unstable. G_r, the antenna's radiation conductance, is chosen to satisfy the oscillation condition.

4.1 Unit Cell Design

Depending on the type of feedback and the transistor configuration, the susceptance looking into the gates can be either capacitive or inductive. The admittance chart of Fig. 6.24 demonstrates the resonance condition for each of these cases. If devices have capacitive susceptance, the midway point between the interconnecting transmission lines is virtually short-circuited. For this condition to be satisfied, two adjacent devices must be out of phase. This condition corresponds to an eigenvector excitation of $(1, -1)$, which results in a voltage null at the plane of symmetry. However, if the devices are inductive, the midway point

SPATIAL POWER-COMBINING OSCILLATORS   269

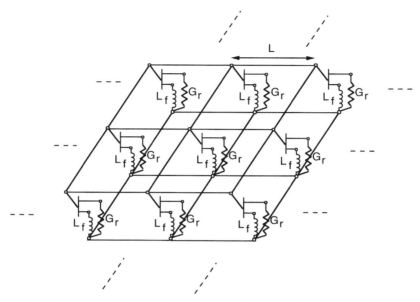

FIGURE 6.23 One possible circuit configuration for a spatial power-combining oscillator employing three-terminal devices.

between the interconnecting transmission lines is virtually open-circuited. This corresponds to an eigenvector excitation of (1,1), which causes a voltage maximum at the plane of symmetry. In order to facilitate the design procedure, one can then identify a unit cell for the circuit. Figure 6.25 shows one possible microstrip realization of the unit cell for the spatial power combiners reported

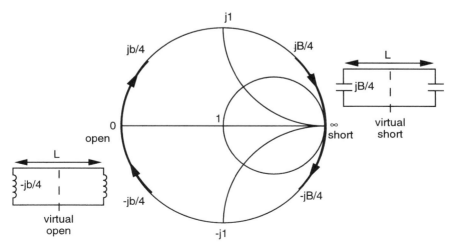

FIGURE 6.24 Admittance chart representing resonance achieved using two inductors or capacitors.

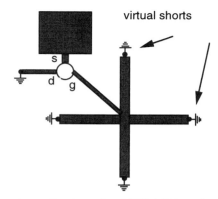

FIGURE 6.25 Microstrip realization of a MESFET-based unit cell for the spatial power-combining oscillator.

in reference [23]. A microstrip line connected to the drain provides the positive feedback to make the transistor unstable.

4.2 Spatial Power-Combining Oscillator Array

Figure 6.26 shows a nine-MESFET spatial power combining oscillator. The circuit was fabricated on Duroid™ substrate with $\epsilon_r = 2.33$ and a thickness of 31 mils and used Fujitsu FSX03LG MESFETs. The measured effective radiated

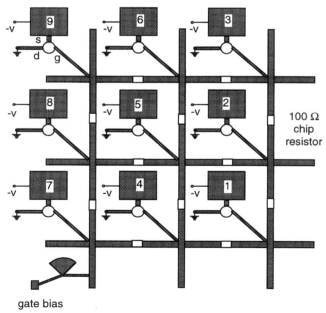

FIGURE 6.26 A nine-MESFET spatial power-combining oscillator.

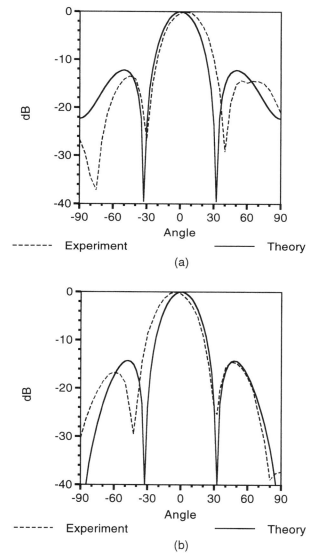

FIGURE 6.27 Predicted and measured E- and H-plane radiation patterns for the nine-MESFET combiner.

power was 4.29 W. Figures 6.27a and 6.27b show the predicted and measured radiation patterns for the nine-device power combiners.

The combiner's sensitivity to single- or multiple-device failures was also reported in reference [23]. In order to perform these experiments, each device was connected through a switch to the power supply, thereby making it possible to turn on and off any individual device in the circuit. Experiments performed verified that any individual device in the nine-device combiner could be turned

TABLE 6.4 Summary of the Results from Switching Off Any Individual Device in the Nine MESFET Power Combining Oscillator

Device Removed	Oscillator Frequency (GHz)	EIRP Drop (dB)
#1	8.036	0.66
#2	8.037	0.91
#3	8.036	1.01
#4	8.035	1.41
#5	8.035	0.84
#6	8.033	1.17
#7	8.039	0.24
#8	8.037	0.82
#9	8.036	1.19

off without disturbing the synchronization. The results are summarized in Table 6.4. It was also reported that any row or column of the nine-device combiner could operate by itself. The summary of the results are given is Table 6.5. In the nine-device combiner, any of the four devices forming a 2×2 combining array can operated by itself. The summary of the experimental results is given in Table 6.6. These results indicate that such circuits are very tolerant of device parameter variations as well as device failures.

Currently, the construction of larger oscillator arrays is underway at the University of Central Florida. Recently a twenty-six-device oscillator array with an EIRP of 19.9 W at 10 GHz was tested.

5 DISCUSSION

In this chapter, several different techniques of constructing multilayer spatial (quasi-optical) array modules were discussed. The major motivation for this work has been its potential to develop low-cost and compact millimeter-wave phased array modules for commercial and military applications.

TABLE 6.5 Summary of the Results from Turning On Any Row or Any Column of the Nine-MESFET Power-Combining Oscillator

Device Number Switched On	Oscillator Frequency (GHz)	EIRP (mW)	Antenna Directivity (dB)	Power/Device (mW)
#1, #2, #3	8.01	325.8	11	8.62
#4, #5, #6	8.03	304.1	11	8.05
#7, #8, #9	8.03	382.8	11	10.14
#3, #6, #9	8.00	382.8	11	10.14
#2, #5, #8	8.03	349.1	11	9.24
#1, #4, #7	8.02	333.4	11	10.83

TABLE 6.6 Summary of the Results Obtained from Turning On Four Adjacent Devices Forming a 2 × 2 Array in the Nine-MESFET Power–Combining Oscillator

Device Number Switched On	Oscillator Frequency (GHz)	EIRP (mW)	Antenna Directivity (dB)	Power/Device (mW)
#4, #5, #7, #8	8.03	961.6	13.24	11.37
#1, #2, #4, #5	8.28	857.0	13.24	10.13
#2, #3, #5, #6	8.28	897.4	13.24	10.62
#5, #6, #8, #9	8.36	897.4	13.24	10.62

Conventional active phased arrays are expensive because relatively few elements can be packaged together. The major cost drivers are the packaging, assembly, and test costs associated with multiple chips per active array element, not the chip costs themselves. Figure 6.28 summarizes the projected cost per element for different frequency bands in active phased arrays versus time. Note that even with significant technical advances and yield improvements, costs on the order of US$150/element or more are projected in the next several years. Additionally, the cost of other components and of integrating the active elements in an array typically doubles the cost of the active array assembly.

These trends taken together indicate that the projected cost of the active array segmant of a system consisting of 500 (22 × 22 array) to 3000 (55 × 55 array) radiating elements will most likely be on the order of US$150,000 to US$1,000,000 as the year 2000 approaches, even with continued improvements in processing and conventional technology. A 22 × 22 array provides a 5-degree beam; a 55 × 55 array provides a 1.5-degree beam. These are array sizes that can easily be related to radar and communications options. Such costs are just too high for very many of these applications. A revolutionary, rather than evolutionary, development is required to provide a cost breakthrough for phased arrays on the order of 5 to 1 at the least.

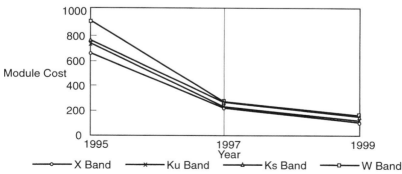

FIGURE 6.28 Active phased arrays projected cost per element for different frequency bands.

Multilayer quasi-optical phased array modules would pack many elements together in a single assembly. Such an array would consist of phase shifter layers cascaded with amplifier layers. The number of modules is compressed by as much as 100:1 compared to conventional phased array modules containing 4 or 8 active array elements each. Cost projections indicate the potential for 3- to 6-fold savings in cost in the millimeter-wave region for such phased arrays. However, quasi-optical array technology requires many developments before its implementation. The amplifiers must be reciprocal, phase shift arrays are required, and separate manifolding systems for various receive and transmit channels may be required. These complications put quasi-optic phased array technology well into the next decade before a practical system can be implemented.

REFERENCES

[1] R. J. Mailloux, "Phased array architecture for millimeter wave active arrays," *IEEE Antennas Propag. Soc. Newsletter,* vol. 28, no. 1, pp. 4–7, Feb. 1986.

[2] T. Ivanov and A. Mortazawi, "A two-stage spatial amplifier with hard horn feeds," *IEEE Microwave Guided Wave Lett.,* vol. 6, no. 2, pp. 88–90, Feb. 1996.

[3] E. Lier and P. S. Kildal, "Soft and hard horn antennas," *IEEE Trans. Antennas Propag.,* vol. 36, no. 8, pp. 1152–1157, Aug. 1988.

[4] T. Ivanov, A. Balasubramaniyan, and A. Mortazawi, "One and two stage spatial amplifiers," *IEEE Trans. Microwave Theory Tech.,* vol. 43, no. 9, pp. 2138–2143, Sept. 1995.

[5] T. Ivanov and A. Mortazawi, "Two stage double layer microstrip spatial amplifiers," *IEEE MTT-S Int. Microwave Symp. Dig.,* pp. 589–592, May 1995.

[6] T. Ivanov and A. Mortazawi, "A double layer microstrip spatial amplifier with increased active device density," *25th Eur. Microwave Conf. Proc.,* pp. 320–322, Sept 1995.

[7] M. Kim *et al.*, "A grid amplifier," *IEEE Microwave Guided Wave Lett.,* vol. 1, no. 11, pp. 322–324, Nov. 1991.

[8] T. Mader *et al.*, "Planar MESFET transmission amplifier," *Electron. Lett.,* vol. 29, no. 19, pp. 1699–1701, Sept. 1993.

[9] R. A. York, "Quasi-Optical Power-Combining Techniques," in *Millimeter and Microwave Engineering for Communications and Radar,* J. Wiltse, ed., vol. CR54, pp. 63–97, SPIE Press, Bellingham, WA, 1994.

[10] P. S. Kildal, "Definition of artificially soft and hard surfaces for electromagnetic waves," *Electron. Lett.,* vol. 24, no. 3, pp. 168–170, Feb. 1988.

[11] T. Ivanov and A. Mortazawi, "Hard horn feeds for quasi-optical amplifiers," *Int. Conf. Infrared and Millimeter Waves, Conf. Digest,* pp. 149–150, Dec. 1995.

[12] J. Hubert, "High-Power Quasi-Optical Ka-Band Amplifier Arrays," *Int. Conf. on Infrared and Millimeter Waves, Conf. Dig,* pp. 145–146, Dec 1995.

[13] J. W. Mink, "Quasi-optical power combining of solid-state millimeter-wave sources," *IEEE Microwave Theory Tech.,* vol. MTT-34, pp. 273–279, Feb. 1986.

[14] K. D. Stephan, "Inter-injection-locked oscillators for power combining and phased arrays," *IEEE Trans. Microwave Theory Tech.,* vol. MTT-34, pp. 1017–1025, Oct. 1986.

[15] Z. Popovic *et al.*, "A 100 MESFET planar grid oscillator," *IEEE Trans. Microwave Theory Tech.,* vol. MTT-39, pp. 193–199, Feb. 1991.

[16] R. A. York and R. C. Compton, "Quasi-optical power combining using mutually synchronized oscillator array," *IEEE Trans. Microwave Theory Tech.,* vol. MTT-39, pp. 1000–1009, June 1991.

[17] J. A. Navaro, Y. H. Shu, and K. Chang, "Broadband electronically tunable planar active radiating elements and spatial power combiners using notch antennas," *IEEE Trans. Microwave Theory Tech.,* vol. MTT-40, pp. 3223–328, Feb. 1992.

[18] H. Kondo *et al.*, "Millimeter and submillimeter-wave quasi-optical oscillator with multi-elements," *IEEE Trans. Microwave Theory Tech.,* vol. MTT-40, pp. 857–863, May 1992.

[19] J. Birkeland and T. Itoh, "A 16-element quasi-optical FET oscillator power combining array with external injection locking," *IEEE Trans. Microwave Theory Tech.,* vol. MTT-40, pp. 475–481, March 1992.

[20] A. Balasubramaniyan and A. Mortazawi, "Two-dimensional MESFET-based spatial power combiners," *Microwave Guided Wave Lett.,* vol. 3, no. 10, pp. 366–368, Oct. 1993.

[21] A. Mortazawi and B.C. De Loach, Jr., "Spatial power combining oscillators based on an extended resonance technique," *IEEE Trans. Microwave Theory Tech.,* vol. MTT. 42, pp. 2222–2228, Dec. 1994.

[22] A. Mortazawi and B. C. De Loach, Jr., "A two dimensional power combining array employing an extended resonance technique," *IEEE Microwave Guided Wave Lett.,* vol. 3, no. 7, pp. 214–216, July 1993.

[23] A. Mortazawi and B. C. De Loach, Jr., "Multiple element oscillators utilizing a new power combining Technique," *IEEE Trans. Microwave Theory Tech.,* vol. MTT. 40, pp. 2397–2402, Dec. 1992.

CHAPTER SEVEN

Planar Quasi-Optical Power Combining

MICHAEL B. STEER
JAMES W. MINK
HUAN-SHENG HWANG
North Carolina State University, Raleigh

1 INTRODUCTION

Conventional power combining makes use of propagating beam modes in free space to combine power from many devices over a distance of many wavelengths. One of the attractive features of these systems are that metallic walls are not required to guide the wave. This is important at millimeter-wave frequencies because conductive metallic losses can become excessive. Much bulk is also avoided by using free space to combine the output of individual active devices rather than using corporate combiners. An alternative to three-dimensional free-space quasi-optical power combining is to confine the beam in two dimensions using the waveguiding structures shown in Fig. 7.1. This structure supports surface modes that are bound by the dielectric in the vertical direction and in the transverse direction support a quasi-optical mode with an essentially Gaussian profile. The lenses are required to periodically refocus the traveling electromagnetic energy. Variations on this principle either have ground planes on the top as well or have no ground planes. The so-called hybrid dielectric slab beam

---

Active and Quasi-Optical Arrays for Solid-State Power Combining, Edited by Robert A. York and Zoya B. Popović.
ISBN 0-471-14614-5 © 1997 John Wiley & Sons, Inc.

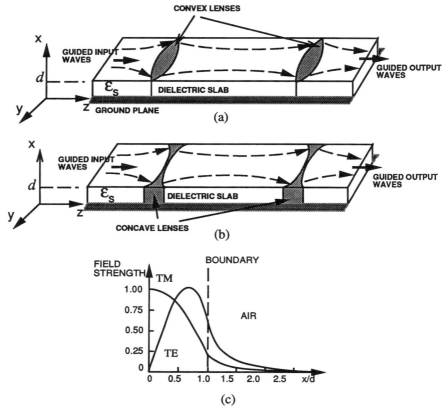

FIGURE 7.1 Two-dimensional slab waveguiding system with (a) convex and (b) concave lenses and (c) the vertical field profile through the slab.

waveguide (HDSBW) bridges the gap between conventional dielectric waveguides used at millimeter-wave frequencies and the slab-type dielectric waveguide used at optical frequencies. In HDSBW power combining, the lateral dimensions are relaxed in just one dimension and passive and active devices can be introduced by selectively metalizing the surface or, utilizing lamination, interior surfaces of the slab.

Both transverse electric (TE) and transverse magnetic (TM) modes are supported with TE mode propagation preferred. In the TE mode the E field is parallel to the ground plane and is compatible with slotline microwave integrated circuit technology. Vivaldi antennas [1] in the conducting plane couple the input and output ports of an active device as shown in Fig. 7.2. The overall view of the amplifier is shown in Fig. 7.2a. It consists of two dielectric sheets separated from each other by a thick highly conducting ground plane. The thick

INTRODUCTION 279

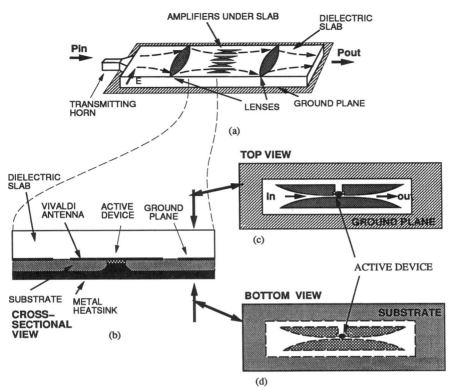

FIGURE 7.2 Two-dimensional QO power-combining amplifier utilizing TE (transverse electric) mode propagation.

ground plane and metal pedestal aid in heat removal. The upper dielectric supports and confines a TE mode with negligible electric field at the ground plane and essentially eliminates metallic losses. In the region of the Vivaldi antennas the transverse electric field drops down from the dielectric slab in which the waves propagate. The Vivaldi antenna gradually narrows the slot and so eventually a large voltage is presented to the input of the device. At the output of the device a large voltage appears across the slotline and a slotline mode propagates toward the end of the antenna where it is lifted up into the dielectric slab. Experimentally it is found that the Vivaldi antenna here provides isolation between the forward traveling and backward traveling waves of at least 20 dB. The phasing of the Vivaldi antennas is adjusted so that the phase delay of the portion of the wave that continues to propagate in the dielectric slab matches (within multiples of 360°) the signal that is amplified. This behavior is similar to that of a traveling wave amplifier described graphically in Fig. 7.3.

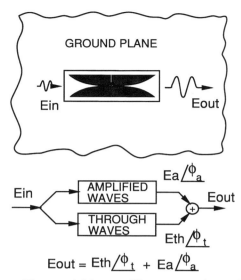

FIGURE 7.3 Wave model for amplifier array underneath the slab.

2 THEORY OF PLANAR QUASI-OPTICAL WAVEGUIDING

The HDSBW in Fig. 7.1 supports TE, with $E_z = 0$, and TM, with $H_z = 0$, slab beam modes, defined with respect to the direction of propagation. The mode families are TE_{nm}, and TM_{nm}; where n and m are the mode indices in the normal (x) and transverse (y) directions. The TM E field and TE H field are respectively derived from the x-directed electric and magnetic vector potentials $\hat{a}_x(\Psi)$ and $\hat{a}_x(\Phi)$ so that the governing equations for the E field are

$$\nabla^2 \Psi + k^2 \Psi = 0 \tag{7.1}$$

$$\vec{E} = \left(\frac{k_0}{k}\right)^2 \vec{\nabla} \times (\vec{\nabla} \times \hat{a}_x \Psi) \tag{7.2}$$

$$\sqrt{\frac{\mu_0}{\epsilon_0}} \vec{H} = jk_0 (\vec{\nabla} \times \hat{a}_x \Psi) \tag{7.3}$$

where Ψ and $(k_0/k)^2 \cdot (\partial \Psi/\partial x)$ are continuous at $x = d$, and $\partial \Psi/\partial x = 0$ at $x = 0$. For the H field, the governing equations are

$$\nabla^2 \Phi + k^2 \Phi = 0 \tag{7.4}$$

$$\vec{E} = -jk_0 (\vec{\nabla} \times \hat{a}_x \Phi) \tag{7.5}$$

$$\sqrt{\frac{\mu_0}{\epsilon_0}} \vec{H} = \vec{\nabla} \times (\vec{\nabla} \times \hat{a}_x \Phi) \tag{7.6}$$

where Φ and $\partial\Phi/\partial x$ are continuous at $x = d$, and $\Phi = 0$ at $x = 0$. The value of k is the wave number in the HDSBW with $k = k_s = \omega\sqrt{\mu_0\epsilon_0\epsilon_s} = k_0\sqrt{\epsilon_s}$.

In the conventional analysis of the fields in a dielectric waveguide, all field components are assumed to be independent of the y coordinate and the TM modes have E_x, H_y, E_y, H_x, E_z components while TE modes have H_x, E_y, H_y, E_x, H_z components. However, in the wavebeam condition the fields are strongly collimated along the z axis but diverge slowly in the y direction. Under this assumption, the field components E_y and H_x of the TM modes, and H_y and E_x of the TE field, are small when compared with other components and so can be neglected.

For the dielectric slab waveguide the TE mode field components can be expressed using orthogonal Gauss–Hermite functions [2]:

$$E_y = -k_0 \sum_{n=0}^{\bar{N}} \sum_{m=0}^{\infty} B_{nm}\beta_n G_n(x) Q_{nm}(y, z) e^{-j\beta_n z} \tag{7.7}$$

$$\sqrt{\frac{\mu_0}{\epsilon_0}} H_x = \sum_{n=0}^{\bar{N}} \sum_{m=0}^{\infty} B_{nm}\beta_n^2 G_n(x) Q_{nm}(y, z) e^{-j\beta_n z} \tag{7.8}$$

$$\sqrt{\frac{\mu_0}{\epsilon_0}} H_z = j \sum_{n=0}^{\bar{N}} \sum_{m=0}^{\infty} B_{nm}\beta_n \frac{dG_n(x)}{dx} Q_{nm}(y, z) e^{-j\beta_n z} \tag{7.9}$$

with

$$G_n(x) = \begin{cases} \sin(\sqrt{k_s^2 - \beta_n^2}\,x) & \text{for } 0 \le x \le d \\ \sin(\sqrt{k_s^2 - \beta_n^2}\,d)\, e^{-\sqrt{\beta_n^2 - k_0^2}(x-d)} & \text{for } d \le x \end{cases} \tag{7.10}$$

$$\bar{N} < \frac{k_{0d}}{\pi}\sqrt{\epsilon_s - 1} - 1/2 \tag{7.11}$$

where B_{nm} is the amplitude of the nmth mode, and β_n is the propagation constant in the waveguide. The value of β_n is determined by the well-known characteristic equation of a conventional dielectric slab waveguide in which fields do not vary in the y direction:

$$\sqrt{\beta_n^2 - k_0^2}\,\tan[\sqrt{k_s^2 - \beta_n^2}\,d] = -\sqrt{k_s^2 - \beta_n^2} \tag{7.12}$$

The function $Q_{nm}(y, z)$ is composed of the Gauss–Hermite function which is here defined as $\text{He}_q(x) = (-1)^q e^{x^2/2}(d^q/dx^q)(e^{-q^2/2})$. Note that there are two definitions for the Gauss–Hermite function and the one we are using has the orthogonal relationship

$$\int_{-\infty}^{\infty} \text{He}_q(Y)\text{He}_r(Y) e^{-Y^2/2}\, dY = \begin{cases} 0 & \text{for } q \ne r \\ \sqrt{2\pi q!} & \text{for } q = r \end{cases} \tag{7.13}$$

$Q_{nm}(y, z)$ implicitly includes the information about the beam size and phase curvature of the guided waves in the slab and is given by

$$Q_{nm}(y, z) = \sqrt{2\pi} \frac{\hat{v}_n}{[1 + \hat{z}_n^2]^{1/4}} \cdot He_m \left\{ \frac{\sqrt{2} \, \hat{v}_n y}{[1 + \hat{z}_n^2]^{1/2}} \right\}$$

$$\cdot \exp\left\{ -\frac{1}{2} \frac{(\hat{v}_n y)^2}{1 + \hat{z}_n^2} - j\left[\frac{1}{2} \frac{(\hat{v}_n^4/\beta_n) z y^2}{1 + \hat{z}_n^2} - \left(m + \frac{1}{2}\right) \tan^{-1}(\hat{z}_n) \right] \right\}$$

(7.14)

where $\hat{z}_n = \hat{v}_n^2 z / \beta_n$. More intuitive expressions for the Gaussian beammodes in the slab are obtained by rearranging $Q_{nm}(y, z)$ so that H_x, E_y, H_z are expressed in terms of the beam width $W_n(z)$ and and phase curvature $R_n(z)$ for the nth beammode. Defining

$$R_n(z) = z(1 + z_{0,n}^2/z^2) \tag{7.15}$$

$$W_n(z) = W_{0,n}(1 + z^2/z_{0,n}^2)^{1/2} \tag{7.16}$$

where $z_{0,n} = \beta_n/v_n^2$ and the beam width at $z = 0$ is

$$W_{0,n} = W_n(z = 0) = \sqrt{2 z_{0,n}/\beta_n} \tag{7.17}$$

Substituting equations (7.15)–(7.17) into equations (7.7)–(7.9) leads to H_x, E_y, H_z expressions that are similar to those for the Gaussian beam modes in a three-dimensional structure. Thus we have

$$E_{y,nm}(x, y, z) = C_{nm} G_n(x) \sqrt{\frac{\beta_n/z_{0,n}}{\frac{W_n(z)}{W_{0,n}}}} \cdot He_m \frac{\sqrt{2 \frac{\beta_n}{z_{0,n}}}}{\frac{W_n(z)}{W_{0,n}}} y$$

$$\cdot \exp\left(-\frac{y^2}{W_n^2(z)}\right) \exp\left(-\frac{j\beta_n y^2}{2 R_n(z)}\right) \exp\left[j\left(m + \frac{1}{2}\right) \tan^{-1}\left(\frac{z}{z_{0,n}}\right) \right]$$

(7.18)

with

$$H_x(x, y, z) = -\frac{\beta_n}{k_0} \frac{E_x}{\sqrt{\epsilon_0/\mu_0}} \tag{7.19}$$

$$H_z(x, y, z) = \frac{-1}{k_0 \sqrt{\epsilon_0/\mu_0}} \frac{1}{G_n(x)} \frac{dG_n(x)}{dx} \tag{7.20}$$

and

$$C_{nm} = -\sqrt{2\pi k_0 \beta_n} B_{nm} \tag{7.21}$$

Similar expressions are obtained for the TM modes [2].

Unlike equation (7.14), equation (7.18) explicitly shows the beam width and phase curvature of the fields in the slab, and the patterns of the fields are easily discerned. The field profiles of the lower-order transverse TE modes are depicted in Fig. 7.4 with the relative field strength measured as the change in the reflection coefficient as a small pyramid of lossy material was moved across the surface of the slab.

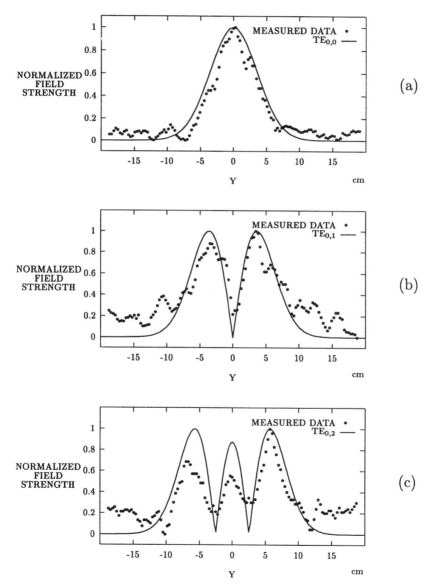

FIGURE 7.4 Measured and calculated field profile of the lowest frequency TE modes: (a) $TE_{0,0}$, (b) $TE_{0,1}$, and (c) $TE_{0,2}$.

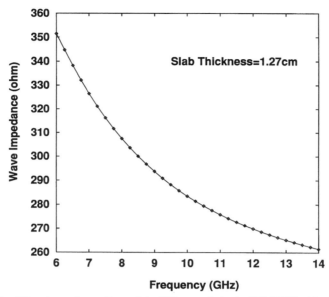

FIGURE 7.5 Wave impedance Z_{TE}, of the $TE_{0,0}$ mode in the HDSBW with $d = 1.27$ cm and $\epsilon_s = 2.57$.

The wave impedance of the nth TE mode is obtained from the propagation constant, β_n. The value of β_n is determined from the nth root of the characteristic equation (7.12). Then the wave impedance of the nth mode in the slab is

$$Z_{TE} = -\frac{E_y}{H_x} = -\frac{k_0}{\sqrt{\epsilon_0/\mu_0}} \beta_n = \omega\mu_0/\beta_n \qquad (7.22)$$

This impedance is plotted in Fig. 7.5 for the lowest-order TE mode, assuming dimensions associated with the experimental planar systems described in the next section.

3 PLANAR QUASI-OPTICAL OSCILLATOR

A planar quasi-optical oscillator is shown in Fig. 7.6 and consists of a curved and a planar reflector [3]. Energy propagates in a quasi-optical mode in a direction perpendicular to the planar reflector which is at the waist of the resonant modes. The curved reflector is circular, approximating the parabolic phase-front of the modes. In this way, energy radiating from one oscillator is coupled into the quasi-optical mode, is reflected by the curved reflector, and illuminates all of the other oscillators, thereby achieving the desired one-to-many coupling. The distance between the planar reflector and the center of the curved reflector is 30.48 cm, the radius of the curved reflector is 60.96 cm, and

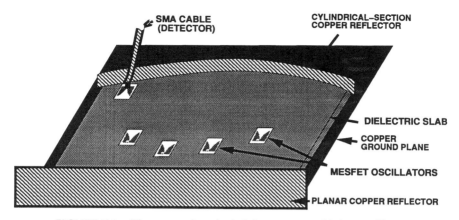

FIGURE 7.6 Planar quasi-optical slab power-combining oscillator.

the thickness and width of the dielectric slab are 1.27 cm and 38.10 cm, respectively. The dielectric is Rexolite ($\epsilon_r = 2.57$, tan $\delta = 0.0006$ at X band). The resonator was characterized in reference [4] using an L-shaped antenna parallel to the ground plane and was inserted into the dielectric through the planar reflector. With this antenna orientation, TE modes are preferentially excited while the excitation of the TM modes is poor. Quality factors, the Q's, of the TE modes are in the range of 1500 to 2250, with a peak Q of 2240 at 11.5 GHz. The published loss tangent of 0.0006 at X band of the Rexolite dielectric [5] corresponds to a Q of 1667 in this frequency range. This is an approximate upper limit on Q of the slab resonator and is modified by additional losses due to the grounded copper plane but by reduced dielectric losses as part of the beam energy is guided outside the slab. With the L-shaped antenna rotated so that it was normal to the ground plane, a TM mode achieves Q's in the range of 150 to 250.

The qth resonant TE_{nm} mode is designated as TE_{nmq}, where q is refers to the number of standing wave patterns along the axis of the resonator. The propagation constant, β_{nq}, of the TE_{0nq} resonant modes of the semiconfocal HDSBW resonator is [2, (27), (29)]

$$\beta_{mq} = \frac{1}{a}\left(q\pi + \left(m + \frac{1}{2}\right)\frac{\pi}{4}\right) \quad (7.23)$$

and this is related to the resonant frequency of the TE_{0nq} mode by [6, (46)]

$$\sqrt{\beta_{mq}^2 - \left(\frac{2\pi}{c}f\right)^2} \tan\left(d\sqrt{\epsilon_r\left(\frac{2\pi}{c}f\right)^2 - \beta_{mq}^2}\right) = -\sqrt{\epsilon_r\left(\frac{2\pi}{c}f\right)^2 - \beta_{mq}^2} \quad (7.24)$$

These equations are numerically solved to obtain resonant frequencies f_{0mq}.

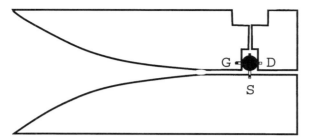

FIGURE 7.7 Single oscillator unit constructed on a Rogers RT/Duroid substrate with $\epsilon_r = 2.33$.

Using a constant relative permittivity of 2.57 [5] for X-band results in a difference between measured and calculated resonant frequencies except in the center of the frequency range, apparently because the relative permittivity, ϵ_r, depends on frequency. The decrease of ϵ_r with increasing frequency [5] yields a relative error in the prediction of the frequencies of resonance of the HDSBW resonator of 0.18% (15 MHz) in the worst case, occurring at the lower limit of X band.

A unit oscillator is shown in Fig. 7.7 and uses a Hewlett Packard ATF-10235 MESFET. The essential element of the oscillator is the end-fire Vivaldi antenna which provides excellent decoupling of forward and backward traveling waves. The design of these oscillating elements was optimized so that oscillation in free space did not occur. This involved optimizing the taper and the drain-gate

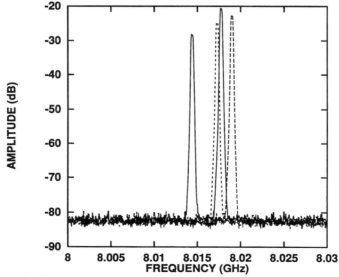

FIGURE 7.8 Oscillator spectrum with, from left to right, 1, 2, 4, and 3 unit oscillators biased.

feedback. Cavity signals were sensed by a Vivaldi antenna on the periphery of the cavity where the fields can be expected to be small. The procedure for establishing the frequency of oscillation is to fix the dimensions of the cavity (which selects a set of possible TE oscillation modes), apply bias to the on-axis oscillator first (which ensures TE_{00} modes), and then apply bias to the other oscillators. In Fig. 7.8 the oscillation behavior with one to four unit oscillators is shown. With four oscillator units the linewidth is < 6 kHz at 30 dB down. Over an extended interval (10s and longer) the center frequency wanders by up to 7 kHz with negligible change in output power level but at all times maintaining the narrow linewidth. Injection locking the power combining oscillator with a signal from a synthesized source 35 dB below the oscillator level reduces the linewidth to < 3 Hz at 30 dB down. The lock-in bandwidth is 350 kHz and the locking bandwidth is 470 kHz. Increasing the power of the injected signal by 3 dB increases the bandwidth to 590 kHz and 700 kHz, respectively.

4 RECTANGULAR WAVEGUIDE TRANSITION

To excite Gaussian beammodes in the slab, an E-plane expansion horn antenna with a dielectric taper inside can be used as shown in Fig. 7.9. The tapered horn conveys the energy from the feeding waveguide to the horn aperture but there is a discontinuity between the horn aperture and slab. Ideal matching requires that the wave impedance and propagation constants match at the interface between the waveguide horn and the dielectric slab.

As the input wave travels from the waveguide to the horn aperture the wave can be viewed as a plane wave in the aperture if the flare angle ϕ of the horn is assumed to be small. That is, if the phase difference, δ', is less than $\lambda/8$, the phase front in the horn aperture can be treated as planar. The reflection coefficient Γ_{TE} at the interface can be obtained by viewing horn and slab as two transmission lines which have different characteristic wave impedances. The

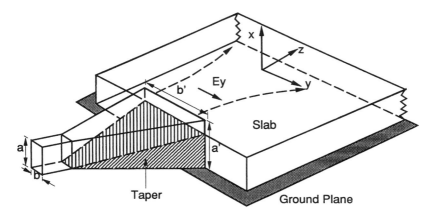

FIGURE 7.9 Rectangular waveguide to HDSBW transition.

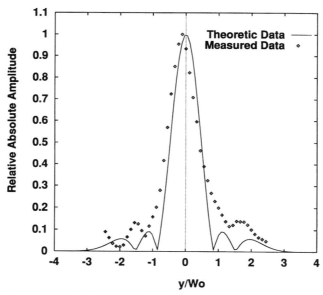

FIGURE 7.10 Calculated and measured fields in the slab system excited by the tapered waveguide transition.

wave impedances of both sides can be obtained from equation (7.22), and the reflection coefficient for TE_{nm} mode becomes

$$\Gamma_{TE} = \frac{Z_{TE,slab} - Z_{TE,horn}}{Z_{TE,slab} + Z_{TE,horn}} = \frac{\beta_{horn} - \beta_{slab}}{\beta_{horn} + \beta_{slab}} \quad (7.25)$$

the field excited in the slab is calculated as $E_{y,slab} = (1 + \Gamma_{TE})E_{y,aperture}$.

The horn excites a number of even-order HDSBW modes with the result that at the interface there is some low-level transverse oscillatory variation to the fields as shown in Fig. 7.10.

5 PLANAR QUASI-OPTICAL AMPLIFIER

An HDSBW system with amplifiers is shown in Fig. 7.2. The HDSBW system uses Rexolite ($\epsilon_r = 2.57$) for the dielectric slab and Macor ($\epsilon_r = 5.9$) for the convex lenses with a radius of 30.48 cm and focal length of 28.54 cm. The slab width is 30 cm. The aperture width of both horn antennas is 9 cm, designed to be the spot size of the slab beam-mode near the aperture. Energy propagates in a TE Gaussian beam mode along the waveguide, passes the lenses, and is refocused in the middle area of the waveguide. Four MESFET amplifiers, each consisting of one MESFET with two Vivaldi antennas (see Fig. 7.2c), are located in this area to amplify the guided energy and are placed underneath the

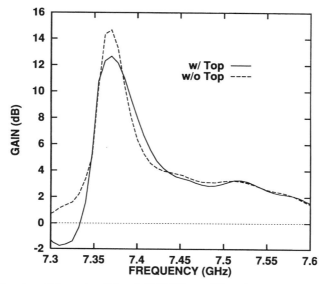

FIGURE 7.11 Amplifier gain of four MESFET amplifiers in the HDSBW system with and without a metallic top cover as $P_{in} = -10$ dBm.

slab so that they do not distort the fields. The design of the amplifiers is discussed in reference [7].

The amplifier gain of the four MESFET amplifiers in the HDSBW system is shown in Fig. 7.11 and the small signal gain was 15 dB and with a metal cover was 12.5 dB, both at a frequency of 7.384 GHz. This gain was measured by taking the ratio of the amplifier's output with the DC bias on and then off. Figure 7.12 shows the gain of the complete system with the amplifiers turned on. The solid line is the reverse transmission $|S_{12}|$ and the dashed line is the forward transmission $|S_{21}|$. $|S_{12}|$ shows that the back scattering to the input port is minimal, and $|S_{21}|$ shows that positive power gain is achieved through the complete system. The input power level was -10 dBm. The loss through the passive HDSBW system is shown in Fig. 7.13. Here the amplifiers are in the system without bias. With the metallic top cover the loss is about 3.5 dB less than without the cover. Sources of loss include input/output mismatch, radiation loss of horns, and insertion loss of amplifiers and lenses which cause field scattering and diffraction.

The saturation characteristic of the amplifier is shown in Fig. 7.14. The measurements were taken at a frequency of 7.384 GHz where the amplifier gain was maximum. Saturation was reached when the input power level was -2.5 dBm for no metallic top and $+2.5$ dBm with the metallic top. The highest power gain (the "real" gain) achieved was 10.5 dB with the top and 8.5 dB without the top. The power-added efficiency is 5.2% with the input power at 5 dBm. With concave lenses the small signal gain increases 2 dB and the power added

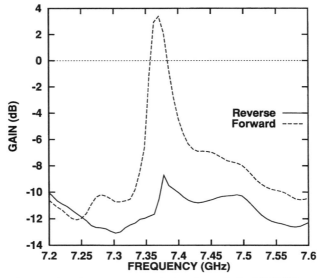

FIGURE 7.12 Forward and reverse transmission through the HDSBW amplifier system at $P_{in} = -10$ dBm.

efficiency is 9%. This efficiency is very low and is largely the result of nonoptimum biasing of the active devices but this is required to ensure stability.

Figure 7.15 shows the power distribution across the top of the HDSBW system for three cases: amplifiers with bias, amplifiers without bias, and no

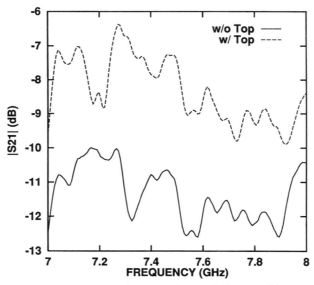

FIGURE 7.13 Measured response of the passive HDSBW amplifier system (no bias) with and without a metallic top cover. Without the cover, the system loss is 10 to 12.5 dB. With the cover, the system loss is 6.5 to 9 dB.

PLANAR QUASI-OPTICAL AMPLIFIER   291

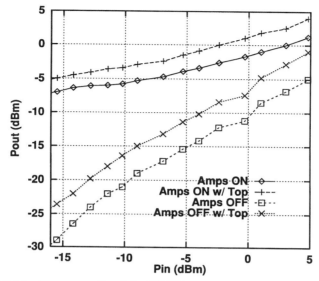

FIGURE 7.14 P_{out} versus P_{in} of the HDSBW amplifier system at 7.389 GHz.

amplifiers in the system. With the amplifiers biased the power distribution spreads out across the slab, whereas with the amplifiers not biased or not in the system the power distribution has the distinctive Gaussian shape. The field distribution is little affected by the ground discontinuities due to the amplifier unit cells. In contrast, with the amplifiers on the top of the slab waveguide as in reference [7], significant scatter is introduced.

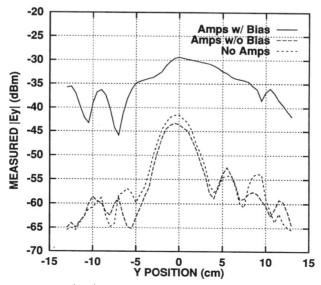

FIGURE 7.15 Measured $|Ey|$ distribution across the top of the HDSBW amplifier system.

ACKNOWLEDGMENTS

The work described here is the result of the efforts of many people. The contributions of J. Harvey, F. K. Schwering, F. Poegel, S. Irrgang, S. Zeisberg, A. Schuenemann, and G. P. Monahan are gratefully acknowledged. The authors also gratefully acknowledge the support of the U.S. Army Research Office through grants DAAL03-89-G-0030 and DAAH04-95-1-0536.

REFERENCES

[1] W. K. Leverich, X.-D. Wu, and K. Chang, "FET active slotline notch antenna for quasi-optical power combining," *IEEE Trans. Microwave Theory Tech.,* vol. MTT-41, pp. 1515–1517, Sept. 1993.

[2] J. W. Mink and F. K. Schwering, "A hybrid dielectric slab-beam waveguide for the sub-millimeter wave region" *IEEE Trans. Microwave Theory Tech.,* pp. 1720–1729, Oct. 1993.

[3] F. Poegel, S. Irrgang, S. Zeisberg. A. Schuenemann, G. P. Monahan, H. Hwang, M. B. Steer, J. W. Mink, and F. K. Schwering, "Demonstration of an oscillating quasi-optical slab power combiner," *1995 IEEE MTT-S Int. Microwave Symp. Din.* pp. 917–920, May 1995.

[4] S. Zeisberg, A. Schunemann, G. P. Monahan, P. L. Heron, M. B. Steer, J. W. Mink, and F. W. Schwering, "Experimental investigation of a quasi-optical slab resonator," *IEEE Microwave Guided Wave Lett.,* vol. 3, Aug. 1993, pp. 253–255.

[5] Robert S. Elliot, *An Introduction to Guided Waves and Microwave Circuits,* Prentice-Hall, Englewood Cliffs, NJ, 1993.

[6] R. E. Collin, *Field Theory of Guided Waves,* McGraw-Hill, New York, 1960.

[7] H. Hwang, G. P. Monahan, M. B. Steer, J. W. Mink, J. Harvey, A. Paollela, and F. K. Schwering, "A dielectric slab waveguide with four planar power amplifiers," *IEEE MTT-S Int. Microwave Symp. Dig.,* pp. 921–924, May 1995.

CHAPTER EIGHT

Grid Oscillators

ZOYA B. POPOVIĆ
University of Colorado, Boulder

WAYNE A. SHIROMA
University of Hawaii, Honolulu

ROBERT M. WEIKLE II
University of Virginia, Charlottesville

1 INTRODUCTION

Chapter 1 outlined the advantages of combining the powers of a large number of solid-state devices in free space, and it discussed how these devices can function as oscillators or amplifiers. Although a large amount of work has recently been done in amplifier combining, the first large-scale, free-space power combiner demonstrated at microwave frequencies was the grid oscillator [1].

1.1 What Are Grid Oscillators?

A grid oscillator is shown schematically in Fig. 8.1, where a metal grating is loaded with either two- or three-terminal solid-state devices. The grating is printed on a dielectric substrate and has a period much smaller than a free-space wavelength. The vertical leads of the grating serve as antennas, and the horizontal leads serve as DC bias lines. If properly designed, the bias lines do not affect the high-frequency performance of the active grid, because the radiated electric field is vertically polarized. A mirror is placed behind the grid to provide the

---

Active and Quasi-Optical Arrays for Solid-State Power Combining, Edited by Robert A. York and Zoya B. Popović.
ISBN 0-471-14614-5 © 1997 John Wiley & Sons, Inc.

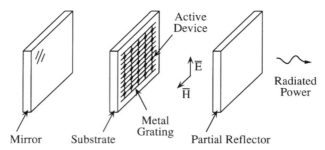

FIGURE 8.1 Schematic of a grid oscillator. A grating loaded with active devices is placed in a Fabry–Perot cavity.

positive feedback necessary for oscillation. A partially transparent reflector is sometimes placed in front of the grid; this configuration is analogous to a Fabry–Perot cavity laser, where the active grid serves as the gain medium.

When DC bias is applied, an oscillation is triggered by transients or noise, and each device oscillates at a different frequency. A noncoherent wave radiates from the grid, reflects off the mirrors, and injection-locks the oscillating devices. At the onset of oscillation, different modes of the cavity compete, as in a laser. Since most grid oscillators use low-Q, unstable Fabry–Perot resonators, the mode with the lowest diffraction loss most likely dominates (Fig. 8.2). After a

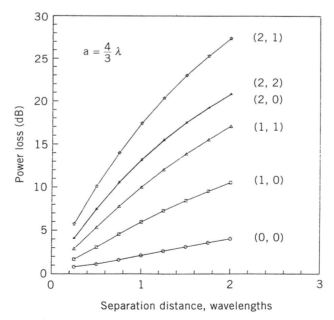

FIGURE 8.2 Diffraction loss per round trip for different modes of a Fabry–Perot cavity with mirrors $\frac{4}{3}\lambda$ on a side for different cavity lengths [38]. The diffraction loss is qualitatively shown by the field-amplitude profiles of the lowest-order (0,0) mode and the higher-order (2,1) mode.

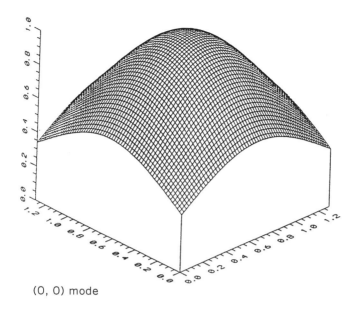

(0, 0) mode

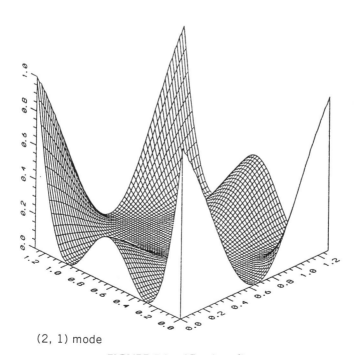

(2, 1) mode

FIGURE 8.2 (*Continued*)

few round trips, the higher-order modes lose most of their power to diffraction, resulting in a single-frequency, self-locked, coherent oscillation.

In some ways the grid oscillator is fundamentally different from the oscillator antenna arrays described in Chapter 4. First of all, since there is no RF ground, there is no guiding mechanism in the plane of the substrate; guiding occurs only in the radiation direction, so the mirror is effectively a ground. This makes it easy for the devices to be unstable, so it is difficult to realize amplifiers in this form. The challenges of grid amplifier design are discussed in the next chapter.

Grid oscillators also differ from oscillator antenna arrays in that the period of the grating is *small compared to a free-space wavelength,* in most cases on the order of $\lambda_0/10$. This means that the antennas connected to each of the active devices are strongly coupled, and standard antenna array theory does not apply. The high spatial density of active devices also gives the possibility of high-power densities and real-estate-efficient power combining. The radiated power grows as n, where n is the number of active devices. Since the period of the grid is small, the grid radiates like an active surface with antenna gain, which means that the power density in the far field grows as the area times n, which is $An = (\sqrt{np})^2 n = p^2 n^2$, where p is the period of the grating. The high power density across the grid also points to a potential problem for high-power oscillators. Namely, since the power is generated in a two-dimensional structure, as the number of devices grows, the radiated power and the heat both grow as the area (e.g., as $(\sqrt{np})^2$ for a square grid of side $\sqrt{np}$). The heatsink can only be placed at the perimeter (which is $4\sqrt{np}$ long for a square), so the relative size of the heatsink decreases as the number of combined devices increases.

Most grid oscillators demonstrated to date have been hybrid, but since they are planar and only one side of the substrate is loaded, they can be fabricated monolithically with single-sided processing. This is especially important at millimeter-wave frequencies, where the grid period is small and a large number of devices must be combined to achieve watt-level power. Grid-oscillator operation is then limited by the yield across a wafer.

1.2 Figures-of-Merit for Grid Oscillators

The ultimate figure-of-merit for free-space power combiners is radiated power. So far, the highest demonstrated radiated power has been 10 W at X band [2]. Since the power from a free-space combiner is not collected in a guiding structure, but instead is radiated into free space, the radiated power must be calculated from far-field antenna measurements. A typical measurement setup is shown in Fig. 8.3, and several figures-of-merit are outlined below.

- *Spectrum.* The spectrum analyzer measures the fundamental oscillation frequency and harmonic content and also verifies the locked behavior of the grid. Self-injection locking occurs for a range of DC biases and mirror positions for most grids demonstrated to date.

INTRODUCTION

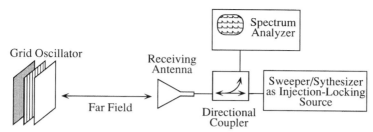

FIGURE 8.3 Typical setup for characterizing a grid oscillator. The sweeper is used as an optional injection-locking source.

- *Effective Radiated Power (ERP).* The power measured on the spectrum analyzer, P_r, is collected by a receiving antenna of gain G_r located a distance r from the grid oscillator. The ERP can be found from Friis's transmission formula:

$$\mathrm{ERP} = P_t G_t = \frac{P_r}{G_r}\left(\frac{4\pi r}{\lambda_0}\right)^2$$

  This is an unambiguous measurement since P_r and G_r can be determined accurately. The ERP, also known as the equivalent isotropic radiated power (EIRP), is the equivalent power that would be radiated by an isotropic source ($G_t = 1$).

- *Radiation Pattern.* The E- and H-plane radiation patterns measured in the far field give an indication of the current distribution across the grids. In most cases, the radiation pattern is similar to that of a uniform current sheet of roughly the same geometric size as the grid. Since the period of the grid is much smaller than the free-space wavelength, this confirms the assumption that all of the oscillators radiate coherently.

- *Polarization.* The radiating leads are oriented in one direction, so all grid oscillators demonstrated to date have been linearly polarized. The level of the cross-polarized component provides useful information about how well the bias lines are isolated from the RF fields, and it is measured by rotating the linearly polarized horn by 90°.

- *Radiated Power.* A full copolar and cross-polar radiation pattern is necessary to calculate the grid directivity so that the radiated power can be backed out from the ERP measurement. This is rarely done, however. In most cases, the directivity and radiated power are estimated from the E- and H-plane radiation patterns. The radiated power is $P_t = \mathrm{ERP}/G_t$.

- *Effective Transmitted Power (ETP).* An alternative figure-of-merit for the radiated power is the effective transmitted power, $\mathrm{ETP} = \mathrm{ERP}/D_t$ [3]. This definition assumes that the grid is a uniformly illuminated aperture, so the directivity is calculated from the physical area of the array, $D_t = 4\pi A_{\mathrm{array}}/\lambda_0^2$.

- *Conversion Efficiency.* The DC-to-RF conversion efficiency is $\eta = P_t/P_{DC}$. Typical conversion efficiencies have been between 10% and 20%.
- *Isotropic Conversion Gain.* An alternative figure-of-merit for the DC-to-RF conversion efficiency is the isotropic conversion efficiency, $\eta_i = ERP/P_{DC} = \eta G_t$ [4]. Since this can be greater than 100%, it is often given in decibels and is called the isotropic conversion gain.
- *Tuning Bandwidth.* The self-locked oscillation frequency can be tuned by changing the bias across the devices or by translating the mirror. Up to 10% tuning bandwidths have been measured, with considerable power variations in the tuning range. It is also possible to tune the frequency by inserting varactor diodes, either in each of the grid unit cells or in a separate loaded grid.
- *Injection-Locking Bandwidth.* The grid can be injection-locked by an external source. This can improve the spectrum in terms of noise and can also be used for frequency modulation [5]. The injection-locking bandwidth is usually measured in terms of the injected signal power and provides information about the Q of the grid.
- *Noise.* As the grid elements lock, their signals add coherently, while the noises add incoherently, assuming they are uncorrelated. This can result in an improvement in the signal-to-noise ratio by a factor of n^2. The reduction of SSB noise in grid oscillators was verified as the number of elements was increased [2].
- *Switching Speed.* In grid oscillators, all of the devices are biased in parallel. This results in a large time constant, which limits the switching speed of the grids. The oscillator switching speed is also limited by the time it takes all of the individual oscillators to lock together. This transient is important for pulsed operation that could potentially lead to high peak powers, and it has not been studied much.
- *Reliability.* One of the potential advantages of grid oscillators is graceful degradation [5]. A systematic study of grid reliability is needed, as well as a figure-of-merit for describing it. For example, it would be useful to know what percentage of devices can fail before the power drops by 3 dB.

2 OVERVIEW

This section summarizes the pioneering work of the earliest grid oscillators. It is important to point out that all of these grids are proof-of-concept prototypes and do not necessarily represent the current state of the art. A good overview of the early transistor-grid oscillators can also be found in reference [6]. Refinements to these original grid designs are described in Sections 4 and 5. Section 2.3 tabulates the important figures-of-merit for all grid oscillators reported in the literature to date.

2.1 Two-Terminal Grid Oscillators

An inductive or capacitive grating lends itself naturally to loading with two-terminal devices. Two-terminal power-generating devices that have been considered as grid loads are Gunn diodes, IMPATT diodes, tunnel diodes, resonant tunneling diodes (RTDs), and Josephson junctions.

One problem with Gunn and IMPATT diodes is their low conversion efficiency which leads to heatsinking problems. Another problem associated with Gunn diodes is that they are inherently resonant and have a relatively high Q of their own. This makes it difficult to lock a large number of mismatched diodes in a low-Q external cavity, such as the ones used in grid oscillators. Limited experiments with up to 100 Gunn diodes on a high-resistivity Si substrate showed evidence of locking and power combining [7]. The output-power dependence on operating temperature was evaluated on a 3×3 X-band grid by measuring radiation patterns for substrate temperatures of 40°C and 50°C at a bias point of 10 V and 2.15 A (Fig. 8.4). These measurements indicate the importance of good heatsinking in grid oscillators.

Mizuno and co-workers [8] used a different type of quasi-optical resonator to combine the output powers of 18 Gunn diodes and 6 MESFETs at X-band. Active devices are placed in a metal grooved mirror that provides good heatsinking and simple biasing (Fig. 8.5a). The output power is coupled through a spherical curved mirror to a waveguide, which reduces the Q of the cavity to about 300. The spacing between elements is $\lambda/2$, with an optimal groove depth of $\lambda/2$ for the diode combiner and $\lambda/4$ for the transistor combiner. The output power was about 20 mW for 15 Gunn diodes. An important property was experimentally confirmed: As more diodes locked to the fundamental cavity

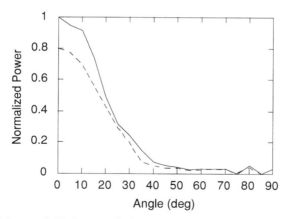

FIGURE 8.4 Measured H-plane radiation patterns of a 9-element Gunn-diode grid oscillator for substrate temperatures of 40°C (solid line) and 50°C (dashed line). The patterns are symmetric about boresight [7].

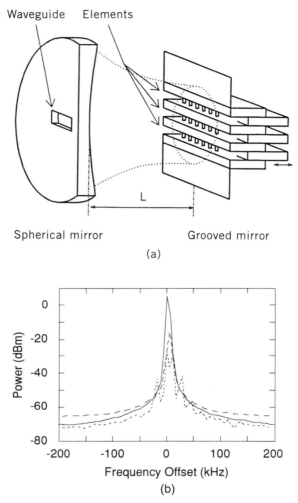

FIGURE 8.5 (a) A quasi-optical combiner using two-terminal devices in a grooved mirror. (b) Output spectrum for a grid with one diode (short-dashed line), three diodes (long-dashed line), and nine diodes (solid line) [8].

mode, the spectrum narrowed and the power increased (Fig. 8.5b). The same type of grooved-mirror resonator was used to combine three Gunn diodes at 50 GHz, giving an output power of 3 mW, 50% more than the power for a single diode [9,10].

Resonant tunneling diodes have been proposed for grid oscillators operating at submillimeter-waves because of their broadband negative resistance. De Lisio *et al.* [11] demonstrated a low-frequency prototype grid using 16 low-power tunnel diodes at L band, yielding an ERP of 1.3 mW and an ETP of 500 μW at 1.86 GHz. At higher millimeter-wave frequencies, Wengler used a grid loaded with Josephson junctions to obtain 0.36 μW at 190 GHz [39] and 0.7 μW at 345 GHz [40].

2.2 Three-Terminal Grid Oscillators

At frequencies up to 100 GHz, three-terminal devices have better conversion efficiency than Gunn diodes. A transistor grid can be loaded in three different ways, depending on which two of the transistor terminals are connected to the radiating grid leads. Figure 8.6 shows two configurations that have been demonstrated to date. In Fig. 8.6a, the fields in the cavity provide direct feedback between the input (gate) and output (drain) of the device, so arrays based on this connection are called "direct-feedback" grids. In contrast, the radiating leads for the grid in Fig. 8.6b are connected to the device's output terminals, and the feedback between input and output occurs through evanescent coupling. Oscillator arrays based on this connection are called "gate-feedback" grids. To distinguish it from the gate-feedback configuration, the direct-feedback grid has also been termed "source-feedback" [12].

Using an equivalent circuit for the grid, De Lisio [13] analyzed the impedance that a device embedded in a grid presents to the cavity fields. For the same transistor and grid, the direct-feedback connection results in a negative resistance over a wider frequency range than the gate-feedback connection (Fig. 8.7). This implies that the direct-feedback grid is capable of oscillating over a broad band, from DC to beyond 12 GHz, except near 9.5 GHz. In the gate-feedback configuration, the opposite is true. The sign reversal in the resistance at 9.5 GHz is related to the resonance of the L-C circuit in the grid equivalent circuit, which is described in Section 3.1.

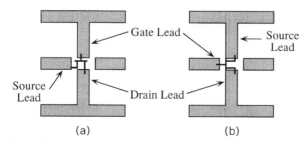

FIGURE 8.6 Two different ways to connect a transistor in an inductive unit cell: (a) The direct-feedback grid, in which the gate and drain leads radiate, and (b) the gate-feedback grid, in which the drain and source leads radiate.

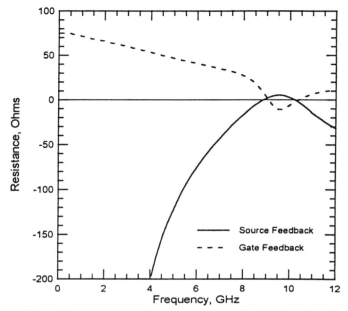

FIGURE 8.7 The resistance presented to a wave in the feedback cavity of a transistor grid oscillator, for the two different configurations of Fig. 8.6 [13].

Direct-Feedback Grid Oscillators Figure 8.8 shows the first transistor-grid oscillator, which was also the first working grid oscillator [1]. The gate and drain leads of the MESFETs are connected to the vertical leads of the grid, and the source leads are connected to the metallized backside of the substrate through vias. The drain bias is provided through the horizontal leads of the grid, and the gates are self-biased. The gate leads act as receiving antennas, and the induced voltage is amplified and re-radiated by the drain leads. This positive feedback is facilitated by a dielectric slab placed in front of the grid. Since the currents radiate in the vertical direction, very little power is radiated in the horizontal polarization. The cross-polarized signal was 20 dB below the copolar at boresight. This grid delivered an ERP of 20.7 W at 9.7 GHz. The power radiated from the grid was 474 mW, which is about 25 times the power typically measured for a single-device microstrip oscillator. Based on a two-dimensional pattern measurement, the directivity was found to be 16.4 dB, which is close to the theoretical directivity of 25 short current elements on an infinite grounded dielectric with a dielectric cover above. When the slab is translated from 4.8 to 7.2 mm away from the grid (a $0.2\lambda_0$ distance), the frequency changed by 1% and the power changed by more than 30 dB. Tilting the slab resulted in beam steering of up to 15% in the far field.

Due to the thin metallized substrate of this grid oscillator, it is possible that microstrip modes that coupled the devices through the source leads were excited. To avoid this problem, Popović *et al.* [5] designed a bar-grid oscillator that

OVERVIEW 303

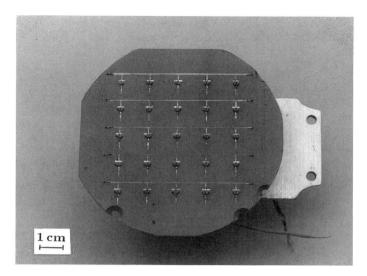

FIGURE 8.8 The first transistor-grid oscillator [1]. Twenty-five MESFETs load a grid printed on 2.35-mm-thick Duroid with $\epsilon_r = 10.5$. The period is $0.4\lambda_0$, or about one guided wavelength at 9.7 GHz.

eliminated the dielectric substrate; the active devices were soldered directly onto metal bars that act simultaneously as bias lines and heatsinks (Fig. 8.9). Adjacent rows of transistors share drain and gate bias bars to minimize the number of bias lines. Each unit cell contains two TEM guides, with the gate and drain leads each feeding a different guide. The bar-grid design has two other advantages over the first grid oscillator: All three-device terminals are placed on a common plane, eliminating the source via, and the gates are no longer self-biased, providing the opportunity for gate-bias tuning.

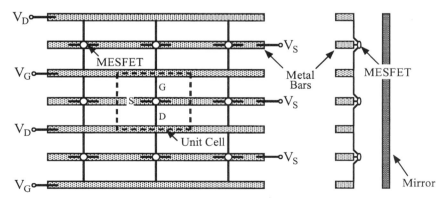

FIGURE 8.9 The bar-grid oscillator [5]. Thirty-six MESFETs are soldered directly onto metal bars, which form short TEM-waveguide sections. The period is $0.6\lambda_0$ at 3.3 GHz.

304 GRID OSCILLATORS

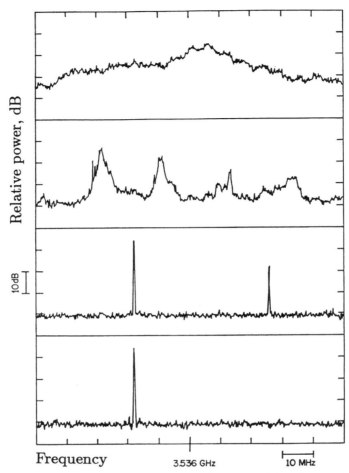

FIGURE 8.10 Self injection-locking sequence of the bar-grid oscillator. (From reference [5], © 1990 IEEE.)

The grid oscillated at 3 GHz, in agreement with the frequency predicted from an equivalent circuit model for a grid unit cell [5]. With two dielectric slabs placed in front of the grid for impedance matching, an ERP of 3W was measured. The directivity was 11.3 dB, based on a two-dimensional pattern measurement. The DC-to-RF conversion efficiency was 22% and the isotropic conversion gain was 4.8 dB. The frequency was tuned with mirror position over a 10% bandwidth, and the gate voltage modulated the frequency by about 250 MHz/V. Once warmed up, this grid operated for several days with essentially no change in power or frequency, and it was stable to within 2 MHz in a 24-hour period. Reliability was also investigated by removing several devices randomly across the grid. When one device was removed, the power dropped by 1-2 dB, depending on the device location. Up to six devices (16%) could be removed while maintaining locked oscillation, but the power dropped as much as 8.5 dB.

The grid has an interesting locking sequence that gives insight into the locking mechanism (Fig. 8.10). Successive spectra are recorded as the gate bias is varied over a 1-V range. The top figure shows the unlocked grid spectrum. Locking starts in the next figure, where different bumps are associated with different grid rows, narrowing as the locking continues. Mixing products are generated due to the competing modes, and locking finally occurs when they all lock to a single mode. This sequence is similar to injection-locking, except that the different modes lock to one of the oscillator modes itself, rather than to an external signal.

The next important development was a 100-element planar grid oscillator [14], shown in Fig. 8.11. The design is similar to the bar-grid oscillator, except that the bars are effectively thinned to printed strips on a dielectric substrate. Since all three transistor leads are on one side of the substrate, this structure can be fabricated monolithically with no backside processing. Figure 8.12 shows the measured radiation pattern, which corresponds to a directivity of 16 dB. This grid produced an ERP of 22W at 5 GHz, with a DC-to-RF conversion efficiency of 20% and an isotropic conversion gain of 8.5 dB. This is an improvement over previous designs since the grid has higher directivity with similar efficiency.

Gate-Feedback Grid Oscillators In the planar grid oscillator discussed above, the MESFET gate lead is parallel to the radiated electric field. Consequently, the gate and radiated field are strongly coupled. Although it provides the direct

FIGURE 8.11 The 100-MESFET planar grid oscillator. The grid is printed on a 0.5-mm-thick substrate with $\epsilon_r = 2.2$ and is placed on top of a 2.5-mm-thick substrate with $\epsilon_r = 10.5$. The period is 8 mm, or about $\lambda_0/8$ at 5 GHz. The ferrite beads at the bias lines act as chokes for lower frequencies. (From reference [14], © 1991 IEEE.)

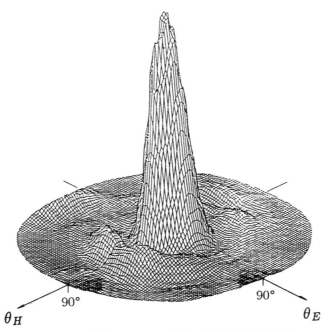

FIGURE 8.12 Measured radiation pattern of the 100-MESFET planar grid oscillator [14]. The vertical scale is linear in power, and the axes are given in spherical coordinates by $\theta_E = \theta \sin \phi$ and $\theta_H = \theta \cos \phi$. (From reference [14], © 1991 IEEE.)

feedback path necessary for oscillation, this configuration may also have drawbacks. Because the output is directly coupled to the MESFET gate, the external circuit provided by the grid may not be a sensitive design parameter in controlling oscillation frequency; the MESFET parasitics and transconductance may be the dominant factors. In addition, the feedback network necessary for oscillation and the output matching needed for optimum power generation are not independent. As a result, it is unlikely that an optimum output impedance can be presented to the grid at the frequency of interest using this configuration.

In the gate-feedback grid of Fig. 8.6b, the gate of the device connects to the horizontal lead which is orthogonal to the radiated electric field. Because the gate and output are no longer directly coupled, the oscillation frequency is more sensitive to the grid structure and the feedback impedance it presents to the gate. The vertical leads connected to source and drain no longer play a major role in the feedback circuit and thus may be tailored primarily for maximum output power.

Several gate-feedback grid oscillators have been demonstrated to date. The first two were proof-of-concept grids, which used Fujitsu MESFET chips (FSC11X) soldered and wire-bonded to grids printed on Duroid substrates [12]. A 4 × 4 X-band oscillator with a period of 9 mm produced 335 mW at 11.6 GHz

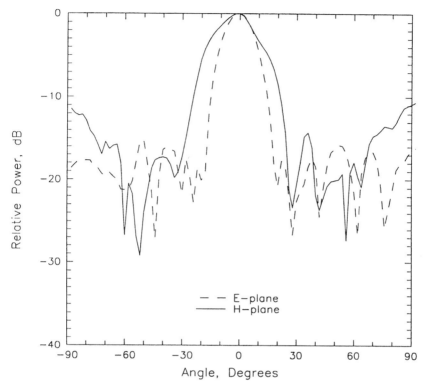

FIGURE 8.13 Measured radiation patterns of the X-band gate-feedback oscillator grid [12].

with a DC-to-RF conversion efficiency of 20%. The radiation patterns are shown in Fig. 8.13. By reducing the period to 5 mm, a 36-element array using the same chip FETs was scaled for Ku band. This grid produced 235 mW at 17 GHz with a conversion efficiency of 7%.

Other gate-feedback grid oscillators include a monolithic heterojunction bipolar transistor (HBT) grid and a 100-element power grid. The HBT grid consisted of 36 devices fabricated on a GaAs substrate, oscillated at 35 GHz, and was the first reported monolithic oscillator grid [15]. The 100-element power grid was designed for maximum output power using a simple, linear load-line analysis which is described in Section 4. This grid produced an output power of 10.3 W at 9.8 GHz with a DC-to-RF conversion efficiency of 23% [2]. To date, this is the highest power produced by a planar transistor-grid oscillator.

2.3 Comparison of Reported Grids

Table 8.1 summarizes all grid oscillators reported in the literature to date. The table is divided into four categories: (a) the prototype grids, (b) grids designed

TABLE 8.1 Comparison of Reported Grid Oscillators

Description	Configuration	Feedback Type	Frequency (GHz)	ERP (W)	Power (mW)	ETP (mW)	Efficiency (%)	Directivity (dB)
First transistor grid oscillator (Popović et al., 1988 [1])	5 × 5 MESFET	Direct	9.7	20.7	474	373	15	16.4
Bar-grid oscillator (Popović et al., 1990 [5])	6 × 6 MESFET	Direct	3.1	3.0	220	621	22	11.3
Planar-grid oscillator (Popović et al., 1991 [14])	10 × 10 MESFET	Direct	5.0	22.0	550	985	20	16.0
Grooved-mirror oscillator (Nakayama et al., 1990 [8])	3 × 2 MESFET	Direct	12.4	—	0.4	—	—	—
Grooved-mirror oscillator (Nakayama et al., 1990 [8])	3 × 3 Gunn diode 5 × 3 Gunn diode	— —	10.2 —	— —	3 20	— —	— —	— —
X-band oscillator (Weikle et al., 1992 [12])	4 × 4 MESFET	Gate	11.6	15.0	335	618	20	16.5
Ku-band oscillator (Weikle et al., 1992 [12])	6 × 6 MESFET	Gate	17.0	3.3	235	91	7	11.5
Ka-band monolithic grid (Kim et al., 1992 [15])	6 × 6 HBT	Base	34.7	0.2	—	28	—	—
Ka-band monolithic grid (Harris et al., 1993 [16])	3 × 2 PHEMT	Gate	37.0	—	5	—	—	—
Ka-band monolithic grid (Unpublished, 1995 [17])	10 × 10 PHEMT	Direct	31.1	—	—	—	—	—

Design (Reference)	Array	Coupling	Frequency (GHz)					
Dipole/bow-tie oscillator (Bundy et al., 1994 [18])	6 × 4 PHEMT	Direct	10.0	0.3	—	54	—	—
Grooved-mirror oscillator (Kondo et al., 1992 [9,10])	3 × 1 Gunn diode	—	55.8	—	3	—	—	—
Tunnel-diode oscillator (De Lisio et al., 1995 [11])	4 × 4 Tunnel diode	—	1.9	0.0013	—	0.52	—	—
Voltage-controlled oscillator (Bundy et al., 1992 [19,20])	6 × 4 MESFET	Direct	4.9 (10% BW)	—	—	—	—	—
Voltage-controlled oscillator (Oak et al., 1995 [21])	4 × 4 MESFET	Gate	12.4 (2% BW)	—	—	—	—	—
Grid amplifier with external feedback (Kim et al., 1993 [22])	10 × 10 HBT	—	6.5–11.5	6.3	—	72	—	—
Dual-frequency oscillator (Shiroma et al., 1995 [23])	5 × 5 PHEMT	Direct Direct	3.9 6.2	0.3 0.3	— —	74 30	— —	— —
Beam-steering oscillator (Li et al., 1994 [24])	4 × 2 MESFET	Gate	11.0	0.2	—	12	—	Scans −6.5° to +5°
Power oscillator (Hacker et al., 1994 [2])	10 × 10 MESFET	Gate	9.8	660	10300	8750	23	18.0
Three-dimensional grid (Shiroma et al., 1994 [25])	5 × 5 × 1 PHEMT 5 × 5 × 2 PHEMT 5 × 5 × 4 PHEMT	Direct Direct Direct	4.9 5.0 5.2	1.8 2.2 8.0	97 152 265	330 388 1350	15 10 12	12.6 11.7 14.8
Dipole/slot oscillator (Shiroma, 1996 [26])	4 × 4 MESFET	Direct	8.6	0.5	90	82	17	7.4

310 GRID OSCILLATORS

for higher frequencies, (c) grids designed for added capabilities, and (d) grids designed for optimizing output power.

3 ANALYSIS TECHNIQUES

A rigorous analysis of the grid oscillator would require an understanding of the complicated interaction between the semiconductor devices, the metal grating, and the radiated and evanescent fields. Determining the driving-point impedance presented to the device at each feed point of the array is thus a nontrivial task. The mutual coupling between the array elements and the unknown behavior of the fields at the edges of the grid make the problem difficult. Even grids of moderate size would generate a large number of unknowns. To make the problem more tractable, it is commonly assumed that the grid is infinite in extent.

If we assume that the devices in each cell in Fig. 8.14 are identical and oscillate at the same frequency and in phase (as they should for a power-combining array), then the radiated fields must satisfy boundary conditions imposed by the array's symmetry. Symmetry planes that run horizontally along the bias lines can be replaced with electric walls. This boundary condition arises because identical currents flow above and below these planes, resulting in a cancellation of the tangential electric field. By a similar argument, the vertical

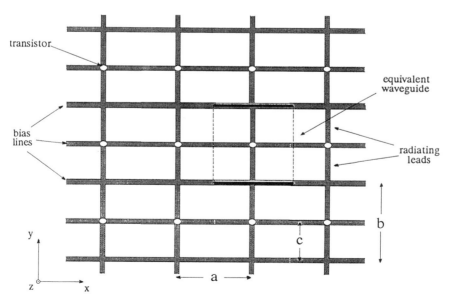

FIGURE 8.14 Diagram of a crossed-dipole array showing planes of symmetry. An active device lies at the center of each unit cell. Horizontal symmetry planes are replaced by electric walls (solid lines) while vertical symmetry planes are replaced by magnetic walls (dashed lines).

symmetry planes lying between adjacent devices can be replaced by magnetic walls. In this way, an infinite array of phase-locked sources can be represented by an equivalent waveguide that has electric walls on the top and bottom and magnetic walls on the sides. In essence, this model allows us to find the coupling between a uniform plane wave and the devices in the array.

Two distinct analysis techniques using the infinite-grid approximation are discussed here, the induced EMF method in Section 3.1 and a full-wave technique in Section 3.2. Experimental results for several grid oscillators designed by these methods follow in Section 3.3.

3.1 The Induced EMF Method for Planar Grids

The induced electromotive force (EMF) method was originally used to calculate the self-impedance of various antennas [27,28]. Eisenhart and Khan [29] extended this method to waveguides to calculate the driving-point impedance of a waveguide mounting structure. The EMF method is primarily used for simple configurations for which the current distribution on the radiating structure can be postulated with reasonable accuracy. In essence, the EMF method is an application of Poynting's Theorem and the conservation of complex power.

A current source placed at the feed point of an antenna delivers a total power of $P = Z_{in}|I_0|^2$, where Z_{in} is the complex (driving-point) impedance seen by the source. According to Poynting's Theorem, the current distribution on the antenna will radiate a total power of

$$-\int_V \vec{E} \cdot \vec{J}^* \, dV = \oint_S \vec{E} \times \vec{H}^* \cdot d\vec{s} + j\omega \int_V (\vec{H}^* \cdot \vec{B} - \vec{E} \cdot \vec{D}^*) \, dV$$

The surface integral represents power radiated by the antenna, and the volume integral on the right represents reactive power in the near field. Equating the above expressions allows the driving-point impedance to be determined:

$$Z_{in} = -\frac{1}{|I_0|^2} \int_V \vec{E} \cdot \vec{J}^* \, dV \qquad (8.1)$$

Modal Expansion for the Equivalent Waveguide The induced EMF method is well suited to analyzing grid structures with simple geometries, such as the one shown in Fig. 8.14. Given an assumed current distribution, the EMF integral (8.1) can be evaluated to find the driving-point impedance seen by a device embedded in the array. Currents on the grid metallization radiate into the equivalent waveguide and couple to the TEM, TM, and TE modes supported by that guide. The transverse electric field in the equivalent waveguide can be expressed as a superposition of these waveguide modes:

$$\vec{E} = e^{TEM}\hat{w}^{TEM} + \sum_{m,n} e^{TM}_{mn}\hat{u}^{TM}_{mn} + \sum_{m,n} e^{TE}_{mn}\hat{v}^{TE}_{mn}$$

where e^{TEM}, e_{mn}^{TM}, and e_{mn}^{TE} are expansion coefficients to be determined. The vectors $\hat{w}^{TEM}$, $\hat{u}_{mn}^{TM}$, and $\hat{v}_{mn}^{TE}$ are basis vectors representing the transverse TEM, TM, and TE modes, respectively [30].

The impressed current distribution on the grid metallization (at $z = 0$) can be expanded in a similar fashion,

$$\vec{J} = j^{TEM}\hat{w}^{TEM} + \sum_{m,n} j_{mn}^{TM}\hat{u}_{mn}^{TM} + \sum_{m,n} j_{mn}^{TE}\hat{v}_{mn}^{TE}$$

The expansion coefficients, j^{TEM}, j_{mn}^{TM}, and j_{mn}^{TE} are determined by *assuming* a current distribution and evaluating the inner products, $\langle \vec{J} \mid \hat{u}_{mn}^i \rangle$. The expansion coefficients for $\vec{E}$ and $\vec{J}$ are related by Maxwell's boundary conditions. The discontinuity in $\vec{H}$ at $z = 0$ is given by $\hat{z} \times (\vec{H}^+ - \vec{H}^-) = \vec{J}$, where the superscripts $+$ and $-$ refer, respectively, to the fields on the positive and negative sides of the surface over which the impressed current flows. In addition, the transverse components of $\vec{E}$ and $\vec{H}$ for a particular waveguide mode are related by that mode's intrinsic wave impedance. These relations allow us to write

$$e^{TEM} = -j^{TEM}(Z^{TEM+} \parallel Z^{TEM-})$$
$$e_{mn}^{TM} = -j_{mn}^{TM}(Z_{mn}^{TM+} \parallel Z_{mn}^{TM-})$$
$$e_{mn}^{TM} = -j_{mn}^{TE}(Z_{mn}^{TE+} \parallel Z_{mn}^{TE-})$$

where "$+$" and "$-$" refer to the intrinsic wave impedances looking in the positive and negative z directions, respectively, and "$\parallel$" indicates that the mode impedances are to be added in parallel. The waveguide mode impedances Z^{TEM}, Z_{mn}^{TM}, and Z_{mn}^{TE} are the standard ones given in reference [31].

Using the EMF integral (8.1) with the modal expansions, we obtain the following expression for the driving-point impedance seen by a device embedded in the array:

$$Z = \frac{1}{|I_0|^2}\left\{(Z^{TEM+} \parallel Z^{TEM-})\mid j^{TEM}\mid^2 + \sum_{m,n}\mid j_{mn}^{TM}\mid^2(Z_{mn}^{TM+} \parallel Z_{mn}^{TM-}) \right.$$
$$\left. + \sum_{m,n}\mid j_{mn}^{TE}\mid^2(Z_{mn}^{TE+} \parallel Z_{mn}^{TE-})\right\} \quad (8.2)$$

To find the driving-point impedance of a particular grid structure, a specific current distribution is assumed. From this assumed current distribution, the current expansion coefficients are determined and (8.2) gives the embedding impedance.

Planar Grid Model The basic configuration of the planar transistor grid is shown in Fig.8.14. To apply the EMF method, a simple, piecewise continuous current distribution is assumed. Currents flowing in the vertical direction are assumed to be uniform along the vertical leads. In addition, the currents must vanish at the sides of the unit cell where there is a magnetic wall. It is further assumed that all currents are uniformly distributed across the widths of the leads. Using the assumed current distributions and the field basis vectors $\hat{w}^{\text{TEM}}$, $\hat{u}_{mn}^{\text{TM}}$, and $\hat{v}_{mn}^{\text{TE}}$, we find the two-port network representing the grid to have the form shown in Fig. 8.15a. The impedance terms for this network are given by the modal sums,

$$Z_0 = \frac{b}{a}(Z^{\text{TEM}+} \parallel Z^{\text{TEM}-}) \tag{8.3a}$$

$$Z_L = \frac{2b}{a} \sum_{m=1}^{\infty} \cos^2\left(\frac{m\pi}{2}\right) \text{sinc}^2\left(\frac{m\pi w}{2a}\right)(Z_{m0}^{\text{TE}+} \parallel Z_{m0}^{\text{TE}-}) \tag{8.3b}$$

$$Z_C = \frac{2}{ab} \sum_{n=1}^{\infty} \frac{1}{k_y^2} \sin^2\left(\frac{n\pi c}{b}\right) \text{sinc}^2\left(\frac{n\pi w_s}{2b}\right)(Z_{0n}^{\text{TM}+} \parallel Z_{0n}^{\text{TM}-}) \tag{8.3c}$$

$$Z = \frac{4}{ab} \sum_{m,n>0}^{\infty} \frac{1}{k_c^2} \cos^2\left(\frac{m\pi}{2}\right) \sin^2\left(\frac{n\pi c}{b}\right) \text{sinc}^2\left(\frac{m\pi w}{2a}\right) \text{sinc}^2\left(\frac{n\pi w_s}{2b}\right)$$
$$\times \left\{ \left(1 - \frac{a}{a-w}\right)^2 Z_{mn}^{\text{TM}+} \parallel Z_{mn}^{\text{TM}-} + \left(\frac{k_x}{k_y} + \frac{k_y a}{k_x(a-w)}\right)^2 Z_{mn}^{\text{TE}+} \parallel Z_{mn}^{\text{TE}-} \right\} \tag{8.3d}$$

where $\text{sinc}(x) = \sin(x)/x$ and $r = b/c$. In Fig. 8.15b, the "T" model for the grid equivalent network is rearranged to incorporate a center-tapped transformer. Using this model, the impedance terms (8.3) can be related to the grid structure in a simple manner. Z_0 is a parallel combination of TEM impedances and

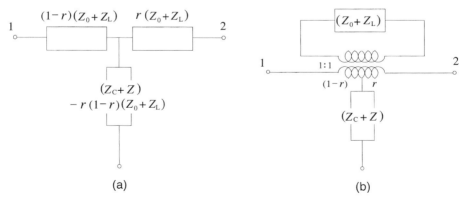

(a) (b)

FIGURE 8.15 (a) Two-port equivalent network for the crossed-dipole grid. (b) Two-port transistor grid model with a center-tapped transformer. (c) Equivalent transmission-line model for the planar transistor grid. Propagating TEM modes are represented with transmission lines and evanescent modes with reactive circuit elements. The transistor shown is connected in the "gate-feedback" configuration. (*Continued*)

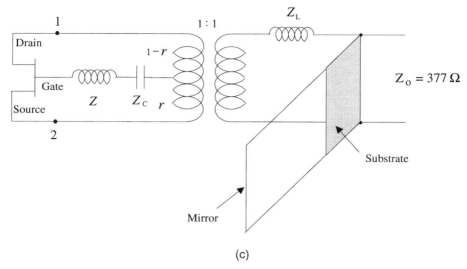

FIGURE 8.15 (*Continued*)

represents energy propagating away from the array. Z_L is a sum over inductive TE modes and models the inductance associated with the vertical leads. Z_C and Z are sums of inductive and capacitive modes and represent evanescent fields produced by currents on the horizontal leads. The parameter r corresponds to the turns ratio for the center-tapped transformer. Replacing the mode impedances by their corresponding lumped-circuit elements, we obtain the grid-transmission line model shown in Fig. 8.15c, where the transistor equivalent circuit is connected to terminals 1 and 2.

3.2 Generalized Full-Wave Analysis

The EMF method presented above is applicable to grid structures for which the current distribution can be reasonably approximated, such as the crossed-dipole geometry. In contrast, the full-wave technique discussed here does not depend on knowing the current distribution *a priori,* making it possible to analyze grids with arbitrary periodic metallization. Additional design flexibility is available with this technique since different metallization patterns are allowed on either side of the dielectric substrate.

Fig. 8.16 illustrates the geometry of the problem. Using the boundary conditions enforced by the electric and magnetic walls of the unit cell, the transverse radiated electric field and surface current can be expanded in Fourier series,

$$E_x(x, y, z) = \sum_{m=1}^{\infty} \sum_{n=1}^{\infty} E_{mnx}(z) \sin\left(\frac{m\pi x}{a}\right) \sin\left(\frac{n\pi y}{b}\right)$$

$$J_x(x, y) = \sum_{m=1}^{\infty} \sum_{n=1}^{\infty} J_{mnx} \sin\left(\frac{m\pi x}{a}\right) \sin\left(\frac{n\pi y}{b}\right)$$

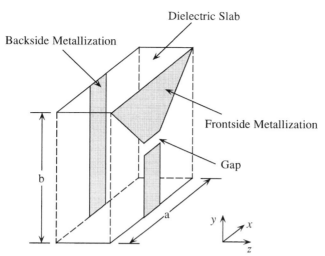

FIGURE 8.16 A unit cell with metallization patterns on the front and back of a dielectric substrate.

where a is the width and b is the height of the unit cell. Similar expressions can be found for the y components. A relationship between the unknown current and electric-field expansion coefficients is then derived by solving the source-free wave equation in the free-space and dielectric regions. Appropriate boundary conditions for the tangential fields are applied at each interface, resulting in an analytical expression relating the Fourier series coefficient of the tangential electric field to those of the surface currents at both dielectric interfaces:

$$E_{fmnx} = f(J_{mnfx}, J_{mnfy}, J_{mnbx}, J_{mnby}) \tag{8.4}$$

The subscripts f and b refer to the frontside and backside dielectric interfaces, respectively. The exact form of (8.4) is complicated and is derived in more detail in references [18] and [32]. Similar expressions are found for the other tangential electric field coefficients E_{fmny}, E_{bmnx}, and E_{bmny}. These relationships are valid for any grid geometry. The unknown surface current for a particular grid is determined from the method of moments using appropriate constraints on the tangential electric field. A time-harmonic voltage generator represents the device, producing a constant electric field across the gap. After the current on the structure is found, the driving-point impedance is calculated as the ratio of the voltage across the gap to the current through the generator driving the gap. The entire analysis is performed over a range of frequencies, yielding a frequency-dependent impedance of the passive grid.

For a grid loaded with two-port active devices, there are two gaps in the metallization, one for each port. In this case, the analysis is performed twice, first with gap 1 driven by a generator while gap 2 is metallized (short-circuited),

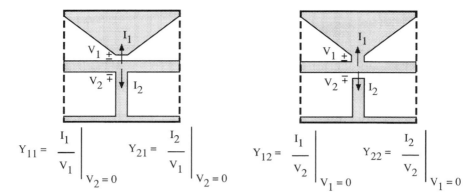

FIGURE 8.17 For a two-port grid, Y parameters are found first by driving gap 1 with a generator V_1 while gap 2 is shorted, then driving gap 2 with generator V_2 while gap 1 is shorted. (From reference [18], © 1994 IEEE.)

and next with gap 2 driven while gap 1 is metallized (Fig. 8.17). Since the current on the entire structure is calculated in each case, two-port Y parameters can be determined and converted to two-port S parameters. By slightly modifying the analytical formulation, two additional ports can be introduced, representing the free space in front and in back of the grid. Adding these extra ports has the advantage of reducing the design cycle, since the front and back mirrors can be adjusted within the circuit simulator rather than repeating time-consuming moment-method simulations. More importantly, the additional ports allow grids to be cascaded together to form quasi-optical subsystems [23].

Once the grid is characterized, appropriate active device models are connected to the grid network and commercially available software can be used to simulate the grid oscillator equivalent circuit. Either linear or nonlinear analyses can be performed, depending on the choice of the device model. Figure 8.18 illustrates the equivalent circuit of a grid oscillator cascaded with a loaded frequency-selective surface. The networks model the grid metallization and dielectric substrate. The free space in front and in back of the grid are represented with transmission lines. Radiation into free space is represented by the resistor.

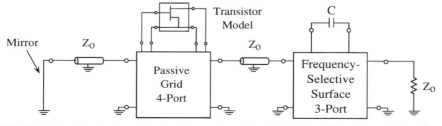

FIGURE 8.18 Equivalent circuit of a transistor-loaded grid oscillator cascaded with a capacitor-loaded frequency-selective surface.

3.3 Verification of the Grid Models

The equivalent circuits obtained from both the EMF and full-wave methods rely on the assumptions of an infinite grid and identical current sources in each cell. The EMF method further assumes a particular current distribution on the grid metallization. As a result, it is natural to question the accuracy of these models. Although both models are intended to calculate the driving-point impedances seen at the transistor terminals, they can also be used to predict scattering parameters for a plane wave normally incident on a grid. For passive grids, these scattering parameters can be measured directly with a network analyzer and a pair of horn antennas.

The EMF method was verified on two passive grids with the devices replaced with shorted or opened gaps. In both cases, the transmission-line model accurately predicted the phase of the reflection coefficient [14]. The full-wave method was verified against measurements on inductive grids, capacitive grids, crossed-dipole arrays, Jerusalem cross arrays, and nested square-loop arrays, with excellent agreement in all cases [18,32]. This theory was further validated on cascaded gratings with passive loads [23].

Comparisons were also made between the EMF and full-wave methods. In reference [33], the reflection coefficient was computed for a metal mesh on a dielectric substrate. Both methods were in good agreement. In reference [32], the driving-point impedance was computed for an inductive grating with periodic gaps. Both methods agreed at low frequencies, but diverged at higher frequencies due to the breakdown in the assumed current distribution used in the EMF method.

For our purposes, the most important validation of the grid model is its effectiveness in predicting the oscillation frequency of an active grid. This is typically done by determining the closed-loop gain of the oscillator circuit, taking into account the saturation of the active device. A useful oscillation test that is easily implemented in a circuit simulator is described in reference [34].

EMF Model   The EMF model has accurately predicted the oscillation frequencies of numerous grids [2,12,14]. Mirror-tuning curves for both direct- and gate-feedback grid oscillators are shown in Figure 8.19. Figure 8.19a shows that there are mirror positions between consecutive tuning ranges for which the grid unlocks. This can be understood by considering the grid as a plane-wave patch radiator with an image due to the mirror. For some patch-image separations, there is a null in the array pattern at boresight, and in this case the devices are not encouraged to lock in phase.

Full-Wave Model   Two examples of grid oscillators designed using the full-wave method are shown in Fig. 8.20. The first grid has dipole radiating elements and is fabricated on 2.54-mm-thick Duroid with $\epsilon_r = 10.5$. Additional dielectric layers (Stycast Hi-K with $\epsilon_r = 10$) are placed between the grid and the

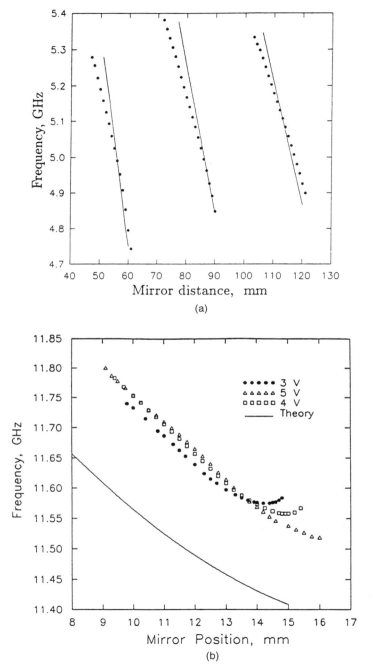

FIGURE 8.19 Mirror tuning curves for the (a) 100-MESFET and (b) gate-feedback grid oscillators. Measurements are indicated by the dots. (From references [12, 14], © 1991, 1992 IEEE.)

ANALYSIS TECHNIQUES

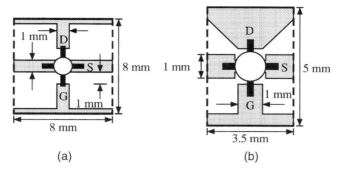

FIGURE 8.20 Grid oscillator unit cells with (a) dipole radiating elements and (b) dipole and bow-tie radiating elements. Both grids are loaded with HP ATF-35576 PHEMTs.

mirror. Table 8.2 summarizes the results for six different configurations. When the mirror is placed directly behind an electrically thin substrate (approximately $0.13\lambda_d$ in the first configuration), the disagreement between simulation and measurement is almost 20%. We believe that in this case, the locking is due to nearest-neighbor coupling, which does not enforce the assumed electric and magnetic wall boundary conditions.

Figure 8.20b shows a very different type of grid oscillator that demonstrates the versatility of the generalized analysis [18]. The unit cell has a dipole and bow-tie radiating element. Linear simulations predict an oscillation frequency of 9.30 GHz when a mirror is placed 5.0 mm behind the dielectric. The measured oscillation frequency for a 6 × 4 grid is 9.95 GHz. Note that the packaged device takes up a large portion of the unit cell, but the theory still predicts the oscillation frequency within 7%.

To validate the theory for a millimeter-wave oscillator, a monolithic 100-PHEMT grid was designed by the University of Colorado, fabricated by Honeywell, and tested by TLC Precision Wafer Technology, Inc. The grid was fabricated on a GaAs wafer and placed on an alumina substrate. The measured and

TABLE 8.2 Measured versus Simulated Oscillation Frequencies for the 5 × 5 PHEMT Grid Oscillator of Fig. 8.20a

Duroid Thickness (mm)	Stycast Thickness (mm)	Measured Frequency (GHz)	Predicted Frequency (GHz)	Percent Error (%)
2.54	0	4.70	3.9	17
5.08	0	3.44	3.1	10
2.54	6.35	5.64	5.4	4
2.54	12.70	3.93	3.8	3
2.54	19.05	4.78	4.6	4
2.54	25.40	5.28	5.1	3

simulated oscillation frequencies were 31.1 and 31.3 GHz, respectively, a difference of less than 1% [17].

4 POWER OPTIMIZATION

For maximum output power, an optimum load impedance should be presented to the devices embedded in the grid oscillator. In general, determining this load impedance is tedious. In addition to requiring a nonlinear transistor model that accounts for device saturation, an accurate design relies on a precise grid model permitting the embedding impedances at each harmonic of the fundamental design frequency to be determined. Unfortunately, nonlinear transistor models are not always available. Approximate methods based on linear circuit theory, however, can often give reasonable results. Hacker used one such approximation to design a 10-W gate-feedback grid at 10 GHz [2].

Hacker's method is based on a simple load-line analysis [35,36]. This method allows one to calculate an approximate load impedance so that the transistor output terminals swing between maximum allowable voltage and current limits. The optimum impedance is given by $R_{opt} = 2V_{DC}/I_{DSS}$, where V_{DC} is the drain-source DC bias voltage and I_{DSS} is the maximum saturated drain current. This resistance must be presented to the transistor transconductance current source as shown in Fig. 8.21.

The power grid oscillator demonstrated by Hacker *et al.* used Fujitsu power FETs (FLK052XP). From the manufacturer specifications, $R_{opt} = 83\ \Omega$. Three requirements must be satisfied to realize a grid oscillator optimized for power. First, the grid dimensions and geometry must be adjusted iteratively until this load impedance is presented to the transistors. Second, for oscillation to occur the internal loop gain of the grid must be greater than unity with 0° phase shift at the frequency of interest. Third, the active device must saturate so that it operates at maximum power-added efficiency [37]. For this to occur, the feedback to the grid must be controlled. Most grid oscillators demonstrated to date

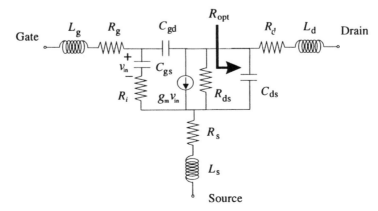

FIGURE 8.21 The optimum load resistance is presented to the current source in the FET equivalent circuit model.

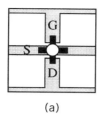 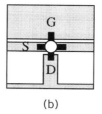

(a)   (b)

FIGURE 8.22 Unit cell of a grid oscillator with (a) dipole radiating elements and (b) dipole/slot radiating elements. Both cells are 6 mm² and are loaded with HP ATF-26884 MESFETs.

have excessive feedback, resulting in output power levels that are suboptimal [2]. Hacker reduced the feedback in his gate-feedback grid by meandering the horizontal gate bias lines, which increased the center-tapped inductance in the planar grid model of Fig. 8.15c.

The feedback can also be controlled by using an asymmetric unit cell (Fig. 8.22b). The drain is connected to a short dipole radiating element, as before. However, the width of the radiating element connected to the gate is extended across the full width of the unit cell, resulting in a slot rather than a dipole radiator. By adjusting the dimensions of this slot, the amount of feedback to the gate, and hence the compression level, can be controlled.

As a demonstration, two 4 × 4 X-band grid oscillators were fabricated with the metallization patterns shown in Fig. 8.22 [26]. Both grids had the same unit-cell dimensions, substrate, and mirror spacing, so the only difference between the two was the radiating element connected to the gate. At the same DC bias level, the dipole/slot-grid oscillator demonstrated 58% more ERP than the dipole-grid oscillator. Figure 8.23 indicates that the asymmetry of the unit cell does not adversely affect the radiation pattern or polarization.

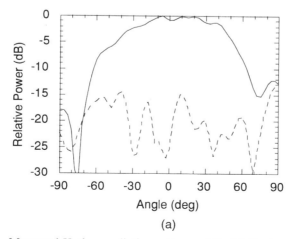

(a)

FIGURE 8.23 Measured H-plane radiation patterns of the (a) dipole and (b) dipole/slot grid oscillators of Fig. 8.22. The solid line indicates the copolar pattern, and the dashed line indicates the cross-polarization pattern [26]. (*Continued*)

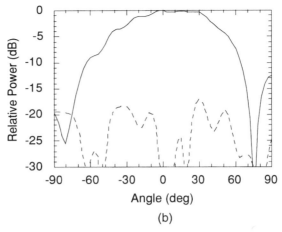

FIGURE 8.23 (*Continued*)

5 CASCADED GRIDS

An attractive feature of the planar grid approach is that separate grids performing different functions can be stacked together, resulting in a component with added capabilities. Sections 5.1 and 5.2 describe techniques for tuning the frequency of a grid oscillator by cascading two different grids. Section 5.3 discusses the advantage of cascading several grid oscillators, effectively forming a three-dimensional grid.

5.1 Voltage-Controlled Grid Oscillator

Previous sections in this chapter described how mirror or DC-bias adjustments can tune the frequency of a grid oscillator. However, mechanically adjusting the mirror is impractical for modulation purposes. DC-bias tuning is more practical, but results in rather modest tuning bandwidths.

Another way to achieve electronic frequency tuning is to cascade a transistor–oscillator grid with a tuning-diode grid (Fig. 8.24). Both grids have the same dimensions and metallization patterns, so each unit-cell waveguide contains one transistor and one tuning diode. The junction capacitance of the diode grid changes as the reverse voltage is varied, causing a tuning of the transistor grid's oscillation frequency. This quasi-optical voltage-controlled oscillator (VCO) was the first demonstration of a component consisting of cascaded grids loaded with semiconductor devices [19,20].

Two types of grid metallization patterns were investigated (Fig. 8.25). A 7×7 dipole VCO had a 7% tuning bandwidth at 2.8 GHz, but the variation in output power was 25 dB. A 7×7 bow-tie VCO had a 10% tuning bandwidth at 6 GHz, with 12-dB power variation. The best performance was obtained from a 6×4 bow-tie grid with a smaller period; this VCO achieved 10% tuning at 5 GHz with less than 2-dB power variation. For comparison, mirror and DC-

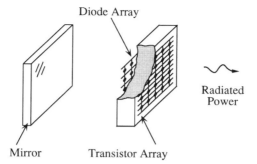

FIGURE 8.24 A voltage-controlled grid oscillator consisting of a transistor grid and a diode grid placed back-to-back on a dielectric substrate [20].

bias tuning (with the varactor grid removed) resulted in less than 4% bandwidths.

Oak and Weikle [21] demonstrated a gate-feedback VCO at X-band, with varactor diodes embedded on the same side of the substrate as the transistors. The 2% bandwidth achieved through varactor tuning was more than twice that obtained through mirror tuning.

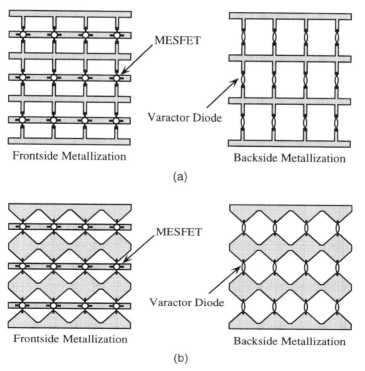

FIGURE 8.25 Two different metallization patterns for the quasi-optical VCO: (a) dipole and (b) bow-tie. The grids are printed on 0.5-mm-thick Duroid substrates with $\epsilon_r = 2.2$, and placed back-to-back [20].

324 GRID OSCILLATORS

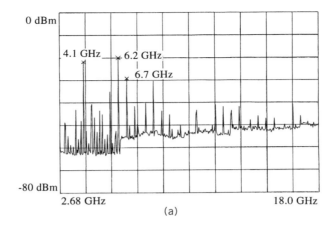

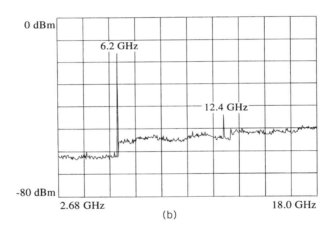

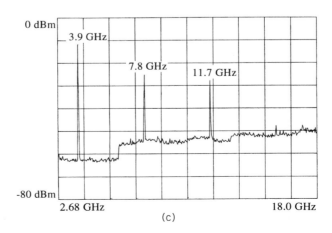

5.2 Dual-Frequency Grid Oscillator

Grid oscillators with thick, high-permittivity substrates often have several *competing* modes, as shown in Fig. 8.26a. This is an unlocked spectrum for a grid oscillator with a relatively thick (15 mm), high-permittivity ($\epsilon_r = 10$) substrate [23]. Adjusting the DC bias or mirror separation usually allows the grid to lock to one or two of these modes, but not all of them. For example, adjusting the gate bias resulted in a locked mode at 6.2 GHz with an ERP of 0.25 W and a cross-polarization of 14 dB (Fig. 8.26b). Another locked, polarized mode was obtained at 6.7 GHz, but with considerably less power. However, the oscillator could not lock to the 4-GHz mode through either bias or mirror tuning.

For the oscillator to operate in the 4-GHz mode, the cavity Q could be increased for this frequency while simultaneously lowering the Q for the unwanted frequencies. The oscillator would then injection-lock to the higher-Q mode alone. To achieve this, a variable-reflectance grid loaded with varactor diodes was placed in front of the oscillator. At -4 V, the diode grid presented reflectivities of 85% and 42% at 3.9 and 6.2 GHz, respectively. As shown in Fig. 8.26c, the oscillator locks at 3.9 GHz with an ERP of 0.26 W and a cross-polarization of 12 dB.

The varactor-diode grid used here was the same one used in the VCO above. The only difference in the two configurations is the separation between the transistor and diode grids. If the diode grid is placed in the *nearfield* of the transistor grid, continuous *frequency tuning* is achieved; if it is placed in the *farfield* of the transistor grid, discrete *mode selection* is achieved.

5.3 Three-Dimensional Grid Oscillator

Grid oscillators have typically demonstrated DC-to-RF conversion efficiencies of about 20%, so a considerable amount of heat is dissipated in the substrate. For poor thermal conductors such as GaAs, this poses a serious problem. Heat removal is even more difficult if large-scale combining is to be achieved, since the surface area of the substrate must be increased to accommodate a larger grid.

Instead of loading all of the devices onto one large grid, an alternate approach is to distribute them over several small grids (Fig. 8.27). Large numbers of devices can be combined while reducing the thermal dissipation over each surface. In effect, this extends the grid to the third dimension. To demonstrate, four identical grid oscillators with 25 transistors each were cascaded [25]. The spacing between grids was optimized experimentally. Dielectric spacers with approximately the same permittivity as the grid substrates were inserted between adjacent oscillators to facilitate the intergrid coupling and to provide

---

FIGURE 8.26 Spectrum of (a) an unlocked grid oscillator with several competing modes, (b) the grid oscillator locked at 6.2 GHz after adjusting the gate bias, and (c) the locked oscillator cascaded with a varactor-diode grid biased at -4 V. The power levels shown represent the power received by a horn antenna in the far field. (From reference [23], © 1995 IEEE.)

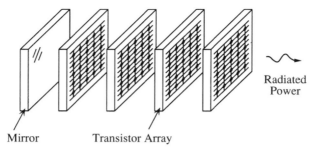

FIGURE 8.27 A three-dimensional power-combining oscillator [25]. Several oscillator grids are cascaded to achieve large-scale combining while relaxing heatsinking requirements.

mechanical support. The 100-element quadruple grid oscillator produced an ERP of 8 W at 5 GHz, 6.5 dB more t
han that of a 25-element single grid. The output power for the quadruple grid was 265 mW, compared to 100 mW for a single grid (Fig. 8.28).

When four grids were combined, the combining efficiency, defined as

$$\eta_c = \frac{\text{output power of an } N\text{-grid combiner}}{N \times \text{output power of 1 grid}}$$

was about 70%. In a two-dimensional grid oscillator, a passive mirror provides the cavity feedback. In a three-dimensional grid oscillator, each grid provides active feedback to its neighbors. Unless this feedback is carefully controlled, this could result in oversaturation of the transistors, leading to reduced output power and conversion efficiency. Future iterations of this design configuration will include mechanisms for feedback control inserted between adjacent grids.

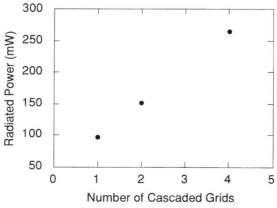

FIGURE 8.28 Measured output power versus the number of cascaded oscillator grids.

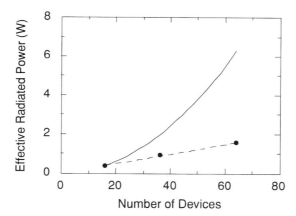

FIGURE 8.29 Measured ERP versus the number of devices. The solid line represents the theoretical ERP versus n^2 relationship, the dots represent the measured data, and the dashed line represents an ERP versus n relationship. The devices are HP ATF-35576 PHEMTs.

6 CONCLUSION

In this chapter, we presented an overview of the current state of the art in grid oscillator power combiners. Future work will continue to focus on optimizing these grids for higher power, higher frequencies, and added functionality. Some system applications of grid oscillators (e.g., self-oscillating mixers) are described in Chapter 12.

In Section 1 we concluded that the ERP grows as the square of the number of devices, assuming that all of the devices across the grid are identical. To investigate this, several grids were fabricated with the same devices, dielectric and radiating structure. Figure 8.29 shows the measured ERP as a function of grid size, and it is seen that the ERP is proportional to the number of devices rather than to the square. This indicates that the individual oscillators are not identical, and as a result the grid might not be equivalent to a uniform current sheet with an effective area equal to the geometrical area.

REFERENCES

[1] Z. B. Popović, M. Kim, and D. B. Rutledge, "Grid oscillators," *Int. J. Infrared Millimeter Waves,* vol. 9, no. 7, pp. 647–654, 1988.

[2] J. B. Hacker, M. P. De Lisio, M. Kim, C.-M. Liu, S.-J. Li, S. W. Wedge, and D. B. Rutledge, "A 10-watt X-band grid oscillator," *1994 IEEE MTT-S Int. Microwave Symp. Dig.* (San Diego, CA), pp. 823–826, May 1994.

[3] M. Gouker, "Toward standard figures-of-merit for spatial and quasi-optical power-combined arrays," *IEEE Trans. Microwave Theory Tech.,* vol. 43, pp. 1614–1617, July 1995.

[4] K. D. Stephan and T. Itoh, "A planar quasi-optical subharmonically pumped mixer characterized by isotropic conversion loss," *IEEE Trans. Microwave Theory Tech.*, vol. 32, pp. 97–102, Jan. 1984.

[5] Z. B. Popović, R. M. Weikle II, M. Kim, K. A. Potter, and D. B. Rutledge, "Bar-grid oscillators," *IEEE Trans. Microwave Theory Tech.*, vol. 38, pp. 225–230, Mar. 1990.

[6] R. M. Weikle II, M. Kim, J. B. Hacker, M. P. De Lisio, Z. B. Popović, and D. B. Rutledge, "Transistor oscillator and amplifier grids," *Proc. IEEE,* vol. 80, pp. 1800–1809, Nov. 1992.

[7] Z. B. Popović, unpublished work.

[8] M. Nakayama, M. Hieda, T. Tanaka, and K. Mizuno, "Millimeter and submillimeter wave quasi-optical oscillator with multi-elements," *1990 IEEE MTT-S Int. Microwave Symp. Dig.* (Dallas, TX), pp. 1209–1212, 1994.

[9] H. Kondo, M. Hieda, M. Nakayama, T. Tanaka, K. Osakabe, and K. Mizuno, "Millimeter and submillimeter wave quasi-optical oscillator with multi-elements," *IEEE Trans. Microwave Theory Tech.,* vol. 40, pp. 857–863, May 1992.

[10] K. Mizuno, private communication.

[11] M. P. De Lisio, J. F. Davis, S.-J. Li, D. B. Rutledge, and J. J. Rosenberg, "A 16-element tunnel diode grid oscillator," *1995 IEEE AP-S Int. Symp. Dig.* (Newport Beach, CA), pp. 1284–1287, 1995.

[12] R. M. Weikle II, M. Kim, J. B. Hacker, M. P. De Lisio, and D. B. Rutledge, "Planar MESFET grid oscillators using gate feedback," *IEEE Trans. Microwave Theory Tech.,* vol. 40, pp. 1997–2003, Nov. 1992.

[13] M. P. De Lisio, *Hybrid and Monolithic Active Quasi-Optical Grids,* Ph.D. thesis, California Institute of Technology, Pasadena, CA, 1996.

[14] Z. B. Popović, R. M. Weikle II, M. Kim, and D. B. Rutledge, "A 100-MESFET planar grid oscillator," *IEEE Trans. Microwave Theory Tech.,* vol. 39, pp. 193–200, Feb. 1991.

[15] M. Kim, E. A. Sovero, R. M. Weikle II, J. B. Hacker, M. P. De Lisio, and D. B. Rutledge, "A 35 GHz HBT monolithic grid oscillator," *Proc. Seventeenth Int. Conf. Infrared Millimeter Waves* (Pasadena, CA), pp. 402–403, Dec. 1992.

[16] H. M. Harris, A. Torabi, R. W. McMillan, C. J. Summers, J. C. Wiltse, S. M. Halpern, and D. W. Griffin, "Quasi-optical power combining of solid-state sources in Ka-band," *1993 IEEE MTT-S Int. Microwave Symp. Dig.* (Atlanta, GA), pp. 159–162, June 1993.

[17] J. Geddes, private communication.

[18] S. C. Bundy and Z. B. Popović, "A generalized analysis for grid oscillator design," *IEEE Trans. Microwave Theory Tech.,* vol. 42, pp. 2486–2491, Dec. 1994.

[19] S. C. Bundy, T. B. Mader, and Z. B. Popović, "Quasi-optical array VCOs," *1992 IEEE MTT-S Int. Microwave Symp. Dig.* (Albuquerque, NM), pp. 1539–1542, June 1992.

[20] T. Mader, S. Bundy, and Z. B. Popović, "Quasi-optical VCOs," *IEEE Trans. Microwave Theory Tech.*, vol. 41, pp. 1775–1781, Oct. 1993.

[21] A. C. Oak and R. M. Weikle II, "A varactor tuned 16-element MESFET grid oscillator," *1995 IEEE AP-S Int. Symp. Dig.* (Newport Beach, CA), pp. 1296–1299, 1995.

[22] M. Kim, E. A. Sovero, J. B. Hacker, M. P. De Lisio, J. J. Rosenberg, and D. B. Rutledge, "A 6.5–11.5 GHz source using a grid amplifier with a twist reflector," *IEEE Trans. Microwave Theory Tech.,* vol. 41, pp. 1772–1774, Oct. 1993.

[23] W. A. Shiroma, S. C. Bundy, S. Hollung, B. D. Bauernfeind, and Z. B. Popović, "Cascaded active and passive quasi-optical grids," *IEEE Trans. Microwave Theory Tech.,* vol. 43, pp. 2904–2909, Dec. 1995.

[24] S. Li and D. B. Rutledge, "Grid oscillator beam-steering array," *1994 IEEE AP-S Int. Symp. Dig.,* pp. 868–871, 1994.

[25] W. A. Shiroma, B. L. Shaw, and Z. B. Popović, "A 100-transistor quadruple grid oscillator," *IEEE Microwave Guided Wave Lett.,* vol. 4, pp. 350–351, Oct. 1994.

[26] W. A. Shiroma, *Cascaded Active and Passive Grids for Quasi-Optical Front Ends,* Ph.D. thesis, University of Colorado, Boulder, CO, 1996.

[27] L. Brillouin, "Origin of radiation resistance," *Radioélectricité,* vol. 3, pp. 147–152, 1922.

[28] P. S. Carter, "Circuit relations in radiating systems and applications to antenna problems," *Proc. IRE,* vol. 20, pp. 1004–1041, June 1932.

[29] R. L. Eisenhart and P. J. Khan, "Theoretical and experimental analysis of a waveguide mounting structure," *IEEE Trans. Microwave Theory Tech.,* vol. 19, pp. 706–719, Aug. 1971.

[30] R. M. Weikle, II, *Quasi-Optical Planar Grids for Microwave and Millimeter-Wave Power Combining,* Ph.D. thesis, California Institute of Technology, Pasadena, CA, 1992.

[31] R. F. Harrington, *Time-Harmonic Electromagnetic Fields,* McGraw-Hill, New York, 1961.

[32] S. C. Bundy, *Analysis and Design of Grid Oscillators,* Ph.D. thesis, University of Colorado, Boulder, CO, 1994.

[33] S. C. Bundy, W. A. Shiroma, and Z. B. Popović, "Design-oriented analysis of grid power combiners," *Workshop on Millimeter-Wave Power Generation and Beam Control* (Huntsville, AL), pp. 197–208, Sept. 1993.

[34] R. D. Martinez and R. C. Compton, "A general approach for the s-parameter design of oscillators with 1 and 2-port active devices," *IEEE Trans. Microwave Theory Tech.,* vol. 40, pp. 569–574, Mar. 1992.

[35] S. C. Cripps, "Old-fashioned remedies for GaAs FET power amplifier designers," *IEEE MTT-S Newsletter,* no. 128, pp. 13–17, summer 1991.

[36] S. C. Cripps, *GaAs FET Power Amplifier Design.* Technical Notes 3.2, Matcom, Inc. (415-493-6127).

[37] K. M. Johnson, "Large signal GaAs MESFET oscillator design," *IEEE Trans. Microwave Theory Tech.,* vol. 27, pp. 217–227, Mar. 1979.

[38] Z. B. Popović, "Grid Oscillators," Ph.D. thesis, California Institute of Technology, Pasadena, CA, 1990.

[39] M. J. Wengler, B. Guan, and E. K. Track, "190-GHz radiation from a quasioptical Josephson junction array," *IEEE Trans. Microwave Theory Tech.,* vol. 43, pp. 984–988, Apr. 1995.

[40] M. J. Wengler, private communication.

CHAPTER NINE

Grid Amplifiers

MICHAEL P. DE LISIO
University of Hawaii at Mānoa, Honolulu

CHEH-MING LIU
Rockwell International Science Center, California

A grid amplifier is a two-dimensional periodic array of closely spaced differential transistor pairs. Hybrid microwave grid amplifiers were the first quasi-optical amplifiers developed. Experiments have shown that the gain and noise figure of a grid amplifier are comparable to those of a single device, while the output power scales with the total number of devices incorporated in the array. Very recently, monolithic grid amplifiers have demonstrated appreciable gain and power combining at millimeter-wave frequencies.

1 INTRODUCTION AND BACKGROUND

Quasi-optical components combine the output powers of many solid-state devices in free space without the conductor losses associated with waveguide or transmission-line combiners. Moreover, the planar nature of quasi-optical components makes them well-suited for monolithic fabrication. Hundreds, possibly thousands, of devices could be incorporated through wafer-scale integration. This creates the potential for microwave and millimeter-wave components with

---

Active and Quasi-Optical Arrays for Solid-State Power Combining, Edited by Robert A. York and Zoya B. Popović.
ISBN 0-471-14614-5 © 1997 John Wiley & Sons, Inc.

332 GRID AMPLIFIERS

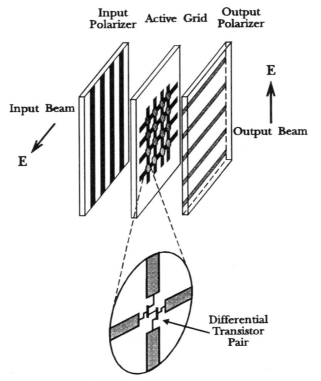

FIGURE 9.1 A grid amplifier. A horizontally polarized input beam is incident from the left. The output beam is vertically polarized and is radiated to the right.

greatly increased power and dynamic range. A wide variety of quasi-optical devices have been successful, including detectors [1], multipliers [2,3], phase shifters [4,5], oscillators [6–8], and mixers [9]. Many of these components are discussed elsewhere in this book. For quite a while, however, a critical component had been missing—the amplifier.

The first quasi-optical amplifier was a 25-element grid built and tested by Moonil Kim at Caltech in 1991 [10]. The grid was a hybrid construction, with packaged MESFETs attached to an etched Duroid board. Figure 9.1 shows the approach. The grid is a two-dimensional periodic array of closely spaced differential transistor pairs. A horizontally polarized input beam is incident from the left, exciting RF currents on the horizontal leads of the grid. These currents drive the inputs of the transistor pair in the differential mode. The output currents are redirected along the vertical leads, producing a vertically polarized output beam to the right. The cross-polarized input and output has two advantages. First, it provides good input–output isolation, reducing the potential for feedback oscillations. Second, the amplifier's input and output circuits can be independently

INTRODUCTION AND BACKGROUND 333

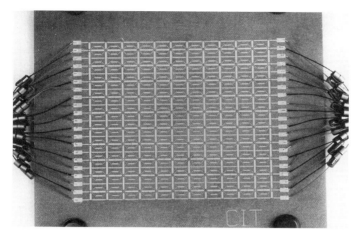

FIGURE 9.2 Photograph of the 100-element HBT grid amplifier. The active area of the array is 8 cm on a side.

tuned using metal-strip polarizers, which also confine the beam to the forward direction. This first grid demonstrated a peak gain of 11 dB at 3.3 GHz.

The next grid to be developed was a 100-element hybrid HBT array, also built and tested by Moonil Kim at Caltech [11]. A photograph is shown in Fig. 9.2. Unlike the previous amplifier, this grid has a completely planar unit cell, making it ideally suited for monolithic integration. The unit cell is illustrated in Fig. 9.3. The active element is a custom-made differential-pair HBT chip, fabricated by Rockwell International. The chip layout is shown in Fig. 9.4. The horizontal leads of the unit cell couple the input beam to the bases of the HBT pair. A thin capacitive gap is etched in the inductive base leads to achieve a better input match. The collectors are attached to the vertical leads, which radiate the output beam. Bias is supplied to the collector and emitter via the thin meandering lines. The bias lines should have a rather large reactance so as not to disturb the input and output fields. Bias to the base is taken from the collector bias through an on-chip self-bias network.

Figure 9.5 shows the measured gain of the amplifier. The peak gain is 10 dB at 10 GHz, with a 10% 3-dB bandwidth. Without bias, the gain is below −8 dB. The input and output return losses were measured to be better than 15 dB at 10 GHz. The measured noise figure was 7 dB, and the saturated output power was 450 mW. These results are consistent with the idea that the gain and noise figure are comparable to those of a single-transistor amplifier, while the power scales with the size of the array. Furthermore, experiments indicate that the amplifier degrades gracefully. Device failure was simulated by wire bonding over the gaps in the base leads of the unit cell, effectively detuning the input. With 10% of the cells disabled, the gain drops by only 1 dB, in accord with the

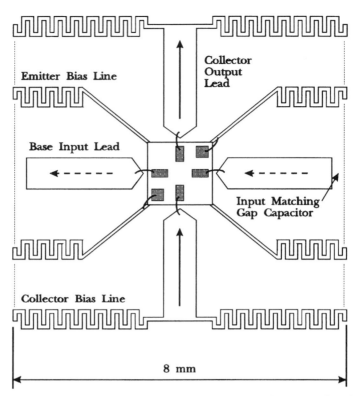

FIGURE 9.3 The HBT grid amplifier unit cell. Arrows indicate the direction of RF currents.

assumption that each cell contributes equally to the radiated field. This implies that a grid amplifier could be quite forgiving to the single-point device failures that would devastate vacuum-tube or conventional solid-state amplifiers.

Finally, the grid amplifier could amplify beams with non-normal incidences while preserving their angle. This is shown in Fig. 9.6. This graph shows the output radiation pattern for three different beams incident at $0°$, $+20°$, and $-20°$. The peaks in the output patterns are within one degree of the incident angles. The beams have similar widths, side lobe levels, and null locations. In another measurement, the grid was found to amplify beams with incident angles up to $30°$ with less than a 3-dB gain reduction. These results indicate that a grid amplifier would work well in a steered-beam application, easing the loss requirements for the steering system.

Following the success of these original grids, a number of other types of quasi-optical amplifiers have been demonstrated. These amplifiers typically have a larger unit cell, usually employing more conventional antennas such as patches and slots. These approaches have been very successful and are discussed

INTRODUCTION AND BACKGROUND   335

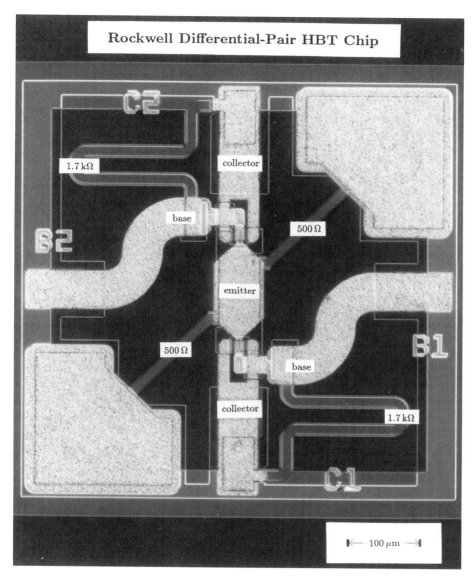

FIGURE 9.4 Photograph of the HBT differential-pair chip.

in Chapters 5 and 6. Progress in grid amplifiers has continued as well, with advances made in bandwidth, noise figure, and output power, as well as the development of gain and stability modeling [12–15]. In addition, gain and power combining has recently been demonstrated at millimeter-wave frequencies with monolithic grid amplifiers [16–18].

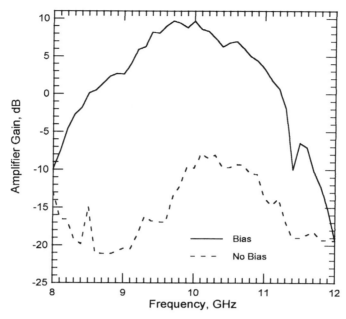

FIGURE 9.5 Gain versus frequency for the 100-element HBT grid amplifier.

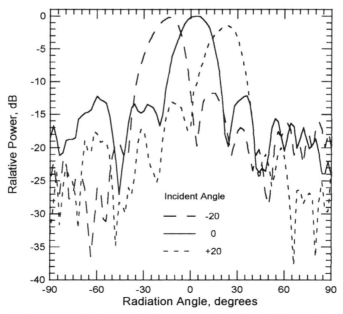

FIGURE 9.6 The output H-plane radiation patterns for three different incident beams.

2 MODELING

An interesting property of grid amplifiers is that, to first order, the size and structure of the unit cell determine the driving-point impedances seen by the device, while the power increases with the total grid area. This allows one to (a) optimize for gain and noise figure by the choice of the unit cell and (b) independently select the grid size to meet the power requirement. The design of early grids was primarily empirical. For the HBT amplifier, a number of smaller 16-element grids were built and tested. The unit cell used for the full-size grid was the prototype that gave the best results. Furthermore, grid amplifiers are often unstable, and thus susceptible to spurious oscillations. The development of accurate gain and stability models is critical if these grids are ever to be designed confidently.

2.1 Gain Modeling

Consider the grid amplifier configuration shown in Fig. 9.7a. The grid is constructed on a substrate with a relative dielectric constant ϵ and a thickness t_s. The input and output polarizers are located a distance t_i and t_o from the grid. To simplify this example, the polarizers are metallic gratings suspended in free space. A simplified unit cell is shown in Fig. 9.7b. The transistors are assumed to be FETs. Horizontal leads are connected to the gates of the transistor pair, and

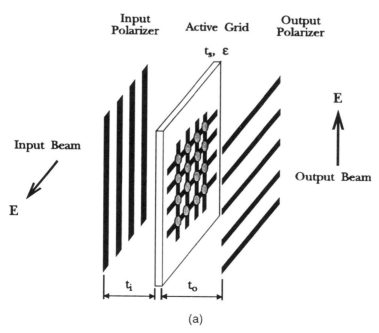

FIGURE 9.7 Simplified amplifier grid configuration (a) and unit cell (b). (*Continued*)

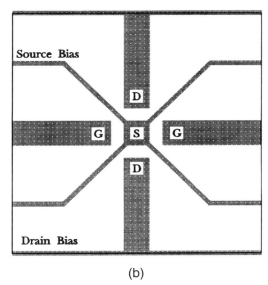

(b)

FIGURE 9.7 (*Continued*)

vertical leads are connected to the drains. Bias to the source and drain is provided along the narrow lines.

The electromagnetic analysis of a large grid array incorporating many elements is a very complicated task. A considerable computational effort may be required even for relatively small arrays. This problem may be simplified by assuming that the size of the array is infinite. Under this assumption, all active elements will be presented with the same driving-point impedance, and edge effects are ignored. One can also exploit the natural symmetries of the grid and define a unit cell [6,19], usually including a single active element. The analysis of the entire grid, then, is reduced to examining a single element in a unit cell.

A rather intuitive transmission-line equivalent circuit for the configuration of Fig. 9.7 is proposed in Fig. 9.8a. Free space is represented by the resistors, with characteristic impedance η_o. For a square unit cell, $\eta_o = 377\ \Omega$. The air gaps between the polarizers and the grid appear as transmission lines with characteristic impedance η_o and appropriate electrical length. The substrate is also modeled by a transmission line. The effect of the dielectric is to reduce the characteristic impedance by a factor of $\sqrt{\epsilon}$, the index of refraction of the material, and to increase the electrical length by this same factor. The polarizers are assumed to be perfect—they present a short circuit to a wave polarized along the direction of the metal strips and they are invisible to a wave polarized orthogonal to the strips. The size and structure of the unit cell will determine the values of the remaining inductive elements. Coupling to the gates of the devices through the short horizontal input leads is represented by the inductance L_i. The vertical output leads attached to the drains are modeled by the inductors labeled L_o. The

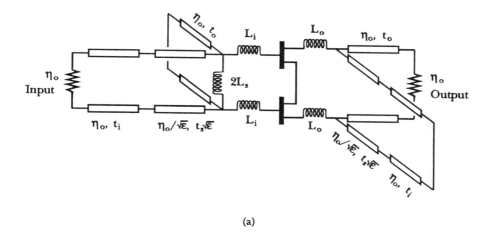

(a)

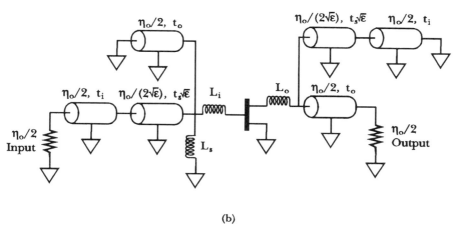

(b)

FIGURE 9.8 Transmission-line equivalent circuit for the simplified amplifier configuration of Fig. 9.7. The circuit's symmetry allows one to bisect the full-cell model (a), resulting in the half-cell model (b).

thin bias lines are primarily horizontally directed. Therefore, they will appear at the input and are represented by the shunt inductance L_s. It is clear that these lines should be made highly reactive to avoid shorting out the input of the grid. Numerical values for these reactive elements can be computed using any number of methods—from very simple quasi-static models to complicated moment method or finite-element techniques.

The circuit in Fig. 9.8a represents the full cell and therefore contains two transistors. The symmetry of this circuit allows one to bisect it along the horizontal symmetry plane with a virtual ground, resulting in the circuit shown in

Fig. 9.8b. This new circuit represents a half cell, containing only a single transistor. This simplifies the analysis because, often, it is the transistor model itself that is the most complicated part of the circuit. The bisection results in the characteristic impedances of free space and all transmission lines becoming half of their usual values, while their effective lengths remain the same.

This model may also be used for beams with non-normal incidences. For a beam with an incidence angle of θ, the length of each transmission line should be scaled by $\cos\theta_i$, where θ_i is the incident angle in the material, as given by Snell's law of refraction. In addition, the characteristic impedance of each transmission line should be multiplied by $\cos\theta_i$ for beams with TM polarization, or $\sec\theta_i$ for beams with TE polarization. The gain should then be scaled by a $\cos^2\theta$ obliquity factor to account for the foreshortening of the grid's input and output area.

The model presented here is by no means intended to be complete. Rather, it is meant to provide a basis for understanding the more complicated equivalent circuit models which will be presented later. These circuits will include the effects of imperfect polarizers, dielectric tuning slabs, and other elements. In addition, the rather simple inductor network representing the coupling between the radiated waves and the device via the unit cell structure could be replaced with a more extensive one. Nevertheless, despite its apparent simplicity, this equivalent circuit has been used to design and model the behavior of a number of different amplifier grids with considerable success.

2.2 Stability Modeling

Stability is a vital concern for any amplifier. This may be especially true for quasi-optical amplifiers, where many oscillation modes may exist due to the large number of devices involved. Many grid amplifiers have, in fact, suffered from spurious oscillations of some kind [10–15]. Often, these oscillations appear at moderate bias levels. If these instabilities cannot be suppressed, the amplifier will be limited to very low bias currents, thereby restricting the gain and output power. Clearly, a stability analysis for grid amplifiers is of paramount importance.

Most of our stability modeling is the result of testing a 16-element hybrid HBT grid amplifier [13,15]. This amplifier was plagued by oscillations, even at very low bias levels. The oscillations were not sensitive to the positions of the polarizers or the thickness of the substrate. On the other hand, the oscillations were very sensitive to changes made in the unit cell itself; shorting the capacitive gaps in the base lead lowered the frequency considerably. In addition, the radiation pattern of the oscillation mode was examined. Most of the power was radiated in the plane of the grid, with relatively little power radiated normal to the grid. The radiation was polarized primarily in the direction of the drain leads. This evidence led to the conclusion that the transistor pair was oscillating in the common mode. Similar results were measured in a larger pHEMT amplifier grid [12,14].

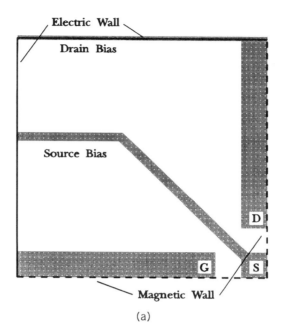

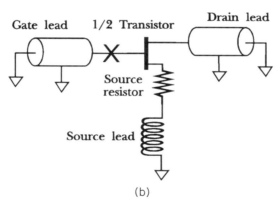

FIGURE 9.9 (a) Quarter unit cell used for analyzing the common-mode oscillation. Boundary conditions are imposed by grid symmetry and common-mode operation of the transistor pair. (b) Common-mode equivalent circuit. The × marks the spot where a circulator is inserted to determine the oscillation frequency.

It is possible to define a unit cell for this common-mode oscillation. The symmetry of the array and the distribution of currents demand that the full unit cell be bounded on all four sides by electric walls, where the tangential electric field must vanish. One can further exploit the symmetry of the full cell and define a quarter-cell, as shown in Fig. 9.9a. The quarter cell is bounded by two electric walls and two magnetic walls, where the tangential magnetic field must vanish. Figure 9.9b proposes an equivalent circuit for the common-mode oscillation. The transistors are assumed to be FETs. The device shown in this quarter-cell circuit is one-half of a transistor, because a full unit cell contains two transistors. The input and output leads appear as transmission lines, and the common-mode source resistors must be included. The source bias line is modeled as an inductance. The length and impedance of the transmission lines are chosen empirically, as is the value of the source inductance.

The circuit in Fig. 9.9b may be analyzed using a variety of methods. One approach that is especially intuitive and useful is the technique suggested by Martinez and Compton [20]. An ideal circulator is added at the node marked with an ×. The reflection coefficient from the circulator is defined as the circular function C. This function may be thought of as a measure of the open-loop gain of the circuit. Classical stability theory dictates that the circuit will be unstable if the magnitude of the circular function is greater than unity at the zero-phase point.

Two approaches have been used to suppress common-mode oscillations. The first is to add a reactive component into the circuit. This tends to add a phase shift to the circular function, rotating the plot in the complex plane and eventually reducing the magnitude of the zero-phase point. Another approach is to add resistive elements, reducing the magnitude of the entire circular function. The reactive approach has been successful for HBT amplifier grids [13,15–17], where a series stabilizing capacitor is added in the base lead. This should not affect the maximum available gain of the device, but may reduce the bandwidth. The resistive approach has been successful in pHEMT arrays [12,14], where a small series resistance is added in the gate leads. This will stabilize the amplifier at the expense of lowering the available gain and increasing the minimum noise figure.

3 A 100-ELEMENT HYBRID pHEMT GRID AMPLIFIER

In recent years, pHEMT technology has developed rapidly. pHEMTs are capable of providing high gain, high output power, high efficiency, and low noise figure at frequencies over 100 GHz, making them the device of choice for millimeter-wave applications [21–24]. It is quite likely that any future applications of active quasi-optical technology will involve pHEMTs. This section will detail the construction, modelling, and performance of a hybrid pHEMT amplifier grid [12,14]. In addition, aspects of the quasi-optical measurements will be discussed.

3.1 Grid Construction

The grid is constructed using 0.1-μm AlGaAs/InGaAs/GaAs pHEMTs fabricated by Lockheed Martin Laboratories. The chip layout is shown in Fig. 9.10. The sources of two pHEMTs are tied together to form a differential pair. Each

A 100-ELEMENT HYBRID pHEMT GRID AMPLIFIER   343

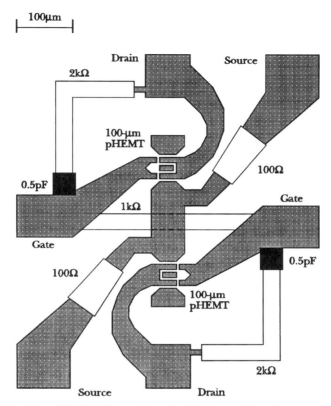

FIGURE 9.10 The pHEMT differential-pair chip layout. The chips were fabricated by Lockheed Martin Laboratories.

transistor has a gate width of 100 μm. Unlike the HBTs in reference [11], which were self-biased, the control terminal can be externally biased. The 1-kΩ resistor allows this gate control bias voltage to pass from cell to cell. This is possible because the gate draws very little bias current. The gate bias resistor passes under the air-bridged source connection. These devices are designed to be used as high as 100 GHz [21]. Therefore, a drain-gate feedback network consisting of the 2-kΩ resistor and 0.5-pF capacitor is included to help stabilize the amplifier at X band. The two 100-Ω source resistors are intended to reduce the common-mode gain. These resistors should not affect the differential-mode performance but will reduce the overall efficiency of the amplifier.

Figure 9.11 shows the unit cell. The differential-pair chip is located at the center of the cell, with the necessary connections made with bond wires. The cell is 7.3 mm on a side. The input beam is coupled to the gates of the pHEMTs through the horizontal input leads, which also supply the gate bias. The output beam is radiated from the vertical drain leads. Both input and output leads are 0.4-mm wide. Bias to the drain and source is provided by the meandering lines,

344 GRID AMPLIFIERS

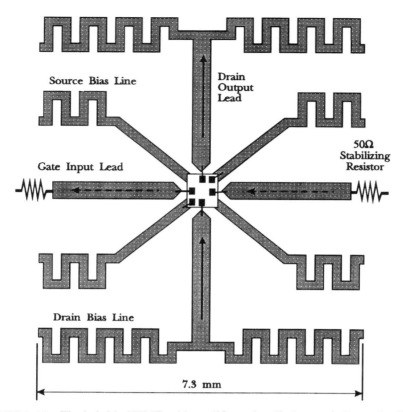

FIGURE 9.11 The hybrid pHEMT grid amplifier unit cell. Arrows indicate the direction of RF currents.

which are 0.2-mm wide. The bias lines are intended to present a rather high impedance to the input wave. The resistors in the gate leads suppress spurious common-mode oscillations and will be discussed later. The entire grid is a 10×10 array of these cells etched onto a Duroid substrate. Figure 9.12 is a photograph of the 100-element array.

Initially, the grid suffered from common-mode oscillations. The frequency was between 8 and 9 GHz, depending on the bias level. The onset of these oscillations was at a drain bias current of 9 mA per transistor—well below the device's usual operating point. The gain of the amplifier at the onset of oscillation was less than 4 dB. Following the approach outlined in Section 2.2, we can model the common-mode instability with a transmission-line circuit. The circular function C of the stability circuit model is plotted in Fig. 9.13. The circuit is theoretically unstable at 8.9 GHz because the magnitude of C is greater than unity and the phase is zero. This is close to where the actual oscillations occurred.

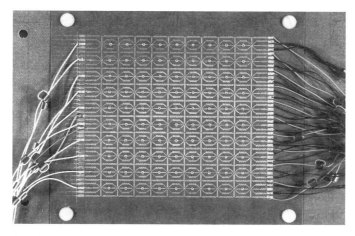

FIGURE 9.12 Photograph of the 100-element pHEMT grid amplifier. The active area of the array is 7.3 cm on a side.

To suppress these oscillations, 50-Ω chip resistors were placed in the radiating gate lead, midway between transistor pairs. With the resistors in place, the magnitude of C is less than unity, so the grid should be stable. After adding the resistors, the grid could be biased to a current of over 15 mA per transistor without oscillations. Unfortunately, these gate resistors will degrade the gain and

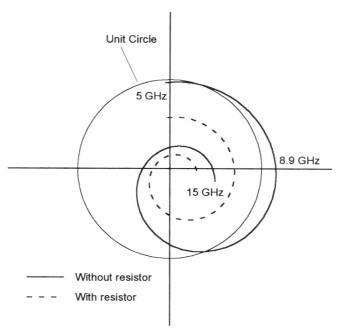

FIGURE 9.13 Simulated circular function for the common-mode oscillation model. Without gate resistors, the grid is unstable at 8.9 GHz. The stabilized amplifier has a gain margin of 3 dB.

345

noise figure of the amplifier. Series stabilizing capacitors like the ones used in reference [13] could not be used because this lead supplies the gate bias. Interestingly, the stability theory indicates that the 100-Ω source resistors originally thought necessary to prevent common-mode oscillations may actually be contributing to the instability. Removal of the source resistors may result in a more stable amplifier, as well as a more efficient grid, although this has not been experimentally verified.

3.2 Gain

Two different techniques have been used to measure the gain of quasi-optical grid amplifiers. The first is a far-field approach as shown in Fig. 9.14 [10,11]. The first step is to calibrate the system. This is done with the two horn antennas copolarized, as shown in Fig. 9.14a. The calibrated power P_c is related to the transmitted power P_t using the Friis transmission equation:

$$\frac{P_c}{P_t} = \left(\frac{G_h \lambda}{4\pi(2R)}\right)^2 = L_c \tag{9.1}$$

where G_h is the gain of the transmitting and receiving horn antenna, and R is the distance between the grid and each horn. Then, the amplifier gain is measured

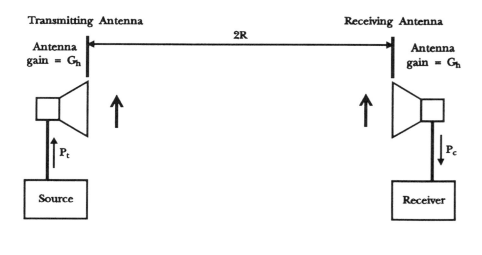

(a)

FIGURE 9.14 Far-field setup used to measure the gain. (a) The calibration step is performed with the horns copolarized. (b) The measurement step is performed with the horns cross-polarized.

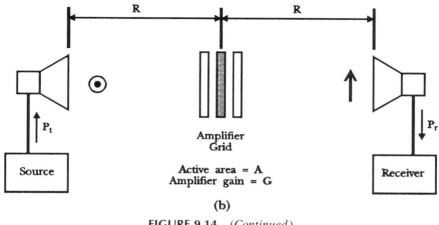

FIGURE 9.14 (*Continued*)

using the setup shown in Fig. 9.14b. In this case, the received power is related to the transmitted power by

$$\frac{P_r}{P_t} = \left(\frac{G_h A}{4\pi R^2}\right) G \left(\frac{G_h A}{4\pi R^2}\right) = GL^2 \quad (9.2)$$

where G is the gain of the grid amplifier and A is the physical area of the active array. This allows us to express the amplifier gain as

$$G = \frac{P_r}{P_c}\left(\frac{\lambda R}{2A}\right)^2 \quad (9.3)$$

This simple formula reveals that the gain of the amplifier can be computed from a relative power measurement and three well-known parameters.

Another method used to measure gain is the lens-focused approach shown in Fig. 9.15 [11]. This is a modified version of the quasi-optical reflectometer reported by D. R. Gagnon [25]. The system is calibrated by placing an absorbing screen at the focal plane, as depicted in Fig. 9.15a. The screen has an aperture cut in it, matching the physical area A of the amplifier grid. The gain is measured in the cross-polarized setup illustrated in Fig. 9.15b. For this method, the amplifier gain is simply the ratio of the power received from the grid P_r to the calibration power P_c.

At this point, it should be stressed that the far-field arrangement of Fig. 9.14 is not how a quasi-optical amplifier is ultimately intended to be used—the high path losses between the grid and the horns would result in a very inefficient system. Instead, a quasi-optical amplifier would be employed in a system where the microwave beam is confined, such as in an overmoded waveguide or a lens-focused system like Fig. 9.15. Measuring the amplifier using the lens-focused

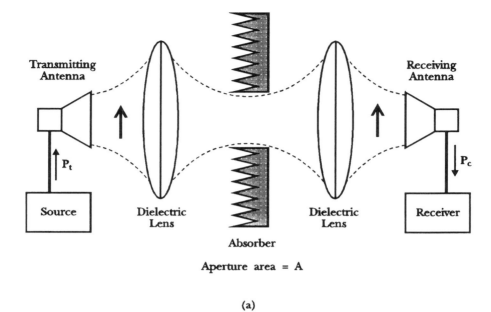

FIGURE 9.15 Lens-focused setup used for gain measurement. (a) The calibration step is performed with copolarized horns and an aperture cut into an absorbing screen. (b) The measurement setup is performed with the horns cross-polarized.

A 100-ELEMENT HYBRID pHEMT GRID AMPLIFIER 349

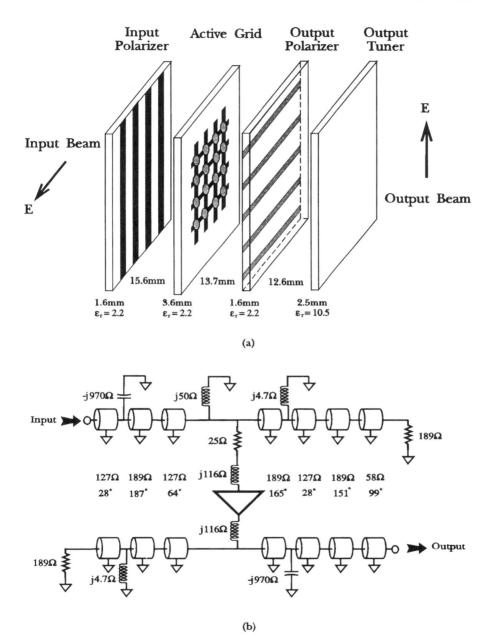

FIGURE 9.16 (a) The assembled amplifier grid tuned for peak gain at 10 GHz. (b) The transmission-line equivalent circuit for the amplifier at 10 GHz.

approach thus has the advantage of being more straightforward. Much of the diffraction losses are eliminated, possibly resulting in a system with appreciable flange-to-flange gain. On the other hand, the far-field method is simpler to set up in the laboratory, because there are no lenses to focus or align. The gain of the 100-element HBT grid amplifier was measured using both approaches, with nearly identical results [11]. For simplicity, the far-field method was used to measure the gain of the 100-element pHEMT grid.

The assembled grid amplifier tuned for 10 GHz is shown in Fig. 9.16a. The grid and polarizers are fabricated on Duroid boards with a relative dielectric constant of 2.2. A high-dielectric Duroid slab is used as an output tuner to increase the gain. The transmission-line equivalent circuit is shown in Fig. 9.16b. This is a half-cell model, so free space is represented by the 189-Ω resistors. The air gaps and substrates appear as transmission lines. The radiating leads are modeled as series inductors, while the meandering drain and source bias lines appear as a shunt inductance at the input surface of the grid. The polarizers are not assumed to be perfect. For a wave polarized along the direction of the metal strips, the polarizer is modeled by a low-impedance inductor. For a wave polarized orthogonal to the strips, the polarizer appears as a high-impedance capacitor. Numerical values for the radiating lead and polarizer reactances are computed by first using the method of moments to approximate the surface current distribution on the leads. Once an estimate for the surface

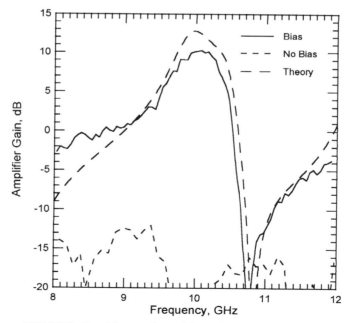

FIGURE 9.17 Measured amplifier gain versus frequency.

current has been obtained, the driving-point impedance is calculated using the induced EMF technique [14]. The numerical value of the reactance associated with the meandering bias lines in empirically determined with the aid of Hewlett-Packard's high-frequency structure simulator (HFSS) [26].

Figure 9.17 shows the measured small-signal gain of the amplifier. The peak gain is 10 dB at 10 GHz. The 3-dB bandwidth is 700 MHz, which corresponds to 7%. The modeled gain is also plotted, and it agrees well with the measured results. Without bias, the gain is below −12 dB throughout the entire frequency range. At peak gain, the difference between the biased and unbiased gain is over 30 dB. An on-off ratio this high is a good indication that the grid is operating properly. The grid amplifier can be tuned to operate at other frequencies simply by changing the positions of the external polarizers and tuning slab. Figure 9.18 shows the gain of the amplifier when tuned for a slightly lower frequency. The peak gain is 12 dB at 9 GHz, with a 3-dB bandwidth of 1.3 GHz. This corresponds to a 15% bandwidth—the highest reported for a quasi-optical amplifier to date.

To further validate the model, the gain can be measured as a function of polarizer position. The amplifier is tuned for 10 GHz, with the positions of the polarizers and tuners given in Fig. 9.16. Figure 9.19 shows the gain at 10 GHz

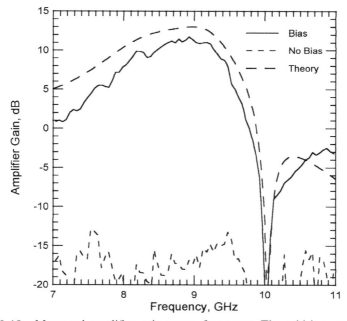

FIGURE 9.18 Measured amplifier gain versus frequency. The grid is tuned for peak gain at 9 GHz.

352 GRID AMPLIFIERS

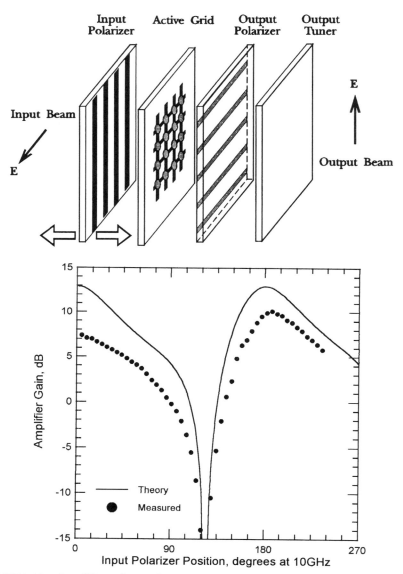

FIGURE 9.19 Amplifier gain as a function of input polarizer position. The grid is tuned for peak gain at 10 GHz.

as a function of input polarizer position. As seen from the model, the position of the input polarizer actually tunes the grid's output circuit. Figure 9.20 shows the gain as a function of output polarizer position, with the output tuner removed for convenience. The output polarizer tunes the input circuit of the amplifier. The measured tuning curves agree quite well with the theoretical predictions, which gives more credibility to the transmission-line model.

A 100-ELEMENT HYBRID pHEMT GRID AMPLIFIER

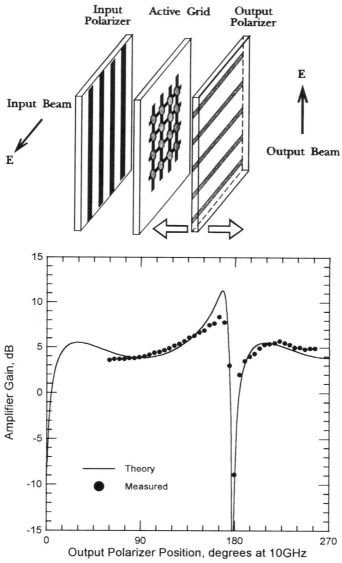

FIGURE 9.20 Amplifier gain as a function of output polarizer position.

3.3 Angular Dependence

The grid amplifier is also an antenna and therefore has a radiation pattern. Figure 9.21 shows the output H-plane radiation pattern at 10 GHz. This measurement is performed by fixing the position of the amplifier and the input horn, and sampling the output pattern, as indicated in the figure. Both horns are in the far field. Limitations in the measurement setup prevented measuring angles

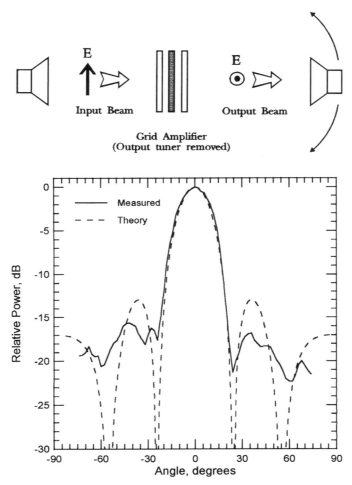

FIGURE 9.21 Output H-plane radiation pattern for the amplifier at 10 GHz. The input beam is normally incident.

greater than 75°. Also plotted is a theoretical pattern generated assuming a uniform array of ten equally spaced elementary dipoles. The main beam of the measured pattern agrees very well with the theory, indicating that the radiation pattern is diffraction limited.

The grid is a multimode device and should therefore be capable of amplifying beams at different incidence angles. The angular dependence of the gain is measured by rotating the grid between two fixed cross-polarized horns. The grid is tuned for 10 GHz, with the output tuning slab removed. Figure 9.22 shows the result at 10 GHz for input TM polarization, output TE. Figure 9.23 shows the

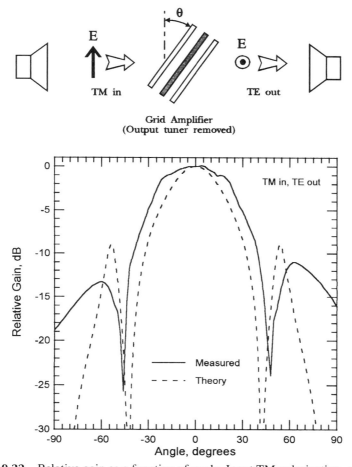

FIGURE 9.22 Relative gain as a function of angle. Input TM polarization, output TE.

result for input TE polarization, output TM. For both orientations, the grid will amplify beams with incidence angles up to 25° with less than a 3-dB gain reduction. The theory predicts the patterns well for small angles, but fails at larger angles where the finite size of the grid causes the simple transmission-line model to break down.

The nulls in the patterns of Figs. 9.22 and 9.23 near ±50° are caused by the input polarizer being farther than a half wavelength from the active surface of the array. The nulls disappear when the polarizer is moved closer to the grid by a half wavelength. This is shown in Fig. 9.24. Theoretically, moving a polarizer a half wavelength should not affect the gain at that frequency. The gain with the closer polarizer spacing, however, is 3 dB less. This can also be seen in the

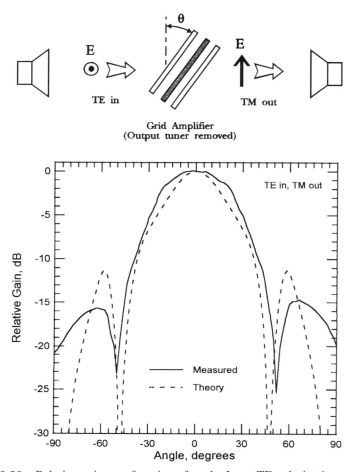

FIGURE 9.23 Relative gain as a function of angle. Input TE polarization, output TE.

input-polarizer tuning curve, Fig. 9.19. This may be evidence of evanescent-wave coupling, a phenomena that occurs in closely spaced quasi-optical systems [27]. Nevertheless, in this configuration, the grid will amplify beams with incident angles up to 30° with less than a 3-dB gain reduction. The patterns approach the theoretical $\cos^2\theta$ obliquity limit.

3.4 Noise

Because the noise from the individual elements are uncorrelated, the noise figure of a grid amplifier should be no worse than that of a single device. The noise figure can be measured quasi-optically in the far field. The setup is similar to the one shown in Fig. 9.14, with the signal source replaced by a noise source

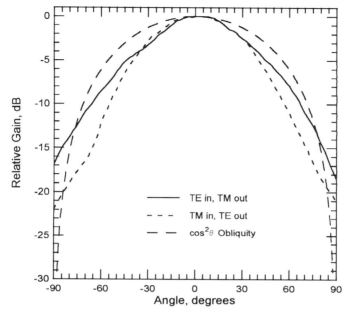

FIGURE 9.24 Relative gain as a function of angle with the input polarizer moved closer by one-half wavelength. The peak gain is reduced by 3 dB.

and the receiver replaced by a noise meter. In order to overcome the large path losses, a noise source with a high excess noise ratio (ENR) is necessary. In addition, a low-noise preamp placed at the receiving antenna will help increase the sensitivity. In the measurements, there will be a noise contribution from the background, and a calibration is made using the setup shown in Fig. 9.14a. The antenna temperature at the receiving horn T_c can be expressed as a contribution from the noise source with temperature T_t and the effective background temperature T_b:

$$T_c = L_c T_t + (1 - L_c) T_b \qquad (9.4)$$

where L_c is the space loss factor given in (9.1). The noise meter measures T_c for two different source temperatures T_t. From these measurements, L_c and T_b can be determined. Next, the amplifier is inserted, as in the arrangement shown in Fig. 9.14b. The temperature of the receiving horn T_r can now be expressed as

$$T_r = GLT_a + GL^2 T_t + (1 - L)(GL + 1)T_b \qquad (9.5)$$

where T_a is the noise temperature of the grid, G is the gain of the amplifier, and L is the space loss factor defined in (9.2). Note that L is related to the measured

358 GRID AMPLIFIERS

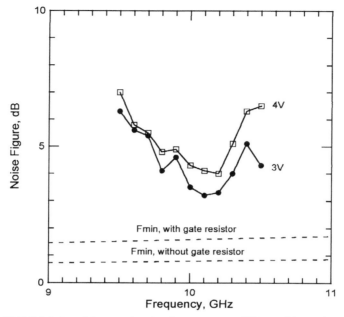

FIGURE 9.25 Measured noise figure at two different bias points.

loss factor L_c through the expression

$$L = \sqrt{L_c}\left(\frac{2A}{\lambda R}\right) \quad (9.6)$$

These measurements are then used to determine G and T_a.

Figure 9.25 shows the measured result for the pHEMT amplifier. The grid is tuned for 10 GHz. The minimum noise figure is 3 dB with 3 V bias. This is a 4-dB improvement over the previous HBT grid [11]. Also plotted is the predicted minimum noise figure F_{min} for the pHEMT. As seen in the figure, the stabilizing gate resistor degrades F_{min} by 0.8 dB. The measured noise figure is only 1.4 dB higher than the predicted minimum noise figure. It should be noted that no attempt was made to noise match the amplifier input. Furthermore, the gain extracted from the noise measurement was nearly identical to the measured gain curve in Fig. 9.17, which adds credibility to the noise measurements. These measurements show that the noise figure of the amplifier is not considerably worse than that of a single device.

3.5 Power

Unlike the noise, the output power of the grid should increase with the number of elements incorporated. Like the gain and noise figure, the power saturation can be measured using the far-field setup shown in Fig. 9.14. The approach proceeds as before. First, the system is calibrated using the configuration shown

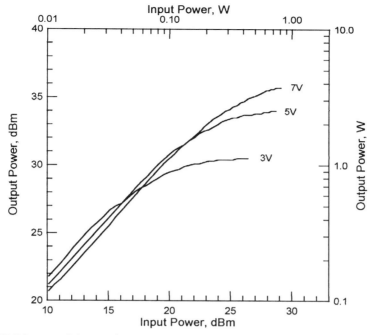

FIGURE 9.26 Measured output power at 9 GHz at three different bias points.

in Fig. 9.14a. If the transmitted power P_t is known, the gain of the horn antennas will be given by the equation:

$$G_h = \sqrt{\frac{P_c}{P_t}}\left(\frac{4\pi(2R)}{\lambda}\right) \tag{9.7}$$

We can then write an expression for the input power to the grid P_i:

$$P_i = P_t\left(\frac{G_h A}{4\pi R^2}\right) \tag{9.8}$$

We can also write a related expression for the grid output power P_o:

$$P_o = P_r\left(\frac{4\pi R^2}{G_h A}\right) \tag{9.9}$$

Note that a high-power transmitter may be required to overcome the path loss and saturate the array. The lens-focused approach shown in Fig. 9.15 would presumably eliminate much of this path loss and therefore ease the source power requirement. However, if the active size of the grid is comparable to the focused beam waist, the central array elements will tend to saturate before the outer

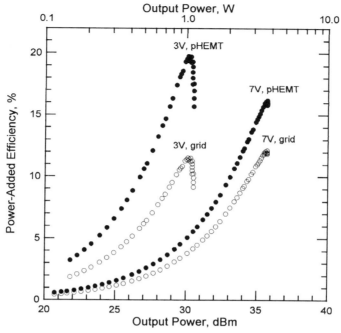

FIGURE 9.27 Power-added efficiency at 9 GHz versus amplifier output power. The solid circles are the efficiency for the pHEMT alone, discounting the DC power lost in the on-chip source resistors.

elements. This reduces the overall efficiency of the amplifier. On the other hand, the far-field technique illuminates the array uniformly, saturating all of the elements together.

The result is shown in Fig. 9.26. The grid is tuned for peak gain at 9 GHz to match the source output. The highest measured output power is 3.7 W at the 7-V bias. This is an 8-dB improvement over the HBT grid amplifier [11] and is the highest reported output power for a quasi-optical amplifier to date. Each pHEMT is providing about 18.5 mW, which is quite reasonable for these particular devices. The total DC power supplied to the grid is 24.5 W. To avoid overheating, the grid is only biased for several seconds at a time.

Figure 9.27 shows the power-added efficiency at two bias levels. The peak overall efficiency is 12%. There is a considerable bias voltage drop across the 100-Ω source resistors in the pHEMT chip. If we discount the DC power dissipated in these resistors, we can calculate an efficiency for the pHEMT alone. The device efficiency peaks at 20% for the 3-V bias at an output power of 1 W. The amplifier gain is plotted as a function of output power for the two bias levels in Fig. 9.28. At the maximum power of 3.7 W, the associated gain is 7 dB. The amplifier's 1-dB compression point is 680 mW for the lower bias and is 2.3 W for the higher bias.

A MONOLITHIC HBT GRID AMPLIFIER

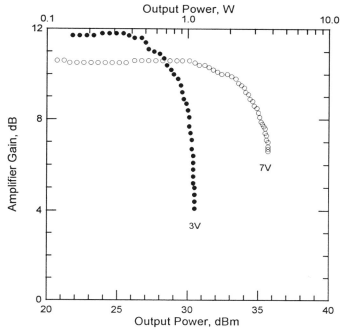

FIGURE 9.28 Associated gain at 9 GHz as a function of amplifier output power.

4 A MONOLITHIC HBT GRID AMPLIFIER

The advantages of quasi-optics become most apparent at millimeter-wave frequencies, where monolithically fabricated arrays could combine the outputs of many solid-state sources. However, most active quasi-optical components have been hybrid constructions operating at lower frequencies. Very recently, millimeter-wave monolithic quasi-optical amplifiers have been demonstrated [16–18,28,29]. This section will discuss the construction and performance of a 36-element monolithic HBT amplifier grid [15–17], the first monolithic quasi-optical amplifier to demonstrate appreciable gain and power combining at millimeter-wave frequencies. The peak gain of the array is 5 dB at 40 GHz. The amplifier's saturated output power is 670 mW.

4.1 Grid Construction

The amplifier arrays were fabricated using the HBT process established at Rockwell International [30]. Under optimum bias, the maximum available gain of an individual transistor is 9 dB at 40 GHz. The unit cell is shown in Fig. 9.29. Each cell contains a differential HBT pair. The bias resistor provides a self-bias to the base, so there are only two external bias connections required. The emitter

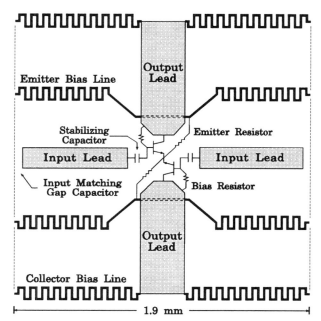

FIGURE 9.29 The monolithic HBT grid amplifier unit cell.

resistors are intended to reduce the common-mode gain and should not affect the differential-mode performance. The capacitors at the base of each transistor are included to stabilize the grid. Without them, the stability theory of Section 2.2 predicts that this grid will oscillate. The cell is 1.9 mm on a side. The horizontal leads couple the input beam to the bases of the HBTs. A narrow capacitive gap cancels the inductance of the input lead. The vertical drain leads radiate the output beam. The meandering lines provide the collector and emitter bias.

Figure 9.30 shows the 36-element array. The grid is monolithically fabricated on a 566-μm GaAs substrate. The thickness of the substrate is chosen to be a quarter-wavelength at 40 GHz. This is necessary to match the low input impedance of the HBT to freespace for maximum gain. The monolithic chip is attached to a Duroid substrate, and external dielectric tuning slabs are used to improve the input and output matching.

4.2 Gain

The gain is measured in the far field. The grid is biased with 16 mA per transistor. No oscillations were observed. To avoid thermal damage, the grid is biased for less than one second at a time. Figure 9.31 shows the measured small-signal gain response. The peak gain is 5 dB at 40 GHz, with a 3-dB bandwidth of 1.8 GHz, corresponding to 4.5%. The zero-bias gain is below

FIGURE 9.30 Photograph of the 36-element monolithic HBT grid amplifier compared with a dime. The active area of the array is 11.4 mm on a side.

−5 dB. At peak gain, the on–off ratio is over 25 dB. Also shown is the modeled gain, which agrees reasonably well with the measurements. The theoretical gain curve includes the skin-effect conductor losses in the input and output leads.

Figure 9.32 shows the relative gain as a function of polarizer position. Theory and experiment are in excellent agreement in Fig. 9.32a, the input polarizer

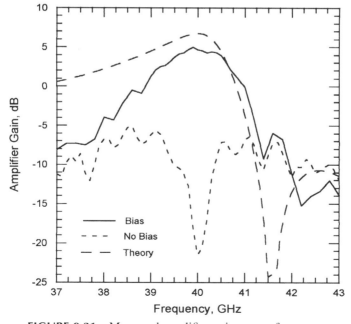

FIGURE 9.31 Measured amplifier gain versus frequency.

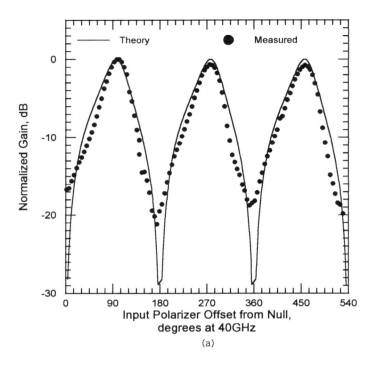

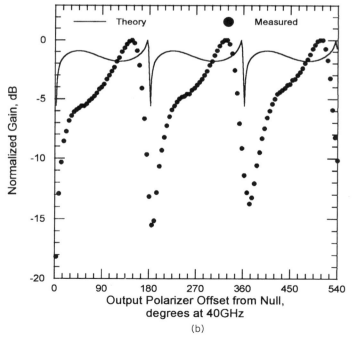

FIGURE 9.32 Relative amplifier gain at 40 GHz as a function of input (a) and output (b) polarizer positions.

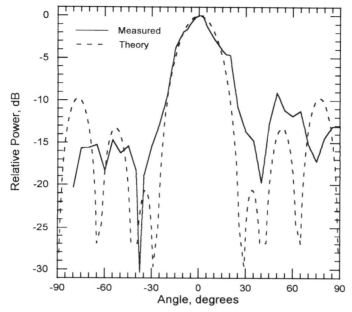

FIGURE 9.33 Output H-plane pattern at 40 GHz.

tuning curve. The agreement in the output polarizer tuning curve, Fig. 9.32b, is not as close. This may be due to the critical matching condition for the input circuit.

4.3 Angular Dependence

The grid's output H-plane radiation pattern is shown in Fig. 9.33. The pattern is measured by fixing the position of the grid and input antenna and rotating the receiving antenna. Also plotted is a theoretical pattern for six elementary dipoles placed in front of a mirror. The mirror is included to model the reflections from the input polarizer. The main lobes of the two patterns agree, which indicates that the pattern is diffraction limited.

4.4 Power

The output power is also measured at 40 GHz in the far field. A 10-W Ka-band traveling-wave tube (TWT) amplifier is used to saturate the grid amplifier. Figure 9.34 shows the power saturation characteristics for the array at three different bias levels. The peak output power is 670 mW for the 7-V bias. The gain and power-added efficiency for the 7-V bias are plotted in Fig. 9.35 as a function of amplifier output power. The associated gain of the grid is 2.5 dB at

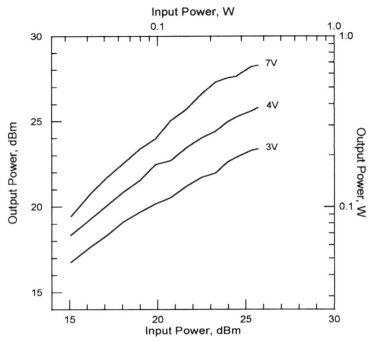

FIGURE 9.34 Measured output power at 40 GHz at three different bias points.

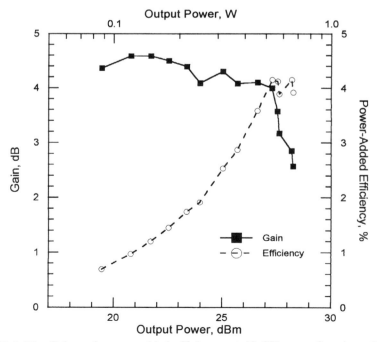

FIGURE 9.35 Gain and power-added efficiency at 40 GHz as a function of output power.

maximum output. The maximum efficiency is 4% at 500-mW output power. At 7 V, 70% of the DC power is actually dissipated in the bias lines and emitter resistors.

5 A MONOLITHIC pHEMT GRID AMPLIFIER

This section details the construction and performance of a 36-element monolithic U-band amplifier grid [14,18]. The active devices are pHEMT differential pairs, monolithically fabricated by Lockheed Martin Laboratories. The grid has a maximum gain of 6.5 dB at 44 GHz. Simply by changing the positions of external polarizers and tuning elements, the grid can operate as high as 60 GHz, with 2.5 dB gain.

5.1 Grid Construction

The grids were fabricated using the pHEMT process developed by Lockheed Martin Laboratories [21]. Figure 9.36 shows the unit cell. Each cell contains two 0.1-μm pHEMTs, with the sources tied together to form a differential pair.

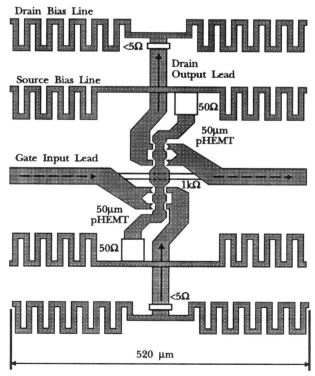

FIGURE 9.36 The monolithic pHEMT grid amplifier unit cell. Arrows indicate the direction of RF currents.

Each transistor has a gate width of 50 μm, distributed among four fingers. The 1-kΩ resistor allows gate bias to flow from cell to cell. This is possible because the gate draws very little DC current. The gate bias resistor passes under the air-bridged source connection. The two 50-Ω source resistors are intended to reduce the common-mode gain. The cell size is 520 μm, intentionally kept small to avoid exciting spurious substrate modes. The input beam is coupled to the gates of the transistor pair through the horizontal input leads. These leads also supply the gate bias. The output beam is radiated from the vertical drain leads. The small resistors in the drains are used for DC probing and diagnostics. The thin meandering lines supply bias to the drain and source. The grid is a 6 × 6 array of these cells fabricated on a 15-mil GaAs substrate. Figure 9.37 is a photograph of the monolithic grid.

The assembled amplifier is shown in Fig. 9.38a. The GaAs chip is mounted on a Duroid board with a relative dielectric constant of 2.33. The polarizers are also fabricated on Duroid. The corresponding transmission-line equivalent circuit is shown in Fig. 9.38b. The air gaps and substrates appear as transmission-lines. The radiating gate and drain leads are modeled as inductors. The polarizers are modeled as low-impedance inductors or high-impedance capacitors, depending on the polarization. The meandering bias lines appear as a shunt inductance at the input surface of the grid. Numerical values for the reactive elements were computed using the same methods mentioned in Section 3.2.

5.2 Gain

The gain is measured using the far-field approach. The input and output antennas are open-ended waveguide with the flanges sawed off. The waveguide has an

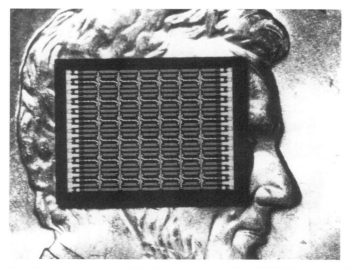

FIGURE 9.37 Photograph of the 36-element monolithic pHEMT grid amplifier compared with a penny. The active area of the array is 3.1 mm on a side.

A MONOLITHIC pHEMT GRID AMPLIFIER 369

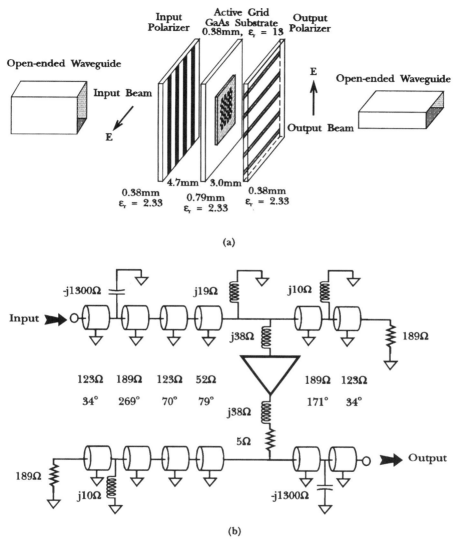

FIGURE 9.38 (a) The assembled grid amplifier. The polarizer positions are adjusted for peak gain at 48 GHz. (b) Transmission-line equivalent circuit at 48 GHz used for gain modeling.

aperture roughly the same size as the active area of the grid, which helps to maximize the power density incident on the array [14]. The grid is biased with a drain voltage of 3 V and a drain current of 8 mA per transistor. To avoid overheating, the grid is only biased for a second at a time. The amplifier is completely stable—no spurious oscillations were observed.

Figure 9.39 shows the measured small-signal gain of the array. The peak gain is 6 dB at 48 GHz. The 3-dB gain bandwidth is 1.7 GHz, which corresponds to

370 GRID AMPLIFIERS

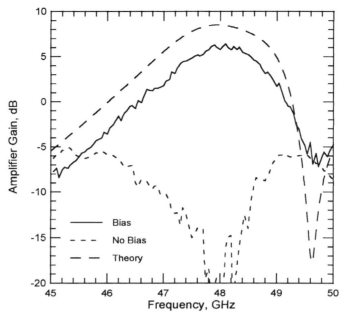

FIGURE 9.39 Measured amplifier gain versus frequency. The polarizer positions are shown in Fig. 9.38a.

3.5%. Without bias, the gain is below −5 dB over the entire frequency range. At peak gain, the on–off ratio is over 30 dB. An on–off ratio this high is a good indication that the amplifier is operating correctly. The gain modeled from the equivalent circuit is also plotted. The agreement is quite good, although the model slightly overpredicts the gain. It should be noted that the theoretical gain curve is generated using a measured value for the device transconductance g_m. Figure 9.40a shows the gain at 48 GHz as a function of input polarizer position. Figure 9.40b shows the gain as a function of output polarizer position. Again, the model agrees with the measured results.

5.3 Tuning Range

By changing the positions of the input and output polarizers, the amplifier can be tuned to operate at other frequencies. Figure 9.41 shows the measured gain with the amplifier tuned for a higher frequency. The peak gain is 2.5 dB at 60 GHz. The modeled gain is also shown, and it agrees with the measurement. For this measurement, a high-dielectric Duroid slab tuner is placed at the output of the amplifier to improve the match. Figure 9.42 illustrates the tuning range of the grid. Gain is plotted as a function of frequency with the amplifier tuned for 44, 48, 54, and 60 GHz. The gain is 6.5 dB with the amplifier tuned for 44 GHz, with a 3-dB bandwidth of 2.0 GHz (4.5%). When tuned for 54 GHz, the grid has a peak gain of 4 dB with a bandwidth of 3.2 GHz (6%).

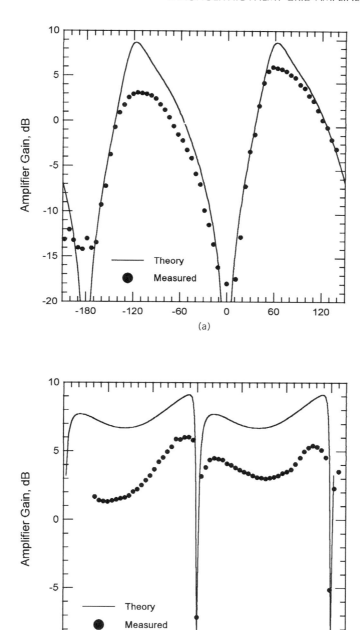

FIGURE 9.40 Amplifier gain at 48 GHz as a function of input (a) and output (b) polarizer positions.

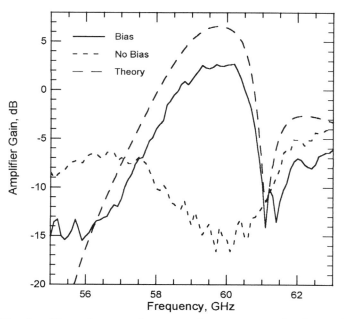

FIGURE 9.41 Amplifier gain versus frequency. An output tuner has been added, and the polarizers are adjusted for peak gain at 60 GHz. The noise floor of the receiver prevents measuring gain below -15 dB.

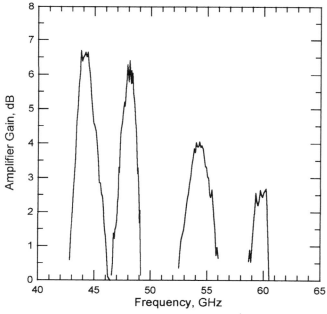

FIGURE 9.42 Tuning range of the amplifier. The 60-GHz measurement includes an output tuner.

6 CONCLUSIONS

The promise of quasi-optics has always been large-scale power combining of active solid-state devices at millimeter-wave frequencies. The goal would be a long-awaited replacement for vacuum tube amplifiers, to be used in medium-power millimeter-wave systems. Today, many pieces of the puzzle have been found. Watt-level power combining has been demonstrated in a microwave hybrid grid amplifier [12]. Monolithic grid amplifiers [16–18], as well as other types of monolithic quasi-optical amplifiers [28,29], have been successful, delivering significant gain and power combining at millimeter-wave frequencies.

Many questions remain unanswered. More extensive electromagnetic modeling of gain and stability is needed. Accurate thermal modeling is also a vital concern. Multiple-element quasi-optical systems are just now being explored. Finally, systems for confining the radiated beam need to be investigated. A promising confinement approach has recently been demonstrated by Ivanov and Mortazawi [31]. This approach uses a hard-horn feed, resulting in a system with a flange-to-flange power gain of 14 dB at 10 GHz.

ACKNOWLEDGMENTS

The authors are deeply indebted to Prof. David Rutledge at the California Institute of Technology, under whose guidance all of the work presented here was performed. We would also like to thank Dr. Moonil Kim at NASA Jet Propulsion Laboratory, who laid the groundwork for quasi-optical amplifiers. Finally, we are grateful to the Rockwell Science Center and Lockheed Martin Laboratories for their generous fabrication of the HBT and pHEMT differential chips and monolithic arrays.

REFERENCES

[1] D. B. Rutledge and S. E. Schwarz, "Planar multimode detector arrays for infrared and millimeter-wave applications," *IEEE J. Quantum Electron.* vol. 17, pp. 407–414, March 1981.

[2] H.-X. L. Liu, L. B. Sjogren, C. W. Domier, N. C. Luhmann, Jr., D. L. Sivco, and A. Y. Cho, "Monolithic quasi-optical frequency tripler array with 5-W output power at 99 GHz," *IEEE Electron. Device Lett.,* vol. 14, pp. 329–331, July 1993.

[3] J.-C. Chiao, A. Markelz, Y. Li, J. Hacker, T. Crowe, J. Allen, and D. B. Rutledge, "Terahertz grid frequency doublers," *6th Int. Symp. Space Terahertz Tech.,* March 1995.

[4] W. W. Lam, H. Z. Chen, K. S. Stolt, C. F. Jou, N. C. Luhmann, Jr., and D. B. Rutledge, "Millimeter-wave diode grid phase shifters," *IEEE Trans. Microwave Theory Tech.,* vol. 36, pp. 902–907, May 1988.

[5] L. B. Sjogren, H.-X. L. Liu, X.-H. Qin, C. W. Domier, and N. C. Luhmann, Jr., "Phased array operation of a diode grid impedance surface," *IEEE Trans. Microwave Theory Tech.,* vol. 42, pp. 565–572, April 1994.

[6] Z. B. Popović, R. M. Weikle, M. Kim, and D. B. Rutledge, "A 100-MESFET planar grid oscillator," *IEEE Trans. Microwave Theory Tech.*, vol. 39, pp. 193–200, March 1990.

[7] R. M. Weikle, M. Kim, J. B. Hacker, M. P. De Lisio, and D. B. Rutledge, "Planar MESFET grid oscillators using gate feedback," *IEEE Trans. Microwave Theory Tech.*, vol. 40, pp. 1997–2003, Nov. 1992.

[8] J. B. Hacker, M. P. De Lisio, M. Kim, C.-M. Liu, S.-J. Li, S. W. Wedge, and D. B. Rutledge, "A 10-watt X-band grid oscillator," *1994 IEEE MTT-S Int. Microwave Symp. Dig.*, pp. 823–826, 1994.

[9] J. B. Hacker, R. M. Weikle, M. Kim, M. P. De Lisio, and D. B. Rutledge, "A 100-element planar Schottky diode grid mixer," *IEEE Trans. Microwave Theory Tech.*, vol. 40, pp. 557–562, March 1992.

[10] M. Kim, J. J. Rosenberg, R. P. Smith, R. M. Weikle, J. B. Hacker, M. P. De Lisio, and D. B. Rutledge, "A grid amplifier," *IEEE Microwave Guided Wave Lett.*, vol. 1, pp. 322–324, Nov. 1991.

[11] M. Kim, E. A. Sovero, J. B. Hacker, M. P. De Lisio, J.-C. Chiao, S.-J. Li, D. R. Gagnon, J. J. Rosenberg, and D. B. Rutledge, "A 100-element HBT grid amplifier," *IEEE Trans. Microwave Theory Tech.*, vol. 41, pp. 1762–1771, Oct. 1993.

[12] M. P. De Lisio, S. W. Duncan, D.-W. Tu, C.-M. Liu, A. Moussessian, J. J. Rosenberg, and D. B. Rutledge, "Modelling and performance of a 100-element pHEMT grid amplifier," *IEEE Trans. Microwave Theory Tech.*, Dec. 1996.

[13] C.-M. Liu, E. A. Sovero, M. P. De Lisio, A. Moussessian, J. J. Rosenberg, and D. B. Rutledge, "Gain and stability models for HBT grid amplifiers," *1995 IEEE AP-S Int. Symp. Dig.*, pp. 1292–1295, 1995.

[14] M. P. De Lisio, "Hybrid and Monolithic Active Quasi-Optical Grids," Ph.D. thesis, California Institute of Technology, Pasadena, CA, 1996.

[15] C.-M. Liu, "HBT Grid Amplifiers," Ph.D. thesis, California Institute of Technology, Pasadena, CA, 1996.

[16] C.-M. Liu, E. A. Sovero, W. J. Ho, J. A. Higgins, and D. B. Rutledge, "A Millimeter-wave monolithic grid amplifier," *Int. J. Infrared Millimeter Waves*, vol. 16, pp. 1901–1910, Nov. 1995.

[17] C.-M. Liu, E. A. Sovero, W. J. Ho, J. A. Higgins, M. P. De Lisio, and D. B. Rutledge, "Monolithic 40-GHz 670-mW HBT grid amplifier," *1996 IEEE MTT-S Int. Microwave Symp. Dig.*, pp. 1123–1126, 1996.

[18] M. P. De Lisio, S. W. Duncan, D.-W. Tu, S. Weinreb, C.-M. Liu, D. B. Rutledge, "A 44-60 GHz monolithic pHEMT grid amplifier," *1996 IEEE MTT-S Int. Microwave Symp. Dig.*, pp. 1127–1130, 1996.

[19] J. B. Hacker and R. M. Weikle, "Quasi-optical grid arrays," in *Frequency Selective Surface and Grid Array*, T. K. Wu, ed., Wiley New York, 1995, pp. 249–324.

[20] R. D. Martinez and R. C. Compton, "A general approach for the S-parameter design of oscillators with 1 and 2-port active devices," *IEEE Trans. Microwave Theory Tech.*, vol. 40, pp. 596–574, March 1992.

[21] D.-W. Tu, S. W. Duncan, A. Eskandarian, B. Golja, B. C. Kane, S. P. Svenson, S. Weinreb, and N. E. Byer, "High gain monolithic W-band low noise amplifiers based on pseudomorphic high electron mobility transistors," *IEEE Trans. Microwave Theory Tech.*, vol. 42, pp. 2590–2597, Dec. 1994.

[22] P. M. Smith, D. W. Ferguson, W. F. Kopp, P. C. Chao, W. Hu, P. Ho, and J. M. Ballingall, "A high power, high efficiency millimeter wave pseudomorphic HEMT," *1991 IEEE MTT-S Int. Microwave Symp. Dig.,* pp. 717–720, 1991.

[23] S. Shanfield, A. Platzker, L. Aucoin, T. Kazior, B. I. Patel, A. Bertand, W. Hoke, and P. Lyman, "One-watt, very high efficiency 10 and 18 GHz pseudomorphic HEMT's fabricated by dry first recess etching," *1992 IEEE MTT-S Int. Microwave Symp. Dig.,* pp. 639–641, 1992.

[24] S. T. Fu, L. F. Lester, and T. Rogers, "Ku-band high power high efficiency pseudomorphic HEMT," *1994 IEEE MTT-S Int. Microwave Symp. Dig.,* pp. 793–796, 1994.

[25] D. R. Gagnon, "Highly sensitive measurements with a lens-focused reflectometer," *IEEE Trans. Microwave Theory Tech.,* vol. 39, pp. 2237–2240, Dec. 1991.

[26] "HP 85180A high frequency structure simulator," Hewlett-Packard, Networks Measurements Division, Santa Rosa, CA.

[27] J. Bae, J.-C. Chiao, K. Mizuno, and D. B. Rutledge, "Metal mesh couplers using evanescent waves at millimeter and submillimeter wavelengths," *1995 IEEE MTT-S Int. Microwave Symp. Dig.,* pp. 597–600, 1995.

[28] J. A. Higgins, E. A. Sovero, and W. J. Ho, "44-GHz monolithic plane wave amplifiers," *IEEE Microwave Guided Wave Lett.,* vol. 5, pp. 347–348, Oct. 1995.

[29] E. A. Sovero, Y. Kwon, D. S. Deakin, A. L. Sailer, and J. A. Higgins, "A PHEMT based monolithic plane wave amplifier for 42 GHz," *1996 IEEE MTT-S Int. Microwave Symp. Dig.,* pp. 1111–1114, 1996.

[30] P. Asbeck, F. Chang, K.-C. Wang, G. Sullivan, and D. Cheung, "GaAs-based heterojunction bipolar transistors for very high performance electronic circuits," *Proc. IEEE,* vol. 18, pp. 1709–1726, Dec. 1993.

[31] T. Ivanov and A. Mortazawi, "A two stage spatial amplifier with hard horn feeds," *IEEE Microwave Guided Wave Lett.,* vol. 6. pp. 88–90, Feb. 1996.

CHAPTER TEN

Beam-Control Arrays

KARL D. STEPHAN
University of Massachusetts, Amherst

Switches and phase shifters are examples of electronic control elements used in RF and microwave system design. In the context of systems using active or quasi-optical arrays, RF energy is typically transmitted in the form of a free-space wave or beam. This chapter describes the application of active grid arrays to the task of beam control. The historical precedents of active beam-control arrays are briefly reviewed in Section 1. The basic principles of operation are described in Section 2. In Section 3 we treat the advantages and limitations of beam-control grids from both electrical and mechanical standpoints. Section 4 gives practical examples of both switching and phase-shifting grid designs. Some concluding remarks about the field are made in Section 5.

1 BACKGROUND

Beam-control arrays incorporate active (nonlinear) devices in periodic structures such as grids. The electromagnetic characteristics of grids have been studied for many years. After Hertz used a grid of parallel wires to demonstrate the polarized nature of radio waves, numerous investigators such as J. J. Thomson and H. Lamb devised quasistatic theories around 1900 which accounted for grid reflection and transmission behavior [1]. At that time, the usefulness of grids in RF and microwave engineering was limited by the difficulty of generating waves that were short enough to be controlled by grids

Active and Quasi-Optical Arrays for Solid-State Power Combining, Edited by Robert A. York and Zoya B. Popović.
ISBN 0-471-14614-5 © 1997 John Wiley & Sons, Inc.

378 BEAM-CONTROL ARRAYS

of practical size. But since about 1970, developments in the generation and control of millimeter-wave energy by solid-state devices have led to more widespread applications of grids to beam control. We will now describe some immediate predecessors of active millimeter-wave beam-control grids.

1.1 Passive Grids for Millimeter-Wave Beams

The canonical situation for beam control by means of a passive grid consists of a plane wave normally incident upon an infinitely thin sheet of perfectly conducting material. The material is perforated in a periodic pattern whose period is considerably smaller than a free-space wavelength at the highest frequency of operation. If the pattern is invariant under a 90-degree rotation, the behavior of the grid is independent of polarization.

Motivated by a need for low-loss quasi-optical frequency diplexers at millimeter wavelengths, Arnaud and Pelow [2] experimentally investigated a variety of grid patterns with respect to their transmission characteristics versus frequency. (An important assumption underlying their work is that the grid period is much smaller than the shortest wavelength considered.) Several of these patterns, their approximate equivalent circuits (explained below), and their qualitative frequency responses are shown in Fig. 10.1.

The left-hand illustration in Fig. 10.1a shows a very simple structure consisting of a square grid of thin conducting strips. The propagation of a linearly polarized plane wave in a vacuum can be considered as propagation along a single-mode transmission line whose characteristic impedance is $[\mu_0/\epsilon_0]^{1/2} = Z_s = 377 \, \Omega$, where μ_0 and ϵ_0 are the permeability and permittivity of free space, respectively. The center illustration of Fig. 10.1a shows the transmission-line equivalent circuit which approximately models the observed frequency behavior of the simple mesh grid. This behavior is shown qualitatively in the right-hand illustration of Fig. 10.1a. The current induced in the grid conductors by the incident field is analogous to the current flowing through the shunt inductor of the equivalent circuit. For a particular grid period and strip width, an equivalent inductance can be found by matching the response function calculated from the equivalent circuit to the experimentally observed response.

While the equivalent circuit of Fig. 10.1a shows some transmission loss at all frequencies, placing two parallel grids a fixed distance apart as in Fig. 10.1b produces a bandpass response which exhibits zero loss at a finite frequency, for lossless grids. The lowest frequency of zero loss corresponds to an intergrid spacing of slightly less than one-half wavelength. This Fabry–Perot structure is the basis of certain active grid systems designed to show low transmission loss in one mode, as we will describe below.

The last grid structure illustrated that has relevance to the active grids described in this chapter is the so-called Jerusalem-cross pattern of Fig. 10.1e. This pattern consists of central crossed strips, each of which terminates in a wide bar which is close to a similar bar in the adjacent cell of the grid. Leaving aside

BACKGROUND 379

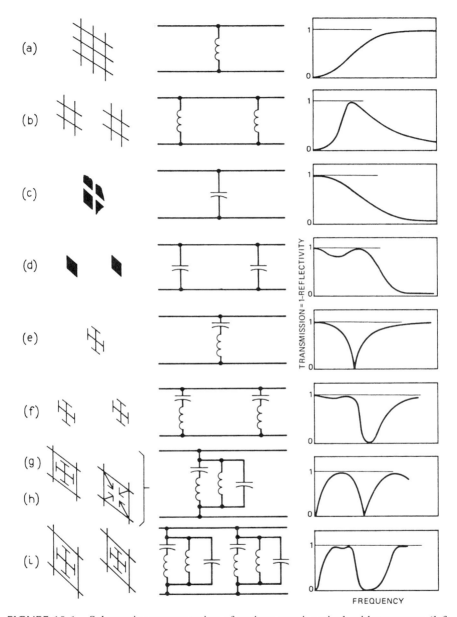

FIGURE 10.1 Schematic representation of various quasi-optical grid structures (left-hand column), their transmission-line equivalent circuits (middle column), and qualitative transmission response (right-hand column). (From reference [2], © 1975 AT&T. All rights reserved. Reprinted with permission.)

the question of how such a structure can be self-supporting (in practice a thin dielectric membrane can support the metal pattern), we note that the central strips provide an inductive reactance, just as in the simple grid structure of Fig. 10.1a. The gaps between the wide bars at the end of each strip in the cell pattern show capacitive reactance. Not surprisingly, the experimental response of this pattern shows a transmission zero (infinite loss) at a finite frequency. At this series-resonant frequency, the equivalent inductance of the central strips resonates with the equivalent capacitance of the gaps between the bars and presents a short circuit to the incoming plane wave, which is completely reflected. This circuit-resonant condition (as distinguished from a Fabry–Perot resonance) results in total reflection and zero transmission through the single grid of Fig. 10.1e at the resonant frequency. If one has already obtained an equivalent inductance value by experiments with simple grids or other independent means, the equivalent capacitance of a particular Jerusalem-cross grid can be found easily from the previously known equivalent inductance and the measured series-resonant frequency.

All the active-grid structures discussed in this chapter are variations on the cases discussed above. An active grid is achieved by inserting an active device into each cell of a passive grid so that the device's electronically controlled impedance affects the grid's response characteristics in a desired way.

1.2 Active "RADANT" Grids for Microwave Beams

In 1981 Chekroun *et al.* [3] described a novel system for controlling the phase of a microwave beam. Their invention, which they termed "RADANT" (a contraction of "radome" and "antenna"), consisted of a set of parallel wires embedded in a low-loss, low-dielectric-constant medium such as plastic foam. In the simplest functional form of the system, the wires are all parallel and lie in a plane perpendicular to the direction of propagation of the incident linearly polarized plane wave. Removing, for example, the horizontal conductors from the simple grid of Fig. 10.1a produces this configuration, which will still show inductive behavior as illustrated for vertically polarized radiation (electric field parallel to wires). If each wire is cut into pieces whose length equals the interwire spacing, and a PIN diode is inserted at the location of each cut, we obtain a single-plane RADANT phase shifter.

First suppose the diodes are forward-biased. A forward-biased PIN diode has a resistance of only a few ohms, so without oversimplifying too much we can consider that a wire with forward-biased diodes resembles a continuous conductor. Except for the absence of horizontal strips, this is the situation of Fig. 10.1a. We can expect the transmitted wave to experience a phase shift corresponding to that which a shunt inductor causes in the equivalent circuit.

Next, suppose the PIN diodes are reverse-biased. A reverse-biased PIN diode exhibits capacitive reactance, just as the gap between the bars of the grid in Fig. 10.1e shows capacitive reactance. If the diode capacitance is chosen so that the series resonance of the wire inductance and diode capacitance occurs well above the frequency of operation, the net equivalent reactance of the wires interrupted by the turned-off PIN diodes is capacitive. The phase shift in this case is opposite in sense to the forward-biased case (e.g., retardation instead of advance). Suppose the system parameters are chosen to permit equal transmission loss in the two cases. Thus the only effect of changing the diode bias from the reverse direction to the forward direction is to change the net transmission phase shift through the system. In this way an electronically controlled phase shifter for plane-wave beams has been achieved.

While the RADANT system exhibited some promising characteristics, obtaining sufficient phase shifts to be useful required many stacked layers of wire planes. Fabricating such three-dimensional structures with discrete components is apparently too costly to be practical. In the years since RADANT was developed, it has become possible to fabricate monolithic diode arrays of useful size and potentially low cost.

2 ACTIVE GRID ARRAYS: BASIC PRINCIPLES OF OPERATION

Although extraordinary fabrication methods can produce metallic grids with control diodes embedded in the grid without any supporting dielectric, most beam-control grids are fabricated on the surface of a dielectric substrate. At this point we should make a distinction between the specific term "mesh," which implies a periodic pattern of conductors running in both horizontal and vertical directions, and the more general term "grid," which includes meshes as well as arrays of parallel wires or strips which never touch. Strictly speaking, the structure shown in Fig. 10.1a is a mesh. As mentioned above, only mesh-type grids are capable of polarization-independent behavior. Since the ability to process dual polarizations is often a valuable property of quasi-optical systems, we will begin our description of beam-control array operation with a mesh structure.

2.1 Passive Square Mesh on a Dielectric Interface

In 1985 Whitbourn and Compton [4] published formulas for the behavior of a thin metal mesh on the planar interface between two dielectric volumes whose indices of refraction are n_1 and n_2, respectively. Figure 10.2 shows the dimensions of a square mesh which has a period of g and a strip width of $2a$. Define a dimensionless frequency

$$\Omega \equiv \frac{g}{\lambda_0} \qquad (10.1)$$

normalized to the grid period, in which λ_0 is the vacuum wavelength corresponding to the actual frequency under consideration. Suppose vertically polarized radiation impinges upon the mesh of Fig. 10.2 from the left. Empirically, it has been found that for a free-standing mesh (vacuum dielectric on both sides) a

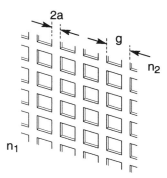

FIGURE 10.2 Periodic conducting mesh with period g and strip width $2a$, on planar interface between media having indices of refraction n_1 and n_2. (From reference [5], © 1993 IEEE.)

broad transmission passband occurs at a normalized frequency Ω_0, which is typically near unity. Physically, at this resonance the inductive reactance of the vertical strips is canceled out by the capacitive reactance of the spaces that separate the horizontal strips. Note that this frequency is considerably higher than the quasistatic limit imposed by Arnaud and Pelow above. At this frequency the mesh period is no longer small with respect to the wavelength, and quasistatic theory no longer applies.

Assuming the free-standing mesh's normalized resonant frequency Ω_0 has been determined, Whitbourn and Compton's formula for the parallel-resonant frequency Ω_0' of a mesh between dielectrics having indices of refraction different from unity is

$$\Omega_0' = \Omega_0 \left[\frac{2}{n_1^2 + n_2^2} \right]^{1/2} \tag{10.2}$$

Indices of refraction greater than unity raise the capacitance of the mesh and so lower the parallel-resonant frequency. By extending the equivalent-circuit concept of Fig. 10.1 to the square mesh between two dielectrics, Whitbourn and Compton found that it is adequately modeled by the circuit shown in Fig. 10.3,

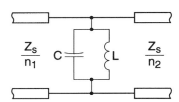

FIGURE 10.3 Equivalent circuit of square mesh shown in Fig. 10.2. (From reference [5], © 1993 IEEE.)

in which the characteristic impedances of media 1 and 2 have been reduced to Z_s/n_1 and Z_s/n_2, respectively, by the appropriate index of refraction. They also obtained formulas for the inductive reactance X_L and capacitive reactance X_C of the mesh equivalent circuit components, which are

$$\frac{X_L}{Z_S} = -\left(\Omega_0' \ln \csc\left[\frac{\pi a}{g}\right]\right)\left(\frac{\Omega}{\Omega_0'} - \frac{\Omega_0'}{\Omega}\right)^{-1} \quad (10.3)$$

and

$$\frac{X_C}{Z_S} = \frac{2}{n_1^2 + n_2^2}\left(4\Omega_0' \ln \csc\left[\frac{\pi a}{g}\right]\right)^{-1}\left(\frac{\Omega}{\Omega_0'} - \frac{\Omega_0'}{\Omega}\right) \quad (10.4)$$

The usefulness of these formulas will become apparent when an active device is inserted into each cell of the mesh, a situation we will now examine.

2.2 Active Mesh: Reflection Design

Except for specialized applications such as attenuators and continuously variable phase shifters, the active devices in a beam-control array will usually be in only one of two states, which we will designate as *on* or *off*. In the case of PIN diodes, *on* corresponds to forward bias and *off* to reverse bias. In the case of varactor diodes, we can define *on* as a high-capacitance state and *off* as a low-capacitance state. A switching function is achieved with binary-state devices by designing the same physical circuit to provide maximum transmission (switch closed) for one device state and minimum transmission (switch open) in the other state. Secondary characteristics such as return loss and power-handling capability are also important considerations, but in this section we will focus on designing the switch for maximum loss (isolation) in the switch-open condition.

While designs can be developed that use PIN diodes in either the *on* or the *off* state for the switch-open condition, the configuration to be described [5] has been found to be straightforward to design. It uses the PIN-diode *off* state to create an equivalent-circuit series resonance that reflects the incident energy.

If we modify the planar mesh of Fig. 10.2 by inserting a small PIN diode in each vertical strip of the mesh, a schematic representation of the result is shown in Fig. 10.4. (We have also destroyed the 90-degree symmetry and thus the polarization independence, but this can be restored by putting diodes in the horizontal strips as we will show below.) In a square mesh with diodes, the diode equivalent circuit appears in series with the equivalent inductance that models the vertical strips. Since the impedance of a PIN diode in its *off* state is primarily capacitive, we show a capacitance C_{diode} in the modified mesh's equivalent

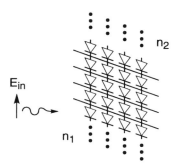

FIGURE 10.4 Mesh of Fig. 10.2 with PIN diodes inserted in each vertical strip. (From reference [5], © 1993 IEEE.)

circuit in Fig. 10.5. As with the passive mesh of Fig. 10.2, the diode mesh of Fig. 10.4 is formed on the interface between two semi-infinite regions having indices of refraction n_1 and n_2.

The diode capacitance now performs the same function as the gap capacitance in the Jerusalem-cross grid of Fig. 10.1e. This means that the mesh with PIN diodes in the *off* state will also exhibit a series resonance when the capacitive reactance of the diodes cancels the inductive reactance of the grid strips. Near the frequency of this resonance, the mesh will present nearly a short circuit to an incoming plane wave, with linear polarization parallel to the strips with diodes. This mode in which the PIN diodes are *off* is termed the reflection mode of the switch, and the desirable property of the switch in this mode is high isolation between the input side and the output side.

For a design that works at a single frequency or a relatively narrow range of frequencies, one must choose the diode capacitance C_{diode} and mesh inductance L so as to resonate at the desired frequency

$$f_S = \frac{1}{2\pi\sqrt{LC_{\text{diode}}}} \qquad (10.5)$$

An approximate value for L can be obtained by finding X_L from Eq. (10.3) and solving for L at f_S. The value of C_{diode} depends upon the electrical characteristics and physical layout of the particular device. The isolation of the switch in this mode is limited by losses in the diode and surrounding structure.

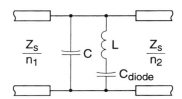

FIGURE 10.5 Equivalent circuit of mesh with PIN diodes in *off* state. (From reference [5], © 1993 IEEE.)

2.3 Active Mesh: Transmission Design

In contrast to its primarily capacitive reactance when reverse-biased, a forward-biased PIN diode exhibits a resistance at RF and microwave frequencies that is inversely proportional to DC bias current, down to a limit determined by parasitic resistances and carrier lifetime. In the active mesh equivalent circuit of Fig. 10.5 the capacitance C_{diode} will thus be replaced by an equivalent resistor of value R_{diode} when the devices are switched from reverse bias to forward bias by an external drive circuit (not shown). While it would be convenient if the inductive reactance of L canceled out the capacitive reactance of the mesh capacitance C near the design frequency f_s, this parallel resonance usually occurs at too high a frequency to be useful. Therefore a slightly more complex approach must be taken to obtain maximum beam transmission with the diodes *on*.

In Fig. 10.6 we show the same active mesh as in Fig. 10.5. Suppose the substrate is no longer semi-infinite, but is approximately one-quarter-wave thick in the dielectric with index of refraction n_2, as shown. On the back side of the substrate a second mesh is shown. This mesh is similar to the active mesh in period and strip dimensions, except that the diodes have been replaced by short circuits. We thus have two partially reflecting grids separated by a layer of dielectric, which forms a Fabry–Perot-like cavity. At a frequency which depends on the grid impedance and the dielectric thickness, a cavity resonance similar to that which gives rise to the passband shown in Fig. 10.1b occurs, and the transmission loss is at a minimum (ideally zero for a zero-loss system).

The resonance of lowest frequency occurs near the frequency for which the front and back grid are separated by approximately a quarter-wavelength in the dielectric. This is in contrast to the usual type of Fabry–Perot cavity, which is approximately an even number of half-wavelengths long. The reason for this difference is that, viewed from within the dielectric, the shunt impedances of the grids are considerably greater than the characteristic impedance of the medium, and so the grids act more as transmission-line perturbations than partially transparent mirrors. The equivalent circuit of the active-grid–dielectric–passive-grid "sandwich" is shown in Fig. 10.7. In the zero-loss case ($R_{diode} = 0$),

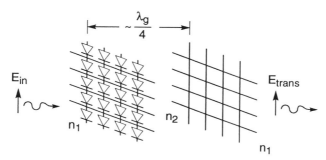

FIGURE 10.6 PIN diode mesh backed by passive mesh on back side of substrate having index of refraction n_2. (From reference [5], © 1993 IEEE.)

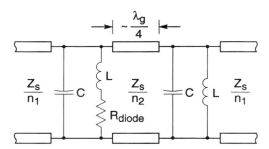

FIGURE 10.7 Equivalent circuit of mesh with PIN diodes *on,* substrate, and similar passive mesh. (From reference [5], © 1993 IEEE.)

perfect transmission would occur in the small-perturbation limit when the wave reflected from the left-hand grid is canceled by the wave reflected from the right-hand grid, which occurs for a quarter-wave spacing. For grids with progressively smaller shunt impedances, the resonant cavity length increases toward a half-wavelength.

In connection with an analysis of quasi-optical bandpass filters, Saleh [6] derived expressions for the transmission and reflection matrices of an arbitrary stack of grids and dielectric spacers, including the effects of polarization. His formulas can be used for the analysis of structures such as the dual-grid system of Fig. 10.6 if sufficiently accurate equivalent circuits can be found for the grids. In this way a grid spacing (i.e., substrate thickness) can be found which places the frequency of maximum transmission at f_s, the center frequency of the overall switch design. While the frequency of maximum isolation with diodes *off* is determined solely by the diode and mesh characteristics (including n_1 and n_2), the frequency of maximum transmission with diodes *on* can be adjusted by means of the substrate thickness, affording a measure of independence between the two design procedures. The factors influencing the choice of mesh design in terms of period and shape, required diode characteristics, and substrate properties will be discussed in more detail below.

2.4 Active Mesh: Phase Shift Design

A single active grid with varactor diodes embedded in the vertical strips can act as a quasi-optical phase shifting element in either reflection or transmission. Lam *et al.* [7] published a paper that described one of the earliest experimental versions of a millimeter-wave monolithic diode-grid phase shifter. If the capacitance C_{diode} in Fig. 10.5 varies with applied bias, the mesh reactance will also vary. This variable shunt reactance is the key to obtaining phase shifts in either transmission through the grid or reflection from the grid. If the grid and its substrate are backed by a plane mirror as shown in Fig. 10.8, all incident power will be reflected (for a zero-loss system), but the phase of the incident power will depend upon the variable capacitance of the diodes in the grid. While varactor diodes provide a continuously variable phase shift, PIN diodes in the same configuration will produce a two-state phase shifter, corresponding to the *on* and

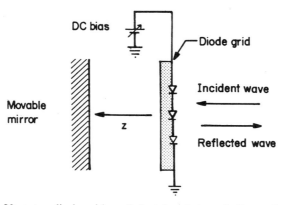

FIGURE 10.8 Varactor-diode grid on dielectric substrate in front of adjustable mirror for quasi-optical reflection phase shifter. (From reference [7], © 1988 IEEE.

off states of the PIN diodes. The particular design process needed for a phase-shifting grid depends on the exact nature of the phase-shifting circuit in which it is used. Suitable circuit design techniques for conventional transmission-line phase shifters are discussed in White [8] and can be adapted to the quasi-optical case in a relatively straightforward way.

One possible configuration is as follows. Suppose the mirror distance z in Fig. 10.8 is adjusted so that at the desired frequency of operation, the combined electrical length of the substrate and the substrate–mirror air gap is one-quarter of a wavelength. The plane-wave equivalent input impedance of such a combination is infinity, across which the shunt impedance of the grid appears. So for this special case, the plane-wave impedance of the grid is also the plane-wave impedance of the entire system. If X represents the net reactance of the mesh with diodes and Z_0 represents the impedance of free space, the phase angle of the plane-wave reflection coefficient is

$$\angle \Gamma = \sin^{-1}\left[\frac{2XZ_0}{Z_0^2 + X^2}\right] \qquad (10.6)$$

As the net reactance X varies over positive and negative values, so will the phase angle of the reflection coefficient. This simplified analysis neglects all losses, which can be a serious limiting factor in quasi-optical phase-shifting grids. In the next section we will discuss the question of loss and other practical limitations of these circuits.

3 ADVANTAGES AND LIMITATIONS OF BEAM-CONTROL ARRAYS

Beam-control arrays offer several advantages over more conventional control components using single-mode media such as waveguide or microstrip line. Among these are:

1. *Multibeam Capability.* The performance of a typical beam-control array degrades only gradually as the incoming beam's angle of incidence deviates from normal. If one considers each beam direction as a separate beam, a beam-control array can thus operate simultaneously on an infinite number of "beams" in this sense. In some imaging applications, for example, the incoming signal takes the form of a ray bundle that emerges from imaging optics. Each ray in the bundle travels in a slightly different direction, since it arises from a different part of the scene to be imaged. A beam-control array can perform the same switching or absorbing function upon all the rays in the bundle at the same time. This cannot be done with a single-mode control element.
2. *High-Power Capability.* While at this writing the author is unaware of any use of quasi-optical beam-control arrays in high-power systems, their potential for this area is considerable. If a high-power microwave beam is switched by an array of many diodes, each diode must switch only a fraction of the total beam power. Distribution of the total power to many diodes should lead to a very large power-handling capability for the array switch as a whole. As a more detailed explanation will show below, the picture is not quite this simple, but it seems that there are no obvious theoretical roadblocks to prevent the development of arrays that can switch kilowatt-level beams.
3. *Beam-Steering and Focusing Operation.* The simplest form of phase-shifting array consists of a grid of identical diodes with identical bias voltages. Any ray transmitted or reflected from the array thus undergoes the same phase shift, regardless of its point of intersection with the array. If the array is segmented into individually addressable "tiles," the phase shift encountered by the incident wave can now be independently controlled at each tile. In principle, a wide variety of electronically controllable quasi-optical elements could be made with such a system. The simplest of these is a phased array which would receive an incoming beam and aim it in any desired direction. Both diverging and converging lenses could also be synthesized with such an array, and the focal length would be under the control of the user. These hypothetical systems would require that each tile be not much larger than a wavelength and capable of a full 360 degrees of phase shift with acceptable loss. Such units do not presently exist, although examples of arrays with more limited capabilities will be shown below.

Having discussed the potential and actual advantages of beam-control arrays, we will now turn to the topic of practical limitations.

3.1 Electrical Limitations

Active-Element Loss White [8] discusses the fundamental limitations of switching and phase-shifting performance of PIN diodes in a lossless network. His analysis applies to any control device, even an FET, as long as it is used in a two-terminal variable-impedance mode. We will now summarize his results and then apply them to the special case of beam-control arrays.

We will first consider the case of a single-pole, single-throw switch. In one switching state the switch transmits energy with an insertion loss IL whose value we wish to determine. In the other state the switch reflects the incoming RF energy. White analyzed the case of a single switching element used to switch energy in a single-mode transmission line. He assumes the energy source to have a total available power P_A. The network which connects the energy source, the diode, and the load comprises a lossless three-port. At a single frequency one can always synthesize a lossless three-port network which will equalize the *on*-state and *off*-state RF power P_D dissipated in the switching element. When the same RF power is dissipated in either state, the average RF dissipated power is minimized.

Assume the switching element is connected to such a lossless three-port. Suppose in the *on* state the element is adequately modeled by a resistance of value R_F. In the *off* state, suppose the element's equivalent circuit is an R-C network consisting of capacitance C_J in series with resistance R_R. Strictly speaking, these values should be measured at the desired frequency of operation f. In the absence of such data, low-frequency values can be used, but the consequent neglect of increased losses due to the skin effect may result in overly optimistic results. Any reactive components that do not depend on the switching state can usually be absorbed into the lossless three-port.

With these component values White defines a switching cutoff frequency f_{CS}:

$$f_{CS} = \frac{1}{2\pi C_J \sqrt{R_F R_R}} \qquad (10.7)$$

which is a characteristic of the switching element only. With a perturbation analysis he shows that the ratio of dissipated RF power to available RF power is

$$\frac{P_D}{P_A} \simeq \frac{2f}{f_{CS}} \qquad (10.8)$$

Defining insertion loss IL in a well-matched system to be the ratio of total power available from the source P_A to power delivered to the load P_L under the conditions described above, energy conservation says that

$$IL = \frac{1}{1 - \frac{P_D}{P_A}} \simeq \frac{1}{1 - \frac{2f}{f_{CS}}} \qquad (10.9)$$

which is accurate within a few percent as long as the operating frequency f is less than a tenth of the switching cutoff frequency f_{CS}.

Extending this analysis to the case of many switching elements in an array is trivial since insertion loss is a dimensionless ratio. The ideal plane-wave case assumes a lossless grid with switching elements embedded in it. From the viewpoint of circuit theory, such a grid is simply a lossless three-port, since a single cell of the grid models the entire array in the transmission-line formalism discussed above. Therefore a grid of identical elements will have the same theoretical loss as a single element. While the ideal minimum-loss three-port may not be realizable in practice, there are various ways [9,10] to obtain the actual three-port model and calculate actual loss using conventional network-analysis methods. Using these models, a design may be optimized for performance in one state at the expense of the other, but in no case will the average loss be less than predicted by equation (10.9).

Turning to the case of phase-shifting beam-control arrays, White performs perturbation analyses similar to the one summarized above. He assumes that the switching element is used in a binary (two-state) phase shifter. Equalizing the losses in both states results again in minimum average dissipated loss. As mentioned earlier, the two basic classes of phase-shifting circuits operate in either reflection or transmission. For a reflection-type phase shifter, White finds that the ratio of dissipated to available power is

$$\frac{P_D}{P_A} \simeq 4\left(\frac{f}{f_{CS}}\right) \sin\left(\frac{\Delta\phi}{2}\right) \qquad (10.10)$$

where $\Delta\phi$ is the desired phase shift change when the switching element changes state. Other things being equal, a larger phase shift will entail a higher minimum loss.

For the transmission phase-shift circuit White finds that the appropriate expression is

$$\frac{P_D}{P_A} \simeq 2\left(\frac{f}{f_{CS}}\right) \tan(\Delta\phi) \qquad (10.11)$$

This expression should be used only for simple transmission circuits with small phase shifts (less than 45°), since transmission phase shifters that provide larger shifts than this usually take the form of a delay-line switching network, which is beyond the scope of this chapter.

Varactor-diode phase shifters whose phase depends continuously upon the applied voltage will show an average loss over the total phase-shift range that is predicted closely by equations (10.10) or (10.11) if representative values for R_F, R_R, and C_J are chosen.

For a given switching element in an optimized lossless three-port, the best theoretical value of any performance parameter which is expressed as a dimensionless ratio (e.g., insertion loss, isolation, return loss, etc.) will be the same whether one element is used in a guided-wave circuit or whether many such devices are used in a quasi-optical beam-control array. In practice the performance in an array will often be worse because of losses to be discussed in the next sections. The point is that the devices used in a beam-forming control array must be intrinsically capable of delivering the desired performance. No circuit-design ingenuity can compensate for active devices whose switching cutoff frequency is inadequate for the task.

Other Losses The calculation of loss in quasi-optical systems is at present an imprecise science. A typical beam-control array consists of active devices interconnected via metallized patterns on a dielectric substrate. For metals of relatively high conductivity such as gold and aluminum, skin depths at millimeter wavelengths are in the range of microns, which is also in the range of typical metallization thicknesses. Experience has shown that measured millimeter-wave conductor loss always exceeds the theoretical value calculated from the bulk DC resistivity, sometimes by a factor of two or more. While the loss tangents of numerous dielectrics commonly used in quasi-optical systems have been measured quite precisely in some cases [11], doped semiconductor substrates show loss in excess of the pure-crystal value, and such loss is sometimes difficult to predict for high-resistivity materials.

Another contribution to loss is scattering. If the surface of the beam-control array is flat and the incoming wavefront is curved, perfect transmission is impossible even if an appropriate matching layer covers the array. Bias leads, joints between array tiles, mechanical support structures, and other large-scale obstacles all scatter incoming radiation in a way that is hard to predict. Modeling such effects is a very challenging problem numerically because the system size is typically many wavelengths in all three dimensions. Experiment is the court of last resort in such cases.

RF Power Limits As mentioned previously, the potential uses of beam-control arrays for controlling high-power microwave and millimeter-wave beams remain largely unexplored at this writing. Nevertheless, something can be said about the theoretical maximum power capability of these systems.

Suppose a beam-control array is illuminated by a Gaussian beam whose waist radius *at the array plane* is w. Following Siegman's treatment of power density in a Gaussian beam [12], we find that the total power P in the beam must equal the integral of the power density $U(r)$ over the plane of the array, where r is the radial distance from the beam axis. This condition leads to the following expression for power density:

$$U(r) = \frac{2P}{\pi w^2} e^{-2r^2/w^2} \qquad (10.12)$$

The power density on the axis of a Gaussian beam is 2.5 times higher than that which would result if the power were distributed uniformly over an aperture of diameter $3w$. Since power is concentrated near the axis, the maximum stress in a high-power application will be applied to devices in that region. If the array has a uniform density of ρ switching elements per unit area, the beam power P_e controlled by each element can be approximated by

$$P_e(r) \simeq \frac{U(r)}{\rho} \qquad (10.13)$$

Once P_e is found, the transmission-line equivalent circuit of the array can be used to find the maximum voltage and current stresses on a particular switching element. White [8] shows that if the maximum allowable voltage stress on an element is V_M and the maximum allowable current stress is I_M (both rms), then the maximum theoretical power-handling capability of the switch is

$$P_M = \frac{V_M I_M}{2} \qquad (10.14)$$

This is the highest instantaneous power the device can theoretically control before breaking down. In practice either the voltage or the current limitation is the determining factor, and one or the other limitation is reached at a level somewhat below P_M.

As an example, we will calculate the theoretical maximum power-handling capability of an array with switching elements whose V_M is 50 V and whose I_M is 20 mA. According to equation (10.14) the maximum peak power per element is 0.5 W. If a square array with a device-to-device spacing of 1.25 mm is used to switch a beam whose waist radius is 2.5 cm, the elements on the beam axis will receive maximum power P_M when the total power P is 314 W. Larger arrays of the same elements could switch beams of higher power.

While it might appear prudent to use the smallest allowable element spacing to distribute RF power over the most elements, another issue comes into play when the elements are PIN diodes: DC control power. Both the RF power-handling capacity and the DC power consumption increase as the square of diode density. This limitation is basically a steady-state thermal one and will be discussed in the next section.

3.2 Mechanical Limitations

Thermal A single PIN diode draws a current of several milliamperes at a voltage drop of approximately one volt when in the forward-bias condition. Even in small-signal applications such as radiometric chopping, the average power dissipated by each diode if switched in a 50% duty cycle is several milliwatts. The total heat load created by an array of hundreds or thousands of diodes can be considerable, and cooling the array becomes an important issue. If the array is used for high-power RF switching, the RF heat load will add to the DC load.

Small transmission arrays (less than about 3 cm^2) can be cooled by a combination of passive convection and conduction through the semiconductor substrate to the edges of the array. Larger arrays than this must generally be mounted to a relatively thick substrate with good thermal characteristics. Unfortunately, most materials that are transparent to short microwaves and millimeter waves (e.g., polytetrafluoroethylene or quartz) are not good thermal conductors, and good thermal conductors (e.g., aluminum or steel) do not transmit RF energy. One possible exception to this rule is beryllia, but its toxic properties make it undesirable for all but the most demanding applications. These limitations mean that if an application involves substantial heat dissipation, it is best to use arrays in the reflection mode rather than the transmission mode. Reflection arrays can be backed by a substantial metallic heatsink.

If ways can be found to avoid large heat loads, the need for a heatsink can be reduced and larger-area transmission-mode arrays will be feasible. Some applications such as T-R switching in radar systems can be designed to require a very small duty cycle ratio of device *on* time to *off* time. Varactor-diode switches are another obvious choice here. The same power-spreading property that makes PIN-diode arrays capable of handling very high power will allow varactor-diode arrays to handle higher power as well, although varactor circuits have power limitations that are discussed below.

Structural The fabrication of a beam-control array is a nontrivial mechanical design problem, especially for larger sizes. If the array is used in the transmission mode, the array substrate (often relatively fragile GaAs) and dielectric matching layers must all be supported rigidly and without significant air gaps between the layers. From a mechanical standpoint it is desirable to fabricate the largest arrays that the foundry can handle, but foundry and yield limitations will prescribe a maximum tile size. If the system requirement demands an array size larger than the maximum tile size, the array must be made up of two or more tiles.

A transmission array made with tiles would have to be mechanically supported with a grid-like structure that would hold the tiles in alignment without causing significant beam blockage. It is much simpler to construct a reflection-type array on a metallic backing plate which can provide mechanical stability and thermal heatsinking as well. Connections for DC biasing can also be made more easily to a reflection-mode system, since bias paths can cross behind areas that are within the beam as long as they run behind the reflection plane at the bottom of the array.

Two other mechanical requirements are for uniformity of dielectric thickness and front-surface flatness.

Whenever beam-control array designs use the dielectric substrate as part of a Fabry–Perot resonator, the resonator frequency accuracy depends directly upon the thickness accuracy of the dielectric. Since the thickness of a high-dielectric substrate an electrical quarter-wave thick will typically be less than 1 mm at millimeter wavelengths, controlling the array resonant frequency to within 1% requires thickness tolerances on the order of microns. This order of substrate thickness accuracy is not customarily supplied by typical IC fabrication processes, but can be achieved by special techniques such as precision lapping.

If the array is used in reflection, it is important to control the heights of tile surfaces relative to one another to within at most about 3% of a free-space wavelength. Larger variations than this will scatter an increasing proportion of the beam because the reflected waves will no longer add in phase.

4 EXAMPLES OF BEAM-CONTROL ARRAYS

4.1 Switching Hybrid PIN-Diode Array for 94 GHz

With P. Goldsmith and F. Spooner, the author developed a prototype version of a transmission-mode beam-control switch array for use in the 94-GHz region [5]. The earliest version of the array was built in hybrid form, but the design process is similar for either hybrid or monolithic versions.

Array Design The goal was to make a switching array of PIN diodes for a 94-GHz beam with a beam waist radius of 0.6 cm. In the reflection mode (with the diodes *off*) the array was to reflect the incoming energy with minimum reflection loss. In the transmission mode (with diodes *on*) the array was to transmit the incoming beam. Low insertion loss in the transmission mode was not a critical requirement since the radiation was transmitted to a quasi-optical load anyway. A more critical parameter was reflection (return) loss in the transmission mode, which should be as high as possible over the frequency range of interest.

The most significant dimension in diode-array design is the cell size g, which in square arrays equals the diode-to-diode spacing. (Certain advantages may be obtained with cells having more than one diode, but will not be discussed here). The upper bound on cell size is the point at which substantial power "leaks through" the mesh openings when the switch is in the reflection mode. This upper bound is poorly defined but is in the neighborhood of $g = \lambda$. A conservative value for g of 0.89 mm was chosen for this array, which amounts to only 0.28 λ at 94 GHz.

The diode *off*-state capacitance C_{diode} together with the equivalent-circuit mesh inductance L must resonate at the design frequency f_s given by equation (10.5). For a given cell size, the mesh inductance cannot be made arbitrarily small since the only way of lowering it is to make the strips wider. But physically the strip width $2a$ cannot exceed the cell size g.

For the hybrid design under study, the diodes were discrete beam-lead devices with an *off*-state capacitance of approximately 17 fF. For an f_s of 94 GHz, equation (10.5) gives the required mesh inductance L to be 169 pH. Experiments with rectangular meshes of the desired cell size failed to achieve values of inductance this low, so a circular metallized area at each mesh strip intersection was added. This shape provided the desired mesh inductance without unduly raising the mesh capacitance. While closed-form analytical expressions for mesh inductance and capacitance are not available for shapes other than a rectangular mesh, recent progress in software designed to characterize the elec-

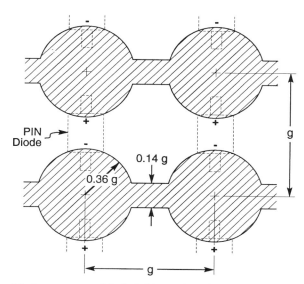

FIGURE 10.9 Mesh pattern used in hybrid quasi-optical PIN diode switch of Stephan et al. ($g = 0.89$ mm). (From reference [5], © 1993 IEEE.)

tromagnetic behavior of planar structures means that the behavior of more complex mesh shapes can now be modeled successfully before expensive experimental work is performed.

Since the mesh was to be used with linear polarization only, diodes were needed in only one of the two mesh axes. The mesh layout with dimensions and beam-lead PIN diode locations is shown in Fig. 10.9. The chrome-gold metallized pattern consists of a matrix of circular dots connected by horizontal strips only. The vertical connections were completed when the PIN diodes were bonded onto the alumina substrate.

Once the reflection design was completed, the transmission design could begin. An optimum Fabry–Perot transmission cavity uses reflectors of similar characteristics at either end of the cavity. To obtain a passive mesh (no diodes) with a surface impedance similar to that of the active mesh with diodes *on*, we rotated the pattern 90° so that the metallized strips connecting the round patches were in the vertical plane, as shown in Fig. 10.10. For vertically polarized radiation this pattern's behavior closely resembled that of the active mesh with forward-biased diodes.

The dielectric thickness required to place the resonator passband near 94 GHz was not achievable with alumina substrates of standard dimensions, so a fused-quartz spacer layer was inserted between the two alumina substrates supporting the meshes. The two 0.25-mm-thick alumina substrates ($\epsilon_r = 9.5$) separated by a 0.15-mm-thick layer of fused quartz ($\epsilon_r = 3.8$) composed a cavity with a total electrical length at 94 GHz of 0.574λ, which indicates that the second longitudinal mode of the cavity is used. (A wider transmission bandwidth would be

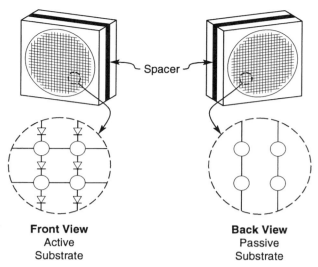

FIGURE 10.10 View of active-mesh alumina substrate, fused-quartz spacer layer, and passive-mesh alumina substrate showing relative orientation of meshes. (From reference [5], © 1993 IEEE.)

available from the lowest-order mode, but alumina substrates of suitable thinness were not available.)

To conserve discrete diodes, the array was laid out in a pattern with a roughly circular outline as illustrated in Fig. 10.11. A total of 464 diodes were bonded to the chrome-gold substrate. This pattern fit the circular cross section of the

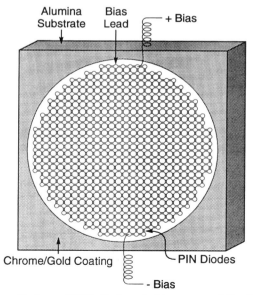

FIGURE 10.11 Overall view of PIN-diode switching array of Stephan *et al.* showing chrome-gold pattern with diodes on alumina substrate. (From reference [5], © 1993 IEEE.)

quasi-optical beam, but the simple series–parallel biasing circuit used meant that diodes in the top and bottom rows received about twice as much bias current as diodes in the middle rows. (Dual-polarization arrays can be biased in a way that avoids this problem, as we will describe below.) Nevertheless, the relatively slight variation in diode resistance over the array did not appear to impair performance significantly, as the following results will show.

Array Tests and Results No standards presently exist for tests of quasi-optical components. While this problem may be solved in the future if such components come into wider use, the short-term solution is to custom-build a quasi-optical test setup such as the one shown in Fig. 10.12. Lenses with a diameter equal to their focal length of 7.5 cm were used to create a beam with a waist radius of about 0.6 cm. Plastic antireflection coatings were applied to the fused-quartz lenses.

For the transmission tests, the 0-dB reference was taken to be the loss of the test setup with no array inserted at the waist of the beam. Residual reflections between the lenses caused a rapid oscillation of the frequency response, which was eliminated by means of the 5% smoothing function of the scalar network analyzer used. The data shown therefore result from taking a moving average over a 2-GHz range centered at each frequency measured.

Fig. 10.13 shows the results of transmission measurements performed on the array. With no DC current through the diodes (*off* state), the array's series resonance occurs at 90.3 GHz. The very low surface impedance presented by the active mesh near that frequency results in transmission loss exceeding 20 dB over a 12-GHz bandwidth. At the design frequency of 94 GHz, *off*-state transmission loss is about 26 dB. In the *on* state with 110-mA total array current, the residual losses caused by the nonzero diode resistance contribute to a minimum insertion loss of 3.7 dB at 93.9 GHz.

The reflection loss is of more interest than the insertion loss in the reflection/absorption application for which this array was designed. In the reflection state an incoming radiometric signal is reflected to the input of a radiometer, and in the absorption state the array should present an ambient-temperature matched

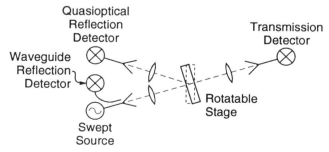

FIGURE 10.12 Quasi-optical reflection and transmission test setup of Stephan *et al.* (From reference [5], © 1993 IEEE.)

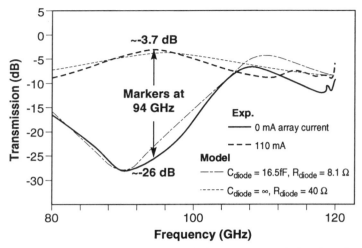

FIGURE 10.13 Measured and modeled switch array transmission loss for array of Stephan *et al.*; DC current of 0 mA and 110 mA. (From reference [5], © 1993 IEEE.)

load to the radiometer. Measurements [13] have confirmed that the equivalent black-body temperature of an array of forward-biased PIN diodes is within a few degrees of the physical temperature for reasonable bias currents, so whether absorption takes place in the somewhat lossy array, or in the passive absorber behind it, is of no great consequence. The reflection loss of the array at an angle of incidence of 17° was measured versus frequency for both states and is shown in Fig. 10.14. Although the equivalent-circuit model of the array assumes nor-

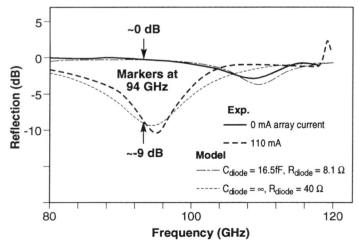

FIGURE 10.14 Measured and modeled switch array reflection loss for array of Stephan *et al.*; DC current of 0 mA and 110 mA. (From reference [5], © 1993 IEEE.)

mal incidence, the observed changes in behavior between normal incidence and slightly off-normal angles of incidence are usually small. The 0-dB reference was a flat aluminum reflector placed at the same physical location as the array. The reflection loss of the array in the *off* state was too low to measure accurately, but was well below 1 dB. The highest reflection (return) loss in the *on* state was 12 dB, falling to 9 dB at the design frequency of 94 GHz. This figure could be improved by means of a passive matching layer and would not affect the low reflection loss of the *off* state.

Quasi-optical network analysis software based on Saleh's formulas was developed to simulate the array's behavior. Component values of the array equivalent circuits were varied so that the simulated response approximated the measured response as shown in Figs. 10.13 and 10.14, with the resulting values shown in the figures. The *off*-state diode capacitance C_{diode} of 16.5 fF agreed closely with the nominal value of 17 fF, but the corresponding series resistance of 8.1 Ω is rather high, as is the *on*-state resistance of 40 Ω. These values are higher than the corresponding low-frequency ones because the skin effect restricts millimeter-wave currents to a relatively thin layer near the surface of the devices, increasing their effective resistance.

4.2 Switching and Phase Shifting Monolithic Varactor-Diode Array at 60 GHz

More recent work by Qin *et al.* [14] has shown that monolithic arrays of Schottky-contact varactor diodes can produce useful switching and phase shifting functions as well. Varactor diodes show a high-impedance state when reverse-biased to near the breakdown voltage, since this reduces their capacitance to C_{min} (about 10 fF for the diodes in this array). Zero bias or a small forward bias raises the device capacitance to C_{max} (about 80 fF in this case), which is the low-impedance state. The series resistance of varactor diodes typically decreases slightly with increasing reverse bias and is not under the direct control of the user.

Varactors have both advantages and disadvantages as control elements when compared to PIN diodes. The single most significant advantage of varactor diodes is their low power consumption vis-à-vis PIN diodes, since the only current drawn is a small reverse-bias leakage current. Their negligible power consumption removes the diode-density limitation imposed by thermal heating, which means varactor-diode arrays can be much more densely packed than PIN-diode arrays. The 60-GHz array under consideration, for example, uses a cell size of 900 μm $\times$ 360 μm (0.18λ $\times$ 0.072λ at 60 GHz). A 2.5-cm^2 array with these dimensions contains some 2000 diodes. A second advantage of varactor diodes is that their reactance is continuously variable with bias. In phase-shifting applications this can in principle allow a single array (or perhaps two) to achieve the complete 360° phase shift required in most phased-array applications.

On the other hand, the continuous variation of capacitance with bias voltage can create problems at the system level. Unless the fabrication process is very tightly controlled, the C-V characteristic of each array is slightly different. This means that the application of a uniform phase-control voltage to a set of phase-control array tiles will result in different phase shifts from each tile. The bias behavior of PIN-diode arrays tends to be much more predictable in this regard, at least when used in the binary *(on–off)* mode. A second potential disadvantage of varactor-diode arrays also arises from the variation of capacitance with voltage. Unlike the limited response bandwidth of PIN-diode RF resistance, which has a natural rolloff in the high kilohertz range, the response of varactor diode capacitance to applied voltage extends well above the microwave or millimeter-wave operating frequency of the device. This fundamental difference will limit varactor-based arrays to the control of relatively low-power beams, since varactors show undesirable harmonic generation and cross-modulation effects at higher power levels. (However, these effects can be exploited in other applications as described in Chapter 11.)

Computer-controlled D/A converters that bias individual rows of diodes can to some extent overcome the problem of C-V variation, and in fact Qin *et al.* used such a control system to achieve some of the results we will now summarize. A single grid of varactor diodes fabricated on a three-inch GaAs substrate was measured for transmission performance at 62 GHz using the test setup shown in Fig. 10.15. This setup differs from the one of Fig. 10.12 in that overmoded waveguides are used to approximate a plane wave at the array. While the electric field in such a setup actually goes to zero at the transition edges, the data so obtained probably do not differ greatly from what would be measured in a true plane-wave situation.

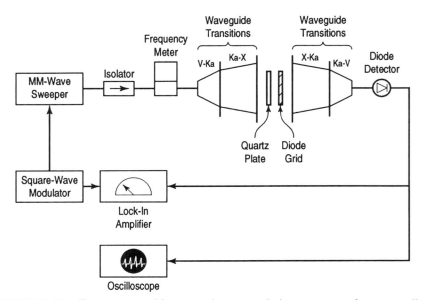

FIGURE 10.15 Test setup used in measuring transmission response of varactor-diode switching array of Qin *et al.* (Adapted from reference [14], © 1995 IEEE.)

As Fig. 10.15 shows, the array was used in conjunction with a quartz plate of unstated thickness. This plate performed a tuning function to produce the reported switch response at 62 GHz that is shown in Fig. 10.16. With the diodes reverse-biased, their capacitance decreases to the extent that the diode impedance dominates the mesh impedance. Proper positioning of the quartz plate produced the minimum insertion loss peak of 2 dB at a bias voltage of about −2.5 V. The insertion-loss maximum of 26 dB (and the resulting *on/off* ratio of 24 dB) is probably a series-resonant phenomenon similar to that used in the PIN-diode array described above. The strip inductance required for series resonance at the stated value of diode capacitance is consistent with expected values. Similar experiments with two stacked grid arrays produced an *on/off* ratio of 42 dB across a 6-GHz bandwidth centered at 63 GHz, with an average minimum insertion loss of 4 dB.

A test setup similar to that in Fig. 10.15 was also used with reflection phase-shift experiments on the same array. The receiving waveguide transitions were replaced with a metal mirror, and the phase of the reflected beam was sampled by means of a waveguide directional coupler installed at the transmitter. The 0-dB reference for loss was not stated explicitly, but the data in Fig. 10.17 show the attractive feature of relatively constant loss across a phase-shift range of some 130°.

4.3 Phase-Shifting Monolithic Varactor-Diode Array at 94 GHz

Recent development work by Choudhury *et al.* [15] of Millitech Corporation has demonstrated the practical utility of diode arrays for quasi-optical beam-switching applications in imaging systems. For beam steering purposes, an imaging system required a reflecting surface in which the reflection phase of

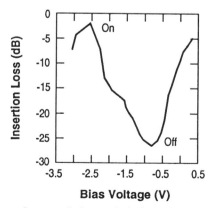

FIGURE 10.16 Measured transmission switch insertion loss versus bias voltage at 62 GHz for varactor-diode array of Qin *et al.* (Adapted from reference [14], © 1995 IEEE.)

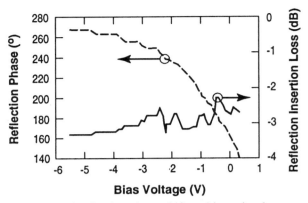

FIGURE 10.17 Measured reflection phase shift and insertion loss versus bias voltage for varactor-diode array of Qin *et al.* (Adapted from reference [14], © 1995 IEEE.)

individual reflecting tiles could be changed by 70° upon command. In this section we describe the design and operation of the monolithic varactor-array tiles designed to perform this function.

Monolithic Varactor Design The performance of a beam-control array using varactors depends critically upon the series resistance and the capacitance ratio (maximum/minimum) of the varactors used. Lower series resistance translates into lower loss, and a large C-V ratio allows a larger phase shift, other things being equal. In the array under discussion, a hyperabrupt doping profile [16] was used for the Schottky-barrier diodes. The doping of a hyperabrupt Schottky-barrier diode decreases in a controlled fashion as distance from the metal–semiconductor junction increases. Nine different layers were grown by molecular-beam epitaxy so that the n-type doping concentration fell in a step-wise fashion from 1.6×10^{17} cm^{-3} at the Schottky junction down to 2.4×10^{16} cm^{-3} just above the heavily doped (5×10^{18} cm^{-3}) n^+ layer to which the lower ohmic contact was made. This GaAs diode structure was backed by a semi-insulating GaAs substrate with a metallic reflecting backshort applied to its opposite side. This carefully designed doping profile yielded a capacitance ratio of 4.9 for a 5.4-μm-diameter discrete diode and a zero-bias capacitance of 32 fF.

A diode with an intrinsically high capacitance ratio can fail to meet performance requirements if the metallic contacting structure shunts the device with too much parasitic capacitance. This problem was largely avoided in this design by means of an air bridge leading to the Schottky contact of the diode, as the SEM photograph of Fig. 10.18 shows. The actual diode structure lies beneath the rounded tip of the lead coming in from the upper right. Low series resistance is provided by the much larger ohmic contact made to the other lead. Low parasitic capacitance between the two leads is insured by keeping them physically separated as much as possible, as well as by the air bridge which minimizes the fraction of electric field lines which enter the high-dielectric substrate near the device.

EXAMPLES OF BEAM-CONTROL ARRAYS   403

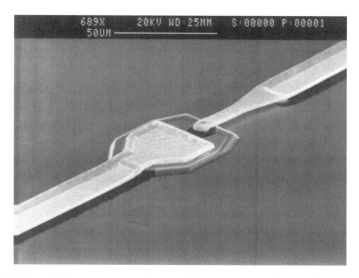

FIGURE 10.18   SEM photograph of monolithic varactor diode used in varactor diode array of Choudhury *et al.* (Millitech Corp.)

Varactor Array Design   Depending on whether single- or dual-polarization phase shifting is required, two different kinds of array layouts can be used. If linearly polarized energy in only one direction is to be phase-shifted, then the diodes can all be parallel to each other and to the direction of polarization, as Fig. 10.19 shows. From the control circuit's point of view, the diodes in the

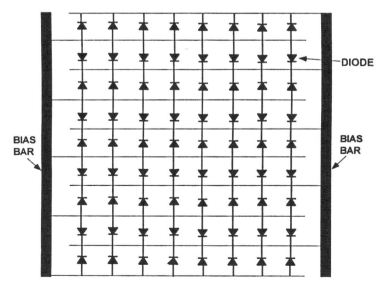

FIGURE 10.19   Layout of varactor-diode array for single-polarization phase shifting. (Millitech Corp.)

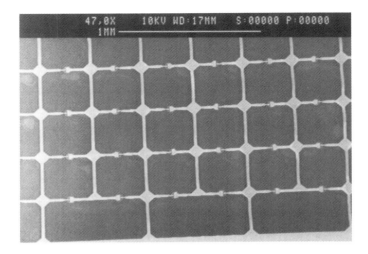

FIGURE 10.20 SEM photograph of single-polarization varactor-diode array with bias bar at bottom. (Millitech Corp.)

entire array are all in parallel. Biasing such a circuit is a simple matter since each tile requires the same voltage as a single diode, typically no more than 15 V. Figure 10.20 is an SEM photo of this kind of array. The wider strip at the bottom is one of the bias bars.

For some applications a single-polarization array is inadequate since the same phase shift must be experienced by two orthogonal polarizations, or by a circularly polarized beam. One layout that has been found to be useful for a dual-polarization array is the "fishnet" pattern shown in Fig. 10.21. This pattern is generated when diodes are inserted in both horizontal and vertical strips of a square metallic grid. As long as the grid appears invariant under a 90° rotation from an RF point of view, its actual angular position with respect to the polarization of incoming energy is arbitrary. Hence the array is laid out at a 45° angle to the perpendicular so that points at equal bias potential can be connected together by bias bars at the edges of the array as shown. From a biasing standpoint the columns of diodes are in series. This means that for a tile with N columns of diodes, the total bias voltage must be N times that required for a single diode. While this is clearly a disadvantage compared to the single-polarization layout, it is a problem that can be overcome.

In either type of layout the usual considerations of strip width and diode spacing apply. As in the arrays previously described, these variables are adjusted together with the substrate thickness to obtain the desired reflection phase shift behavior. The designer of this particular array took advantage of a commercial electromagnetic structure modeling program to obtain the embedding impedance seen by the varactor diode in its unit cell. This information was used together with knowledge of the diode characteristics to model the design of the array before the expensive steps of fabrication and layout were taken.

FIGURE 10.21 Layout of varactor-diode array for dual-polarization phase shifting. (Millitech Corp.)

Array Results While development work is still in progress at this writing, some preliminary data have been obtained from 12.5-mm^2 tiles using the dual-polarization layout shown in Figure 10.21. Figure 10.22 shows test results obtained from a quasi-optical test setup designed to measure the reflection phase shift of varactor-diode beam-control tiles. The setup uses a horn-and-lens

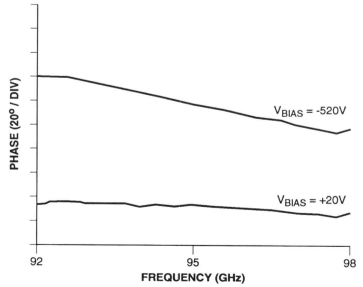

FIGURE 10.22 Reflection phase versus frequency of dual-polarization varactor-diode with bias voltages of −520 V and +20 V. (Millitech Corp.)

405

combination similar to that shown for the reflection measurement of Fig. 10.12, except that the incident and reflected signals were downconverted and measured by a vector network analyzer. In Fig. 10.22 the reflection phase is plotted as a function of frequency for two array bias voltages, $+20$ V and -520 V. The plot shows that a phase shift greater than $70°$ is obtained over most of the 92 to 98-GHz frequency range. Further development work will concentrate on improving yield and devising arrays that are controllable by lower voltages.

5 CONCLUSION

Beam-control arrays are just now beginning to find commercial applications. As more uses are found for quasi-optical and active arrays in microwave and millimeter-wave systems, we can expect beam-control arrays to be included in these systems where conventional control techniques will not fill the bill. Future developments in this area may include:

1. *Beam-Control Arrays Using Three-Terminal Devices.* Currently many conventional microwave control circuits use FETs as gate-controlled variable resistors. The development of HEMTs and other three-terminal devices whose characteristics will make them usable as control elements at millimeter wavelengths will permit the design of beam-control arrays with transistors instead of diodes. As always, biasing connections will present challenges to the designer of quasi-optical arrays, but techniques developed to overcome these problems in amplifier and oscillator arrays will be useful in the design of control arrays as well.
2. *Low-Frequency Controlled-Reflection and Controlled-Transmission Surfaces.* In this context "low frequency" means below 30 GHz. The possibility of electrically controlling the microwave reflectivity of a surface, not necessarily planar, is an attractive one for certain applications. If sufficiently thin, flexible, and inexpensive beam-control reflectors can be designed, they might find wide application in the area of electronic countermeasures.
3. *Smart Antennas.* Quasi-optical beam control arrays may find applications in high-bandwidth millimeter-wave communications systems. Spectrum crowding in the 3 to 7-GHz terrestrial point-to-point microwave communications band led to tighter and tighter antenna specifications to avoid interference. In the future a similar phenomenon may occur at millimeter wavelengths within the interior space of a single office. Intelligent communications systems that direct their transmitted beams automatically will need compact, inexpensive beam steering, which beam-control arrays may provide.

ACKNOWLEDGMENTS

The author would like to thank Debabani Choudhury, Ellen L. Moore, and G. Richard Huguenin of Millitech Corporation, Prof. Paul F. Goldsmith of Cornell University, and Prof. Erik L. Kollberg of Chalmers University for their assistance and cooperation.

REFERENCES

[1] T. Larsen, "A survey of the theory of wire grids," *IRE Trans. Microwave Theory Tech.,* vol. 10, pp. 191–201, May 1962.

[2] J. A. Arnaud and F. A. Pelow, "Resonant-grid quasi-optical diplexers," *Bell System Tech. J.,* vol. 54, no. 2, pp. 263–283, Feb. 1975.

[3] C. Chekroun, D. Herrick, Y. Michel, R. Pauchard, and P. Vidal, "RADANT: new method of electronic scanning," *Microwave J.,* vol. 24, pp. 45–47, 50, 52–53, Feb. 1981.

[4] L. B. Whitbourn and R. C. Compton, "Equivalent circuit formulas for metal grid reflectors at a dielectric boundary," *Appl. Optics,* vol. 24, pp. 217–220, Jan. 15, 1985.

[5] K. D. Stephan, F. H. Spooner, and P. F. Goldsmith, "Quasi-optical millimeter-wave hybrid and monolithic PIN-diode switches," *IEEE Trans. Microwave Theory Tech.,* vol. 41, no. 10, pp. 1791–1798, Oct. 1993.

[6] A. A. M. Saleh, "An adjustable quasi-optical bandpass filter—part I: theory and design formulas," *IEEE Trans. Microwave Theory Tech.,* vol. 22, pp. 728–739, July 1974.

[7] W. W. Lam, C. F. Jou, H. Z. Chen, K. S. Stolt, N. C. Luhmann, and D. B. Rutledge, "Millimeter-wave diode-grid phase shifters," *IEEE Trans. Microwave Theory Tech.,* vol. 36, pp. 902–907, May 1988.

[8] J. F. White, *Microwave Semiconductor Engineering,* Van Nostrand Reinhold, New York, 1982, pp. 143–175.

[9] R. C. Compton and D. B. Rutledge, "Circuit modeling of planar meshes with discrete loads," in *Diffraction Phenomena in Optical Engineering Applications SPIE* vol. 560, J. E. Harvey and D. M. Byrne, eds., Society of Photooptical Instrumentation Engineers, Bellingham, WA, 1985, pp. 29–32.

[10] L. B. Sjogren and N. C. Luhmann, Jr., "An impedance model for the quasi-optical diode array," *IEEE Microwave Guided Wave Lett.,* vol. 1, pp. 297–299, Oct. 1991.

[11] M. N. Afsar, "Dielectric measurements of millimeter-wave materials," *IEEE Trans. Microwave Theory Tech.,* vol. 32, pp. 1598–1609, Dec. 1984.

[12] A. E. Siegman, *An Introduction to Lasers and Masers,* McGraw-Hill, New York, 1971, p. 312.

[13] G. R. Huguenin, private communication.

[14] X. Qin, W.-M. Zhang, C. W. Domier, N. C. Luhmann, Jr., W. Berk, S. Duncan, and D. W. Tu, "Monolithic millimeter-wave beam control array," *1995 IEEE Microwave Theory Tech. Symp. Dig.,* pp. 1669–1672.

[15] D. Choudhury, J. J. Nicholson, R. F. Dec, and E. L. Moore, "Development of a monolithic diode beam steering array for concealed weapon detection system," *Digest of the Twentieth International Conference on Infrared and Millimeter Waves,* Dec. 1995, pp. 289–290.

[16] S. M. Sze, *Physics of Semiconductor Devices,* 2nd ed. Wiley-Interscience, New York, 1981, p. 114.

CHAPTER ELEVEN

Frequency Conversion Grids

JUNG-CHIH CHIAO
University of Hawaii at Mānoa, Honolulu

The demand for solid-state local oscillators at submillimeter wavelengths has been steadily increasing for applications in radio astronomy and remote sensing of the atmosphere. The interest for terahertz applications has fostered a strong need for submillimeter-wave receivers, mixers, and sources, especially tunable high-power sources used as the local oscillators for heterodyne submillimeter-wave receivers. Traditional high-power sources in the terahertz region such as gas lasers and vacuum-tube oscillators are not suitable for this purpose due to their large size, high-voltage supplies, and small tuning ranges. However, frequency multipliers such as Schottky diode multipliers can be used to generate the required terahertz frequencies from lower-frequency solid-state tunable signal sources such as Gunn-diode oscillators.

1 MOTIVATION

1.1 Application

Microwave remote-sensing techniques and recent developments on submillimeter-wave heterodyne radiometric systems have made possible their use in limb sounders at submillimeter wavelengths to study the upper atmosphere of Earth [1]. The Antarctic ozone hole discovered in 1985 [2] shows that it is necessary and urgent to monitor the upper atmosphere in order to detect the

---

Active and Quasi-Optical Arrays for Solid-State Power Combining, Edited by Robert A. York and Zoya B. Popović.
ISBN 0-471-14614-5 © 1997 John Wiley & Sons, Inc.

change of the stratospheric ozone layer which shields life from solar ultraviolet radiation but is depleted by pollution. Observations in the terahertz region have been successful in determining the density of many species such as hydrogen chloride, hydrogen fluoride, and many radicals such as atomic oxygen, chlorine monoxide, and hydroxyl which enter in many atmospheric chemical reactions including the catalytic cycles of HO_x, NO_x, and ClO_x which lead to the destruction of ozone [3]. Technology needs include local oscillators for radiometers and mixers carried by a satellite to monitor ozone depletion and tropospheric chemistry up to 2.5 THz.

Astronomers note that interstellar medium emits energy mostly in the submillimeter-wave spectral region [4] which is very important for the investigation of a wide range of astronomical topics including star-forming molecular cloud regions, the interstellar medium, the composition of planetary atmospheres, galaxies, and cosmic background radiation [5]. Space-borne observation will be required to study these subjects at submillimeter-wave ranges due to the interference with the terrestrial atmosphere. This presents severe technical challenges on submillimeter-wave local oscillators for space-borne heterodyne receivers and focal-plane arrays. For example, the frequency coverage of the NASA SMMM (SubMillimeter Moderate Mission) is from 400 GHz to 1.2 THz, and the minimum output power requirement of the local oscillator at 1 THz is 50 μW. The challenge is to provide a small, lightweight, reliable device requiring low-voltage power supply but generating enough output powers.

The terahertz technologies are also used on plasma diagnostics [6] including thermal imaging, density probing, and backscattering measurement. In fusion research, measurements of plasma electron-density profiles have been obtained by a submillimeter-wave imaging system, using a bow-tie antenna/bismuth microbolometer detector array [7]. Terahertz coherent systems were also used for plasma heating and high-energy accelerators in fusion researches [8].

1.2 Sources

The realization of a simple tunable local oscillator with adequate output powers at submillimeter wavelengths presents severe technical difficulties. Traditional choices such as free-electron lasers, gas lasers, vacuum-tube oscillators, and reflex klystrons are not suitable for space applications due to their large sizes, high-voltage supplies, requirement for coolers, and small tuning ranges. High-frequency carcinotrons or back-wave oscillators have been successfully developed at frequencies up to 1 THz [9]. However, they both are mechanically cumbrous and require high-power supplies or complicated phase lock systems. Therefore, all-solid-state local-oscillator sources would be more appropriate for most of the applications. Several solid-state components theoretically capable of generating oscillations at submillimeter wavelengths have been under development such as TUNNETT [10], quantum-well diodes (QWD) [11], and resonant-tunneling diodes (RTD) [12]. For example, an output power of 0.2 μW at 420 GHz and an oscillation at 712 GHz using RTDs were obtained by Brown *et al.* [13]. It was also reported that it is possible to achieve fundamental oscillation up to 1 THz. A recent developing competitive approach is to use a photomixer of low-temperature-grown GaAs to mix two visible IR lasers to generate radiation in the frequency range from 100 GHz to 3.8 THz [14]. At 1 THz, an output power of 0.8 μW was observed.

Though these devices are potential submillimeter-wave local-oscillator sources, they are still unable to generate enough power for use and the power falls off dramatically at higher frequencies. Therefore, frequency multipliers or upconverters like Schottky diode multipliers pumped by high-power, lower-frequency tunable solid-state sources provide an attractive alternative to generate the required terahertz frequencies. The advantages of using a combination of frequency multipliers and solid-state sources, such as IMPATT, GaAs-Gunn, or InP-Gunn devices, will result from the favorable features of low-frequency sources like lower noise, larger tuning range, lower cost, and higher reliability.

2 WAVEGUIDE MULTIPLIERS

Current diode multipliers have mostly been single-diode structures typically consisting of a Schottky varactor diode placed in a crossed-waveguide mount with a whisker contact shown in Fig. 11.1 [15]. The input power is fed through the input waveguide, passed a low-pass filter, and coupled into the diode chip

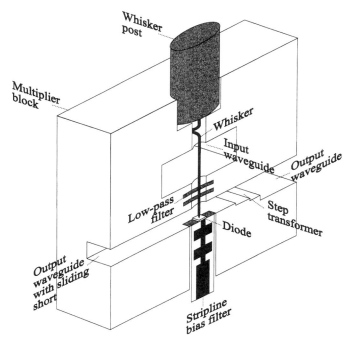

FIGURE 11.1 A typical crossed-waveguide multiplier using one diode [15]. Only half of the structure is shown.

which is mounted on the surface of the output waveguide. One of the diodes on the surface of the chip is contacted by a very thin wire (whisker) which acts like an antenna coupling harmonic power into the output waveguide. Sliding back-shorts in the input and output waveguides allow impedance tuning at the input and output frequencies. The step transformer transforms the impedance of the waveguide to match the diode impedance. A stripline bias filter is used to provide the diode DC bias. The structure design requires matched terminations at the fundamental frequency and the desired harmonic frequency; open-circuited terminations at the higher harmonics; and optimum reactive terminations at the idler frequencies [16].

Frequency multiplication results from the nonlinear impedances of diodes. There are usually two types of diode frequency multipliers: The varistor multipliers utilize the nonlinear $I-V$ relationship with a forward bias, and the varactor multipliers make use of the nonlinear $C-V$ relationship with a reversed bias. Single-varistor multipliers have been well developed with low conversion efficiencies. Resistive diode doublers have been demonstrated at 300 GHz with a conversion loss of 18 dB [17]. Page [18] showed that the frequency multiplication efficiency of a purely resistive multiplier is limited to $1/n^2$, where n is the harmonic number. The advantages of using varistors include that the impedance matching is easier as compared to varactors and the intermodulation distortion is smaller [19].

Manley and Rowe [20] first showed that it is possible to approach 100% conversion efficiency with a purely nonlinear reactive device. Penfield and Rafuse [21] also predicted that an ideal varactor will be able to convert all of the pump power to any higher harmonics with proper termination, bias and pump power. Therefore, varactors have been widely used for frequency multipliers. To date, multipliers using varactors in crossed waveguides have achieved impressive conversion efficiencies. Archer [22] reported a doubler with 10% efficiency at any output frequency in the range 100–260 GHz. The maximum output power is 6 mW at 260 GHz. Faber *et al.* [23] demonstrated a doubler with 35% efficiency at 98 GHz which is close to the theoretical prediction. Erickson [24] demonstrated a doubler with a peak efficiency of 35% at 160 GHz using a balanced configuration of two series Schottky varactors.

At submillimeter wavelengths, considerable effort has been made to increase output powers and output frequencies by using scaled-down versions of millimeterwave waveguide mounts with smaller active-area diodes. Takada *et al.* [25] reported a doubler with 5-mW output power at 300 GHz and a tripler with 500 μW at 450 GHz. Erickson [24] has built a doubler with a maximum output power of 4 mW at 330 GHz and a tripler with output powers of 700 μW and 550 μW at 474 GHz and 498 GHz, respectively. Zimmermann *et al.* [26] demonstrated triplers, quadruplers, quintuplers, and sextuplers with output frequencies in the range of 450–750 GHz. A tripler with 250-μW output power at 630 GHz, a quadrupler with 30-μW output power at 750 GHz [27], and a chain of a doubler and a tripler with 380-μW output power at 690 GHz [28] were reported. Above 800 GHz, Rothermel *et al.* [29] reported an octupler producing 0.66 μW at 800 GHz when pumped by 20 mW at 100 GHz and a seventh harmonic with an output power of 1 μW at 805 GHz when pumped by 18 mW at 115 GHz. Rydberg *et al.* [30] have demonstrated a Schottky varactor diode frequency tripler with a measured output power more than 120 μW at 803 GHz. Erickson and Tuovinen [31] presented a waveguide tripler with an output power of 110 μW at 800 GHz. Zimmermann *et al.* [32] have demonstrated an output power of 60 μW at 1 THz by using a cascade of two whisker-contacted Schottky varactor frequency triplers.

Fig. 11.2 shows the state-of-art output powers for frequency multipliers with output frequencies from 50 GHz to 1 THz [33]. Each data point represents a

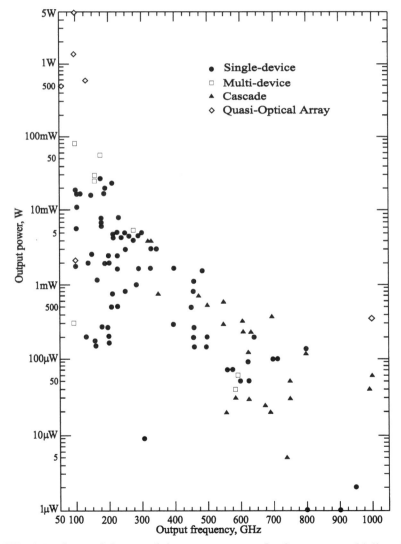

FIGURE 11.2 State of the art of the output powers for frequency multipliers [33].

maximum output power at an operating frequency. Obviously the output powers strongly depend on the pumping powers, so some results do not have high output power due to the availability of pumping sources. Also, some results may not have impressive output powers but have high efficiencies or are historic milestones. In most of the cases, the maximum output power is limited by the power saturation of the devices.

Though waveguide multipliers are highly developed and are capable of producing more output power at submillimeter wavelengths using two or more diodes in series, there are some limitations.

At submillimeter wavelengths, the losses of waveguide components are too high for sensitive radiometers [28] and the machining for single-mode waveguides becomes more complicated and expensive as frequencies increase. Besides the mounting structure parameters (the losses and the embedding impedances at the fundamental, idler, output, and higher harmonic frequencies), the varactor parameter deciding the efficiency is the dynamic cutoff frequency f_c:

$$f_c = \frac{1}{2\pi R_s(f)} \left(\frac{1}{C_{j,\min}} - \frac{1}{C_{j,\max}} \right) \tag{11.1}$$

where $R_s(f)$ is the series resistance of diode, and $C_{j,\min}$ and $C_{j,\max}$ are the minimum and maximum diode junction capacitances with different biases. A large capacitance ratio $C_{j,\max}/C_{j,\min}$ with a small $C_{j,\min}$ will result a high cutoff frequency. However, a small $C_{j,\min}$ is usually achieved by reducing the device active area which will also limit the saturation power. Therefore, increasing operating frequencies for a solid-state device inherently limits the output power.

Power-combining techniques then become essential to overcome this problem. Conventional power-combining techniques utilizing waveguides and striplines have serious limitation at higher frequencies though they have reached high combining efficiencies at microwave and millimeter-wave frequencies [34,35]. These structures have high losses at submillimeter wavelengths due to ohmic dissipation, are narrowband due to resonant-cavities, and usually require many hours of manual labor to assemble. Therefore, spatial power-combining techniques have been applied to submillimeter-wave frequency multipliers. Frerking et al. [36] used a quasi-optical Fabry–Perot interferometer to diplex the input and output frequencies and a varistor GaAs Schottky diode to make a frequency doubler with an output power of 50 μW at 600 GHz. Archer [37] used quasi-optical filtering and tuning elements in the outputs of triplers and quadruplers to reduce losses and demonstrated a peak efficiency of 7.5% at 265 GHz and 5% at 340 GHz. Steup [38] demonstrated a 580-GHz quadrupler consisting of four whisker-contacted diodes and a quasi-optical antenna array in the output which acts as a tunable power combiner to produce an output power of 29 μW with a pump power of 120 mW. A corner-cube doubler has been constructed by Lyons et al. [39] to generate 2 μW at 952 GHz with an input power of 22 mW. These methods demonstrated the low-loss feature of spatial power combining, but they didn't produce more power due to the power saturation of a single diode or a small number of diodes. One approach to overcome the low power of solid-state devices in the submillimeter-wave band is to combine a large number of devices together.

3 QUASI-OPTICAL GRID MULTIPLIERS

3.1 Concept

The grid multiplier concept is shown in Fig.11.3. The input beam at the fundamental frequency enters from the left. The first element is a pair of dielectric tuning slabs that act to transform the impedance of the input wave to one appropriate for the multiplier grid. Typically inductive reactance is needed to cancel capacitance of the diodes in the grid. In addition, the free-space wave impedance, 377 Ω, is inconveniently high and needs to be reduced. Next the beam passes through a low-pass filter that passes the fundamental frequency, but reflects harmonics. Then the beam hits the diode grid. A diode grid is a periodic circuit loaded with diodes. The grid acts as a nonlinear surface which results from the nonlinearity of $I-V$ or $C-V$ characteristics of diodes and generates harmonics. This harmonic beam radiates both forward and backward, but the backward beam reflects off the low-pass filter. The forward beam passes through the band-pass filter, and then through another pair of tuning slabs. One important feature is that the tuning slabs are outside the filters, so the input and output can be tuned independently. The entire structure is quite compact, only a few wavelengths thick. The design is also suitable for cascading, so even higher harmonics could be produced. The multiplication process preserves the beam shape. Therefore, a focused beam could be used so that different sizes of multiplier grids could be cascaded.

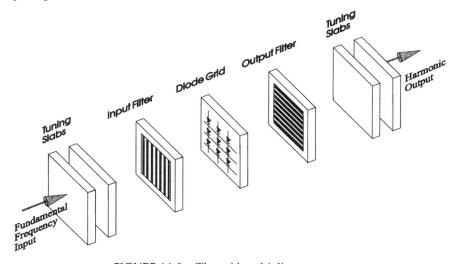

FIGURE 11.3 The grid multiplier concept.

3.2 Advantages

A grid of many planar devices quasi-optically coupled in free space does not require the construction of single-mode waveguides and can potentially overcome the power limit of conventional single-diode multipliers. By integrating a large number of devices together into the grid, very large powers can be achieved [40]. Because the input power is distributed onto the entire grid, each device only handles a certain amount of input power which will increase the total saturation power and the dynamic range. To increase the power, we simply increase the number of devices and the size of the grid. Also because the power is combined in free space, losses associated with waveguide walls and feed networks are eliminated. Conduction loss, dielectric loss, and radiation loss which limit the operating frequencies using traditional planar-waveguide components will be reduced in a quasi-optical system. Quasi-optical systems use mirrors, dielectric slabs, metal-patterned grids, or Fabry–Perot interferometers for tuning the impedances, so the impedance tuning is easier than most of the power combiners based on microstrip lines or waveguides. The impedance tuning is independent of the active grid itself, so there is no need to change the design or refabricate the active devices for optimizing the performance of the entire system. The tuning elements are easier to be replaced or redesigned without changing the active grids. Furthermore, both active and passive quasi-optical circuits are suitable for monolithic manufacture using planar photolithographic techniques and existing semiconductor fabrication technologies. The fabrication is simpler and costs less without transmission lines or waveguides.

Although quasi-optical systems look more like optical systems, they can be modeled with reasonable accuracy using simple transmission-line and lumped-element components by assuming unit cells in a large array. This provides two major advantages. First, unit cells decide the performance of the grid. Design changes such as power requirements or operating frequencies can be done by simply changing the number of unit cells or scaling the unit cells, unlike microstrip circuits or waveguides which require a new iteration of analysis and layout. Second, the transmission-line models allow the well-developed commercial computer-aided design tools for conventional microwave circuits such as Hewlett-Packard Microwave and RF Design Systems (MDS) and Spice to be applied to the design and modeling of quasi-optical circuits. Also, the unit-cell feature makes it possible to use the EMF technique, which was developed by Eisenhart and Khan [41] to derive the impedance of a waveguide mounting structure and later extended by Weikle [42] to determine the characteristics of the grids. Conventional numerical techniques such as the finite-element methods, which use iterative approaches, require large amounts of computer time and memory. This method reduces the computing time and effort significantly.

3.3 Achievements

Lam *et al.* [43] first presented the idea using diode grids, designed for electronic beam steering, as frequency multipliers and predicted a second-harmonic output power of 560 mW at 130 GHz with a conversion efficiency of 35%. Using this quasi-optical array approach on frequency multipliers, a second-harmonic conversion efficiency of 9.5% and a pulsed output power of 0.5 W were achieved at 66 GHz by Jou *et al.* [44]. It was also shown that the experimental results agree well with the predictions using simple transmission-line models and the unit-cell method. Liu *et al.* [45] have demonstrated a frequency tripler consisting of 3100 Schottky-quantum-barrier varactor diodes to produce 5-W pulsed output powers at 99 GHz and a tripler consisting of 3000 multi-quantum-barrier varactor diodes to produce 1.25-W pulsed output powers at 99 GHz. Terahertz quasi-optical grid frequency doublers [46] have been investigated by using 6×6 Schottky diode grids which were originally designed as sideband generators. A peak output power of 330 μW was measured at 1 THz for 2.42-μs 500-GHz input pulses with a peak power of 3.3 W without any impedance tuning.

4 66-GHz FREQUENCY DOUBLER GRID

Millimeter-wave frequency doublers were demonstrated by Jou *et al.* [44] using monolithic grids loaded with 760 Schottky barrier varactor diodes fabricated on 2-cm^2 gallium-arsenide wafers, as shown in Fig. 11.4. The diodes are connected with vertical strips and biased by horizontal strips. The unit-cell size is 500 μm and the dipole antennas have a width of 20 μm. All the diodes have the same orientation. The average series resistance is 27 Ω and the minimum capacitance is 18 fF at a reverse breakdown voltage of -3 V. The barrier height is 0.6 $\pm$ 0.05 V and the dynamic cutoff frequency, f_c, is 340 GHz.

The doubler configuration is similar to the one shown in Fig. 11.3 except the filters consist of a wire polarizing grid with a half-wave plate designed for the fundamental frequency. The power arrives at the diode grid, and the nonlinear

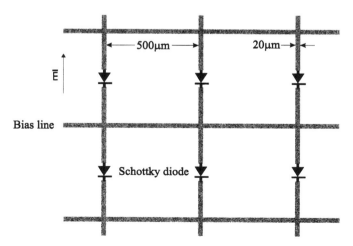

FIGURE 11.4 Doubler grid dimensions. The incident electric field is vertically polarized.

capacitance of the diodes generates harmonics. The half-wave plate provides separation for the fundamental from the second harmonic because it rotates the fundamental polarization by 90°, but does not alter the second harmonic polarization. The polarizers then select the desired frequencies.

The doubler measurement, shown in Fig. 11.5, is performed under far-field illumination condition, so the equivalent-circuit model based on the transmission-line analysis of plane wave could be applied. The power incident on the grid can be calculated accurately using antenna gain formulas. The pump source is a 50-kW pulsed magnetron operating at 33 GHz. The doubling efficiency, η_2, is calculated by

$$\eta_2 = \left(\frac{4\pi r_1 r_2}{A_d}\right)^2 \left(\frac{P_r}{P_t G_1 G_2}\right) \qquad (11.2)$$

where P_t is the transmitted power by the source, P_r is the measured power by the output horn, A_d is the area of the diode grid, r_1 is the distance between the source and the grid, r_2 is the distance between the grid and the output horn, and G_1 and G_2 are the antenna gains of the transmitting and receiving horns, respectively. The calibration can be done by measuring the transmitted and received powers using calibrated detector diodes without the grid in the path.

The multiplier analysis is based on a transmission-line model assuming plane-wave illumination. The substrate, tuners, and filters are represented as sections of transmission line. Their characteristic impedances are equal to the wave impedances in the dielectric. The circuit embedding impedance is calculated as the parallel combination of the impedances looking out to the left and right of the grid. The effective impedance and the frequency multiplication efficiency can be predicted from the large-signal nonlinear analysis [15,48] (to be described later) using the unit-cell method. The optimum efficiency is estimated by sampling the fundamental and second harmonic embedding impedances for maximum output power. The complex conjugate of the optimum embedding impedance is taken as the diode-grid impedance. In experiments, the maximum efficiency can be reached by adjusting the tuning elements to match the grid impedance. In the analysis, the total output power is found by multiplying the computed output power of single diode by the number of diodes on the grid. The doubling efficiency of the grid is calculated by dividing the output power by

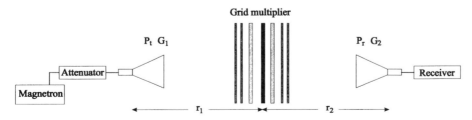

FIGURE 11.5 Millimeter-wave multiplier measurement.

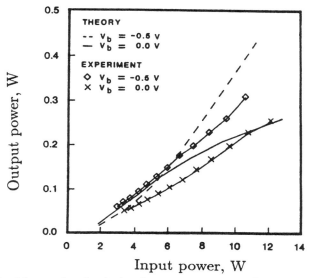

FIGURE 11.6 Measured and calculated output powers as a function of input power.

the input power available to the diodes. Fig.11.6 shows the measured and computed second harmonic output power as a function of input power. The computed second harmonic powers compare reasonably well with the measured results at bias levels of -0.5 V and 0 V, which implies the transmission-line model is quite accurate.

Figures 11.7a and 11.7b show the measured 66-GHz output power and the corresponding doubling efficiency under three different bias conditions, respectively. The highest output power is 0.5 W at a bias voltage of -0.5 V, and the highest efficiency is 9.5% with a 2.5-W input power.

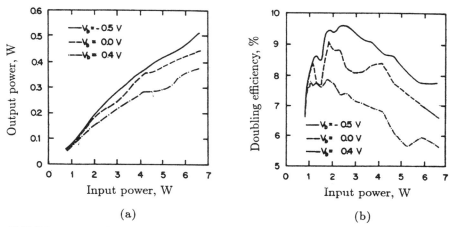

FIGURE 11.7 (a) Measured output power at 66 GHz and (b) doubling efficiency as a function of input power at 33 GHz.

419

5   99-GHz FREQUENCY TRIPLER GRID

Two major requirements need to be satisfied to build a frequency tripler: higher dynamic cutoff frequencies and a reactive idler circuit to recirculate the second harmonic power. To increase the operation frequencies and the frequency multiplication efficiencies, the cutoff frequency of the varactors needs to be increased. A large cutoff frequency requires small values of series resistance and $C_{j,\min}$. A small $C_{j,\min}$ is usually achieved by using a small active area device which, however, limits the output power. Staecker *et al.* [47] show significant improvements in both device cutoff frequency and power handling ability by epitaxially stacking single-quantum-barrier structures into a multi-quantum-barrier structure. Liu *et al.* [45] have demonstrated a grid frequency tripler consisting of 3100 Schottky-quantum-barrier varactors (SQBVs) to produce 5-W pulsed output powers at 99 GHz and a tripler consisting of 3000 multi-quantum-barrier varactors (MQBVs) to produce 1.25-W pulsed output powers at 99 GHz.

Fig. 11.8a shows the epitaxial profile of the MQBV device. Three intrinsic AlGaAs barriers are sandwiched between GaAs active regions. In the SQBV structure, the resistive ohmic contact of the MQBV is replaced with an *in situ* Al Schottky contact formed on a layer of AlGaAs and a δ-doped layer is employed under this AlGaAs layer to compensate the built-in voltage of the Schottky contact to make a symmetric *C–V* curve with the maximum capacitance at zero bias. Figure 11.8b shows the grid layout where two back-to-back devices share the same active area. The back-to-back method doubles the number of barriers in series while the current flows between the two stacks via the n^+ layer and provides an efficient way to cancel second-harmonic currents. The unit-cell size

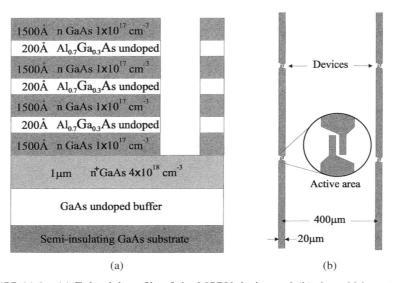

FIGURE 11.8   (a) Epitaxial profile of the MQBV device and (b) the grid layout [45].

is 400 μm and the dipole antenna width is 20 μm. The dipole antenna structure is tapered down at the device to reduce gap capacitance. These devices have a symmetric C–V characteristic to the zero bias because of the back-to-back symmetric structure. The array will only generate odd harmonics of the fundamental and does not require a reactive idler circuit to recirculate the second harmonic power. Furthermore, no bias line is required since these varactors operate at zero bias. A $C_{j,\max}/C_{j,\min}$ ratio of 4 is obtained when the bias changes from zero to ±2 V.

These triplers are measured under the far-field illumination conditions and the measurement setup is the same as that for grid doublers. Cutoff waveguide filters, which have a cutoff frequency higher than the fundamental frequency, are used to prevent contamination of the detected signal from the pump source.

An output power of 1.25 W with a maximum efficiency of 0.7% has been achieved with a tripler consisting of 3000 MQBVs which have a cutoff frequency of 150 GHz. Fig.11.9a shows the measured and calculated efficiencies as a function of input power for the MQBV array. An output power of 5 W with a maximum efficiency of 2% has been achieved with a tripler consisting of 3100 SQBVs which have a cutoff frequency of 485 GHz. Figure 11.9b shows the results for the SQBV array. The simulated output efficiencies are calculated by the same large-signal nonlinear analysis program developed by Siegel and Kerr [15,48] using the measured low-frequency characteristics. The discrepancy between the simulated and measured values is primarily due to diffraction losses in the matching circuits. It is observed that the closer the tuning slabs are to the grid, the smaller the discrepancy is. Due to the measurement setup, the minimum spacing is limited to 0.5 cm, so the diffraction effect on each element has

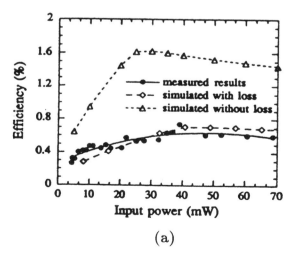

(a)

FIGURE 11.9 The measured and calculated efficiencies as a function of input power for the (a) MQBV and (b) SQBV grids. (*Continued*)

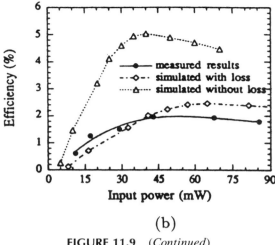

(b)

FIGURE 11.9 (*Continued*)

been individually measured while maintaining all the other elements at the positions giving the maximum output signals. A linear extrapolation of losses for each element can be performed and the result shows that 40% and 25% of losses come from the input and output matching elements, respectively. Based on the measured losses, simulation values are corrected. The good agreement in Figs. 11.9a and 11.9b shows that the diffraction loss is a reasonable assumption.

The better performance of SQBV can be explained by its higher cutoff frequency than that of MQBV which may be reduced by the parasitic resistance resulting from the ohmic contact and the larger capacitance that exists without the reactive Schottky contact present [45].

6 THz FREQUENCY DOUBLER GRID

The proof-of-principle tests of grid multipliers at millimeter wavelengths have demonstrated the advantages of quasi-optical multipliers. A reasonable approach to establish a terahertz local oscillator consisting of a multiplier chain and an oscillator is to work on the end-multiplier first and then work backward to decide the configurations of each cascaded multiplier and the pumping oscillator. A joint effort between Caltech and the University of Virginia (UVA) has been made to achieve this purpose. This is the first attempt to use the state-of-the-art terahertz planar Schottky diodes developed successfully at UVA [49] on quasi-optical grids. These diode grids, which were originally designed as sideband generators (to be described later), have been fabricated and the testing for use as sideband generators is presently underway; nevertheless, they can be tested as frequency multipliers to verify the feasibility of using diode grids for terahertz frequency multiplication. It should be emphasized that these diode grids were originally designed for sideband generators; therefore, some design features may not be optimal or proper for use as multipliers. For example, flipping the diode orientation above and below the center row of the grid is inappropriate for multipliers because the cancellation of electric fields from the two halves in the far field creates an undesired null in the center of the output beam.

6.1 Planar Schottky Diodes

Schottky barrier diodes for submillimeter-wave applications are typically fabricated as circular anode metallization on GaAs substrate with anode diameters from microns to submicrons for whisker contacts [49]. This type of structure provides benefits of simpler fabrication, less parasitic shunt capacitance with the whisker contact, and higher efficiency for input power to be coupled into diodes. However, whisker-contacted diodes are costly to assemble and hard to combine power. Planar (whiskerless) Schottky barrier diodes greatly simplify the system implementation, especially for space-based applications which require high reliability. The planar diodes also allow integration of a large number of devices, easier fabrication for antiparallel diode pairs, and integration with antenna structures or other devices such as filters, oscillators, and amplifiers on the same chips.

However, three major issues need to be solved for using planar Schottky diodes. First, the large shunt capacitance caused by the coplanar contact pads will reduce the cutoff frequencies and limit the operating frequencies below 100 GHz. Garfield et al. [50] developed a novel planar diode structure to overcome this problem. These devices have low parasitic element values while still maintaining excellent current–voltage characteristics by etching a channel through the active GaAs layer.

Second, the formation of submicron anodes on a planar diode is more difficult than a whisker-contacted diode. However, fabrication processes have been greatly improved by the Semiconductor Devices Laboratory at the University of Virginia [51]. The surface channel structure provides the benefit to form the anode on an essentially planar surface, therefore, makes it easier to define the anode formation and the anode-to-finger alignment. Also the structure avoids expensive and troublesome proton bombardment which is required for conventional planar structures.

Third, the LO coupling efficiency is low for a planar diode [49]. This problem can be solved by designing an integrated planar antenna structure with the diodes on the same chips and using outside impedance tuners to optimize the coupling efficiency. Quasi-optical grid is a good candidate to use the planar diodes.

Figure 11.10a shows an SEM picture of the planar Schottky diode located between the metal fingers of the bow-tie structure. Figures 11.10b and 11.10c show the top view and the cross section of the planar Schottky diode. The anode

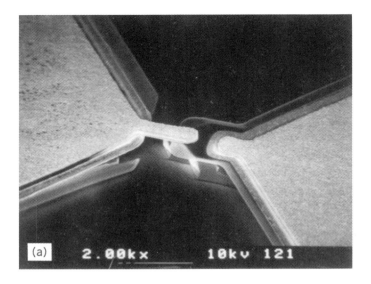
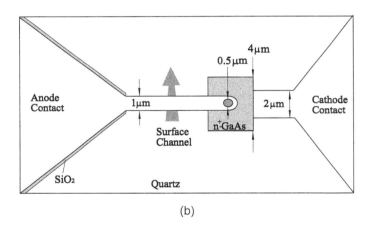
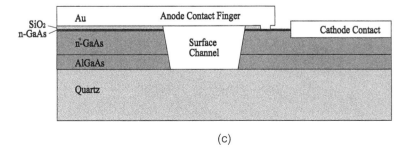

FIGURE 11.10 (a) The SEM picture, (b) the top view and dimensions, and (c) the cross section of the planar Schottky diode.

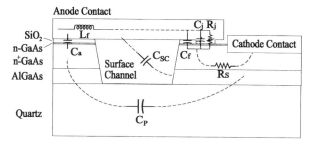

FIGURE 11.11 The cross section of the planar Schottky diode and the equivalent-circuit elements.

has a diameter of 0.5 μm. The n-GaAs layer has a thickness of 0.1 μm and a doping concentration of 4×10^{17} cm^{-3}. The n^+-GaAs layer has a thickness of 3 μm and a doping concentration of 5×10^{18} cm^{-3}. The AlGaAs layer has a thickness of 1.5 μm. A surface channel was etched away underneath the anode contact finger to reduce the shunt capacitance.

Major circuit elements in a planar Schottky diode are shown in Fig. 11.11. The surface channel eliminates the conducting path between the anode contact capacitance, C_{SC}, and the cathode contact. Therefore, instead of having a bigger capacitance, C_{SC}, it is replaced by the series combination of C_a and C_p. Because C_p is much less than C_a, the series capacitance is greatly reduced. The shunt C_{SC} contribution is not significant because it is now distributed across air ($\epsilon_r = 1$) instead of GaAs ($\epsilon_r = 13.1$). C_p is also reduced by using quartz as substrate instead of semi-insulating GaAs. Shunt capacitances from the contact finger, C_a and C_f, are further reduced by reducing the width of the finger and etching the surface channel as close as possible to the anode. The width of the finger is 1 μm and the distance between anode and channel is 2 μm. The finger inductance, L_f, can be calculated and helpful in tuning out the parasitic capacitances.

One of the distinct advantages in planar Schottky diodes is that the ohmic contacts are very close to the anodes, so the electrical path length is greatly reduced. Therefore, the series resistance, R_s, and series inductance will be much less. A thin layer of n-GaAs and a highly doped n^+-GaAs layer should also reduce the series resistance.

The resistive losses at ohmic contacts are further reduced when the planar Schottky diodes are used in a quasi-optical grid. Because the AC current paths are limited to the very small dimensions of the unit cells unlike the single-chip diode integrated with whisker-contact which the AC current has to flow through the entire structure, the RF-resistive losses can be reduced.

6.2 The 6 × 6 Diode-Grid Arrays

These grid multipliers were fabricated by monolithic technology on a 30 μm-thick fused-quartz substrate. Fig. 11.12 shows the bow-tie-shaped metal pattern used for the unit cell. The Schottky diode junction is located at the center of the

426 FREQUENCY CONVERSION GRIDS

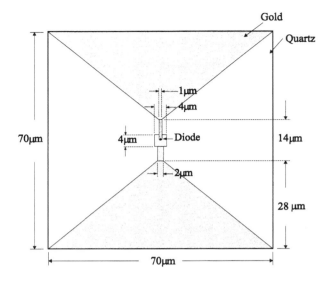

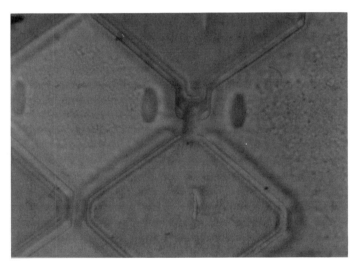

FIGURE 11.12 The unit cell of the grid multiplier.

unit cell. The size of a unit cell is 70 μm × 70 μm. Fig. 11.13 shows the entire 6 × 6 array. The active area is 420 μm × 420 μm. Three diodes are in series from the top or bottom bias line to the center bias line. The adjacent bow ties are not connected except the center row. The diodes in the top and bottom halves of the grid have opposite orientations. This is because the grid was originally designed as a sideband generator. When the grid is used as a frequency multiplier, the orientation change causes an undesired null in the middle of the beam.

FIGURE 11.13 The grid frequency doubler. It is a 6 × 6 array.

6.3 Design Approach

Each diode in the array is presented with an embedding impedance which is determined by the metal pattern repeated throughout each unit cell. By assuming an infinite grid with a uniform plane wave normally incident upon the grid, the symmetry allows us to replace the walls of the unit cell in the grid with electric and magnetic walls to form an equivalent unit-cell waveguide. Therefore, the analysis of the entire grid can be reduced to a simpler analysis of the equivalent unit-cell waveguide. For a TEM incident wave with vertically polarized electric field, the unit-cell waveguide has magnetic walls ($\overline{H}_{\text{tangential}} = 0$) on the sides and electric walls ($\overline{E}_{\text{tangential}} = 0$) on the top and bottom. The propagating mode in the unit-cell waveguide is TEM and the evanescent modes couple to the currents flowing in the metal patterns. The AC current paths are limited in the unit cell by symmetry.

The equivalent unit-cell waveguide incorporated with a device can be presented as a simple transmission-line model to decide the performance of the entire grid, once we know the equivalent circuits for the device and the embedding impedance of the metal pattern. The embedding impedance of the metal pattern can be solved either by EMF techniques or by finite-element methods. Because of the assumption of the grid to be infinite extended, the edge effects of the grid are ignored.

Embedding Impedances—EMF Analysis For simple geometric metal patterns like strips or bow ties, the embedding impedance presented to the terminals of the diode in the unit cell can be solved by the EMF analysis, first described by Eisenhart and Kahn [41], and later developed by Weikle and Hacker [42,52,53] using Galerkin's moment method to find the current distribution on the metal patterns.

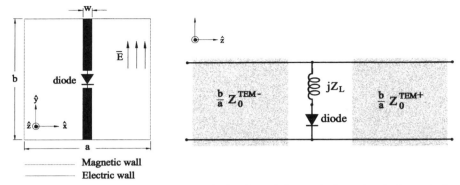

FIGURE 11.14 A unit cell with a vertical strip and the equivalent-circuit elements for the unit-cell waveguide.

For a unit cell with a vertical strip, the equivalent element in the transmission-line model is a shunt inductor, as shown in Fig.11.14, with reactance Z_L:

$$Z_L = \frac{2b}{a} \sum_{m=1}^{\infty} \cos^2\left(\frac{m\pi}{2}\right) \operatorname{sinc}^2\left(\frac{m\pi w}{2a}\right)(Z_{m0}^{TE+} \| Z_{m0}^{TE-}) \quad (11.3)$$

where $Z_0^{TEM} = \sqrt{\mu/\epsilon}$, $Z_{mn}^{TE} = \omega\mu/k_z$, and k_z is the propagation constant and is given by

$$k_z = \sqrt{\omega^2 \mu \epsilon - \left(\frac{m\pi}{a}\right)^2 - \left(\frac{n\pi}{b}\right)^2} \quad (11.4)$$

For a bow-tie metal pattern in a unit cell, the embedding impedance resembles a shunt section of transmission line, as shown in Fig.11.15. The shunt

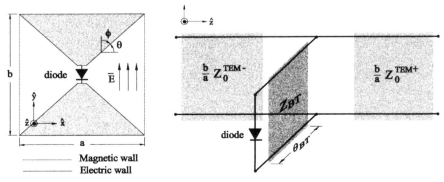

FIGURE 11.15 A unit cell with a bow-tie pattern and the equivalent-circuit elements for the unit-cell waveguide.

transmission line has a characteristic impedance, Z_{BT}, and an electrical length, θ_{BT}, which are determined by

$$Z_{BT} = \sqrt{\frac{Z}{Y}} \tag{11.5}$$

$$\theta_{BT} = \sqrt{ZY} \tag{11.6}$$

where

$$Z = \frac{1}{ab} \sum_{\substack{m=1 \\ n=0}}^{\infty} 2\epsilon_{0n} \frac{k_c^2}{k_x^2} A_{mn}^2 (Z_{mn}^{TE+} \| Z_{mn}^{TE-}) \tag{11.7}$$

$$A_{mn} = \frac{\int_0^b \int_0^\phi \frac{\cos(k_x y \tan \psi)}{\sqrt{\sin^2 \phi - \sin^2 \psi}} \cos k_y y \, d\psi \, dy}{\int_0^\phi \frac{d\psi}{\sqrt{\sin^2 \phi - \sin^2 \psi}}} \tag{11.8}$$

$$Y = \frac{1}{ab} \sum_{\substack{m=0 \\ n=1}}^{\infty} 2\epsilon_{m0} \frac{k_c^2}{k_y^2} B_{mn}^2 (Y_{mn}^{TM+} + Y_{mn}^{TM-}) \tag{11.9}$$

$$B_{mn} = \frac{\int_0^a \int_0^\theta \frac{\cos(k_y x \tan \xi)}{\sqrt{\sin^2 \theta - \sin^2 \xi}} \cos k_x x \, d\xi \, dx}{\int_\theta^\theta \frac{d\xi}{\sqrt{\sin^2 \theta - \sin^2 \xi}}} \tag{11.10}$$

and $Y_{mn}^{TM} = \omega\epsilon/k_z$, $k_x = m\pi/a$, $k_y = n\pi/b$, $k_c^2 = k_x^2 + k_y^2$, and

$$\epsilon_{mn} = \begin{cases} 1, & \text{if } m = n \\ 2, & \text{otherwise} \end{cases} \tag{11.11}$$

The bow-tie metal pattern is chosen to achieve a broader-band operation. Because the embedding impedance of the bow-tie pattern resembles a shunt section of lower-impedance transmission line with an electrical length of a fraction of a wavelength, it has a broader bandwidth than a strip pattern if the junction capacitance is small. However, choosing the bow tie will also degrade the reflection losses at the operating frequencies. A combination of a strip and a bow-tie pattern can be a trade-off option for bandwidth and return losses. Due to the planar diode structure, the contact fingers are required to reduce the parasitic capacitances. Therefore, we can use the innate contact fingers as part of the strip patterns to help tuning out the diode capacitances along with the bow tie to achieve a wider bandwidth with reasonable return losses.

430 FREQUENCY CONVERSION GRIDS

Embedding Impedances—Three-Port Extraction The EMF analysis can provide a gross approximation to the actual impedance for a unit-cell pattern with a combination of a short-strip and a short-ended bow tie. Intuitively, it is expected the narrow strips of contact-fingers to add some inductance in series with the diode, and the short-ended bow tie to have a shorter electrical length than that calculated by the EMF method in the shunt transmission line. To verify this assumption, the Hewlett-Packard High-Frequency Structure Simulator (HFSS), a 3-D finite-element electromagnetic–wave solver, is used with the three-port embedding-impedance extraction technique [52] to extract the embedding impedance of this complex geometry of the unit cell.

Because of the symmetry between the left and right, the unit cell can be reduced to a half-piece by placing a vertical magnetic wall in the center of the unit cell. In order to simplify the analysis and shorten the computation time, the half-cell can be further reduced to a quarter-piece by assuming the anode- and cathode-finger widths the same. This symmetry allows us to place a horizontal electric wall in the center, and the driving point of diode will be exposed at the edge of the quarter-piece unit-cell waveguide. Fig. 11.16 shows the quarter-piece structure used in HFSS for three-port embedding-impedance extraction. Port 1 and port 2 are the front- and back-ports of the unit-cell waveguide, and port 3, which presents the driving point of the diode, is connected to a short section of a rectangular coaxial waveguide. The short section of the rectangular coaxial waveguide connected to the driving point is well-defined with a known propagation constant and a known waveguide impedance.

Simulations over the frequency range of interest are done to solve the s parameters. Post-processing on the s parameters is performed to remove the

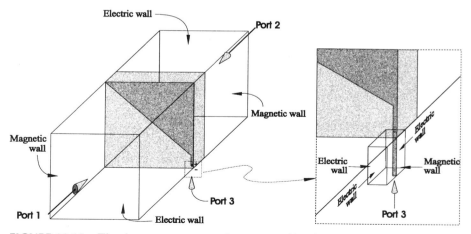

FIGURE 11.16 The three-port quarter-piece waveguide with proper boundary conditions and the detail view of the rectangular coaxial waveguide connected to the driving point of the diode.

effect of adding the coaxial waveguide like de-embedding the phase shift and renormalizing s-parameter matrix to the free-space impedance. The final s-parameter matrix can be compared with the s-parameter matrix of our intuitive circuit model whose element values can be first approximated by the EMF analysis. Then the element values can be fine-tuned until the s-parameter matrices agree with the HFSS results.

A more convenient way to solve the embedding impedance is to utilize the calibration method used for a one-port network analyzer. A mirror is placed behind the grid which means short-circuited the port 2 in the circuit model and the structure becomes a two-port network with the s-parameter matrix **S**:

$$\mathbf{S} = \begin{pmatrix} s_{11} & s_{13} \\ s_{31} & s_{33} \end{pmatrix} \tag{11.12}$$

We can now apply the calibration method for a one-port network analyzer. A series of three electrically distinct loads are placed across the port 3 connected to the driving point of the diode. This reduces the structure to a one-port with port 1 excited by a TEM incidence onto the grid. Typically these loads would be a short circuit, an open circuit, and a matched load. Simulations of the structure over the frequency range of interest are performed for each of the three loads solving three one-port s-parameter files, e_m, e_s, e_o corresponding to the matched termination, short circuit, and open circuit, respectively. These calibration s-parameters can then be used to find the two-port s parameters of the grid using

$$s_{11} = e_m \tag{11.13}$$

$$s_{33} = \frac{e_o + e_s - 2e_m}{e_o - e_s} \tag{11.14}$$

$$s_{13} s_{31} = \frac{2(e_o - e_m)(e_s - e_m)}{e_s - e_o} \tag{11.15}$$

and reciprocity ($s_{13} = s_{31}$) is applied to solve s_{13}. The **S** matrix then is compared with the intuitive circuit model to adjust the element values.

Linear Analysis for Grid Multipliers Simulations at frequencies from 200 GHz to 1.5 THz have been done. Two quarter-piece waveguides are used to simulate (a) the top-half structure, where the anode contact finger has a width of 1 μm, and (b) the bottom half-structure, where the cathode contact finger has a width of 2 μm. The length of the anode finger is 7 μm and the cathode finger is shortened to 5 μm responding to the real physical lengths.

Fig.11.17 shows (a) the unit-cell structure and (b) the corresponding equivalent-circuit elements. The total inductance contributed by the contact fingers is 4.9 pH and the shunt section of the bow-tie transmission line has an impedance of $Z_{BT} = 121$ Ω and an electrical length of $\theta_{BT} = 30.6°$ at 1 THz.

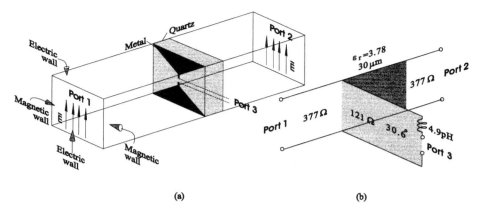

FIGURE 11.17 (a) The unit-cell waveguide for the diode grids and (b) the equivalent circuit.

Figure 11.18 shows the s-parameters plotted on a Smith Chart from 200 GHz to 1.5 THz for HFSS simulation results and the results from the equivalent circuit of Fig.11.17b with a short-circuit load at port 3.

DC I–V Relationship The RF values of the equivalent-circuit elements for the planar Schottky diodes strongly depend on the operating frequencies, pumping powers, temperatures, and biasing conditions. Some models have been developed [50]; however, they can only provide limited information. Due to the lack of information, DC element values are often used to allow us to initiate the RF

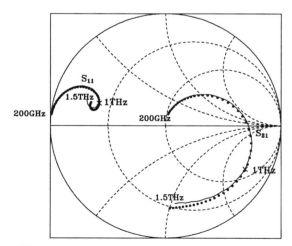

FIGURE 11.18 The s-parameters plotted on a Smith Chart from 200 GHz to 1.5 THz for HFSS simulations (●) and the results from the equivalent circuit (solid lines) in Fig. 11.17b, with a short-circuit load at port 3.

design and analysis. These values, of course, can only provide us an approximation of the circuit performance at RF and will vary significantly due to the skin effect, local temperature, and the change of barrier heights with different biases or pumping powers.

Fig. 11.19 shows a typical *I–V* curve. The solid line indicates the fitted *I–V* characteristic for diodes:

$$I = I_S(e^{V/\eta V_T} - 1) \tag{11.16}$$

where I_S is the saturation current, V_T is the thermal voltage, and η is the ideality factor of diode. The ideality factor can be calculated by $\eta = \Delta V/(\ln(10)V_T)$, where ΔV is the voltage change for a decade change in current. The fitted *I–V* characteristic is plotted with the ideality factor $\eta = 1.1925$ and the saturation current $I_S = 1.3 \times 10^{-16}$ A.

The diode junction resistance can be expressed by

$$R_j = \frac{1}{\partial I/\partial V} \approx \frac{\eta V_T}{I} \tag{11.17}$$

assuming that the diode is strongly forward-biased. For a forward current 0.4 mA with $V = 0.86$ V, the DC junction resistance is estimated to be 75 Ω.

The series resistance, R_s, can be derived from the DC *I–V* curves, given by

$$I = I_S(e^{(V-IR_s)/\eta V_T}) \tag{11.18}$$

where R_s may vary in different *I–V* regions. The R_s is measured with a current of 1.5 mA; and an average value, 14 Ω, is used in the equivalent circuit.

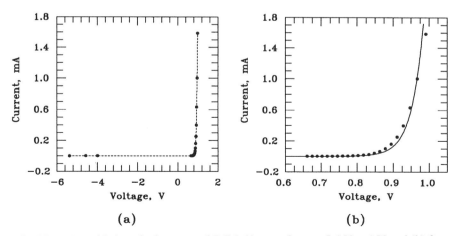

FIGURE 11.19 (a) A typical measured DC *I–V* curve from -5.4 V to 1 V and (b) from 0.65 V to 1 V. The measured data are shown by ●. The solid curve indicates a fitted *I–V* characteristic for diodes.

C–V Relationship In most of the cases, Schottky diodes are reverse-biased to increase the frequency multiplying efficiencies since the diodes are designed as varactors. However, these diodes are originally planned for use as sideband generators and designed to be used under forward bias. Thus, these diodes have very thin epitaxial layer and the capacitance does not change significantly with reverse bias. Therefore, the frequency multiplication in these diode grids possibly comes from varistor multiplication.

The junction capacitance at zero-bias, C_{j0}, can be calculated by

$$C_{j0} = \frac{\epsilon_r \epsilon_o \pi r^2}{d}\left(1 + \frac{bd}{r}\right) \qquad (11.19)$$

where ϵ_r is the relative permittivity of GaAs, ϵ_o is the permittivity in vacuum, r is the diode radius, d is the depth of the depletion layer at zero-bias, and b is the edge-effect factor calculated numerically [54]. For GaAs, the edge-effect factor is 1.5. Louhi [55] suggested that a second-order correction term should be added to decide the junction capacitance for submillimeter wavelengths:

$$C_{j0} = \frac{\epsilon_r \epsilon_O \pi r^2}{d}\left(1 + b_1\frac{d}{r} + b_2\frac{d^2}{r^2}\right) \qquad (11.20)$$

where $b_1 = 1.5$ and $b_2 = 0.3$. The depletion depth, d, can be expressed as a function of bias

$$d = \sqrt{\frac{2\epsilon_r \epsilon_O}{qN_e}(V_{bi} + V_r)} \qquad (11.21)$$

where q is the electronic charge; N_e is the doping density in the depletion region, which is 4×10^{17} cm^{-3}; V_{bi} is the built-in potential; and V_r is the reverse bias applied to the diode which is zero when calculating C_{j0}. The V_{bi} is defined as the potential difference between the top of the Schottky barrier and the edge of the depletion region where no bias is applied and can be determined by

$$V_{bi} = X_m - X_s \qquad (11.22)$$

where X_m and X_s are the work functions of the metal and the doped GaAs, respectively.

The junction capacitance of the diode at zero bias, C_{j0}, is estimated to be 0.4 fF. There is a parasitic capacitance between the contact finger and the active area, C_f, in Fig. 11.11, in parallel with the C_j. C_f can be decided by

$$C_f = \frac{\epsilon_r \epsilon_O A}{d} \qquad (11.23)$$

where ϵ_r is the relative dielectric constant of SiO$_2$, d is the height of the SiO$_2$ layer under the finger, and A is the contact area decided by the finger width and

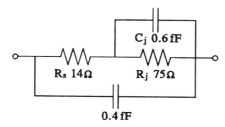

FIGURE 11.20 The equivalent circuit of the terahertz planar Schottky diode.

the distance between the diode and the edge of the surface channel. C_f is estimated to be 0.2 fF. Therefore, the total junction capacitance at zero bias is estimated to be 0.6 fF.

The equivalent circuit used for the planar Schottky diode, shown in Fig. 11.20, is incorporated with the embedding impedance of the grid and the estimated reflectance and transmittance of the diode grids are calculated. The reflection for the input frequency, 500 GHz, is -3.1 dB and the transmission for the second-harmonic frequency, 1 THz, is -7 dB without impedance tuning or filters.

6.4 Measurements

The measurements use the free-electron laser (FEL) as the input source in the Quantum Institute at the University of California, Santa Barbara. The free-electron laser is driven by high-quality, relatively low-energy electron beams from a recirculating electrostatic accelerators and capable of generating kilowatts of polarized radiation tunable from 120 GHz to 4.8 THz with a pulse width varied from 1 μs to tens of microseconds [56].

The measurement setup is shown in Fig. 11.21. The free-electron laser pulsewidth is 2.42 μs with a period of 1.3 s. All the optical components were aligned using an He–Ne laser. Part of the input power at the fundamental frequency in the incident beam is split off into a reference detector. A pyroelectric detector is used as the reference detector. The radiation from the FEL is then focused by a parabolic mirror. The input power is varied by inserting plexiglass attenuators in front of the beam splitter. The harmonic radiation from the grid is collimated by an f/1 parabolic mirror. A metallic-mesh Fabry–Perot interferometer in the collimated beam is used to measure the frequency content. Then the collimated beam is refocused onto a liquid-helium-cooled germanium bolometer by another f/1 parabolic mirror. The bolometer has a theoretical responsivity of 10^5 V/W with a 200-μs response time. A winston cone is placed as close as possible to the input window to collect the radiation and guide it to the small-area detector. The acceptance angle of the winston cone is $\pm 14°$.

An 8-layer metal-mesh filter is used as a low-pass filter on the input side. It has an attenuation more than 60 dB at 1 THz. A circular cutoff-waveguide array filter is used as a high-pass filter on the output side of the grid. The waveguide

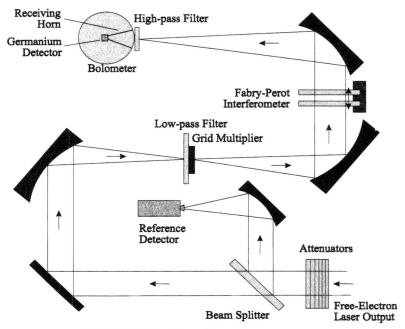

FIGURE 11.21 Experimental arrangement.

array filter is attached to the window of the bolometer. The circular waveguides have a diameter of 200.7 μm and a length of 1.02 mm. The transmittance at 1 THz is 0.76 and the attenuation is more than 60 dB at 500 GHz.

First, the FEL is tuned to the fundamental frequency, f_0, to measure the total incident energy on the focal plane with an electrically calibrated photoacoustic detector (for which an absolute calibration of energy exists), manufactured by Thomas Keating Ltd.. Then the responsivities of the second and the third harmonic energy are decided by tuning the FEL to $2f_0$ and $3f_0$ and measuring the bolometer responses through the high-pass filter, relative to the responses of the photoacoustic detector. Because the response time of the germanium bolometer is much longer than the pulse width, the output voltage response of the bolometer is actually corresponding to the detected energy. Therefore, an absolute calibration of the peak energy responsivity is established. Unfortunately, because the pyroelectric detector is not sensitive enough to detect the higher harmonic pulses generated by the multiplier grids, it is not able to decide the output pulse widths. Therefore, to be conservative, the input pulse width, 2.42 μs, is used to calculate the input peak power and output peak power by dividing the detected peak energy by this pulse width.

The diode grids are glued on the edges of microscope glass slides and suspended in air. Bonding wires connect the bias lines on the substrate to the contact pads on the slides for DC bias.

Time-Domain Responses Fig. 11.22 shows time responses of the bolometer voltage as a function of input power level. Each curve is an average of four pulses. The solid lines show the outputs with the multiplier grid in the path and with four different input power levels which have 0-dB, 3.8-dB, 5.8-dB, and 7-dB attenuations. The dashed line was measured without the multiplier in the path and 0-dB input power attenuation. This measurement shows that the harmonics are radiation from the multiplier grid instead of harmonic contamination from the free-electron laser. The 0-dB attenuation reference was actually set with enough sheets of plexiglass in front of the beam splitter so that there was no output signal detected by the bolometer without the multiplier in the path. Then extra attenuation was added to measure the time responses for different input signal levels.

Another way to show that the terahertz radiation comes from the grid rather than the free-electron laser is to rotate the grid by 90° and measure the cross-polarized signal. Fig. 11.23 shows the time responses. The solid line was measured with the electric field parallel to the diodes, and the dashed line was measured with the electric field perpendicular to the diodes. This result shows that: (1) The bow-tie metal structures on the grid work as linearly polarized antennas and couple the signal into the diodes. The harmonic signal is not generated by pumping high energy on GaAs in the device. (2) The harmonic signal does not come from the free-electron laser.

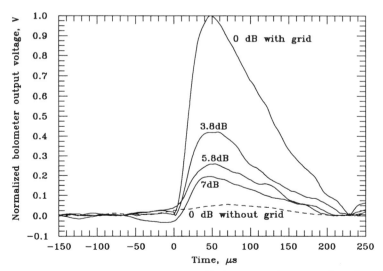

FIGURE 11.22 Normalized time responses of bolometer voltage as a function of input power level. The dashed line was measured with 0-dB attenuation of the input power but without the multiplier in the path. The solid lines were measured with the multiplier in the path and with different attenuations of the input power.

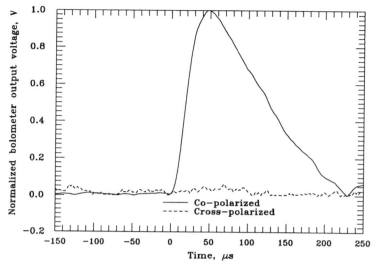

FIGURE 11.23 Normalized time responses of bolometer voltage with different grid orientations. The solid line indicates the electric field parallel to the the diode, and the dashed line indicates the electric field perpendicular to the diode.

Frequencies A Fabry–Perot interferometer is used to measure the signal frequencies. The metal-mesh plates could be positioned in 0.7-μm increments. Fig. 11.24 shows (a) the fundamental frequency from the free-electron laser after passing the low-pass filter but without the high-pass filter and (b) the output frequency from the grid after passing the high-pass filter. Figures 11.24a and 11.24b only show parts of the measured curves. The average distance between two peaks in the output-frequency measurement is 301 μm over 14 peaks. This indicates that the output frequency is 1.00 THz. No higher harmonics have been detected. The average distance between two peaks in the input-frequency measurement is 603 μm over 9 peaks. This indicates that the input frequency is 500 GHz.

Powers Figure 11.25a shows the power dependence of the multiplier with normal incidence. This grid has diodes with an anode diameter of 0.56 μm. The dashed line indicates a square-power relationship. A peak output power of 330 μW at 1 THz is achieved when the diode grid is pumped by a peak input power of 3.3 W at 500 GHz. With low input power, the data do not follow the square-power relationship, possibly due to measurement noise. It should be noticed that these diodes have not yet saturated. To increase the input power in order to investigate the saturation conditions, however, some of the diodes were damaged when the input power reached 13 W. These damaged diodes are open-circuited, verified by a curve tracer.

The data acquisition method is improved to investigate the power dependence with low input power. Each pulse is integrated to reduce noise effect. Because some of the diodes got damaged in the first grid, a new device was used.

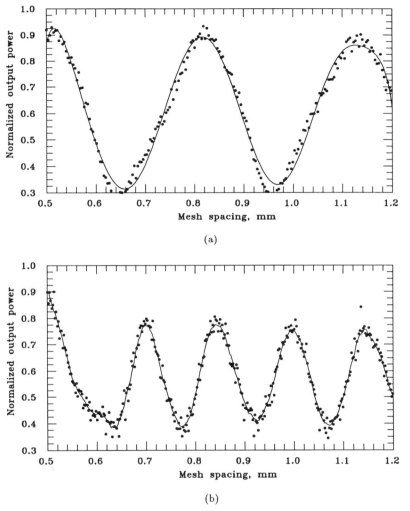

FIGURE 11.24 Normalized output power as a function of the metal-mesh spacing in the Fabry–Perot interferometer. (a) Measurement of 500-GHz input signal and (b) measurement of 1.00-THz output signal.

Figure 11.25b shows the power dependence of the second multiplier with normal incidence in a linear scale. This grid has diodes with an anode diameter of 0.5 μm. The solid line indicates a square-power relationship. With an input power of 800 mW, this multiplier generated a peak output power of 45 μW. These diodes have not saturated yet, and some of the diodes were damaged when the input power was increased to 4 W. It should be noticed that the data with low input power follows the square-power relationship after reducing the noise effect.

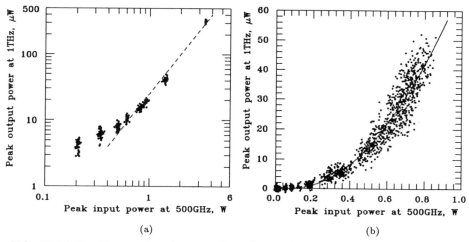

FIGURE 11.25 Measured peak power dependence of (a) the first and (b) the second multiplier grid. The dashed line indicates a square-power relationship. The diodes were not biased.

Patterns The diodes in the top and bottom halves of the grid have opposite orientations because the grid was designed for a sideband generator. When the grid is used as a frequency doubler, this causes an undesired null in the middle of the beam. The output pattern was measured to verify the existence of the null. The measurements were done by rotating the grid itself.

There are three main factors to decide the output patterns: the antenna pattern of the shortened bow-tie with contact-fingers; the effect of flipping the diode orientations across the center row; and the finite angle of the receiving horn at the bolometer.

Figure 11.26 shows the structure of the shortened bow-tie with contact fingers for far-field pattern simulations using HFSS. The antenna driving points are fed by a coaxial rectangular waveguide with the center and the outer conductors connecting to two fingers. The center conductor connecting to the finger in the bottom is bent to avoid the short circuit on the outer conductor of the coaxial waveguide. The modes in the coaxial waveguide are excited by the port in the end of the center conductor. There is a short section of a rectangular cavity (with electric wall in the end) attached to the coaxial waveguide to force the port fields radiating only into the coaxial waveguide, but not the free space. There is a small opening in the substrate to allow the connection of the coax to the antenna. The impedance of the coaxial waveguide can be specified by adjusting the permittivity or the permeability of the material filled in the waveguide to match the antenna impedance. The big rectangular enclosure of the entire structure is specified as absorbing boundary to allow the wave propagating into free space.

Three infinitesimal current elements on one side are assumed to have an opposite phase from three elements on the other side of a linear array to calcu-

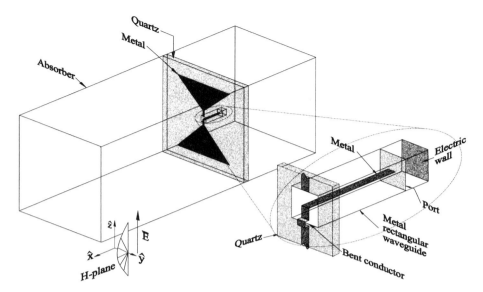

FIGURE 11.26 The simulation structure of the shortened bow-tie with contact fingers for the antenna far-field patterns.

late the array factor, shown in Fig. 11.27. The spacing between elements is 70 μm. The normalized array factor is determined by

$$A(\theta) = \left[\frac{\sin(\frac{3k_0 d}{2} \sin \theta)}{\sin(\frac{k_0 d}{2} \sin \theta)}\right]^2 \cdot \left[\sin\left(\frac{3k_0 d}{2} \sin \theta\right)\right]^2 \qquad (11.24)$$

where θ is the angle, d is the spacing, and the propagation constant k_0 equals $2\pi/\lambda_0$ where λ_0 is the wavelength, 300 μm.

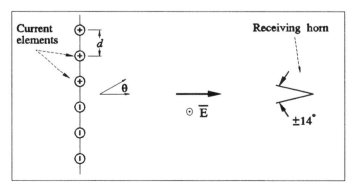

FIGURE 11.27 Calculation for the array factor.

The receiving horn in the bolometer has an acceptance angle of $\pm 14°$. Therefore, taking both the antenna array factor and the finite acceptance-angle effect into account, the array-factor pattern is calculated by

$$P(\phi) = \int_{\phi-14°}^{\phi+14°} A(\theta)\, d\theta \qquad (11.25)$$

Combining both the antenna pattern for the shortened bow-tie with fingers and the array factor with the effect of finite receiving angle, the calculated output pattern and the measured pattern are shown in Figure 11.28. The solid curve shows the measured output pattern with an input power of 300 mW. Peaks appear at 42° and 26° from the center with a power of 3.2 times and 2.3 times bigger than the power in the null, respectively. The accuracy of this measurement is limited by the uncertainty in the position of the rotating axis. The asymmetry of the output pattern may be caused by the off-axis rotating effect or by the fact that the diodes in one array are not completely identical. It shows only 10% of the total radiated power is accepted at normal incidence.

Biasing Power dependence was measured under different biasing conditions to verify varistor frequency multiplication. Figure 11.29 shows the measured power dependence with a bias of 0 V and 0.5 V. The frequency doubling effi-

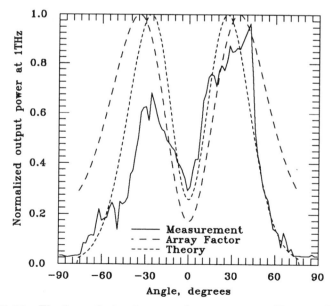

FIGURE 11.28 The theoretical and measured output patterns. The array factor with the effect of limited receiving angle, without the bow-tie antenna pattern, is also shown. The null is caused by the opposite orientations of the diodes in two halves of the grid.

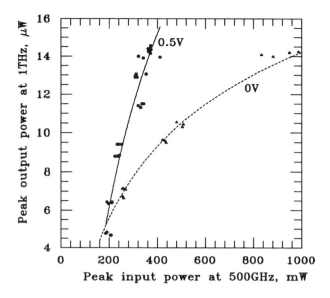

FIGURE 11.29 Measured power dependence with a bias of 0 V and 0.5 V.

ciency with a 14-μW output power was increased 2.6 times with a bias of 0.5 V. It should be noticed that these diodes become saturated with high input power when the diodes are biased at 0 V. This measurement verifies that the frequency multiplication results from varistors.

Peak output power was measured as a function of bias with a peak input power of 400 mW (Fig. 11.30). The maximum output power is 17 μW with a bias

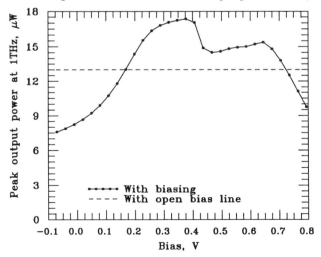

FIGURE 11.30 Measured peak output power as a function of bias. The peak input power is 400 mW. The dashed line was measured with open-circuited bias lines.

of 0.375 V. Comparing with the output power of 13 μW which was measured with open-circuited bias lines, it seems biasing only makes a small improvement in the output power. One possible reason is that these diodes are self-biased when the bias lines are open-circuited.

Tuning In previous measurements, the input or output tuners are not used so that neither the input nor output impedances are matched to achieve the optimal harmonic generations. It is concluded from the past experiences that the impedance tuners are essential to achieve high efficiencies [40]. A pair of dielectric slabs were used as impedance tuners in Jou's [44] and Liu's multipliers [45]. The slabs behave in a manner similar to the double stub tuners in a coaxial line or a waveguide. The slab thickness should be a quarter-wavelength at the fundamental frequency for the input tuner and a quarter-wavelength at the second harmonic frequency for the output tuner.

Figure 11.31 shows one example of tuning results for the terahertz multipliers. The dashed line indicates the average output power of 23 μW which was measured without the tuning slabs. The highest output power is 30.5 μW which is 1.3 times bigger than that without the tuning slabs. It is also observed that the closer the tuning slabs are placed to the array, the larger the second harmonic power is. This is caused by diffraction effect.

6.5 Nonlinear Analysis

The classic theory for nonlinear analysis of frequency multipliers and mixers was studied by Penfield and Rafuse [21] based on ideal varactors at microwave frequencies. Burckhardt [57] further discussed the cases in which the varactors have series resistances and nonideal characteristics. However, for complicated cases at millimeter or submillimeter wavelengths, where the harmonics are terminated with different embedding impedances, the simple analysis fails to predict the performances. For arbitrary nonlinearity, the most convenient analysis method would be numerical. The equivalent circuit is divided into linear and nonlinear subcircuits, as shown in Fig. 11.32, to perform harmonic-balance analysis [15]. The linear subcircuit is analyzed in the frequency domain to reach the solutions that satisfy the external circuit equations, and the nonlinear subcircuit is analyzed in the time domain to reach current and voltage solutions that satisfy the diode conditions. Fourier transform is used between these two domains and iterations are performed until both domains reach reasonable convergence. One popular form of the harmonic-balance analysis is the multiple-reflection technique introduced by Kerr [58]. In his technique, the multiplier circuit is divided into linear and nonlinear subcircuits separated by an inserted transmission line with an arbitrary characteristic impedance. By choosing the length of the transmission line to be an integral number of wavelength at the fundamental frequency, which would also be integral numbers of harmonic wavelengths, the steady-state responses will be the same as without the transmission line. The separation allows us to treat the frequency domain and the time domain separately, and the conditions of the transmission line should be satisfied in both domains.

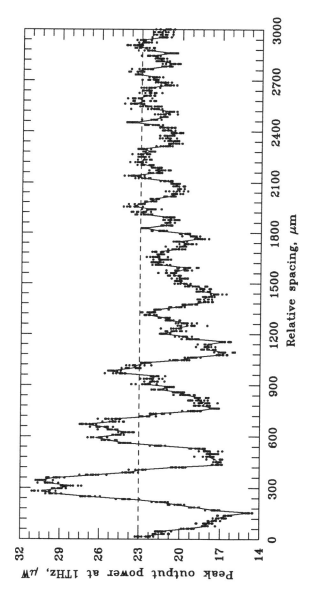

FIGURE 11.31 Measured output power as a function of the relative spacing between the grid and the tuning slabs.

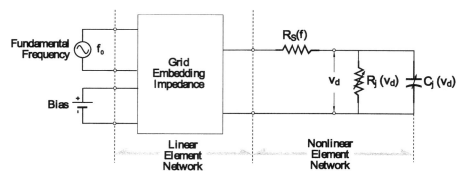

FIGURE 11.32 Equivalent circuit of a frequency conversion device based on diodes using harmonic-balance analysis [15].

Using this technique, Siegel and Kerr [15,48] developed a computer program to perform a full nonlinear large-signal analysis of Schottky diodes, used in multipliers or mixers, with frequency-dependent series-resistances and arbitrary C–V and I–V characteristics. This program calculates current and voltage waveforms based on given diode parameters and embedding impedances at the fundamental and harmonic frequencies, and it derives theoretical efficiency and input impedance at a given operating point where the input power and bias are specified. The program was modified (by Tolmunen and Frerking [59] and by Choudhury *et al.* [60]) to incorporate the embedding impedances and the estimated parameters of the planar Schottky diodes.

Commercial software is also available to perform the harmonic-balance analysis. The Hewlett-Packard Microwave and RF Design Systems also performs the nonlinear analysis. The software provides a list of symbolically defined devices, database-defined devices, and nonlinear models of diodes which allow us to perform the harmonic-balance analysis more accurately.

Figure 11.33 shows the equivalent circuit of the grid frequency doubler in the harmonic-balance analysis. The diode parameters can be specified as a varistor, a varactor, or a combination of both with the parasitic impedances. The measured DC I–V and low-frequency C–V curves or the theoretical values can be used. The embedding impedance and transmission-line equivalent circuit can be incorporated. The input and output filters can also be included. Two circulators are used to isolate the fundamental and harmonic frequencies. Figure 11.34 shows the comparison of measurements and calculations. The number of harmonics included in the calculations is six.

7 SIDEBAND GENERATOR

Hacker [52] proposed a terahertz sideband generator grid used as a grid mixer to upconvert a tunable low-frequency signal, the IF, to the terahertz region by

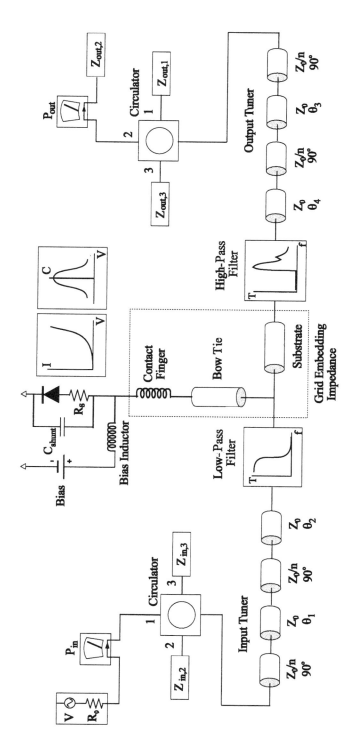

FIGURE 11.33 Equivalent circuit of the grid frequency doubler in the harmonic-balance analysis

448 FREQUENCY CONVERSION GRIDS

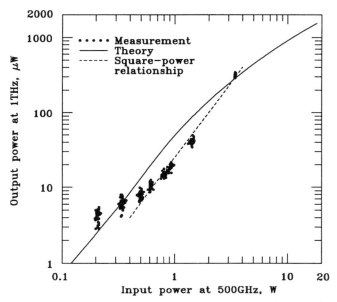

FIGURE 11.34 Comparison of the theoretical and measured power dependences.

mixing it with a single-frequency local oscillator. Figure 11.35 shows the sideband generator grid concept. This terahertz sideband generator grid acts as a nonlinear surface to convert a variable 1-GHz to 20-GHz IF signal onto a 1.6-THz LO signal which incidents normally on the diode grid. A mirror behind the grid is used to tune the impedance to match the free space for a better

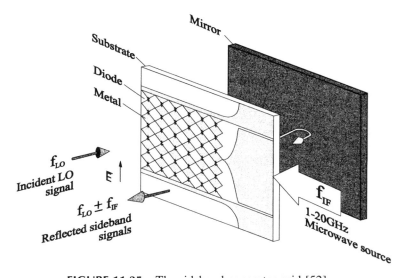

FIGURE 11.35 The sideband generator grid [52].

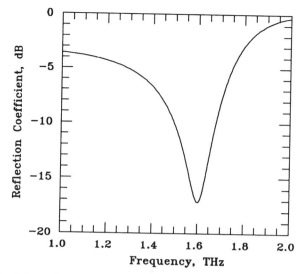

FIGURE 11.36 Theoretical reflection coefficient for the diode grid used as a sideband generator.

coupling efficiency. The IF signal is generated by a microwave sweeper and fed to the grid using a coplanar waveguide. The center row of the grid is connected to the center conductor of the coplanar waveguide, and the edges of the grid are connected to the outer conductors. The diode orientation is flipped over the central row to allow DC biasing.

The flat mirror behind the grid tunes the reflection coefficient null to the desired frequency. The mirror provides a shunt short-circuited free-space transmission line in the equivalent circuit, and the position of the mirror decides the electrical length. By optimizing the dimensions and the metal pattern of the unit cell as well as the mirror position behind the diode-grid, the reflection coefficient can be optimized for the incident signals at the operating frequency. Figure 11.36 shows theoretical reflection coefficient from 1 THz to 2 THz. The null of the reflection coefficient can be tuned between 1.5 THz and 1.9 THz. The electrical length between the grid and the mirror for the null at 1.6 THz is 23°.

8 CONCLUSION

Millimeter-wave frequency multiplier grids demonstrated the advantages of using quasi-optical power-combining techniques for high-frequency signal generation. Linear and nonlinear analysis for frequency conversion grids has been well developed and reaches reasonable agreement with the experiments. The terahertz doubler grids demonstrated the feasibility of using multiplier grids as submillimeter-wave local oscillators, although the diode orientation designed for sideband-generator applications results in a null in the center of the output beam when used as frequency multipliers. A grid with diode orientation appropriate for multiplier applications should improve the output pattern and increase the output power without changing the unit-cell design or the planar diode structures. A new batch (12×12 array) with correct diode orientation is being fabricated as this chapter been written. Increasing the size of grid not only increases the output powers, but also reduces the diffraction losses in the matching circuits and filters.

ACKNOWLEDGMENTS

The author would like to thank Dr. Neville Luhmann at UC Davis, Dr. Jonathan Hacker at Bellcore, Dr. Robert Weikle at UVA, Dr. Tom Crowe at UVA, Dr. Jim Allen at UC Santa Barbara, Dr. Peter Siegel at NASA-JPL, and Dr. David Rutledge at Caltech for their contributions.

REFERENCES

[1] J. W. Waters and P. H. Siegel, "Applications of millimeter and submillimeter technology to earth's upper atmosphere: results to date and potential for the future," *The 4th International Symposium on Space Terahertz Technology,* Los Angeles, CA, March 1993.

[2] J. Farman, B. Gardiner, and J. Shanklin, "Large losses of total ozone in Antarctica reveal seasonal ClO_x/NO_x interaction," *Nature,* vol. 315, p. 207, 1985.

[3] P. B. Hays and H. E. Snell, "Atmospheric remote sensing in the terahertz region," *Proceedings of the 1st International Symposium on Space Terahertz Technology,* p. 482, 1990.

[4] T. G. Phillips, "Developments in submillimeterwave astronomy," *The 19th International Conference on Infrared and Millimeter Waves in Sendai, Japan,* 1994.

[5] S. Gulkis, "Submillimeter wavelength astronomy missions for the 1990s," *Proceedings of the 1st International Symposium on Space Terahertz Technology,* pp. 454–457, 1990.

[6] N. C. Luhmann, "Instrumentation and techniques for plasma diagnostics: an overview," in *Infrared and Millimeter Waves,* vol. 2, K. J. Button, ed., Academic Press, New York, pp. 1–65, 1979.

[7] P. E. Young, D. F. Neikirk, P. P. Tong, and N. C. Luhmann, "Multi-channel far-infrared phase imaging for fusion plasma," *Rev. Sci. Instrum.,* vol. 56, pp. 81–89, 1985.

[8] P. F. Goldsmith, "Coherent systems in the terahertz frequency range: elements, operation and examples," *Proceedings of the 3rd International Symposium on Space Terahertz Technology,* pp. 1–23, 1992.

[9] G. Kantorowicz and P. Palluel, "Backward wave oscillator," in *Infrared and Millimeter Waves,* vol. 1, K. J. Button, ed., Academic Press, New York, pp. 185–212, 1979.

[10] G. I. Haddad and J. R. East, "Tunnel transit-time (TUNNETT) devices for terahertz sources," *Proceedings of the 1st International Symposium on Space Terahertz Technology,* pp. 104–126, 1990.

[11] M. A. Frerking, "The submillimeter mission heterodyne instrument," *Proceedings of the 2nd International Symposium on Space Terahertz Technology,* pp. 17–31, 1991.

[12] D. P. Steenson, R. D. Pollard, R. E. Miles, and J. M. Chamberlain, "Power-combining of resonant tunnelling diode oscillators at W-band," *The 19th International Conference on Infrared and Millimeter Waves in Sendai, Japan,* 1994.

[13] E. R. Brown, C. D. Parker, A. R. Calawa, M. J. Manfra, C. L. Chen, L. J. Mahoney, W. D. Goodhue, J. R. Söderströmt, and T. C. McGill, "High frequency resonant-tunneling oscillators," *Microwave Optical Tech. Lett.,* vol. 4, no. 1, p. 19, 1991.

[14] E .R. Brown, K. A. McIntosh, K. B. Nichols, and C. L. Dennis, "Photomixing up to 3.8 THz in low-temperature-grown GaAs," *Appl. Phys. Lett.,* vol. 66, no 3, pp. 285–287, Jan. 16, 1995.

[15] Peter H. Siegel, "Topics in the Optimization of Millimeter-Wave Mixers," Ph.D. Dissertation, Columbia University, 1984.

[16] A. Räisänen, "Frequency multipliers for millimeter and submillimeter wavelengths," *Proceedings of the IEEE,* vol. 80, no. 11, pp. 1842–1852, Nov. 1992.

[17] A. Mardon, "Nonlinear resistance and nonlinear reactance devices for harmonic generation," in *Millimetre and Submillimetre Waves,* F. A. Benson, ed., Iliffe Book Ltd., London, pp. 179–191, 1969.

[18] C. H. Page, "Frequency conversion with positive nonlinear resistors," *J. Natl. Bur. Stand.,* pp. 179–182, vol. 56, April 1956.

[19] M. V. Schneider, "Metal–semiconductor junctions as frequency converters," in *Infrared and Millimeter Waves,* vol. 6, K. J. Button, ed., Academic Press, New York, pp. 209–275, 1982.

[20] J. M. Manley and H. E. Rowe, "Some general properties of nonlinear elements—Part I. General energy relations," *Proc. IRE,* vol. 44, no. 7, pp. 904–913, July 1956.

[21] P. Penfield, Jr. and R. P. Rafuse, *Varactor Applications,* MIT Press, Cambridge MA, 1962.

[22] J. W. Archer, "Millimeter wavelength frequency multipliers," *IEEE Trans. Microwave Theory Tech.,* vol. MTT-29, no. 6, pp. 552–557, June 1981.

[23] M. T. Faber, J. W. Archer, and R. J. Mattauch, "A high efficiency frequency doubler for 100GHz," *IEEE MTT-S Int. Microwave Symp. Dig.,* pp. 363–366, 1985.

[24] N. Erickson, "High efficiency submillimeter frequency multipliers," *IEEE MTT-S Int. Microwave Symp. Dig.,* pp. 1301–1304, 1990.

[25] T. Takada, T. Makimura and M. Ohmori, "Hybrid integrated frequency doublers and triplers to 300 and 450 GHz," *IEEE Trans. Microwave Theory Tech.,* vol. 28, pp. 966–973, Sept. 1980.

[26] R. Zimmermann, R. Zimmermann, and P. Zimmermann, "490 GHz solid state source with varactor quadrupler," *The 13th International Conference on Infrared and Millimeter Waves Digest*, pp. 77–78, 1988.

[27] R. Zimmermann, R. Zimmermann, and P. Zimmermann, "All-solid-state radiometer at 557 GHz," *Proceedings of the 21st European Microwave Conference*, Stuttgart, pp. 253–256, Sept. 1991.

[28] R. Zimmermann, R. Zimmermann, and P. Zimmermann, "All-solid-state radiometers for environmental studies to 700 GHz," *Proceedings of the 3rd International Symposium on Space Terahertz Technology*, pp. 706–723, 1992.

[29] H. Rothermel, T. G. Phillips, and J. Keene, "A solid-state frequency source for radio astronomy in the 100 to 1000 GHz range," *The International Journal of Infrared and Millimeter Waves*, pp. 83–100, 1989.

[30] A. Rydberg, B. N. Lyons and S.U. Lidholm, "On the development of a high efficiency 750 GHz frequency tripler for THz heterodyne systems," *IEEE Trans. Microwave Theory Tech.*, vol. 40, no. 5, pp. 827–830, May 1992.

[31] N. Erickson and J. Tuovinen, "A waveguide tripler for 720–880 GHz," *Proceedings of the 6th International Symposium on Space Terahertz Technology*, pp. 191–198, 1995.

[32] R. Zimmermann, T. Rose, and T. Crowe, "An all solid-state 1 THz radiometer for space applications," *Proceedings of the 6th International Symposium on Space Terahertz Technology*, pp. 13–27, 1995.

[33] Jung-Chih Chiao, "Quasi-Optical Components for Millimeter and Submillimeter Waves," Ph.D. Dissertation, California Institute of Technology, 1996.

[34] K. J. Russell, "Microwave power combining techniques," *IEEE Trans. Microwave Theory Tech.*, vol. 27, no. 5, pp. 472–478, May 1979.

[35] K. Chang and C. Sun, "Millimeter-wave power combining techniques," *IEEE Trans. Microwave Theory Tech.*, vol. 31, no. 2, pp. 91–107, Feb. 1983.

[36] M. A. Frerking, H. M. Pickett, and J. Farhoomand, "A submillimeter wave quasi-optical frequency doubler," *IEEE MTT-S Int. Microwave Symp. Dig.*, pp. 108–109, 1983.

[37] J. W. Archer, "A novel quasi-optical multiplier design for millimeter and submillimeter wavelengths," *IEEE Trans. Microwave Theory Tech.*, vol. 32, no. 4, pp. 421–427, April 1984.

[38] D. Steup, "Whisker-contacted diode-multipliers as quasioptical SMMW-arrays," *Int. J. Infrared Millimeter Waves*, vol. 14, no. 12, pp. 2519–2532, 1993.

[39] B. N. Lyons, I. Sheridan, W. M. Kelly, U. S. Lidholm, and A. Räisänen, "Experimental and theoretical evaluation of a quasi-optical submillimeter-wave multiplier," *Proceedings of MIOP'90*, Stuttgart, pp. 369–374, 1990.

[40] J. Hacker, M. P. DeLisio, M. Kim, C.-M. Liu, S.-J. Li, S. W. Wedge, and D. B. Rutledge, "A 10 watt X-band grid oscillator," *IEEE MTT-S Int. Microwave Symp. Dig.*, pp. 823–826, 1994.

[41] R. L. Eisenhart and P. J. Khan, "Theoretical and experimental analysis of a waveguide mounting structure," *IEEE Trans. Microwave Theory Tech.*, vol. 19, no. 8, pp. 706–719, Aug. 1971.

[42] R. M. Weikle, II, "Quasi-Optical Planar Grids for Microwave and Millimeter-Wave Power Combining," Ph.D. Dissertation, California Institute of Technology, 1992.

[43] W. W. Lam, C. F. Jou, N. C. Luhmann, Jr., and D. B. Rutledge, "Diode grids for electronic beam steering and frequency multiplication," *Int. J. Infrared Millimeter Waves*, vol. 7, pp. 27–41, 1986.

[44] C. F. Jou, W. W. Lam, H. Z. Chen, K. S. Stolt, N. C. Luhmann, Jr., and D. B. Rutledge, "Millimeter-wave diode-grid frequency doubler," *IEEE Trans. Microwave Theory Tech.*, vol. 36, no. 11, pp. 1507–1514, Nov. 1988.

[45] H-X. L. Liu, L. B. Sjogren, C. W. Domier, N. C. Luhmann, Jr., D. L. Sivco, and A. Y. Cho, "Monolithic quasi-optical frequency tripler array with 5-W output power at 99 GHz," *IEEE Electron Device Lett.*, vol. 14, no. 7, pp. 329–331, July 1993.

[46] J.-C. Chiao, A. Markelz, Y. Li, J. Hacker, T. Crowe, J. Allen, and D. Rutledge, "Terahertz grid frequency doublers," *Proceedings of the 6th International Symposium on Space Terahertz Technology*, pp. 199–206, 1995.

[47] P. W. Staecker, M. E. Hines, F. Occhiuti, and J. F. Cushman, "Multi-watt power generation at millimeter-wave frequencies using epitaxially-stacked varactor diodes," *IEEE MTT-S Int. Microwave Symp. Dig.*, pp. 917–920, 1987.

[48] P. H. Siegel and A. R. Kerr, "The measured and computed performance of a 140–220 GHz Schottky diode mixer," *IEEE Trans. Microwave Theory Tech.*, vol. 32, no. 12, pp. 1579–1590, Dec. 1984.

[49] T. W. Crowe, R. J. Mattauch, H. P. Röser, W. L. Bishop, W. C. B. Peatman, and X. Liu, "GaAs Schottky diodes for THz mixing applications," *Proc. IEEE*, vol. 80, no. 11, pp. 1827–1841, Nov. 1992.

[50] D. G. Garfield, R. J. Mattauch, and W. L. Bishop, "Design, fabrication and testing of a novel planar Schottky barrier diode for millimeter and submillimeter wavelengths," *Proceedings of IEEE Southeastcon'88 in Knoxville, TN*, pp. 154–160, April 1988.

[51] W. L. Bishop, T. W. Crowe, and R. J. Mattauch, "Planar GaAs Schottky diode fabrication: progress and challenges," *Proceedings of the 4th International Symposium on Space Terahertz Technology*, pp. 415–429, 1993.

[52] J. B. Hacker, "Grid Mixers and Power Grid Oscillators," Ph.D. Dissertation, California Institute of Technology, 1994.

[53] J. Hacker and R. Weikle, "Quasi-optical grid arrays," in *Frequency Selective Surface and Grid Array*, T. K. Wu, ed., Wiley, New York, pp. 249–296, 1995.

[54] J. A. Copeland, "Diode edge effect on doping-profile measurements," *IEEE Trans. Electron Devices*, vol. 17, no. 5, pp. 404–407, May 1970.

[55] J. T. Louhi, "The capacitance of a small circular Schottky diode for submillimeter wavelengths," *IEEE Microwave Guided Wave Lett.*, vol. 4, no. 4, 1994.

[56] S. J. Allen, K. Craig, B. Galdrikian, J. N. Heyman, J. P. Kaminski, K. Campman, P. F. Hopkins, A. C. Gossard, D. H. Chow, M. Lui and T. K. Liu, "Materials science in the far-IR with electrostatic based FELs," *FEL 94*, Stanford, CA, August 1994.

[57] C. B. Burckhardt, "Analysis of varactor frequency multipliers for arbitrary capacitance variation and drive level," *Bell Syst. Tech. J.*, vol. 44, no. 4, pp. 675–692, 1965.

[58] A. R. Kerr, "A technique for determining the local oscillator waveforms in a microwave mixer," *IEEE Trans. Microwave Theory Tech.*, pp. 828–831, Oct. 1975.

[59] T. J. Tolmunen and M. A. Frerking, "Theoretical efficiency of multiplier devices," *Proceedings of the 2nd International Symposium on Space Terahertz Technology*, pp. 197–211, 1991.

[60] D. Choudhury, A. V. Räisänen, R. P. Smith, S. C. Martin, J. E. Oswald, R. J. Dengler, M. A. Frerking, and P. H. Siegel, "Frequency tripler with integrated back-to-back barrier-N-N^+ (bbBNN) varactor diodes in a novel split-waveguide block at 220 GHz," *IEEE MTT-S Int. Microwave Symp. Dig.*, pp. 771–774, 1994.

CHAPTER TWELVE

Quasi-Optical Subsystems

ZOYA B. POPOVIĆ
University of Colorado, Boulder

GERALD JOHNSON
Lockheed Martin Corporation, Colorado

1 INTRODUCTION

Depending on who uses it, the words "component," "subsystem," or "system" can mean very different levels of complexity. In this chapter, we consider the analog front end of a communication link or radar to be a subsystem. We examine the potential advantage of active quasi-optical components for some applications. We also describe subsystems in which several quasi-optical active components are designed in an integrated fashion, the goal of which is to provide a system's function. An example is a quasi-optical amplifier fed by several quasi-optical oscillators in a beam-forming transmitter [1].

Quasi-optical components have been considered as an alternative to more conventional technologies for some applications. The motivation for starting research in this area was the lack of high-power solid-state sources at millimeter-wave frequencies. Several watts at Ka band have been demonstrated with active array amplifiers (Chapter 5), but the real potential of the quasi-optical approach also lies in functionality and the possibility of low-cost integration with the rest of the system.

In a quasi-optical transmitter, the output power amplifiers are integrated with the planar (or conformal, if desired) antenna array, saving space and loss in the

---

Active and Quasi-Optical Arrays for Solid-State Power Combining, Edited by Robert A. York and Zoya B. Popović.
ISBN 0-471-14614-5 © 1997 John Wiley & Sons, Inc.

feed network. The output power is obtained by free-space power combining, which eliminates a passive power-combining network. All of the amplifiers are biased in parallel with a simple circuit. The feed to the amplifier is provided from a quasi-optical source, also a two-dimensional planar device. Other functions, such as modulation, polarization control, switching, isolation, tuning, filtering, and frequency conversion can be added by cascading appropriate planar active or passive quasi-optical surfaces with the output planar antenna array. This modular property of quasi-optical subsystems may have advantages for low-cost communication or radar applications. The reliability of quasi-optical components has not been systematically investigated, but limited data show that it should be very good since a relatively large percentage of the devices in an array can fail with graceful power degradation. This assumes that the bias and control connections are made so that a single device failure will not impact adjacent cells.

On the receiver end of a quasi-optical subsystem, an antenna array is populated with low-noise amplifiers. In applications such as radar, wide dynamic range becomes very important. Each amplifier element in the array receives the incoming signal independently, and the amplified powers are incident on a quasi-optical mixer after combining in free-space. The noise figure of the entire array is approximately the same as the noise figure of a single element, provided that the individual element noises are uncorrelated. The output power, however, is a coherent combination of the individual signal powers, and as a result the dynamic range of the receiver is increased as compared to a receiver with a single amplifier. Other functions, such as angle diversity, can be added to the receiver by simply adding several planar mixers.

The above potential benefits from using quasi-optics may not be sufficient for widespread system applications. A quasi-optical subsystem will also need to provide linearity at high power, reasonable bandwidths, flexible modulation types, possibly very high modulation rates, beam-steering and beamforming, flexible polarization control, high operating power efficiency, and rugged, compact packaging. In the concluding section of this chapter, we discuss some of these issues as they relate to specific system examples.

Quasi-optical components or subsystems are not (yet) in production. Here we give a brief description of what such a subsystem might consist of. As an example, the diagram of a quasi-optical transceiver subsystem is sketched in Fig. 12.1. A solid-state power-combining source generates the carrier frequency for the transmitter and also serves as the local oscillator or self-oscillating mixer for the receiver. A planar array of transmit/receive (T/R) active antennas performs the power amplification for the transmit and the low-noise amplification for the receive. Frequency modulation is achieved through the gate bias of either the oscillator or amplifier. Alternatively, an independent modulator can be integrated in cascade with the amplifier for phase modulation. In this approach, both the input and output waves are focused, and the source/mixer is located at the focal point of the amplifier to reduce diffraction loss and thus increases system efficiency.

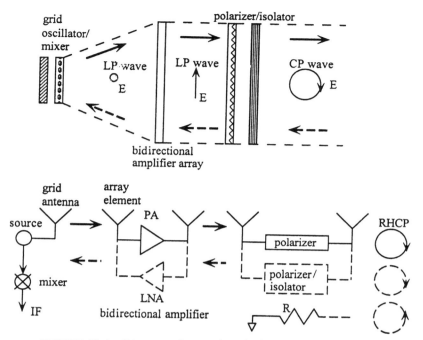

FIGURE 12.1 Diagram of a quasi-optical transceiver subsystem.

The close proximity of the oscillator and amplifier also implies compact transceiver design. The input/output isolation, essential for maintaining amplifier stability, is achieved by using orthogonally polarized input/output antennas. A quasi-optical isolator can be placed in front of the amplifier array to protect the amplifier from high-level reflections. In addition to its important isolation function, this device is a circular polarizer, thus reducing propagation fading effects. Large-scale and low-loss spacial power combining allows this approach to overcome the limited power available from millimeter-wave devices. In addition, reliability is improved since degradation is graceful. By using a grid oscillator as the signal source, we are applying two-level solid-state power combining: the first level at the grid oscillator, and the second at the amplifier array. The quasi-optical approach lends itself to a great deal of flexibility, since individual grids, or arrays, serve different functions and can be cascaded into subsystems.

In addition to serving as a free-space power combiner, the constrained lens amplifier in Fig. 12.1 may also perform beam forming, beam steering, and beam switching over a wide scan angle range with a relatively small cluster of feed horns. The feed may be provided by any one horn in the cluster to produce a single beam in a particular direction, or the horns may simultaneously feed the array for forming multiple beams. These functions are necessary for active array applications in airborne satellite communication terminals and satellite cross-links. Such applications currently call for millimeter-wave phased array development in the 20 to 60-GHz frequency range using active transmit–receive (T/R) modules with phase shifters at each element in the array. The small physical dimensions at millimeter-wave frequencies lead to several unexpected difficulties not encountered at microwaves. For instance, a 4000-element phased array at 40 GHz has an aperture of only 24×24 cm^2, yet there needs to be RF distribution, phase control, and DC power distribution lines to all elements—tasks that approach a total of 36,000 lines [2]. A quasi-optical lens amplifier array could conceivably replace the conventional phased array approach at millimeter-wave frequencies. X-band lens arrays described in Chapter 5 distribute and combine the RF signal in free space, perform beam scanning with a cluster of feed horns, and distribute DC power in a simple fashion with a minimum of distribution lines.

In the following sections of this chapter, some details of quasi-optical subsystems demonstrated to date are described and their potential advantages over conventional approaches are outlined. These subsystems include: a transmitter with beam-forming and beam-switching capabilities; a two-level power combining system consisting of an oscillator/power combiner which feeds an amplifier/combiner; a low-noise receiver with an input amplifier and self-oscillating quasi-optical subharmonic mixers; a quasi-optical isolator; and quasi-optical digital phase modulators.

2 TRANSMITTING QUASI-OPTICAL SUBSYSTEMS

In this section, a quasi-optical transmitter using a lens amplifier is described. The transmitter consists only of quasi-optical components. Several grid oscillator feeds generate the signal which is then amplified by a lens amplifier. The lens amplifier radiates a collimated beam which can be steered in discrete steps with no accompanying on-substrate phase shifters. This type of beam steering is implementationally simple, but only a finite number of discrete beams are achievable. It is shown, however, that this approach has a broader bandwidth compared to constant phase-shifter phased arrays. In contrast, a quasi-optical technique which allows continuous beam-steering has been presented in Chapter 4. Here we also present beam-forming results using several grid-oscillator feeds and a linear lens amplifier.

2.1 Two-Level Power Combining

The measured results of the linear and two-dimensional lens arrays presented in Chapter 5 were accomplished with a horn feed located at the focal distance or along the focal arc. A grid oscillator, such as the ones described in Chapter 8, can perform the feed function in a more compact planar form. The challenge is to design a grid oscillator that will oscillate within the passband of the lens amplifier and provide good illumination of the input side for a good-quality output beam. The details of the design and measurements given here illustrate the issues that arise when several quasi-optical components are integrated into a subsystem.

The grid oscillators are the primary level of this two-level system in that they generate the signal. The lens amplifier array is the second level of the system, performing focusing, amplifying, and re-radiation.

A planar grid oscillator was designed to feed the two-dimensional lens amplifier array. The theory presented in reference [3], based on a full-wave analysis of a single unit cell of the structure, was used. For a given geometry, the metallization is divided into a number of rooftop basis functions and a voltage generator is inserted across the gap in place of a 1-port active device. The moment method is then used to determine the current on the entire structure, and the ratio of the voltage across the gap to the current through the gap is the driving-point impedance seen by the active device. This analysis is extended to a two-port device, such as a transistor, yielding a two-port characterization for the passive part of the grid (the metal geometry, the dielectric, and radiation into free space). An appropriate model for the active device is then connected to this linear network to analyze the behavior of the grid oscillator.

In order to match the narrow-band frequency response of the two-dimensional lens amplifier, a grid oscillator operating at 10 GHz is required. In the design, small-signal s parameters were used to model the Avantek ATF-35576 PHEMTs, the devices selected for fabricating the oscillator. Simulations indicate that a structure with one dipole and one bow-tie radiating element on a 2.54-mm-thick Duroid substrate with a relative permittivity of 10.5 results in an oscillation frequency near 10 GHz. The unit cell of the oscillator is shown in Fig. 12.2a, and the entire grid oscillator is shown in Fig. 12.2b.

According to the simulations, a mirror placed 3 mm behind the dielectric results in an oscillation frequency of 10 GHz. The fabricated grid requires a grid-to-mirror spacing of 5 mm to tune the frequency to 10.0 GHz. We observe a gate-bias tuning bandwidth of 225 MHz with less than 3 dB change in output power. The grid oscillator can be electrically tuned over the operating bandwidth of the lens amplifier array. The 28-element grid oscillator has an effective radiated power of 22 dBm measured in the far field. It is interesting to note here that in this grid oscillator, 20 out of 28 devices (71%) are edge devices, and yet the theory based on an infinite grid and a unit cell was adequate for its design.

For broadside radiation, the grid oscillator is mounted along the optical axis of the two-dimensional lens amplifier array in an anechoic chamber. The best

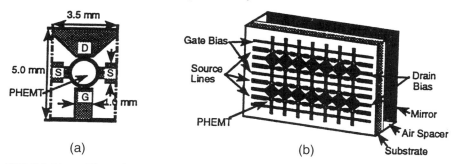

FIGURE 12.2 The unit cell (a) of a grid oscillator (b) designed to feed a quasi-optical lens amplifier array. (From reference [18], © 1994 IEEE.)

output power is produced when the grid oscillator is 15 cm away from the lens. Since the effective size of the grid oscillator is approximately 2.0 cm by 2.5 cm, the optimum feed distance is expected to be closer than the design focal length.

The grid oscillator generates 150 mW of effective radiated power (ERP) at 10.25 GHz. When the oscillator feeds the lens at its optimum feed point, the lens ERP is 525 mW, showing an insertion gain of 5.4 dB. The isolation of the lens is tested by increasing the gate bias, resulting in 2 mA of total drain current. The isolation ratio at 10.25 GHz is 21 dB. A cross-polarization ratio of 23 dB for the lens is found by polarization mismatching the grid oscillator feed and the receive horn with respect to the input and output of the lens, respectively. E-plane and H-plane patterns of the lens fed at broadside are shown in Fig. 12.3. The main lobe 3-dB beamwidths of the E- and H-plane patterns are approximately 11°, which is in good agreement with a theoretical value of 11.3° for a uniformly driven array.

The H-plane pattern side lobe level is better than -12 dB, but the E-plane pattern exhibits a significant lobe at $+20°$. It is believed that the cause of this lobe is an asymmetry in the near-field radiation pattern of the grid oscillator, as well as an asymmetry in the output patch pattern at 10.25 GHz. The measured far-field pattern of the grid oscillator shows an asymmetry in the E-plane at around 20° and a symmetrical H-plane pattern. To verify the lens with a symmetrical feed, the grid oscillator was replaced with a horn and the E-plane radiation patterns of the lens at 10.25 GHz and 9.7 GHz, shown in Fig. 12.4a, indicate that the lobe at $+20°$ is reduced by 5 dB at 10.25 GHz and essentially

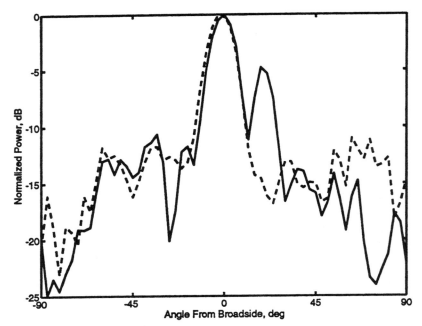

FIGURE 12.3 E-plane and H-plane patterns of the patch–patch 24-element lens amplifier fed at broadside. (From reference [18], © 1994 IEEE.)

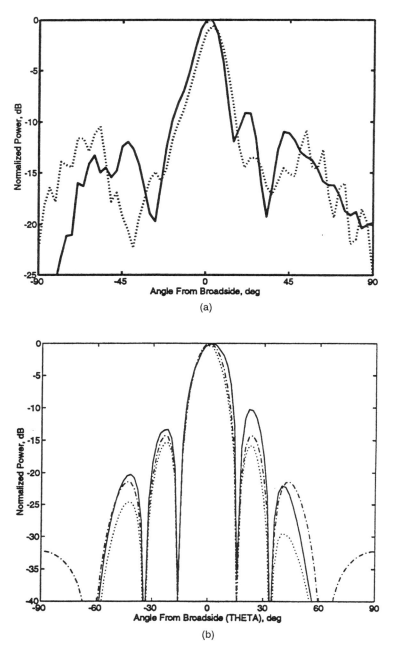

FIGURE 12.4 Measured E-plane radiation patterns of the lens at 10.25 GHz and 9.7 GHz using a horn feed. (a) The resulting patterns indicate that the asymmetry in the grid oscillator feed is mostly responsible for the significant side lobe in the patterns. (From reference [18], © 1994 IEEE.) (b) The theoretical E-plane radiation pattern showing the patch pattern's effect as a function of frequency, assuming uniform amplitude among the elements.

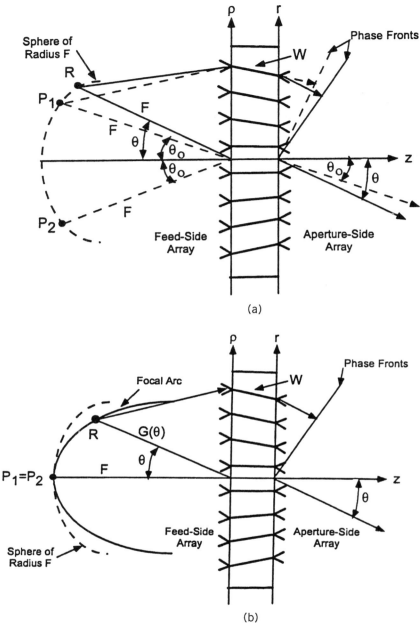

FIGURE 12.5 (a) The lens amplifier has perfect focal points at positions P_1 and P_2 along a sphere of radius F and focal angle θ_0. (b) The corrected distance, $G(\theta)$, as a function of feed angle for the lens. (c) The calculated focal arc for $F/D = 2.0$, which the ratio used in the power-combining transmission lens amplifier arrays described in Chapter 5, and $F/D = 1.2$, which is the ratio used in the low noise receiver array from Chapter 5. These focal arcs determine the feed location(s) for the lens amplifiers when performing beam-steering and beamforming.

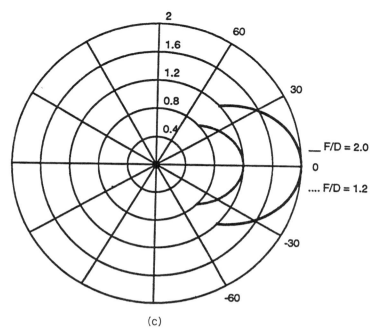

(c)

FIGURE 12.5 (*Continued*)

disappears at 9.7 GHz. Therefore, it appears that there is an asymmetry in the grid oscillator feed that is mostly responsible for the significant side lobe in the patterns. The grid oscillator radiation pattern may be asymmetrical because of a dielectric strip placed across the bias lines on one side of the grid which was required to get the oscillator to reliably lock to the desired mode. To check the frequency dependence of this significant sidelobe being caused by the output patch pattern, we calculated the theoretical array factor multiplied by the output patch pattern.

The patch pattern is found to be frequency dependent, with the radiation pattern relatively symmetrical at 9.75 GHz but steered to $+20°$ at 10.25 GHz. The theoretical E-plane radiation pattern showing the patch pattern's effect as a function of frequency is shown in Fig. 12.4b. The simulation assumes uniform amplitude among the elements. The solid line represents the array's pattern at 10.25 GHz, showing the significant side lobe at $+20°$ and in the asymmetrical pattern. As seen in the measured results, operation at 9.75 GHz (dotted line) reduces the significant side lobe by 5 dB for a nearly symmetrical pattern. For comparison, the array factor assuming isotropic radiation of all elements is shown as a dash–dot line. This model adequately explains the frequency dependence of the E-plane pattern.

2.2 Beam-Steering Using a Lens Amplifier

Since the lens is no longer fed at its focal point at angles other than 0°, lens defocusing due to path length errors in its feed reduces the output power of the lens. Consider the diagram in Fig. 12.5a in which the lens has perfect focal

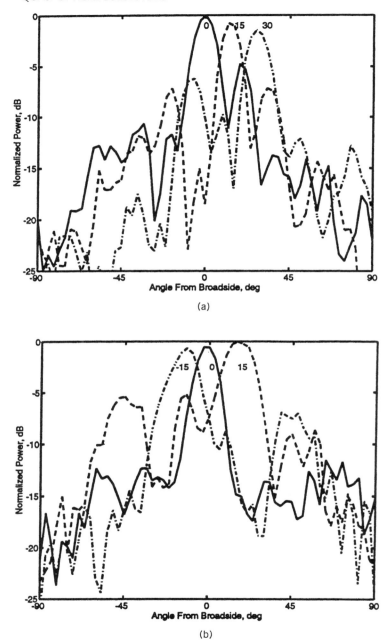

FIGURE 12.6 Measured beam steering in the (a) E- and (b) H-planes using a lens amplifier. (From reference [18], © 1994 IEEE.)

points at positions P_1 and P_2 along a sphere of radius F and focal angle θ_0. A feed location G can be chosen at a distance different than F to minimize the peak error for any scan angle θ [4]. Figure 12.5b illustrates the corrected distance, $G(\theta)$, as a function of feed angle for the lens.

The focal arc minimizes defocusing errors as a function of angle and is instrumental in beam forming at angles other than the focal point angle θ_0. Figure 12.5c shows the calculated focal arc for $F/D = 2.0$, which the ratio used in the power-combining transmission lens amplifier arrays described in Chapter 5, and $F/D = 1.2$, which is the ratio used in the low-noise receiver array from Chapter 5. These focal arcs determine the feed location(s) for the lens amplifiers when performing beam steering and beamforming.

Beam steering is performed by positioning the grid oscillator at various angles off of broadside along the focal arc. Figure 12.6 shows the measured beam steering in the E- and H-planes. The beam can be steered in the E-plane by 30° with less than 2 dB power variation in the main lobe peak. The asymmetric side lobe in the broadside-fed E-plane pattern is reduced when the lens is fed from the +15° and +30° feed points, as expected. Grating lobes are observed for scan angles greater than 24° in the H-plane beam-steering measurement due to the large interelement spacing. These measured patterns compare favorably with the theoretical patterns for E-plane and H-plane shown in Fig. 12.7. These patterns include the output patch pattern at 10.25 GHz to accurately model the measured patterns at this frequency. No grating lobes form in the E-plane pattern due to the triangular lattice of this array. This is in contrast to the case of the linear array described in Chapter 5, even though both arrays have the same interelement spacing. The H-plane pattern shows grating lobes 7 dB below the peak due to the triangular lattice not suppressing these lobes in the $\phi = 90°$, or H-, plane. Therefore, beam scan range is much more limited in the H-plane than the E-plane.

To illustrate the advantage of using a triangular grid for grating lobe suppression, consider a 6×5 rectangular array with interelement period $d_x = d_y = 0.75\lambda_0$ in an arrangement shown in Fig. 12.8a. If the elements of the array are uniformly illuminated and isotropic, a scan angle of only 15° in either E- or H-planes will produce a grating lobe, illustrated as the dash-dot line in Fig. 12.9. To improve the scan performance, a triangular lattice similar to that in Fig. 12.8b may be selected. For comparison purposes, the same interelement period is used, but the second and fourth rows of the array have only five elements and these rows are offset by $d_x/2$ with respect to their neighboring rows. The pattern of this 28-element triangular-lattice array may be scanned to 50° in the E-plane without grating lobe formation as shown in Fig. 12.9 as the solid line. Therefore, the triangular grid offers improved scan performance for a given interelement spacing.

The triangular grid array allows substantially wider interelement spacings for a given scan angle requirement than does the rectangular lattice. This is important for active arrays, where circuits are integrated on the same substrate as the antennas. Larger interelement spacing translates to larger unit cells, which

466 QUASI-OPTICAL SUBSYSTEMS

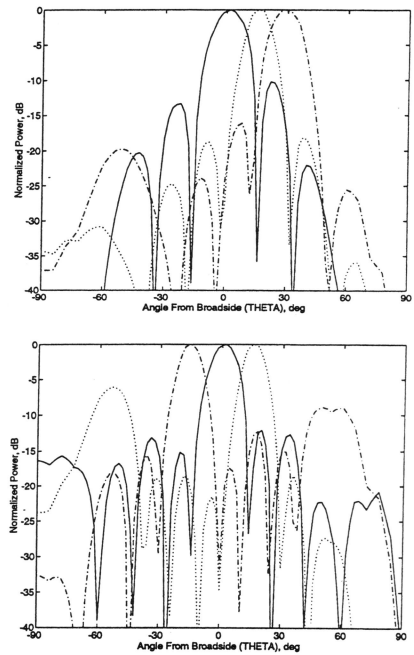

FIGURE 12.7 Theoretical patterns for *E*- and *H*-planes. These patterns include the output patch pattern at 10.25 GHz to accurately model the measured patterns at this frequency.

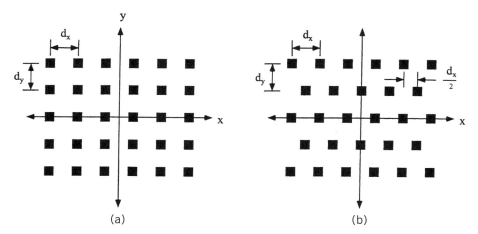

FIGURE 12.8 (a) A 6 × 5 rectangular array with interelement period $d_x = d_y = 0.75\lambda_0$ and (b) a triangular lattice array. For comparison purposes, the same interelement period is used, but the second and fourth rows of the triangular array have only five elements and these rows are offset by $d_x/2$ with respect to their neighboring rows.

means that there is more area available for the active circuits and less possibility of circuit-antenna coupling. As an example of scan properties of triangular and rectangular array lattices, we consider the hypothetical requirement for a main lobe scan angle of 60° from broadside for every plane of scan without grating lobe formation.

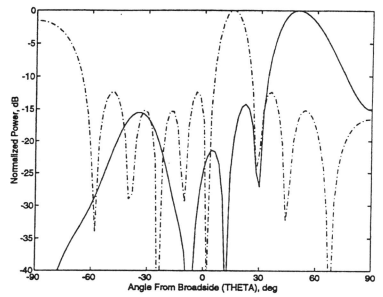

FIGURE 12.9 E- or H-plane scanning patterns will produce a grating lobe for the rectangular lattice array, illustrated as the dash–dot line. A triangular lattice with the same interelement period is used, and the pattern without grating lobe formation is shown in solid line. The triangular lattice offers improved scan performance for a given interelement spacing.

A square lattice needs to have an interelement spacing of $d_x = d_y = 0.536\lambda$. Hence, the area occupied per element location is $d_x \cdot d_y = (0.536\lambda)^2 = 0.287\lambda^2$. For an equilateral-triangular lattice, the requirement is satisfied by $\lambda/d_y = \lambda/\sqrt{3}d_x = 1.866$ or $d_y = 0.536\lambda$, $d_x = 0.309\lambda$. The area per element in this case is $2d_x d_y = 2(0.536)(0.309) = 0.332\lambda^2$. For the same amount of grating-lobe suppression, the number of elements saved using the triangular lattice over the rectangular lattice is about 14%. The relaxation of the area requirements for the triangular lattice is employed in our active lens amplifier arrays to accommodate the space required for the amplifier circuits and delay lines.

The bandwidth of scanned array antennas depends upon the characteristics of the components used in the array. The type of elements, feed networks, and the mechanism used to determine the phase relationship between elements in the array may all contribute to bandwidth limitations. One set of components may dominate the bandwidth constraints of the array, or they may be comparable. A discussion on the bandwidth limitations due to the radiating elements is presented in Chapter 5. Since the lens arrays use a free-space feed, bandwidth limitations due to the array feed are not considered.

To illustrate an advantage of lens amplifiers, we consider the bandwidth of a scanning system for a given scan angle. As mentioned above, a popular method of scanning arrays is with constant phase-type phase shifters. The phase shift is independent of frequency. A phase-steered array will establish the desired scan angle (θ_0, ϕ_0) only at one frequency f_0. Since the phase distribution is designed to remain fixed independent of frequency, the array becomes phased to receive at a different incidence angle (θ_1, ϕ_1) at a different frequency f_1. For a fixed scan plane in ϕ, the amount of beam squint with frequency is given by $f_1 \sin \theta_1 = f_0 \sin \theta_0$ [11]. For a small change in frequency, the change in scan angle is given by $\Delta\theta_0(\text{radians}) = -(\Delta f/f_0) \tan \theta_0$. This shows that the amount of beam squint depends upon the original scan angle as well as on the percent frequency change. At the broadside angle $(\theta_0 = 0°)$, there is no beam squint scanning regardless of the amount of change in frequency. When the desired scan angle increases, the amount of beam squint increases for a given frequency variation. Therefore, the bandwidth limitation of an array steered with constant-phase-type phase shifters must be specified in terms of the desired maximum scan angle θ_0 and some maximum squint angle $\Delta\theta_{0,\text{max}}$. For instance, if the maximum desired scan angle is $\pm 60°$, and the maximum allowable beam squint angle is $\pm 1°$, then the bandwidth $\Delta f/f_0$ is only about 1%. Thus if $\theta_{0,\text{max}} = \pm 60°$ and $\Delta\theta_{0,\text{max}}$ is defined in degrees, then the approximate relationship to bandwidth is $\Delta\theta_{0,\text{max}}(\text{degrees}) \simeq \Delta f/f_0(\text{percent})$. The extent to which one specifies the maximum allowable squint angle may be dependent on the beam width of the main lobe or the resolution of the measurement system that would detect the squint.

In a lens array, the electrical delay length l_{mn} may be related to a time delay ΔT_{mn} in a transmission medium with a phase velocity $v(f)$ that may be frequency dependent according to

$$l_{mn} = v(f)\Delta T_{mn} = \frac{c}{\epsilon_{\text{eff}}(f)} \Delta T_{mn}$$

An ideal TEM transmission medium is dispersionless and will have an effective dielectric constant $\epsilon_{\rm eff}$ that is independent of frequency, yielding a frequency invariant electrical delay length l_{mn}. Such a delay mechanism is called *true-time delay beam steering*. However, the planar lens arrays utilize either microstrip or coplanar waveguide (CPW) for delay lines and amplifier matching networks, which are quasi-TEM media that exhibit some dispersion, which will yield an error in the delay as a function of frequency, which in turn will set a frequency limit on the lens array due to the time-delay network.

2.3 Beam Forming Using a Lens Amplifier

In the two-level power combiner from Section 2.2, a second 28-PHEMT grid oscillator was built to the same specifications as the first grid oscillator feed for the amplifier. With two oscillators, beam forming is demonstrated in both E- and H-planes as the grid oscillators are moved along the focal surface of the lens. Beam forming and beam switching can be achieved with a dual grid-oscillator feed, as shown in Fig. 12.10a.

When both grids are on, the output radiated pattern of the lens amplifier is a superposition of the patterns for the two individual feeds, which makes beam-forming possible. Figure 12.10b shows examples of measured dual-beam patterns in the E- and H-planes of the lens. In the E-plane, the two grids were

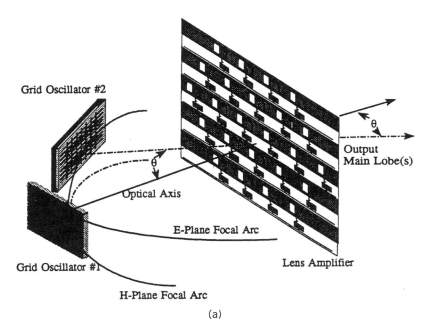

FIGURE 12.10 (a) Beam forming and beam switching can be achieved with a dual grid-oscillator feed positioned on the focal surface of the lens amplifier. (b) Measured E-plane (solid line) and H-plane (dotted line) beam-forming patterns for a dual-grid oscillator feed. (From reference [1], © 1995 IEEE.) (*Continued*)

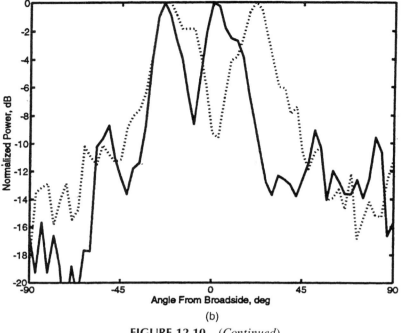

FIGURE 12.10 (*Continued*)

placed at 0° and 24° along the focal arc; and the amplifier pattern, shown as a solid line, shows equal-power beams at 0° and 24°. Two beams at 21° in the H-plane are shown as a dotted line in Fig. 12.10b. The individual grid oscillators can also be bias-switched one at a time, and so the amplifier pattern can be switched between a finite number of beams at different angles. In this case, the switching speed is limited by the oscillator settling time. By keeping the oscillators biased at the threshold of oscillation and pulsing the gate bias, a switching frequency of 5 kHz was measured (0.2-msec switching rate). It was verified that the grid oscillator that is off does not affect the pattern of the lens.

3 RECEIVING QUASI-OPTICAL SUBSYSTEMS

Recently, there has been much interest in using quasi-optical techniques in front ends of communication systems. A number of stand-alone quasi-optical components have been demonstrated, and the next step is to demonstrate entire transmitters and receivers. The motivation for the work presented here is to demonstrate a quasi-optical front end for a receiver useful for radio communications [5]. This receiver consists of a lens amplifier antenna array and a self-oscillating grid mixer, as shown in Fig. 12.11. The lens amplifies an incoming plane wave at the carrier frequency and then focuses it to a focal point where a mixer can be placed. By reciprocity to a transmitting lens amplifier, in a receiving am-

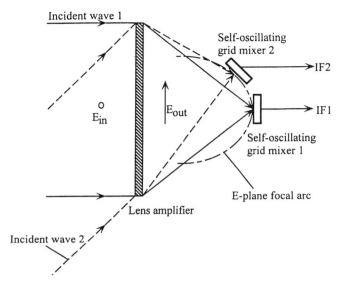

FIGURE 12.11 A quasi-optical receiver consisting of a lens amplifier antenna array and self-oscillating grid mixers. (From reference [5], © 1996 IEEE.)

plifier, there is a multitude of focal points which allow reception of waves incoming from different directions. A receiving lens amplifier with three self-oscillating quasi-optical grid mixers positioned along a focal arc of the amplifier confirms this property. The three mixers receive signals independently from three directions, taking advantage of the angle-preserving nature of the amplifier and providing angle diversity. This is potentially useful in radio communications for eliminating multipath fading nulls, since the probability of a fade of three signals incident from different angles is very low [6]. For a BPSK modulated signal and a standard Rayleigh multipath fading channel, the probability of error is proportional to $(L/(S/N))^L$, where n is the level of diversity. The probability of error for different diversity levels is graphically illustrated in Fig. 12.12 [7], and it shows the advantage of using several diversity levels in communication system operating under Rayleigh multipath fading conditions. The result for an additive white Gaussian noise channel (no fading) is shown in the figure for comparison.

3.1 Self-Oscillating Grid Mixer

One type of quasi-optical mixer that has been demonstrated is a diode-loaded grid [8]. A receiver using this type of component requires a separate LO. Alternatively, one can use a self-oscillating mixer, in which a single device generates the LO and mixes it with the RF signal [9]. A quasi-optical self-oscillating mixer using two transistors has been reported in reference [16].

Grid oscillators can also be used as self-oscillating mixers. A 25-element grid (the same as the one described in Chapter 8) with a free-running oscillation

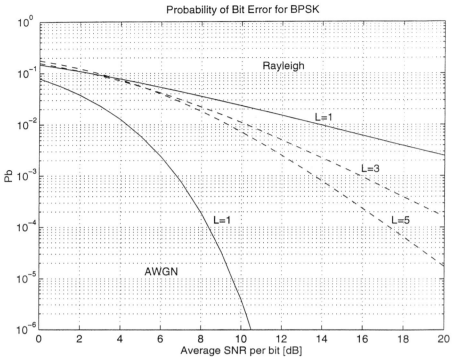

FIGURE 12.12 The probability of error for different diversity levels [7] shows the advantage of using several diversity levels in communication system operating under Rayleigh multipath fading conditions. The result for an additive white Gaussian noise channel (no fading) is shown for comparison.

frequency of 5.36 GHz was used to demonstrate this. A vertically polarized RF signal incident on the grid is received by the vertically oriented drain and gate leads of the grid oscillator and mixes with the LO. The IF signal is detected from the horizontally oriented DC bias lines. The horizontal lines contain very little of the RF and LO signals, so this scheme provides a simple method of isolating these signals from the IF. Since the grid does not provide RF–LO isolation, the oscillator will work as a self-oscillating mixer only if the RF signal is outside of its injection-locking range. The grid can also serve as a subharmonic self-oscillating mixer by using the second harmonic as the LO. This type of mixer has a distinct advantage in millimeter-wave receivers where it may be difficult to build an LO at the fundamental frequency.

The operation of the grid as both a harmonic and subharmonic self-oscillating mixer is shown in Fig. 12.13. The IF power is plotted as the difference between the RF and LO frequencies is varied. Since the second harmonic level of the grid oscillator is approximately -30 dBc, the IF power is much lower for the subharmonic case. The shape of the curve is possibly due to the strong nonlinearities in the circuit. As expected, the bandwidth over which the

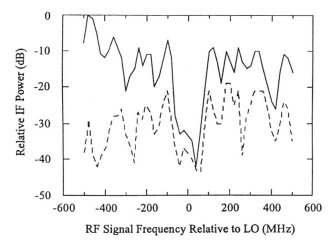

FIGURE 12.13 The measured IF power as a function of frequency difference between the RF incident wave and grid oscillator fundamental (solid line) and second harmonic (dashed line) LO signals. (From reference [5], © 1996 IEEE.)

IF power is negligible corresponds to the regime where the RF signal injection-locks the grid oscillator, which in this case is below 200 MHz. This self-oscillating grid mixer was not specifically designed for low conversion loss or low noise, but rather to provide a simple demonstration of the use of grid oscillators as mixers.

3.2 Receiving Lens Amplifier

The lens amplifier was designed for transmission with maximum gain and with good feed efficiency. It is designed to be fed from focal points situated along a focal surface in the near field of the array. The reciprocal properties of the passive lens design of the amplifier allows it to receive an incident plane wave and focus it down to a focal point at which a mixer can be placed. The amplifier can therefore also be used as the front amplifier in a receiver. The operation of a lens amplifier in a transmitter and receiver is shown in Fig. 12.14a. An example of the measured power gain and isolation of a lens amplifier used as both a transmit and receive lens is shown in Fig. 12.14b.

In the angular diversity receive system presented here, the lens amplifier is used to focus incident plane waves down to three mixers placed along the focal arc of the lens amplifier. The measured relative RF power received at these three focal points as a function of incident angle of a plane wave is shown in Fig. 12.15. For these measurements, an X-band transmit horn was moved along a half-circle from $-40°$ to $40°$ with respect to the optical axis of the lens. This provides a plane wave with variable angle of incidence to the lens amplifier. A receive horn was located at $-30°$, $0°$, and $30°$ along the E-plane focal arc of the lens amplifier. Figure 12.15 shows that the maximum received power for the

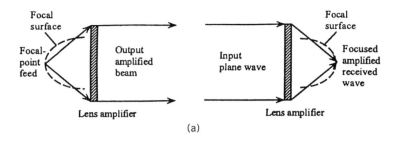

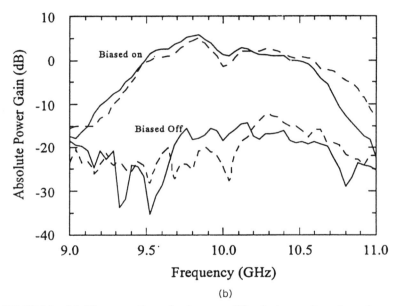

FIGURE 12.14 (a) The operation of a lens amplifier in transmit and receive modes. (b) Measured power gain and isolation of a lens amplifier used as a transmit lens (dashed lines) and receive lens (solid lines). (From reference [5], © 1996 IEEE.)

three different receive horn locations occurs when the transmit horn is located collinearly with the receive horn. Plane waves with different angles of incidence are hence focused to focal points at different angles along the focal arc.

3.3 Quasi-Optical Receiver with Diversity

As shown in Fig. 12.11, the amplifier is fed from the far-field by an RF source. The RF signal is amplified, rotated in polarization, and focused onto the self-oscillating grid mixer, which in this case operates as a subharmonic mixer. With one mixer, reception is limited to a narrow sector where the source can transmit from. To increase the reception range, additional mixers can be positioned along the focal arc of the amplifier. This also allows independent reception for differ-

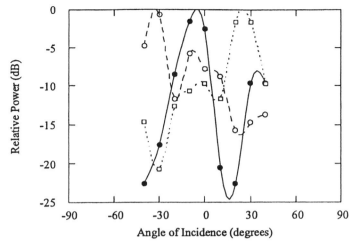

FIGURE 12.15 The measured relative RF power received at three focal points as a function of incident angle of a plane wave. (From reference [5], © 1996 IEEE.)

ent incidence angles of the RF, or angle diversity, which has the benefit of reducing multipath nulls in communication systems with slow fading.

Two additional grid oscillators, identical in design to the first one and with free-running oscillation frequencies within 1% of each other, were placed at angles of $-30°$, $0°$, and $30°$ along the E-plane focal arc of the amplifier. The E-rather than H-plane was chosen since the grating lobes are further from broadside in this case. Figure 12.16 shows the variation in IF power from each

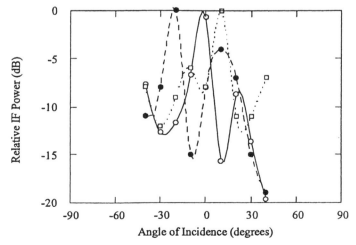

FIGURE 12.16 The measured variation in IF power from each mixer as the source is scanned in the far field. The mixers were positioned at $-30°$ (dashed line), $0°$ (solid line), and $30°$ (dotted line). (From reference [5], © 1996 IEEE.)

mixer as the source is scanned in the far field. It is clear that the reception scan range is increased compared to the previous case, and that the received IF signals from different directions are relatively independent. It is interesting to compare Fig. 12.15 and Fig. 12.16. The dependence of the IF power on scan angle is very similar in shape to the scan patterns of the lens amplifier at the carrier frequency.

4 SOME OTHER COMPONENTS FOR QUASI-OPTICAL SUBSYSTEMS

In this section, we describe several quasi-optical components which were developed as additions to existing ones and with systems requirements in mind. In communications, the carrier polarization is often required to be circular, but all quasi-optical components demonstrated to date have had linear polarization (with the exception of reference [12]). In grid oscillators, linear polarization allows RF isolation of the bias lines; and in most amplifiers, orthogonal linear polarization of the input and output waves provides stability. A quasi-optical linear-to-circular polarizer was designed to be cascaded with an oscillator or amplifier [13]. To be practical, the polarizer needs to have less than 3-dB loss in transmission and an axial ratio better than 2 dB. This polarizer was also used as part of a quasi-optical isolator which was designed to protect an amplifier from high-level reflections which could cause instabilities. Finally, every communication system requires modulation. AM or FM modulation can be achieved by modulating the bias of the transistors in oscillators or amplifiers. A digital phase modulator cascadable with other quasi-optical components which has potential of high modulation speed is also presented here.

4.1 A Quasi-Optical Linear-to-Circular Polarizer

The quasi-optical linear-to-circular polarization converter consists of four capacitively loaded dipole grids, as shown in Fig. 12.17. The dipoles are 10 mm long and periodically spaced 13 mm apart. A 100-pF chip capacitor is soldered across each of the 1-mm gaps. Each grid contains an array of dipoles oriented

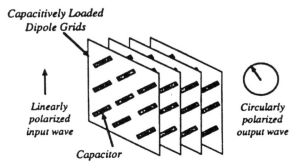

FIGURE 12.17 The quasi-optical linear-to-circular polarization converter consists of four capacitively loaded dipole grids. (From reference [13], © 1995 IEEE.)

45° with respect to a vertically polarized incident plane wave. The field component parallel to the dipoles is phase-shifted 90° relative to the orthogonal component, resulting in a circularly polarized transmitted wave. To achieve low transmission loss for both field components, four identical grids spaced 5.5 mm apart are used.

Simulation and measurement of the transmission coefficient for the component parallel to the strips are shown in Fig. 12.18a. The axial ratio was measured by rotating a receive horn in the far field and is shown in Fig. 12.18b. This linear-to-circular polarization converter has a measured axial ratio of 1.3 dB and a 1.1-dB transmission loss at 8.4 GHz [13]. The axial ratio is better than 3 dB in an 8% bandwidth.

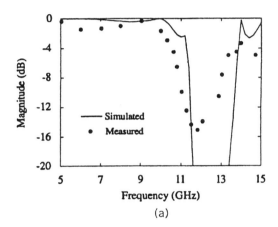

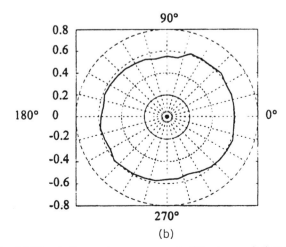

FIGURE 12.18 (a) Simulation and measurement of the transmission coefficient for the component parallel to the strips. (b) The measured axial ratio of the polarizer obtained by rotating a receive horn in the far field. (From reference [13], © 1995 IEEE.)

4.2 A Quasi-Optical Isolator/Directional Coupler

To protect the amplifier against high-level reflections, an external isolator can be used. A quasi-optical isolator was designed to improve the isolation and stability of polarization-isolated quasi-optical amplifiers and provide additional functions such as tuning and linear-to-circular polarization conversion [14]. The isolator, shown in Fig. 12.19, consists of multiple loaded gratings and is cascadable with a quasi-optical amplifier. The vertically polarized output wave from a quasi-optical amplifier is first incident on a pair of grids, which do not introduce any significant transmission loss. Both grids consist of horizontal printed strips, and the second one is in addition loaded with 390-Ω resistors. The vertically polarized wave then passes through a linear-to-circular polarization converter and is right-hand circularly polarized at the output. In the event of a reflection (e.g., the transmitter facing a large conductive object), the right-hand circularly polarized wave reflects back toward the transmitter as a left-hand circularly polarized wave. The left-hand circularly polarized wave becomes horizontally polarized upon passing through the polarization converter and is then absorbed in the resistor-loaded grid.

It is possible to view the linear-to-circular polarization converter as a four-port device by considering the horizontal and vertical polarizations as separate ports at the input and output. In this case, since the horizontally and vertically polarized waves have half the power of the input wave and are 90° out of phase, this device can technically be considered a 3-dB 90° hybrid coupler. However, the ports at the output are actually coupled together to form a circularly polarized wave, and the horizontally polarized port at the input is terminated by the absorbing surface, so effectively this device has only one input and one output port.

To characterize the isolator performance, two X-band horns are placed side-by-side. One of the horns provides the incident vertically polarized wave from port 1 of an HP8510 network analyzer. The other horn is connected to port 2 and is used to measure either the copolarized or cross-polarized signal reflected

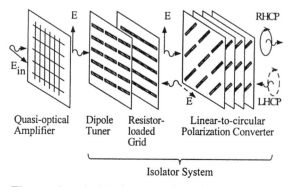

FIGURE 12.19 The quasi-optical isolator consists of multiple loaded gratings and is cascadable with a quasi-optical amplifier. (From reference [14], © 1996 IEEE.)

through the isolator. The isolator system is inserted at the plane of an absorbing aperture in the far field of the horn antennas. A metallic mirror is used as a variable load behind the linear-to-circular polarization converter. Measurements of S_{21} are made for a range of mirror positions. The best isolation was found at 8.83 GHz, where the linear-to-circular polarization converter has a measured axial ratio of 2.5 dB and a transmission loss of 1.1 dB. At 8.83 GHz, the copolarized and cross-polarized isolation for the maximum reflected signal are 9 dB and 19 dB, respectively.

4.3 Quasi-Optical Modulators

Digital phase modulation is a widely used method in today's communication systems [15, 16]. Two quasi-optical transmission-type digital phase modulators are presented in this section. Two two-bit digital modulators can be stacked to form a QPSK modulator. The modulators are designed to be cascadable with other quasi-optical components and to operate in transmission mode [17]. The main design goals were to achieve low transmission loss, (0, 90) and (0, 180) degree phase bits, and equal wave amplitudes for the two phase states. Another consideration is speed, which is discussed at the end of the section.

The 0–90° phase modulator is a printed 10 × 10 grid, has a period much smaller than the free-space wavelength, and is loaded with p–i–n diodes. Each unit cell of the grid contains two diodes, oppositely oriented with respect to the horizontal bias lines (Fig. 12.20a). The diodes are connected to vertical lines of different widths, which represent different reactive loading to an incoming vertically polarized plane wave. When one of the diodes is forward-biased and the other one is reverse-biased, the phase change of the wave transmitted through the grid is much smaller than when the diodes are oppositely biased. Away from resonance, when the transmission loss of a single grid is low, the phase change is also small, so a single grid is not useful for a digital quadrature phase modulator, where 90° phase steps are required. In order to achieve low transmission loss with simultaneous large phase variation between the two states, we use a cascade of two grids. Effectively, this achieves impedance matching to the incoming plane wave. Two grids with unit cells as shown in Fig. 12.20a have been fabricated using HP HSMP-3892 surface mount p–i–n diodes.

An HP8510 network analyzer was used for the grid modulator measurements. The phase of the transmission coefficient was measured for the two bias states of the diodes. The grids exhibit a slightly lower operating frequency than designed, probably due to underestimated diode package inductance in the model obtained from the manufacturer. This also resulted in relatively high transmission loss of about 5 dB, while the simulations predicted less than 1 dB at 10 GHz. At 9.1 GHz, a 90° relative phase shift was measured with less than 1 dB of relative amplitude change between the two phase states.

Another approach for transmission-type digital phase modulators is to use active antenna arrays. The unit cell of a four-element array is shown in

480 QUASI-OPTICAL SUBSYSTEMS

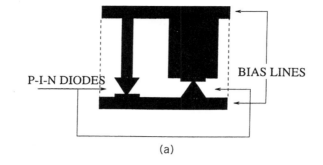

(a)

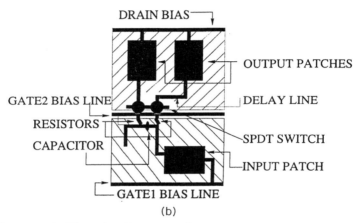

(b)

FIGURE 12.20 (a) The unit cell of a 0–90° phase modulator is loaded with two p–i–n diodes oppositely oriented with respect to the horizontal bias lines. (b) The unit cell of a four-element active antenna modulator array. The input signal is received by patch antennas and coupled to a PHEMT SPDT active switch circuit. The outputs of the switch are connected to two output patch antennas with feed lines which are designed to have a length difference of 180 electrical degrees at 10 GHz. (From reference [17], © 1996 IEEE.)

Fig. 12.20b. The input signal is received by patch antennas and coupled to a PHEMT SPDT active switch circuit with gain to overcome transmission loss due to antenna mismatch. The outputs of the switch are connected to two output patch antennas with feed lines which are designed to have a length difference of 180 electrical degrees at 10 GHz. PHEMT ATF-35576 transistors are used in the SPDT switches, which are designed with two transistors in parallel, so that when one transistor is biased in pinched-off mode the other is biased as an amplifier. The transistors in the switch are DC isolated by a capacitor. The amplifiers are designed to be unconditionally stable, with resistors placed in the gates for stability. To simulate the switch, the pinch-off S parameters of the transistor were measured in a 50-Ω environment. A unit cell was simulated using Hewlett-Packard's MDS microwave circuit simulator with the measured S parameters.

In the active antenna array modulator measurements, an HP70820A transition analyzer was used. The calibration was performed for one state of the amplifier switch, and then the relative phase and amplitude for the other switch state was measured. The measured relative phase and amplitude from the array from 10.1 GHz to 10.3 GHz show between 160 and 175 degrees of relative phase shift between the two states with 2.5 to 3.5 dB of amplitude difference. The bandwidth of this modulator is only a few percent, due to the narrowband nature of the patches.

The modulation speed of the modulator is low because all of the devices are biased in parallel, which results in a large time constant of the switching network. In order to increase the modulation speed, each of the unit cells would need to be addressed individually. This could in principle be done with a more complicated biasing network or using another wave to switch the devices in the modulators on and off.

5 WHAT NEEDS TO BE DONE?

The previous three sections described results which demonstrate integration of several quasi-optical active components into subsystems. These experiments are very crude and should be taken as merely an effort to show the functionality of quasi-optical components and the possible advantages they might bring to a system. In this concluding section, we present an example application and a short discussion on what needs to happen to transition quasi-optics out of the laboratory basic research stage into development and production phases. We attempt to outline required advances for this to happen.

Quasi-optics will be pressed into service when one of two things happen: (a) Quasi-optics prove to be the enabling technology to implement a very important system function, or (b) adequate performance is demonstrated at a cost lower than competing technologies. One of these two things will more likely happen when the application exploits the inherent strengths of quasi-optical technology. In either case, development risks must be sufficiently low not to jeopardize the planned cost or performance. The five most promising advantages are: low cost; high power-added efficiency; small size and weight; high dynamic range; and reliability. Quasi-optics will most likely excel at millimeter-wave frequencies.

5.1 Application Example—Space Communications

One of the most exciting developments today is that of low-earth orbit communication satellite systems. In the near future, these will be the backbone of worldwide communications for voice and high-speed data transmission. These multi-satellite networks are envisioned to operate nearly autonomously, relaying information between hand-held or portable ground terminals anywhere on the earth by hopping from satellite to satellite. Due to the large volume of data, sophisticated satellite cross-links will be required to maintain this connectivity. Since size, weight, and available power are very costly in a spacecraft, only small and power-efficient subsystem designs will survive.

To acquire and track adjacent satellites, steerable phased array transmit antennas are attractive. Multiple, inertialess scanning transmit beams may be required. Since the earth's atmosphere provides a high degree of attenuation due to oxygen absorption at 60 GHz, interference from terrestrial signals will not be a problem. Also at this frequency, a very small aperture is needed for high antenna gain, reducing the costly transmit power, making this an ideal frequency for cross-link operation.

In the case of a communication link of this type, dynamic range is usually not an issue. Thus of the five advantages noted above, all but one is of critical importance. Typical performance parameters for a transmitter of this type might be 5 W of RF power, 25 dB of antenna gain, and perhaps as much as 1 GHz of signal bandwidth. An integral modulator providing a high-speed QPSK modulation format could be used. A transmitter power-added efficiency of greater than 50% is desired. A mission life of 5 to 10 years may be expected, with immunity to the damaging effects of space radiation and atomic oxygen, after surviving the harsh launch environment.

The quasi-optical transmitter should consist of an integrated package, consisting of a stable oscillator, high-speed modulator, phase control for beam steering, driver amplifier, and output transmitter that also serves as an active lens. The modulator may be driven with fiber-optic data lines to accommodate the high data rate and minimize interference. A good thermal design must be implemented internal to the package, perhaps using a diamond heat spreader to cool the high-power monolithic output stage. The package must meet rugged requirements for space qualification, including electromagnetic compatibility, long-life performance stability, and graceful degradation in case of device failures. The entire assembly should fit in a space about the size of a 12-ounce soft drink can.

5.2 Work To Be Done

So how do we get from where we are today to the successful operation described above? First, sufficiently convincing demonstrations must be provided to those who control the means to release the funding to make and manage the necessary investments. Wise and focused use of these funds must be made to complete the development cycle.

Most experimental work today has been with open structures. Except for the radiating aperture, an operational transmitter subsystem must be completely enclosed for environmental and EMI/EMC protection. Additional challenging design tasks, such as packaging and production engineering, must be done, but these do not represent a major technical risk.

A most subtle and difficult technological obstacle that must be routinely accomplished is that of cascading multiple quasi-optical components with very low internal reflections. This is needed to yield power-efficient subsystems with good amplitude flatness and low group delay. This will require excellent modeling tools based on good theoretical design and optimization algorithms and might well be the most challenging area in the maturation of quasi-optical subsystems.

There is not a market for any electrical component today that cannot be integrated with a variety of other components to make up a functioning subsystem. So it is with quasi-optics, and the challenging solutions to functional integration of this new technology will be critical to its ultimate widespread applicability.

REFERENCES

[1] J. Schoenberg, T. Mader, B. Shaw, and Z. B. Popović, "Quasi-optical antenna array amplifiers," *IEEE 1995 Int. Microwave Theory Tech. Symp. Dig.,* pp. 605–609, May 1995.

[2] H. Steyskal, R. J. Mailloux, J. P. Turtle, "EHF active phased array development in the USA," *23rd European Microwave Conference Proceedings,* Sept. 1993, pp. 80–83.

[3] S. C. Bundy, "Analysis and Design of Grid Oscillators," Ph.D. Dissertation, University of Colorado, Boulder, 1994.

[4] J. S. H. Schoenberg, "Quasi-Optical Constrained Lens Amplifiers," Ph.D. Dissertation, University of Colorado, Boulder, 1995.

[5] W. A. Shiroma, E. Bryerton, S. Hollung, and Z. B. Popović, "A quasi-optical receiver with angle diversity," *1996 IEEE MTT-S Int. Symp. Dig.,* San Francisco, June 1996.

[6] W. C. Jakes, *Microwave Mobile Communications,* Wiley, New York, 1974, pp. 311.

[7] Prof. Peter Mathys, University of Colorado, private communication.

[8] J. B. Hacker, R. M. Weikle II, M. Kim, M. P. DeLisio, and D. B. Rutledge, "A 100-element planar Schottky diode grid mixer," *IEEE Trans. Microwave Theory Tech.,* vol. 40, no.3, pp. 557–562, March 1992.

[9] Y. Tajima, "GaAs FET applications for injection-locked oscillators self-oscillating mixers," *IEEE Trans. Microwave Theory Tech.,* vol. 27, no. 7, pp. 629–632, July 1979.

[10] V. D. Hwang, T. Itoh, "Quasi-optical HEMT and MESFET self-oscillating mixers," *IEEE Trans. Microwave Theory Tech.,* vol. 36, no. 12, pp. 1701–1705, Dec. 1988.

[11] R. Johnson, J. Jasic, *The Antenna Engineering Handbook,* 2nd ed., McGraw Hill, New York, 1984, Chapter 16, pp. 225–259.

[12] T. Mader, J. Schoenberg, L. Harmon, and Z. B. Popović, "Planar MESFET transmission amplifier," *IEE Electronics Letters,* vol. 29, no. 19, pp. 1699–1701, Sept. 1993.

[13] W. Shiroma, S. Bundy, S. Hollung, B. Bauernfiend, and Z. B. Popović, "Cascaded active and passive quasi-optical grids," *IEEE Trans. Microwave Theory Tech.*, vol. 43, no. 12, pp. 2904–2909, Dec. 1995.

[14] S. Hollung, M. Marković, W. Shiroma, and Z. B. Popović, "A quasi-optical isolator," *IEEE Microwave Guided Wave Lett.*, vol. 6, no. 5, pp. 205–207, May 1996.

[15] M. J. Vaughan, W. Write, and R. C. Compton, "Active antenna elements for millimeter-wave cellular communications," *1995 Conf. Proc. Int. Symp. Signals, Syst. Electron.*, San Francisco, pp. 9–12, Oct 1995.

[16] C. W. Pobantz, J. Lin, and T. Itoh, "Active integrated antennas for microwave wireless systems," *1995 Conf. Proc. Int. Symp. Signals, Syst. Electron.*, San Francisco, pp. 1–4, Oct 1995.

[17] M. Marković, S. Hollung, Z. B. Popović, "Quasi-optical phase modulators," *1996 IEEE MTT-S International Symp. Digest,* San Francisco, pp. 1247–1250, June 1996.

[18] J. S. H. Schoenberg, S. C. Bundy, and Z. B. Popović, "Two-level power combining using a lens amplifier," *IEEE Trans. Microwave Theory Tech.,* vol. MTT-42, no. 2, pp. 2480–2485, Dec. 1994.

CHAPTER THIRTEEN

Commercial Applications of Quasi-Optics

RICHARD C. COMPTON
Millimeter-Wave Wireless Laboratory, Cornell University, New York

MEHRAN MATLOUBIAN
Hughes Research Laboratories, California

MARK J. VAUGHAN
Endgate Corporation, California

1 INTRODUCTION

Quasi-optical arrays provide the promise of an efficient means of generating the powers necessary for many microwave and millimeter-wave applications [1,2]. These arrays can be designed to serve as direct replacements for existing components such as traveling-wave tubes. However, the flexibility in topology of quasi-optical arrays makes possible a range of applications which are quite different from more traditional RF systems.

In this chapter the key power and device issues for quasi-optical array applications will be discussed. Then a number of applications will be presented. These applications fall into three categories, oscillator arrays, amplifier arrays, and imaging/receiver arrays.

Active and Quasi-Optical Arrays for Solid-State Power Combining, Edited by Robert A. York and Zoya B. Popović.
ISBN 0-471-14614-5 © 1997 John Wiley & Sons, Inc.

2 POWER REQUIREMENTS

The starting point for a particular application is to calculate the necessary transmit power. For a communication link, the Friis formula allows the received power P_R to be calculated for a given transmit power P_T, wavelength λ, and separation distance r [3]:

$$P_R = P_T G_T \frac{\lambda^2}{(4\pi r)^2} G_R \qquad (13.1)$$

where G_T and G_R are the gains of the transmitter and receiver. A key measure of the signal quality is the carrier-to-noise ratio (C/N), which is given by

$$\frac{C}{N} = \frac{P_R}{kTB} * N_F \qquad (13.2)$$

where k is the Boltzmann constant and B is the bandwidth of the receiver which is at a temperature T. The noise figure N_F is a measure of the noise added by the receiver in the conversion stage required to shift the receive signal down to a frequency suitable for demodulation. A typical link budget for calculating the received C/N is shown in Table 13.1.

For radar applications, the received power P_R is calculated using the radar equation [3]

$$P_R = P_T G_T^2 \frac{\lambda^2}{(4\pi)^3 r^4} \sigma \qquad (13.3)$$

where σ, the radar cross section, is the total power scattered by the target divided by the flux density incident on the target.

2.1 Modulation Schemes

The transmitted signal can be represented as

$$\begin{aligned} u(t) &= A(t) \sin(\omega(t)t + \phi(t)) \\ &= A(t) \cos \phi(t) \sin \omega(t)t + A(t) \sin \phi(t) \cos \omega(t)t \end{aligned} \qquad (13.4)$$

TABLE 13.1 Typical Link Budget—60 GHz

Transmit power	10 dBW
Transmitter gain	12 dB
Path loss 100m $\lambda^2/(4\pi r)^2$	−108 dB
Receive antenna gain	6 dB
Noise figure	−6 dB
Carrier level C	−86 dBW
Noise (50 MHz) N	−127 dBW
C/N	41 dB

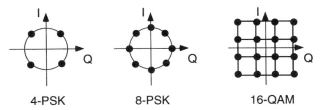

FIGURE 13.1 Constellation diagrams for common linear modulation schemes.

Information is carried by varying the amplitude $A(t)$, phase $\phi(t)$, or frequency $\omega(t)$. In traditional digital radio links amplitude and phase modulation is achieved by using a balanced modulator to impress a baseband digital stream onto a carrier. This technique, called *linear modulation,* achieves good power efficiency as well as efficient use of the spectrum. Nonlinear modulation techniques, such as frequency modulation, usually require more spectrum for a given data rate but place less stringent requirements on performance of the transmitter and receiver [4].

Linear digital modulation schemes, such as quadrature amplitude modulation (QAM) and phase shift keying (PSK), are widely used in modern radios. These modulations may be applied to a low-power signal which is then boosted by a quasi-optical amplifier array. In multichannel transmitters undesirable intermodulation products favor the use of a separate amplifier for each channel [2]. QAM and PSK modulations can also be applied directly to a high-power oscillator. In QAM and PSK modulation, in-phase and quadrature carriers are added to produce a signal

$$u(t) = I(t) \sin \omega t + Q(t) \cos \omega t \qquad (13.5)$$

The allowed amplitudes are often plotted in I/Q space as shown in Fig. 13.1.

The transit power P_T in the Friis formula is an average power. Because the signal is modulated, the peak power requirements are significantly larger than the average power. The transmitter must be able to accommodate this peaking and so the average transmit power must be backed-off so that the peak signal does not overly saturate the transmitter. The amount of backoff depends on the modulation scheme employed. For some common modulations the backoff is as follows: 4 dB for 4-PSK/QAM and 13 dB for 16QAM [5].

2.2 Spectral Efficiency

To minimize interference, specific channels are prescribed with tightly defined masks. Such strict filtering requirements place stringent requirements on the out-of-band emissions and can necessitate filtering after the final output sage (Fig. 13.2).

The bandwidth occupied by a given transmission is determined by the shape of the individual digital pulse. Assuming that the pulses are separated by time

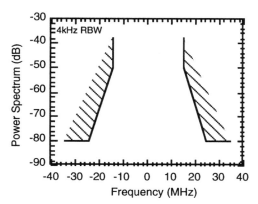

FIGURE 13.2 FCC mask for a 30-MHz wide channel operating in the 6-GHz band. Transmission must lie within the shaded areas. The vertical axis is normalized to the total power.

T, pulses which reach a maximum at t_0 should be zero for neighboring pulses ($t = t_0 + \kappa T$, κ nonzero integer). A commonly used pulse shape is the cosine rolloff

$$p(t) = \frac{\sin(\pi t/T)}{\pi t/T} \frac{\cos(\pi \alpha t/T)}{1 - (2\alpha t/T)^2} \quad (13.6)$$

which spans a bandwidth $W = (1 + \alpha)/T$. For a practical system, α is usually chosen to be around 0.5. We define the spectral efficiency η as the ratio of bit rate R_b to bandwidth W:

$$\eta = \frac{R_b}{T} = \frac{1}{WT} \log_2 M = \frac{\log_2 M}{1 + \alpha} \quad (13.7)$$

where M is the number of allowed values for each pulse. As the spectral efficiency increases, the required carrier-to-noise ratio also grows (Fig. 13.3).

Imperfections in the transmitter and channel means that in practice, larger carrier-to-noise ratios than showed in Fig. 13.3 are required. These impairments are usually factored in as power penalties and can arise from such factors as carrier phase noise and nonlinearities in the transmitter output stage [5].

3 DEVICE TECHNOLOGIES

As discussed in the earlier chapters, semiconductor devices play a vital role in quasi-optical arrays. Semiconductor devices are used in grid oscillators and amplifier arrays, beam steering and phase shifting arrays, and receiving and imaging arrays. In this section a brief overview of different device technologies

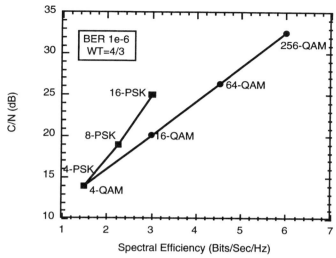

FIGURE 13.3 Required C/N ratios for a range of modulation schemes [5].

is presented. This section is not intended to be a complete device review but rather to give the reader a better understanding of trade-offs involved in selecting the right device technology for a particular application. A number of references are given at the end of the chapter to provide the reader with additional sources of information on each device technology. Three different device technologies will be reviewed: diodes, heterojunction bipolar transistors (HBTs), and field effect transistors (FETs).

3.1 Diodes

Diodes used in quasi-optical applications can generally be divided into two categories. Some are used as oscillators or sources (Gunn diodes, IMPATTs, RTDs), and the others as nonlinear elements or control devices (Schottky diodes, varactors, PINs). The first group of diodes are capable of directly generating power at microwave and millimeter-wave frequencies. Diodes are easier to fabricate than transistors and have higher yields and uniformity. However, diode amplifiers are inefficient and are not well suited to multistage high-gain applications.

Gunn effect diodes, also known as TEDs (transferred electron devices), rely on the bulk property of semiconductors to generate their negative resistance [6,7]. The Gunn effect is observed in semiconductors such as GaAs and InP but not silicon. The property required from a semiconductor to be used as a Gunn diode is an upper conduction band separated from the main lower-energy conduction band by an energy less than the bandgap of the semiconductor (this is necessary so that electron transfer to the upper conduction band occurs before impact ionization). In addition, the electrons in the upper conduction band have a larger effective mass and lower velocity than the electrons in the lower conduction band. As the field applied to the semiconductor increases, a level is reached where electrons transfer to the upper conduction band, resulting in a drop in electron velocity and current with further increase of applied voltage. As a result, a negative differential resistance is obtained. This negative resistance can be used in a circuit for generation of microwave power. Depending on the doping density of the semiconductor, the length of the sample, and the external circuit used with the diode, different modes of Gunn diode operation can be obtained [6]. The operating frequency of Gunn diodes is limited by the energy relaxation time, which corresponds to the time it takes for the electrons in the upper conduction band to return back to the lower conduction band. InP-based Gunn diodes are capable of higher frequency operation than GaAs Gunn diodes because the energy relaxation time of InP is lower than GaAs by a factor of two. InP Gunn diodes are capable of operating up to a frequency of approximately 200 GHz. Gunn diodes with good power performance have been reported with fundamental operating frequencies of 140 GHz [8]. In addition, Gunn diodes with frequencies above 170 GHz have been reported using second and third harmonics [9]. They are capable of low noise operation and are commonly used as local oscillators. Gunn diodes have been used in a number of quasi-optical arrays [10,11].

IMPATT (IMPact Avalanche Transit Time) diodes, as the name suggests, rely on a combination of impact ionization and transit time of electrons across the semiconductor to achieve negative resistance [6,12]. IMPATTs are similar to PN-junction type diodes but are designed specifically (by location of the junction relative to a drift region) to achieve negative resistance over a particular frequency range. IMPATT diodes can be made from silicon as well as GaAs and InP. They are capable of operation over a wide bandwidth and can generate high CW as well as pulsed power at millimeter-wave frequencies. In fact, IMPATT diodes have produced the highest output power of any semiconductor device at millimeter-wave frequencies [13]. But IMPATTs, with typical efficiencies of less than 10%, have much lower efficiencies than HEMTs. Among IMPATT diodes, the Si device is the most mature technology and has demonstrated high reproducibility and reliability. Si IMPATTs with operating frequencies of over 200 GHz have been demonstrated. GaAs and InP IMPATTs have also recently demonstrated good power performance at millimeter-wave frequencies, but because they have lower thermal conductivities than silicon they require good heatsinking. For quasi-optical applications, IMPATT diodes are ideal for generating very high powers at millimeter-wave frequencies. IMPATT diodes have been integrated with antennas to generate 27 mW of output power at 43 GHz [14], and an array of IMPATT diodes has been created which generates 2.2 W of output power at 60 GHz [15].

Resonant tunneling diodes (RTDs) have set records for the highest frequency of operation of any semiconductor device at room temperature. Frequencies of over 700 GHz have been obtained from these diodes [16]. The structure of RTDs consists of alternating thin layers of two types of semiconductors, with different bandgaps, forming quantum wells in the lower bandgap material. The electrons occupy a set of discrete energy levels in the quantum wells. The current flow through the diode depends on tunneling between quantum wells. The probability of tunneling changes as the bias voltage varies the lineup of the discrete energy levels in adjacent wells [17]. RTDs are capable of generating broadband negative resistance, from DC to the cutoff frequency of the diode. This makes design at millimeter waves difficult because the diodes have a tendency to oscillate at lower frequencies. This also has resulted in low output power from these devices at millimeter-wave frequencies compared to Gunn diodes and IMPATTs. (Some researchers have attributed the low output power of RTDs to bias oscillation [18].) RTDs are well suited for high-power quasi-optical arrays at submillimeter-wave frequencies. Compared to IMPATTs and Gunn diodes, RTD technology is relatively immature.

Schottky diodes are probably the simplest semiconductor devices used for microwave and millimeter-wave applications. They consist of two contacts to a semiconductor; one contact is ohmic, while the other is rectifying. For quasi-optical applications, Schottky diodes have been used as mixers [19], frequency multipliers [20,21], phase shifters [22], and modulators [23]. Schottky diodes with cutoff frequencies of more than 2 THz have been fabricated. Due to their simplicity, Schottky diodes can be fabricated with high yield and uniformity.

PN-junction diodes are also commonly used as varactors for high-frequency applications. They typically have greater capacitance variation with DC bias than Schottky diodes and are more efficient as frequency multipliers [24]. In addition, PN-junction diodes are used as tuning elements in VCOs as well as for phase shifting and beam-steering applications. Novel multiple quantum-well varactor diodes have recently demonstrated excellent performance as frequency multipliers. By stacking multiple quantum wells the power handling capability of the diode increases while maintaining a high cutoff frequency [25,26]. Monolithic arrays of Schottky-quantum-barrier varactors (SQBV) have been used as triplers to generate 5 W of pulsed output power at 99 GHz [25].

A somewhat modified varactor diode is the PIN diode. An added intrinsic region between the *P*-type and *N*-type regions allows these diodes to be used for switching applications with high isolation in the off-state, but unlike varactor diodes, their capacitance is relatively constant with variation of reverse bias. PIN diodes are also used for signal modulation and phase shifting applications [27].

3.2 HBTs

Heterojunction bipolar transistors (HBTs) are almost as old as the bipolar transistor itself [28,29]. In conventional silicon bipolar transistors the emitter injection efficiency (and the current gain of the device) depends on the ratio of the emitter doping to the base doping levels. To achieve current gain, the emitter must be doped more heavily than the base, thereby limiting the base doping and

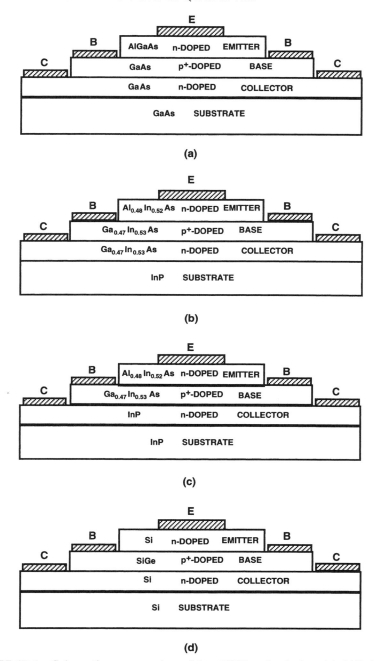

FIGURE 13.4 Schematic cross section of four HBT technologies: (a) AlGaAs/GaAs HBT, (b) AlI-nAs/GaInAs single heterojunction HBT, (c) AlIn As/GaInAs/InP double heterojunction HBT, and (d) Si/SiGe HBT. The schematics are simplified, and not all the layer structures are shown.

resulting in high base resistance. In HBTs the emitter consists of a wider bandgap material than the base. This difference in bandgap between the emitter and base allows a current gain that is not limited by the base doping. The base can be heavily doped to decrease the base resistance and improve the High-frequency performance of the device. In effect, HBTs give the device designer more flexibility to engineer a device for high-frequency performance [30].

HBTs have recently emerged as a promising technology for microwave, and potentially millimeter-wave, power applications [31–34]. There are a number of different HBT structures, four of which are shown in Fig. 13.4. All of these consist of a wide bandgap emitter and a narrower bandgap base. The AlGaAs/GaAs HBT is the most mature power HBT technology [31,32,35–37]. It has exhibited high power and large power-added efficiency and has been used in quasi-optical arrays at microwave [38] as well as millimeter-wave frequencies [39]. A variation of this technology uses GaInP emitters instead of the AlGaAs emitters [40–43]. This results in better etch selectivity between the emitter and the base layers and simplifies device fabrication.

Two different InP-based HBT structures are shown in Figs. 13.4b and 13.4c. Both the base and the collector of the HBT in Fig. 13.4b are GaInAs, while the structure in Fig. 13.4c has a GaInAs base but an InP collector (hence it is referred to as a double-heterojunction HBT, or DHBT). The single-heterojunction HBT (SHBT) has a lower base-collector breakdown voltage due to the low bandgap of the GaInAs collector, but typically it also has higher frequency performance than the DHBT. The SHBT is primarily used for digital circuit applications [44]. Using InP as the collector increases the base-collector breakdown voltage (as well as V_{ceo}) and improves the output conductance of the device. The InP-based power HBT technology is less mature than GaAs-based power HBTs, but recently state-of-the-art power results have been obtained using InP-based materials systems [34,45].

Si/SiGe HBT technology (Fig. 13.4d) also appears promising for microwave applications [46,47]. f_T and f_{max} values of over 100 GHz have been demonstrated using this technology. Si/SiGe HBTs are attractive for realization of low-cost microwave circuits due to the availability of large-diameter silicon substrates. One limitation of MMICs using SiGe technology has been the high losses of silicon substrates. Several approaches have been taken to overcome this problem. In one approach, high-resistivity silicon substrates are used [48], but large-diameter, high-resistivity silicon substrates are not available and are presently too expensive. In another approach, a top low-loss spin-on dielectric is added for fabrication of transmission lines [49]. This approach has the advantage that the SiGe HBT can be fabricated using existing mass-production silicon fabrication lines. One problem with the present Si/SiGe HBT technology has been the low value of V_{ceo}. SiGe HBTs with an output power of approximately 100 mW have been reported at microwave frequencies [50,51]. This power can be improved by increasing the breakdown voltage [52].

III–V-based HBTs have demonstrated f_T and f_{max} values of over 200 GHz [53]. But these HBTs have low breakdown voltages (typically less than 2 V), and are not suitable for millimeter-wave power applications. Research on bandgap engineering of base-collector junctions has resulted in a record combination of breakdown voltage and cutoff frequency for HBTs [54]. By proper scaling of the material structure and minimizing device parasitics, it should be possible to achieve breakdown voltages of more than 10 V and f_{max} values of over 150 GHz in the near future [55].

3.3 FETs

Field effect transistors (FETs) have progressed remarkably in performance over the last 25 years: from the early developments of GaAs MESFETs with approximately 1-μm gatelengths [56] to present advanced InP-based HEMTs with 0.1-μm gatelengths and f_{max} values of 600 GHz [57]. Figure 13.5 shows a comparison of several FET technologies. In a GaAs MESFET structure, shown in Fig. 13.5a, all the layers are made using GaAs with different doping concentrations. Typically the same doped GaAs layer acts as the channel as well as the Schottky layer. The mobility of the electrons in a MESFET channel is lower than in undoped GaAs due to scattering of the electrons by ionized donors. To reduce the MESFET gate length, the thickness of the channel also has to be reduced to allow effective control of the current flow through the channel by the gate. But as the channel thickness is reduced, the doping has to be increased to keep the channel resistance and drain current constant. This criteria limits the cutoff frequency of GaAs MESFETs to approximately 100 GHz [58–60]. MESFETs have demonstrated good power performance, but their frequency of operation is typically limited to less than 30 GHz. GaAs MESFET technology is very mature and MESFETs have been used in a number of quasi-optical oscillator arrays [61,62].

In an AlGaAs/GaAs HEMT structure (Fig. 13.5b) performance is improved over conventional GaAs MESFETs by using a wider bandgap AlGaAs layer as the Schottky layer and a lower bandgap GaAs layer as the channel. The Schottky layer is doped while the channel is undoped. The electrons transfer from the wider bandgap AlGaAs layer into the lower bandgap GaAs channel, forming a two-dimensional electron gas (2-DEG) near the AlGaAs/GaAs interface. In effect, by incorporating an AlGaAs/GaAs heterojunction, a conduction-band edge discontinuity is created that physically separates the donors in the wider bandgap AlGaAs from the mobile electrons in the lower bandgap GaAs channel. This separation significantly reduces the ionized impurity scattering of the electrons in the channel and leads to higher carrier mobility and higher f_T values [63].

By adding indium to the GaAs channel to form GaInAs channels (Fig. 13.5c), the device characteristics are further enhanced. The electron mobility, conduction band discontinuity (ΔE_c), and 2-DEG carrier density all increase with increasing indium content in the channel. However, the indium content is limited by the fact that GaInAs is not lattice matched to the GaAs substrate, and the layer is increasingly strained as the indium content is increased. Due to the lattice mismatch between the GaInAs channel and the GaAs substrate, the AlGaAs/GaInAs HEMTs are commonly referred to as GaAs pseudomorphic

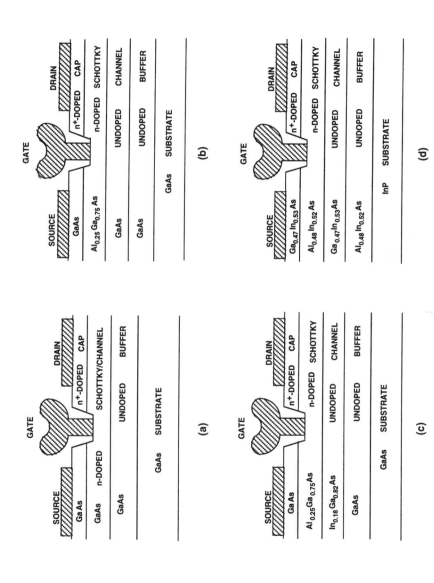

FIGURE 13.5 Schematic cross section of four FET technologies: (a) GaAs MESFET, (b) AlGaAs/GaAs HEMT, (c) AlGaAs/InGaAs pHEMT, and (d) AlInAs/GaInAs HEMT. The schematics are simplified and not all the layer structures are shown.

HEMTs (GaAs pHEMTs). To avoid relaxation of the crystal structure and creation of dislocations in the channel, the indium content of AlGaAs/GaInAs/GaAs HEMT structures is typically limited to a maximum of approximately 22%. GaAs-based pHEMTs have demonstrated excellent low-noise performance [64–66] as well as state-of-the-art power performance at microwave and millimeter-wave frequencies [67–71]. GaAs-based pHEMTs are gradually replacing MESFETs for some microwave applications and have been the primary device of choice for millimeter-wave power applications. Quasi-optical arrays of pHEMT amplifiers with 100 elements have been demonstrated [72,73].

The InP-based HEMT technology (Fig. 13.5d) has the advantage over GaAs pHEMTs that GaInAs channels with higher indium content can be grown on InP substrates. GaInAs with 53% indium ($Ga_{0.47}In_{0.53}As$) is lattice-matched to InP substrates, and instead of an AlGaAs Schottky layer, $Al_{0.48}In_{0.52}As$ (also lattice matched to InP) is used. Compared with the GaAs-based pHEMTs, the $Al_{0.48}In_{0.52}As/Ga_{0.47}In_{0.53}As/InP$ materials system exhibits higher electron mobility, a higher electron peak velocity, higher conduction-band discontinuity (leading to higher 2-DEG density), and higher transconductance [74]. InP-based HEMTs consequently exhibit higher f_T and f_{max} values than GaAs pHEMTs of the same gatelength. (*Note:* To further increase the mobility of the channel, the indium content can be increased beyond 53% [75]. This results in an InP-based pseudomorphic HEMT.) The unique properties of the InP-based HEMT materials system have established this as the leading transistor for millimeter-wave low-noise applications [76–78]. These devices have set records for the fastest transistors operating at room temperature, with the highest reported f_T of 343 GHz [75] and the highest reported f_{max} of 600 GHz [57]. They have also set the world record for the fastest monolithic integrated circuit operating at a frequency of 213 GHz [79].

For power applications, InP-based HEMTs offer a number of advantages over GaAs-based HEMTs. The thermal conductivity of InP is 40% higher than GaAs, allowing a lower operating channel temperature for the same power dissipation. In addition, the higher electron densities in the channel, coupled with the higher electron velocity, can lead to higher current densities. Despite these advantages, until recently, very little work had been done on InP-based HEMTs for power applications [80–83]. This has been primarily due to the low gate-to-drain breakdown voltage and low Schottky barrier height of low-noise InP-based HEMTs. But by proper device layer design it is possible to overcome the drawbacks of InP-based HEMTs for power applications and achieve state-of-the-art results [84]. Due to their very high-gain, power-added efficiency, and thermal conductivity, InP-based HEMTs are well suited for millimeter-wave quasi-optical arrays.

3.4 Device Comparison

Choosing the right device technology for an application depends on several factors, including system specifications, architecture, and system cost. For example, for low-noise applications such as receiving and imaging arrays, InP-based

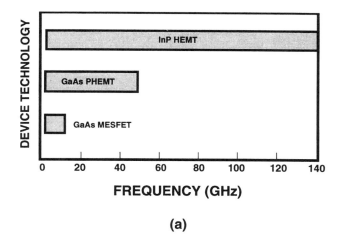

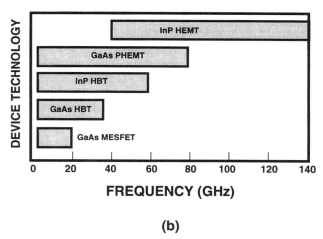

FIGURE 13.6 Comparison of transistor technologies for (a) low-noise and (b) power applications. For low-noise applications, InP-based HEMTs offer the best performance at microwave and millimeter wave-frequencies. At microwave frequencies, depending on the required noise figure, GaAs MESFETs or pHEMTs can often meet the system requirements.

HEMTs have the best performance of any existing transistor technology. But at microwave frequencies the noise figure of low-noise GaAs-based pHEMTs are adequate for most applications (Fig. 13.6a). In fact, low-noise GaAs-based pHEMTs have been inserted in a number of commercial applications including receivers for direct broadcast satellite (DBS) systems.

Choosing the device technology for power applications (for amplifiers or oscillators) is more complicated. It depends on the frequency range of interest, output power, gain, efficiency, bandwidth, linearity, and phase noise. IMPATT and Gunn diodes are capable of generating higher powers than transistors but they have lower efficiencies. RTDs can extend into the submillimeter-wave frequency range but have low output powers. As mentioned previously, another approach for power generation at millimeter-wave frequencies is frequency multiplication using diode arrays [25]. Although this approach generates high powers, the conversion efficiency is low.

Fig. 13.6b shows a comparison of different power transistor technologies. For microwave power applications, both FETs and HBTs have demonstrated good power performance. GaAs-based pHEMTs have become the device of choice among FETs for microwave power applications and are gradually replacing GaAs MESFETs for most applications. While GaAs-based HBTs are more mature than InP-based HBTs, recently, InP-based HBTs have also demonstrated state-of-the-art power performance [54]. In comparison to FETs, HBTs typically have lower gain, but they do not require e-beam lithography for fabricating sub-0.25-μm T-gates. Also, the device characteristics (breakdown voltage, input impedance, etc.) are primarily determined by material properties and are not dependent on processing. For FETs the gate-recess depth and profile have a great impact on the breakdown voltage of the device. HBTs can be fabricated with much higher uniformity and yield compared to FETs, which is important for quasi-optical arrays when a large number of elements is desired. HBTs also have lower phase noise and higher linearity than FETs.

4 OSCILLATOR ARRAYS

In the preceding chapters, various aspects and types of quasi-optical oscillator arrays (and active antennas with which such arrays can be constructed) have been described. Different configurations are presented, and techniques for designing them with increased output power, frequency, and efficiency are provided. For use in a practical application, a configuration must be chosen based on the specific system requirements, and then certain other issues such as the acceptable noise levels, desired modulation scheme, and perhaps special radiation pattern requirements must be addressed. These aspects of quasi-optical oscillator array utilization are described here.

4.1 Array Phase Noise

The phase noise of an array of oscillators is an important parameter, particularly in high-data-rate communication links where the receiver must be able to recover the carrier [5]. Unfortunately, to make quasi-optical arrays compact and inexpensive, the oscillator elements generally end up having low quality-factors, which result in relatively high individual phase-noise levels. In order to understand and hence be able to reduce the noise levels in an array, a matrix equation for the phase noise has been derived which accounts for the effect of the coupling to a first-order approximation [85].

For an array consisting of N elements which are locked to each other via a network described by the coupling coefficients [10–86] $\kappa_{mn} = \lambda_{mn} e^{-j\Phi_{mn}}$, the magnitudes of the phase deviations with frequency ω_p of each element are given by the matrix $\Delta\Phi$ in the formula

$$(I - \Lambda)\Delta\Phi = \Gamma \tag{13.8}$$

where I is the $N \times N$ identity matrix, Λ is an $N \times N$ matrix with elements

$$\Lambda_{mn} \equiv \frac{\lambda_{mn}}{D_m} \cos(\phi_{0m} - \phi_{0n} + \Phi_{mn}) \tag{13.9}$$

Γ is an N-element column matrix defined by

$$\Gamma_m \equiv \frac{-n_m(\omega_p)}{AD_m} \tag{13.10}$$

and

$$D_m \equiv \sum_{n=1}^{N} \lambda_{mn} \cos(\phi_{0m} - \phi_{0n} + \Phi_{mn}) + 2j\omega_p L_m \tag{13.11}$$

In these expressions, ϕ_{0m} specifies the average steady-state phase of oscillator m, $n_m(\omega_p)$ describes the phase noise of oscillator m when operating independently, A is the amplitude of the individual oscillators, and L_m is the equivalent-circuit inductance (proportional to the square of the oscillator's Q-factor).

Because the phase noise of each element is a function of the phase noise of all the oscillators to which it is coupled, determining the total phase noise of the array requires careful summing of the noise terms. The expression obtained from (13.8) for the phase noise of each oscillator in the array, $\Delta\phi_m$, must be written in the form

$$\Delta\phi_m(\omega_p) = a_{m1}n_1(\omega_p) + a_{m2}n_2(\omega_p) + \ldots + a_{mN}n_N(\omega_p).$$

Then, assuming that the uncoupled phase noises of the individual oscillators are uncorrelated, the phase noise of the array can be computed from

$$|\Delta\phi_t|^2 = \frac{1}{N^2} \sum_{n=1}^{N} |n_n(\omega_p)|^2 \left[\sum_{n=1}^{N} a_{mn}\right]^2.$$

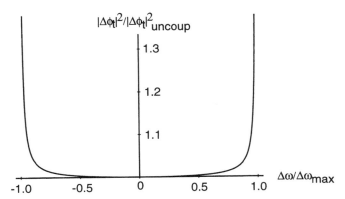

FIGURE 13.7 Phase noise for a two-oscillator array, normalized to that of two uncoupled (but otherwise synchronized) oscillators, as a function of the free-running frequency difference $\Delta\omega$ [85].

The above formulas have been studied for a variety of different coupling schemes. The two-element case is particularly useful because it yields results which are applicable to larger arrays, but obtained from simpler expressions. Figure 13.7 shows a plot of the total phase noise from the two oscillators, as a function of the difference in their free-running frequencies normalized to the maximum frequency separation which still allows locking ($\Delta\omega_{max}$). This shows that as the frequencies of the oscillators come close to unlocking, the phase noise increases dramatically. Other general results are that, as expected, higher-Q oscillators yield lower phase-noise arrays, and that the coupling coefficient phase, Φ, ought to be close to an integer multiple of 2π.

In a linear array of N oscillators with nearest neighbor coupling, the array phase noise generally decreases as more oscillators are added. In the ideal limit of identical oscillators with coupling coefficients having phases which are an integer multiple of 2π, the noise is proportional to $1/N$. However, in less than ideal conditions, the noise does not decrease as rapidly, and can even increase after a point, though never to the level of N uncoupled oscillators.

In some arrays the oscillators are arranged such that the coupling network forms a closed loop. This has a benefit in reducing the number of possible modes in the system (by causing mode degeneracy [87]), and as a result, the phase noise in these arrays is proportional to $1/N$ even when the coupling coefficient phases are not a multiple of 2π.

To reduce the noise in an array of oscillators to practical levels, in nearly all cases the array must be synchronized to an external, low phase-noise source. The synchronization can be achieved either by using the array in a phase-locked loop (PLL) or through injection-locking. In the latter case, the locking signal can be applied to either the whole array (convenient when the signal is radiated at the array and picked up by input antennas or transmission lines in each element) or just a single element which is coupled to the rest of the array (easiest technique when the signal is applied directly through a wired connection).

Comparisons of the phase-noise improvement resulting from each of these techniques can be made using the formulas given above. External injection-locking is modeled in the equations by specifying the coupling between the low-noise source and the injection-locked elements to be unilateral (i.e., $\lambda_{mn} \neq \lambda_{nm} = 0$).

When modeling injection-locking of the full array, the results indicate that there is only a slight variation in the total phase noise as the strength of the coupling between the array oscillators is varied. Hence, the coupling scheme can be designed strictly to satisfy other requirements. Furthermore, the array can be modeled as a single oscillator provided that the free-running frequencies are well within the locking range.

With the injection-locking signal applied just to a single element, that oscillator's phase noise is reduced as much as those in a fully injection-locked array, but the other oscillators have slightly more phase noise. The oscillators located furthest away from the injection-locked one have the greatest phase noise, though this is typically not significantly greater.

Figure 13.8 shows how a PLL can be used in place of injection locking to synchronize an array to an external source and reduce the array noise. An antenna is used to sample the output signal from the array and feed it back to a phase-detector (mixer). The oscillator array replaces the voltage-controlled oscillator (VCO) of the standard PLL, so it must have a control line which can vary the output frequency. This is generally not a problem, however, since the operating frequency is usually a monotonic function of the array's dc bias. The reference oscillator, like in the injection-locking case, is a low-phase-noise, low-power source.

The ability of the PLL to reduce the phase noise of the array is just as good as injection locking. In fact, the first-order PLL ($F(s) = 1$) is described by the same nonlinear differential equation as is injection locking. As mentioned previously, the array can usually be represented by a single oscillator when analyzing injection locking, so it follows that the same applies to the PLL analysis, and consequently the phase-noise improvement ought to be comparable.

The PLL approach has a few advantages over injection locking. First, the signal provided to the array is a low-frequency control signal—no RF

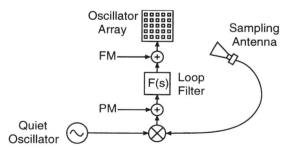

FIGURE 13.8 Use of a PLL to reduce the phase-noise and provide modulation of an oscillator array [85].

connection is needed. Phase and frequency modulation are easily implemented by adding the modulation signals at appropriate places in the loop (as shown in the figure) rather than having to directly modulate the reference oscillator. Finally, the loop filter can be tailored to optimize the system for the desired modulation and noise performance.

4.2 Modulation

In radar or communication applications the array output must be modulated to superimpose information onto the carrier. It can be modulated in amplitude, phase (or equivalently, frequency), or a combination of the two.

Perhaps the simplest way to modulate an oscillator array is to switch it on/off. In communication language this would be binary amplitude-shift keying, while in radar this would be a pulsed system. Davidson *et al.* [15] investigated the pulsed operation for an array of four high-power impatt oscillator operating at 60 GHz. As the array was switched on, the frequency chirped (see Fig. 13.9) due to the finite ramp time of the current. For high-power arrays, this represents a particularly significant problem because of the difficulties in rapidly switching large currents.

For analog or higher-level digital amplitude modulation, class-C (high-efficiency) oscillator elements [88,89] are ideal. In class C a transistor is switched on and off in such a way that the amplitude of the output depends linearly on the drain bias voltage (as illustrated in Fig. 13.10). Many AM radio stations use class-C amplifiers in their transmitters for this reason.

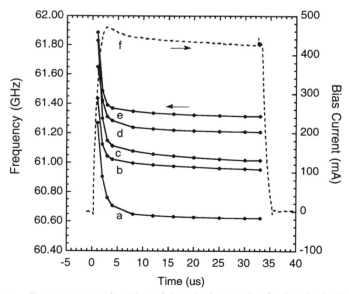

FIGURE 13.9 Frequency as a function of time during a pulse for four locked oscillators (a) and for each of the free-running oscillators (b–e). Also shown is the current for one of the diodes (f) [15].

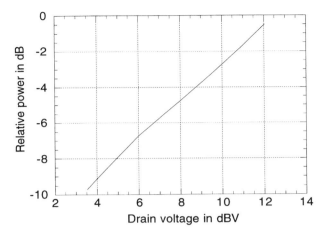

FIGURE 13.10 The output power measured from a class-C oscillator element is proportional to the drain bias voltage squared. Drain bias voltage is in decibels normalized to 1 V. The power is linear over 10 dB of variation [89].

Like with most other FET oscillator elements, the frequency of the class-C oscillator can be modulated by varying the gate bias. In some oscillators a varactor is also integrated into the circuit to explicitly create a voltage-controlled oscillator (VCO), but in all of these the frequency is a nonlinear function of the applied voltage. While this open-loop control can be used to create a simple frequency-modulated source (see the 28-GHz binary FSK link in reference [90], for example), to gain accurate phase or frequency modulation control, feedback becomes important.

The PLL configuration diagrammed in Fig. 13.8 contains this feedback and provides a simple way of modulating an oscillator array (in addition to substantially reducing the phase noise). A phase-modulating signal can be applied in the loop just after the phase detector, and if only digital PSK is needed, the loop filter can be eliminated because the resulting first-order loop is just as capable of tracking a step in phase as any higher-order loop [91]. For frequency modulation the signal would be added into the loop after the filter, which could be a simple op-amp integrator. In either case, only a baseband signal is actually applied to the array—no high-frequency feed is needed.

Unfortunately, for phase modulation this technique only allows shifts of $\pm 90°$. To obtain the full 360° needed for standard QPSK (quadrature PSK), a concept for a "quadrature" PLL has been devised [92]. This technique is quite similar to the one described above, except that the PLL feedback path is split into two branches, as shown in Fig. 13.11.

The signal received from the array via the sampling antenna is split into two paths, each of which is mixed with the reference signal from a low phase-noise source. One of the reference signals is shifted by 90°, however. The baseband feedback signals are then mixed with the digital modulating signals, PM_I and PM_Q, and the final control signal sent to the array is a sum of these products. In

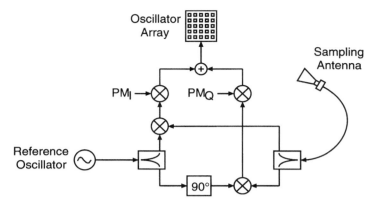

FIGURE 13.11 A "quadrature" PLL can be used to obtain a full 360° of PSK from an array. "In-phase" and "quadrature" phase-modulating signals are applied in separate feedback paths [92].

effect, each of the feedback paths allows phase shifting by ± 90°, and since they are in quadrature, the full 360° swing is obtainable.

Mathematically, the necessary values for PM_I and PM_Q to obtain an output phase θ are

$$PM_I = A \cos \theta \quad \text{and} \quad PM_Q = A \sin \theta \quad (13.12)$$

where A is an arbitrary constant. In other words,

$$\theta = \arg(PM_I + jPM_Q) \quad (13.13)$$

This technique has yet to be applied with an actual array, but simulations and experiments in reference 92 have validated the approach using an HP sweep oscillator in place of the array. In the experiments a 7-GHz carrier was modulated at rates up to 1.5 Mbps.

The quadrature PLL technique could be used to generate any level of PSK, though for any higher level modulation than (4-level) QPSK, QAM is more efficient. However, modulation of the oscillators' amplitude independently of the frequency (and phase) is difficult. A technique to produce 16-QAM that requires only phase modulation and uses combinations of QPSK-modulated elements to create the AM is described next.

As indicated in Fig. 13.12a, five (or a multiple thereof) oscillator elements must be pointing in the same direction for this technique. Broadside arrays with a multiple of five rows could implement this scheme by having some of the rows independently modulated.

Four of the elements are QPSK-modulated by the same signal. The resulting, combined QPSK constellation for them is indicated by x's in Fig. 13.12b—

OSCILLATOR ARRAYS   505

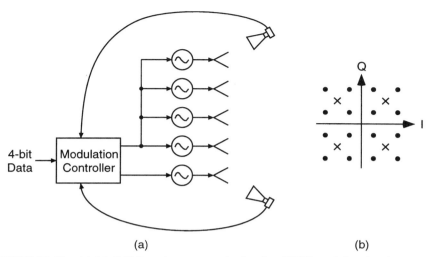

FIGURE 13.12 (a) 16-QAM can be generated using five QPSK-modulated active antennas, four of them modulated together. (b) The combination of these four yield the constellation of x's, but when combined with the fifth, the 16-QAM constellation of dots is achieved.

because the power from four elements is combined, the amplitude is twice that for a single element (referred to as one unit). When the QPSK signal from the fifth element is added in, it splits each of the points in the constellation into four points, located along a circle one unit away from the original points. This final, combined constellation is represented in the figure with dots and is identical to a square 16-QAM constellation (see Fig. 13.1).

Higher levels of QAM could also be generated using extensions of this method. For 64-QAM, three sets of QPSK-modulated elements are needed, in multiples of sixteen, four, and one. 256-QAM requires a total of 85 elements.

4.3 Omniazimuthal Arrays

Communication applications fall roughly into two categories. Point-to-point links transmit from one fixed transceiver to another and use high-gain antennas to minimize power requirements as well as impairments due to multipath and co-channel interference. In point-to-multipoint applications a base station serves multiple subscribers, such as in a cellular link.

Although standard planar, broadside oscillator arrays are well suited for point-to-point systems, they must be modified for multipoint applications. For example, a lens/horn combination could be used to collect the radiated power, which would then be re-radiated with an omnidirectional antenna. Such systems lose the flexibility of adaptive sectoring and/or steering however. Alternatively, quasi-optical arrays can be tailored specifically to produce omniazimuthal beams.

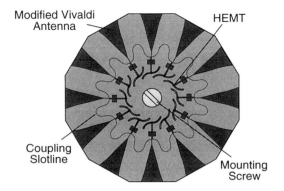

FIGURE 13.13 Twelve-element 28-GHz CPW oscillator array for multipoint applications. The lighter shaded areas represent metal on this 15-mil-thick Rogers TMM3 duroid [94].

Figure 13.13 shows a drawing of a 12-element, omnidirectional 28-GHz oscillator array [93]. The oscillators in this array are integrated with end-fire antennas, which are oriented along the periphery of the circular substrate pointing in different directions. To minimize the amount of destructive interference in the array radiation pattern, the radius, and consequently the antenna length, should generally be made as small as possible. Here the antenna is a shortened version of a Vivaldi tapered slot antenna (TSA) [94].

Unlike in linear, broadside TSA arrays in which the elements are pointing in the same direction, adjacent antennas in this circular array are pointing 30° apart. This, in conjunction with the fact that the TSAs in the circular array can only be spaced as close as 1.1 wavelengths at the periphery due to the need to fit the active circuitry into the center, results in nulls in the array pattern.

One solution to avoid these is to turn the antennas around and have them face a reflector located in the center of the array, as shown in Fig. 13.14. This reflector can have a polygonal horizontal cross-section so that each antenna points at a flat surface, and consequently the images of the antennas, which are pointing outward, can be placed as close together as desired.

In addition to eliminating nulls in the azimuthal pattern, this reflector approach allows the designer to tailor the elevation pattern. As demonstrated in the mechanically steered antenna of reference 95, such a central reflector can be made with a parabolic vertical profile (similar to an hourglass) to provide substantial focusing in the elevation direction. Alternatively, a simpler reflector structure can be made with flat sides on the polygonal cylinder and metal plates above and below to limit the elevation beamwidth. This is the type of reflector illustrated in Fig. 13.14. Optimization of the reflector and active antenna position for a 28 GHz design is demonstrated in reference 96, and a sample array using this reflector is described in reference [97].

As mentioned previously, an advantage of using an omnidirectional array in a multipoint transmitter is the ability to sector the pattern, or transmit different

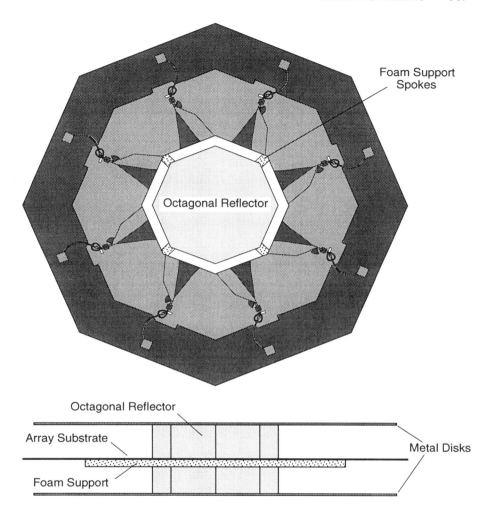

FIGURE 13.14 Plan and side views of an omni-azimuth array configuration containing end-fire antennas pointed at a metal octagonal–cylindrical reflector. Patch antennas near the outer edge of the substrate are for receiving an injection-locking signal [97].

signals in different directions. A technique to provide narrow beamwidths and minimize the side lobes of the individual active antenna radiation patterns in the reflector-type array is shown in Fig. 13.15. Rectangular sheets of metal are placed on each side of the active antenna, vertically between the top and bottom disks. These vanes not only limit the azimuthal beamwidth, but also improve the elevation pattern as well [96].

To modulate and synchronize the elements of such an array independently, each could be incorporated into a PLL circuit. This array of radiating PLL elements would be synchronized to a single external oscillator, but each PLL would have a separate input for the sector's modulation signal.

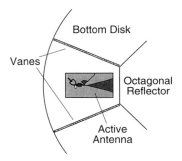

FIGURE 13.15 Overhead view of active antenna and reflector with top plate removed, showing location of vertical metal vanes used for creating sectored patterns.

5 AMPLIFIER ARRAYS

Quasi-optical amplifier arrays have direct applicability to radar and communication applications which presently use traveling wave tubes. For most applications these arrays will be driven close to saturation, where the efficiency and output power is maximized. Near saturation the output power is no longer a linear function of the input and the output phase becomes power-dependent.

5.1 Nonlinearities in Arrays

The performance of some modulation schemes, such as single channel QPSK and analog FM, are not adversely affected by output nonlinearities. However, multichannel links, AM radio, and QAM links, are seriously affected by nonlinearities. Several studies have investigated performance degradation due to nonlinearities in TWTs [98–101]. Deviations from nonlinear behavior produce intermodulation products. The most undesirable are the third-order products which generate spurious signals within the communication band. Third-order products are measured using a two-tone setup in which the input is driven with closely spaced equal signals at f_1 and f_2, while looking for $2f_1 - f_2$ and $2f_2 - f_1$ components at the output. For an input signal

$$u(t) = A(t) \sin(\omega(t)t + \phi(t)) \quad (13.14)$$

the output is of the form

$$u'(t) = G[A(t)] \sin(\omega(t)t + \phi(t) + F[A(t)]) \quad (13.15)$$

where $G[A(t)]$ and $F[A(t)]$ are the AM/AM and AM/PM conversion characteristics. Characterization of the amplifier nonlinearities allows prediction of the AM/AM and AM/PM conversion which can then be used to calculate the bit error rate [99–101].

5.2 Heat Dissipation

Overheating resulting from the conversion of DC bias into thermal energy necessitates careful thermal design. Heatsinks attached to the backside of a

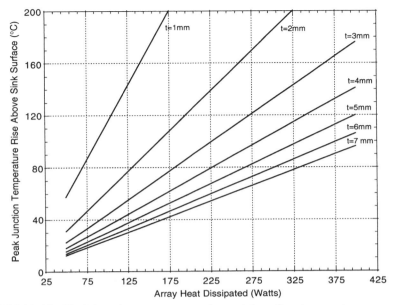

FIGURE 13.16 Peak junction temperature above sink surface temperature versus total power dissipated for AlN substrates of various thicknesses, t. The array has 100 elements and is assumed to be 1.89 in.^{2} mounted on a 3 in.^{2} aluminum nitrate substrate [102].

thinned array can be used for most oscillator array designs, but heatsinking becomes more challenging for amplifier arrays which usually require RF access on both sides.

When the RF input signal is interrupted to an amplifier array, all of the DC bias power (P_{DC}) is converted to heat (P_{heat}). The heat term is given by $P_{heat} = P_{out}(1 - 1/G)/\eta_{pae}$ where G is the array gain, P_{out} is the array's RF output power, and η_{pae} is the array power-added efficiency. For a 35-GHz array assuming $P_{out} = 100$ W, $G = 7.5$ dB, $\eta_{pae} = 25\%$, $P_{heat} = 329$ W. This heat raises the temperature of the array transistors, which should be kept below their specified maximum temperature, typically 150° C for GaAs-based devices (Fig. 13.16) [102].

6 IMAGING AND RECEIVER APPLICATIONS

The millimeter-wave region of the spectrum is also of interest for imaging applications. For a given antenna size, millimeter wavelengths offer higher resolution than microwave-based systems. Furthermore, millimeter waves propagate further than infrared and optical signals in fog and rain conditions. For these reasons, millimeter-wave systems are being developed for aircraft, airport security, plasma imaging, and vehicular radar applications [103]. Imaging arrays offer the promise of realtime images without the need for mechanical or electrical scanning across the scene. These arrays perform the function of a

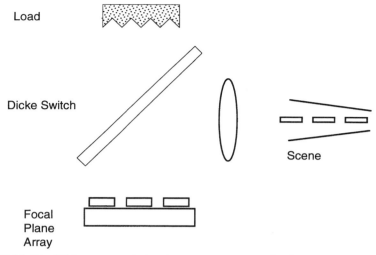

FIGURE 13.17 Dicke array switches between two states. In the transparent state the load is focused onto the receiving focal plane array. In the reflective state, the scene is reflected on the receiver.

camera in which imaging pixels/elements are placed in the focal plane of the optical system. High-quality images require a large numbers of pixels for which quasi-optical arrays are well suited.

The relatively low received powers, particularly in passive systems, makes sensitivity a major design parameter. A number of undesirable effects can degrade the sensitivity. Because of the high-gains in the receiver, drift can become a major problem. Nevertheless, high-contrast images are possible; however, the gain fluctuations must be accounted for by continual calibration. One common scheme is to use a Dicke switch to toggle (10–1000 Hz) the receiver input between the scene and a known reference load. This allows real-time compensation for variations in receiver gain.

Rather than building an expensive switch for each array pixel, Stephan, Spooner, and Goldsmith [104] developed a monolithic quasi-optical PIN diode array to switch the entire imaging array. In reflection the quasi-optical switch reflects the image of the scene onto the receiving array with as little loss as possible. In transmission it appears transparent to the reference load (Fig. 13.17). The array uses a grid of diodes to switch the array between the inductive (shorted or reflecting) and capacitive (open or transmission) state. To achieve an acceptable series inductance and to accommodate the PIN diodes, circular patches were added to the strips. Good transmission properties in the off state is achieved with a second passive array in which the diodes are replaced by shorts. The passive array is placed a quarter wavelength away from the PIN array. More details are given in the beam-control array chapter, Chapter 10. A 94-GHz array PIN diode array has been built by Millitech for detecting concealed metallic and nonmetallic contraband. A photograph of the array is shown in Fig. 13.18 [105].

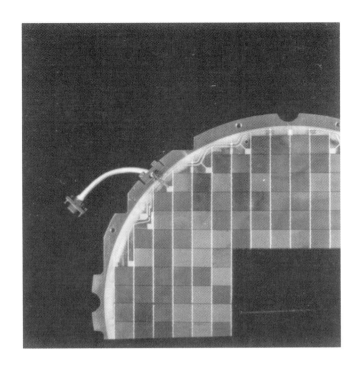

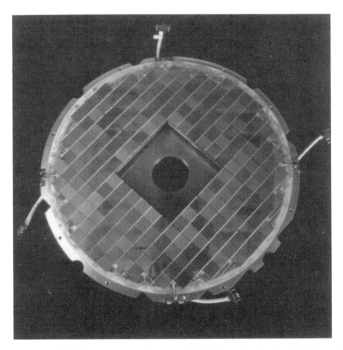

FIGURE 13.18 PIN diode array consisting of 204 tiles. Each tile is 12.7 mm^2 consisting of a 6 × 5 PIN diode subarray. (Photograph courtesy of Millitech Inc., South Deerfield, MA.)

There are several other quasi-optical receivers elements, which can also be integrated into arrays. Stephan *et al.* [106] developed a quasi-optical balanced mixer element, in which the LO and RF are fed in orthogonal polarizations. The element had a typical conversion loss of 6 dB at 10 GHz with a directivity of 6 dBi. In the self oscillating mixer, a pair of HEMTs are used to form a balanced local oscillator [107] which is integrated into the receiving antenna. In reference 108, a FET oscillator provides LO for an integrated mixer diode and also drives an inverted patch antenna. The reader is also referred to the related articles [109–111].

7 REALIZATION OF QUASI-OPTICAL ARRAYS

For commercial applications the cost of the quasi-optical array becomes increasingly important. There has been great debate over the monolithic and hybrid approaches for realization of microwave circuits and which approach is least expensive [112]. This argument (MMICs versus hybrids) can also be extended to quasi-optical arrays. In the MMIC approach all the devices, as well as the antennas, are fabricated on the same substrate. The drawback of this is that a large area of the semiconductor substrate is "wasted" (covered with passive components and antennas), and the device yield has to be high to achieve good array yields. For example, a 100-element millimeter-wave HEMT transistor array with a device yield of 99% (99% HEMT yield for 0.1-μm gate-length HEMTs is considered to be very good) will only have fully worked elements in 37% of the arrays. Considering the area of wafer that such an array will occupy, a yield of 37% is totally unacceptable for realization of low-cost arrays. HBTs can have much higher yields than 0.1-μm gate-length HEMTs, but as mentioned earlier, HBTs have lower gain at millimeter-wave frequencies. Even if HBTs are used, the issue of "wasted" semiconductor real estate still remains. Recently there has been considerable interest in circuit compaction for MMICs in order to reduce the size and therefore the cost of the chips, but these approaches will not necessarily allow reduction of the size of quasi-optical arrays, where element-to-element spacing is limited by the antenna size. One approach to reducing the area occupied by the passive components on the semiconductor is to use a top-layer dielectric such as BCB [113].

In the hybrid approach the antenna is fabricated on a low-cost substrate, and only the active devices are fabricated on the more expensive semiconductor substrate. The active devices are then mounted on the antenna substrate either in packaged form or by direct wire-bonding. The hybrid approach offers several advantages. It allows the array designer to use a substrate with a different dielectric constant than GaAs or InP substrates, which can improve the performance of the antennas [114]. Different device technologies (HBTs, HEMTs, PINs) can easily be integrated on the same antenna substrate to perform different functions. Also, the devices can be prescreened before mounting so high array

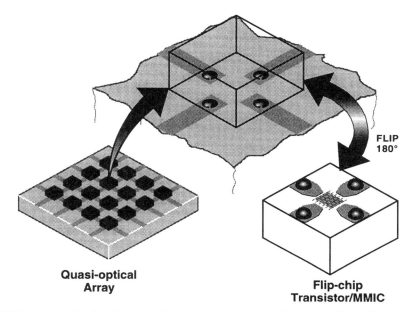

FIGURE 13.19 Realization of a 16-element quasi-optical array by flip-chip mounting of discrete transistors or small MMIC chips on a low-cost dielectric substrate.

yields can be achieved. In addition, the typical cost of discrete devices is low compared to that of MMICs. The obvious disadvantage of this approach is that it requires assembly of the array, which can be time-consuming and costly. Plus, at millimeter-wave frequencies the package parasitics and wire-bonds must be considered in the design.

Flip-chip mounting of components offers a low-cost technique for realization of quasi-optical arrays [115]. In this approach (Fig. 13.19), bumps are fabricated on the device pads, and the device can then be mounted using automated assembly machines onto the antenna substrate. This way the devices can be placed accurately [116, 117] and the array assembled at low cost. In addition, flip-chip connections have high yield, are reproducible, and have low inductance. An SEM photograph of a flip-chip HBT with bumps is shown in Fig. 13.20. Quasi-optical amplifier arrays can also be fabricated by flip-chip mounting the individual amplifier chips onto the antenna substrate rather than wire-bonding them [38]. Large arrays can be assembled by flip-chip mounting smaller MMIC subarrays. The flip-chip approach is probably limited to less than 100 GHz, as the spacing between the elements becomes too small at higher frequencies. In fact, at higher frequencies the element spacing is probably small enough that the array can be made sufficiently compact to merit monolithic fabrication, and then the primary issue becomes achieving high yields for the transistors.

514 COMMERCIAL-APPLICATIONS OF QUASI-OPTICS

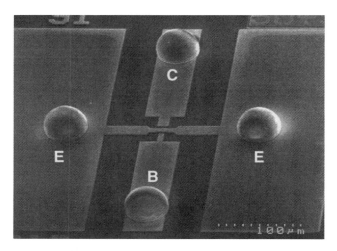

FIGURE 13.20 SEM photograph of a flip-chip HBT. The device has four bumps: two on the emitter pads, one for the base, and one for the collector.

In general, to create quasi-optical arrays or systems, careful consideration must be made of the system requirements and performance needed. The device technology, while keeping the system cost in mind, has to be chosen to meet the system requirements. Then depending on the device technology, the quasi-optical system can be realized using flip-chip or MMIC technologies (or a combination of both).

REFERENCES

[1] S. A. Booth, "Cable television without the wires," *Popular Mechanics,* pp. 58–59, 111, March 1994.
[2] M. K. Nezami, "Design techniques minimize IMD in MMDS transmitters," *Microwaves & RF,* pp. 73–80, May 1995.
[3] J. D. Kraus and K. R. Carver, *Electromagnetics,* McGraw Hill, New York, 1973.
[4] J. G. Proakis and M. Salehi, *Communication Systems Engineering,* Prentice-Hall, Englewood Cliffs, NJ, 1994.
[5] T. Noguchi, Y. Daido, and J. A. Nossek, "Modulation techniques for microwave digital radio," in *Microwave Digital Radio,* L. J. Greenstein and M. Shafi, eds., IEEE Press, New York, 1988.
[6] S. M. Sze, *Physics of Semiconductor Devices,* Wiley 1981, Chapter 11.
[7] M. Shur, *GaAs Devices and Circuits,* Plenum Press, New York, 1987, ch. 4 and 5.
[8] J. D. Crowley, C. Hang, R. E. Dalrymple, D. R. Tringali, F. B. Fank, L. Wandinger, and H. B. Wallace, "140 GHz indium phosphide Gunn diode," *Electronics Lett.,* vol. 30, no. 6, pp. 499–500, March 1994.

[9] A. Rydberg, "High efficiency and output power from second- and third-harmonic millimeter wave InP-TED oscillators at frequencies above 170 GHz," *IEEE Electron Device Lett.*, vol. 11, no. 10, pp. 439–441, Oct. 1990.

[10] R. A. York and R. C. Compton, "Quasi-optical power combining using mutually synchronized oscillator arrays," *IEEE Trans. Microwave Theory Tech.*, vol. 39, no. 6, pp. 1000–1009, June 1991.

[11] A. Mortazawi and T. Itoh, "A periodic planar Gunn diode power combining oscillator," *IEEE Trans. Microwave Theory Tech.*, vol. 38, no. 1, pp. 86–87, Jan. 1990.

[12] S. Y. Liao, *Microwave Devices and Circuits*, Prentice-Hall, Englewood Cliffs, NJ, 1980, Chapter 6.

[13] Y. E. Ma, E. M. Nakaji, and M. F. Thrower, "V-band double-drift read silicon IMPATTs," *1984 IEEE MTT-S Int. Microwave Symp. Dig.*, pp. 167–168.

[14] N. Camilleri and B. Bayraktaroglu, "Monolithic millimeter-wave IMPATT oscillator and active antenna," *IEEE Trans. Microwave Theory Tech.*, vol. 36, no. 12, pp. 1670–1676, Dec. 1988.

[15] A. C. Davidson, F. W. Wise, and R. C. Compton, "A 60-GHz IMPATT oscillator array with pulsed operation," *IEEE Trans. Microwave Theory Tech.*, vol. 41, no. 10, pp. 1845–1850, Oct. 1993.

[16] E. R. Brown, J. R. Söderström, C. D. Parker, L. J. Mahoney, K. M. Molvar, T. C. McGill, "Oscillations up to 712 GHz in AlAs/AlSb resonant tunneling diodes," *Appl. Phys. Lett.*, vol. 58, pp. 2291–2293, 1991.

[17] S. M. Sze, *High Speed Semiconductor Devices*, Wiley, 1990, Chapter 8.

[18] M. Reddy, R. Y. Yu, H. Kroemer, M. J. W. Rodwell, S. C. Martin, R. E. Muller, and R. P. Smith, "Bias stabilization for resonant tunnel diode oscillators," *IEEE Microwave Guided Wave Lett.*, vol. 5, no. 7, pp. 219–221, July 1995.

[19] J. B. Hacker, R. M. Weikle, M. Kim, M. P. DeLisio, and D. B. Rutledge, "A 100-element planar Schottky diode grid mixer," *IEEE Trans. Microwave Theory Tech.*, vol. 40, no. 3, pp. 557–562, March 1992.

[20] S. Nam, T. Uwano, and T. Itoh, "Microstrip-fed planar frequency-multiplying space combiner," *IEEE Trans. Microwave Theory Tech.*, vol. 35, no. 12, pp. 1271–1276, Dec. 1987.

[21] C. F. Jou, W. W. Lam, H. Z. Chen, K. S. Stolt, N. C. Luhmann, and D. B. Rutledge, "Millimeter-wave diode-grid frequency doubler," *IEEE Trans. Microwave Theory Tech.*, vol. 36, no. 11, pp. 1507–1514, Nov. 1988.

[22] W. W. Lam, C. F. Jou, H. Z. Chen, K. S. Stolt, N. C. Luhmann, and D. B. Rutledge, "Millimeter-wave diode-grid phase shifters," *IEEE Trans. Microwave Theory Tech.*, vol. 36, no. 5, pp. 902–907, May 1988.

[23] L. B. Sjogren, H. X. L. Liu, F. Wang, T. Liu, X. H. Qin, W. Wu, E. Chung, C. W. Domier, and N. C. Luhmann, "A monolithic diode array millimeter-wave beam transmittance controller," *IEEE Trans. Microwave Theory Tech.*, vol. 41, no. 10, pp. 1782–1789, Oct. 1993.

[24] N. Camilleri and T. Itoh, "A quasi-optical multiplying slot array," *IEEE Trans. Microwave Theory Tech.*, vol. 33, no. 11, pp. 1189–1195, Nov. 1985.

[25] H. X. L. Liu, L. B. Sjogren, C. W. Domier, N. C. Luhmann, D. L. Sivco, and A. Y. Cho, "Monolithic quasi-optical frequency tripler array with 5-W output power at 99 GHz," *IEEE Electron Dev. Lett.*, vol. 14, no. 7, pp. 329–331, July 1993.

[26] A. Rahal, R. G. Bosisio, C. Rogers, J. Ovey, M. Sawan, and M. Missous, "A W-band medium power multi-stack quantum barrier varactor frequency tripler," *IEEE Microwave Guided Wave Lett.*, vol. 5, no. 11, pp. 368–370, Nov. 1995.

[27] K. D. Stephan, F. H. Spooner, and P. F. Goldsmith, "Quasioptical millimeter-wave hybrid and monolithic PIN diode switches," *IEEE Trans. Microwave Theory Tech.*, vol. 41, no. 10, pp. 1791–1798, Oct. 1993.

[28] W. Shockley, U.S. Patent No. 2,569,347, issued 25 September 1951.

[29] H. Kroemer, "Theory of a wide-gap emitter for transistors," *Proc. IRE,* vol. 45, pp. 1535–1537, Nov. 1957.

[30] H. Kroemer, "Heterostructure bipolar transistors and integrated circuits," *Proc. IEEE,* vol. 70, no. 1, pp. 13–25, Jan. 1982.

[31] M. A. Khatibzadeh, B. Bayraktaroglu, and T. Kim, "12 W monolithic X-band HBT power amplifier," *Proc. Monolithic Microwave Circuits Symp.*, pp. 47–50, 1992.

[32] D. Deakin, W. J. Ho, E. A. Sovero, and J. Higgins, "Power HBT for 44 GHz operation," *1993 IEEE GaAs IC Symp. Dig.*, pp. 371–373.

[33] F. Ali, A. Gupta, M. Salib, B. W. Veasel, D. E. Dawson, "A 2 Watt, 8–14 GHz HBT power MMIC with 20 dB gain and 40% power-added efficiency," *IEEE Trans. Microwave Theory Tech.*, vol. 42, no. 12, pp. 2635–2641, Dec. 1994.

[34] M. Hafizi, P. A. Macdonald, T. Liu, and D. B. Rensch, "Microwave power performance of InP-based double heterojunction bipolar transistors for C- and X-band applications," *1994 IEEE MTT-S Dig.*, pp. 671–674.

[35] F. Ali, M. Salib, and A. Gupta, "A 1-W X–Ku band HBT MMIC amplifier with 50% peak power-added efficiency," *IEEE Microwave Guided Wave Lett.*, vol. 3, no. 8, pp. 271–272, Aug. 1993.

[36] P. K. Ikalainen, S. K. Fan, and M. A. Khatibzadeh, "20 W linear, high efficiency internally matched HBT at 7.5 GHz," *1994 IEEE MTT-S Digest*, pp. 679–680.

[37] M. Salib, F. Ali, A. Gupta, B. Bayraktaroglu, and D. Dawson, "A 5–10 GHz, 1-watt HBT amplifier with 58% peak power-added efficiency," *IEEE Microwave Guided Wave Lett.*, vol. 4, no. 10, pp. 320–322, Oct. 1994.

[38] M. Kim, E. A. Sovero, J. B. Hacker, M. P. DeLisio, J. C. Chiao, S. J. Li, D. R. Gagnon, J. J. Rosenberg, and D. B. Rutledge, "A 100-element HBT grid amplifier," *1993 IEEE MTT-S Dig.*, pp. 615–618.

[39] C. M. Liu, E. A. Sovero, W. J. Ho, and D. B. Rutledge, "A 40-GHz monolithic grid amplifier," *IEEE Microwave Guided Wave Lett.*, to be published.

[40] M. J. Mondry and H. Kroemer, "Heterojunction Bipolar Transistor using a (Ga,In)P emitter on a GaAs base, grown by molecular beam epitaxy," *IEEE Electron Dev. Lett.*, vol. 6, no. 4, pp. 175–177, April 1985.

[41] F. Ren, C. R. Abernathy, S. J. Pearton, J. R. Lothian, P. W. Wisk, T. R. Fullowan, Y. K. Chen, L. W. Yang, S. T. Fu, R. S. Brozovich, and H. H. Lin, "Self-aligned InGaP/GaAs heterojunction bipolar transistors for microwave power application," *IEEE Elect. Dev. Lett.*, vol. 14, no. 7, pp. 332–334, July 1993.

[42] W. Liu, A. Khatibzadeh, T. Henderson, S. K. Fan, and D. Davito, "X-band GaInP/GaAs power heterojunction bipolar transistor," *1993 IEEE MTT-S Dig.*, pp. 1477–1480.

[43] K. Riepe, H. Leier, U. Seiler, A. Marten, and H. Sledzik, "High-efficiency X-band GaInP/GaAs HBT MMIC power amplifier for stable long pulse and CW operation," *1995 IEDM Tech. Dig.,* pp. 795–798.

[44] J. F. Jensen, M. Hafizi, W. E. Stanchina, R. A. Metzger, and D. B. Rensch, "39.5 GHz static frequency divider implemented in AlInAs/GaInAs HBT technology," *1992 GaAs IC Symp. Dig.,* pp. 101–104.

[45] M. Chen, C. Nguyen, T. Liu, and D. Rensch, "High-performance AlInAs/GaInAs/InP DHBT X-band power cell with InP emitter ballast resistor," in *Proceedings of IEEE/Cornell Conference on Advanced Concepts in High Speed Semiconductor Devices and Circuits,* pp. 573–582, Aug. 1995.

[46] U. König, A. Gruhle, and A. Schüppen, "SiGe devices and circuits: Where are advantages over III/V?" *1995 IEEE GaAs IC Symp. Dig.,* pp. 14–17.

[47] D. L. Harame, J. H. Comfort, J. D. Cressler, E. F. Crabbè, J. Y. C. Sun, B. S. Meyerson, and T. Tice, "Si/SiGe epitaxial-base transistors—Part I: materials, physics, and circuits," *IEEE Trans. Electron Devices,* vol. 42, no. 3, pp. 455–468, March 1995.

[48] J. F. Luy, K. M. Strohm, and E. Sasse, "Si/SiGe MMIC technology," *1994 IEEE MTT-S Dig.,* pp. 1755–1757.

[49] M. Case, P. Macdonald, M. Matloubian, M. Chen, L. Larson, and D. Rensch, "High-performance microwave elements for SiGe MMICs," *Proceedings of IEEE/Cornell Conference on Advanced Concepts in High Speed Semiconductor Devices and Circuits,* pp. 85–92, Aug. 1995.

[50] U. Erben, M. Wahl, A. Schüppen, and H. Schumacher, "Class-A SiGe HBT power amplifiers at C-band frequencies," *IEEE Microwave Guided Wave Lett.,* vol. 5, no. 12, pp. 435–436, Dec. 1995.

[51] L. Larson, M. Case, S. Rosenbaum, D. Rensch, M. Chen, P. Macdonald, M. Matloubian, D. Harame, J. Malinowski, B. Meyerson, M. Gilbert, and S. Maas, "Si/SiGe HBT Technology for low-cost monolithic microwave integrated circuits," *1996 IEEE Int. Solid-State Circuits Conf. (ISSCC) Dig.,* pp. 80–81.

[52] K. D. Hobart, F. J. Kub, N. A. Papanicolaou, W. Kruppa, and P. E. Thompson, "Si/Si$1\text{-}x$Gex heterojunction bipolar transistors with high breakdown voltage," *IEEE Electron Dev. Lett.,* vol. 16, no. 5, pp. 205–207, May 1995.

[53] S. Yamahata, K. Kurishima, H. Ito, and Y. Matsuoka, "Over-220-GHz-f_T-and-f_{max} InP/InGaAs double-heterojunction bipolar transistors with a new hexagonal-shaped emitter," *1995 IEEE GaAs IC Symp. Dig.,* pp. 163–166.

[54] C. Nguyen, T. Liu, M. Chen, H. C. Sun, and D. Rensch, "AlInAs/GaInAs/InP double heterojunction bipolar transistor with a novel base-collector design for power applications," *1995 IEDM Tech. Dig.,* pp. 799–802.

[55] Chanh Nguyen, private communication.

[56] J. V. DiLorenzo, and D. D. Khandelwal, *GaAs FET Principles and Technology,* Artech House, Boston, 1982.

[57] P. M. Smith, S. M. J. Liu, M. Y. Kao, P. Ho, S. C. Wang, K. H. G. Duh, S. T. Fu, and P. C. Chao, "W-band high efficiency InP-based power HEMT with 600 GHz f_{max}," *IEEE Microwave Guided Wave Lett.,* vol. 5, no. 7, pp. 230–232, July 1995.

[58] J. V. DiLorenzo and W. R. Wisseman, "GaAs power MESFET: design, fabrication, and performance," *IEEE Trans. Microwave Theory Tech.*, vol. 27, no. 5, pp. 367–378, May 1979.

[59] J. M. Golio, "Ultimate scaling limits for high-frequency GaAs MESFET's," *IEEE Trans. Electron Devices*, vol. 35, no. 7, pp. 839–848, July 1988.

[60] M. B. Das, "Millimeter-wave performance of ultrasubmicrometer-gate field-effect transistors: a comparison of MODFET, MESFET, and PBT structures," *IEEE Trans. Electron Devices*, vol. 34, no. 7, pp. 1429–1440, July 1987.

[61] Z. B. Popovic, R. M. Weikle, M. Kim, and D. B. Rutledge, "A 100-MESFET planar grid oscillator," *IEEE Trans. Microwave Theory Tech.*, vol. 39, no. 2, pp. 193–200, Feb. 1991.

[62] J. B. Hacker, M. P. DeLisio, M. Kim, C. M. Liu, S. J. Li, S. W. Wedge, and D. B. Rutledge, "A 10-watt X-band grid oscillator," *1994 IEEE MTT-S Digest*, pp. 823–826.

[63] L. D. Nguyen, L. E. Larson, and U. K. Mishra, "Ultrahigh-speed modulation-doped field-effect transistors: A tutorial review," *Proc. IEEE*, vol. 80, no. 4, pp. 494–518, April 1992.

[64] K. L. Tan, R. M. Dia, D. C. Streit, L. K. Shaw, A. C. Han, M. D. Sholley, P. H. Liu, T. Q. Trinh, T. Lin, and H. C. Yen, "60-GHz pseudomorphic $Al_{0.25}Ga_{0.75}As/In_{0.28}Ga_{0.72}As$ low-noise HEMT's," *IEEE Electron Dev. Lett.*, vol. 12, no. 1, pp. 23–25, Jan. 1991.

[65] T. Katch, N. Yoshida, H. Minami, T. Kashiwa, and S. Orisaka, "A 60 GHz-band ultra low noise planar-doped HEMT," *1993 IEEE MTT-S Dig.*, pp. 337–340.

[66] K. L. Tan, R. M. Dia, D. C. Streit, T. Lin, T. Q. Trinh, A. C. Han, P. H. Liu, P. D. Chow, and H. C. Yen, "94-GHz 0.1 μm T-gate low-noise pseudomorphic InGaAs HEMT's," *IEEE Electron Dev. Lett.*, vol. 11, no. 12, pp. 585–587, Dec. 1990.

[67] S. Bouthillette, A. Platzker, and L. Aucoin, "High-efficiency 40 watt PsHEMT S-band MIC power amplifiers," *1994 IEEE MTT-S Dig.*, pp. 667–670.

[68] B. Kraemer, R. Basset, P. Chye, D. Day, and J. Wei, "Power pHEMT module delivers 12 watts, 40% PAE over the 8.5 to 10.5 GHz band," *1994 IEEE MTT-S Dig.*, pp. 683–686.

[69] P. M. Smith, C. T. Creamer, W. F. Kopp, D. W. Ferguson, P. Ho, and J. R. Willhite, "A high-power Q-band pHEMT for communication terminal applications," *1994 IEEE MTT-S Dig.*, pp. 809–812.

[70] R. Lai, M. Biedenbender, J. Lee, K. Tan, D. Streit, P. H. Liu, M. Hoppe, and B. Allen, "0.15 μm InGaAs/AlGaAs/GaAs HEMT production process for high performance and high yield V-band power MMICs," *1995 GaAs IC Symp. Dig.*, pp. 105–108.

[71] H. Wang, Y. Hwang, T. H. Chen, M. Biedenbender, D. C. Streit, D. C. W. Lo, G. S. Dow, and B. R. Allen, "A W-band monolithic 175-mW power amplifier," in *1995 IEEE MTT-S Dig.*, pp. 419–422.

[72] M. P. De Lisio, C. M. Liu, A. Moussessian, D. B. Rutledge, and J. J. Rosenberg, "A 100-element MODFET grid amplifier," *1995 IEEE AP-S Int. Symp.*

[73] M. P. De Lisio, S. W. Duncan, D. W. Tu, C. M. Liu, A. Moussessian, J. J. Rosenberg, and D. B. Rutledge, "Modelling and performance of a 100-element pHEMT grid amplifier," *IEEE Trans. Microwave Theory Tech.*, submitted.

[74] U. K. Mishra, A. S. Brown, M. J. Delaney, P. T. Greiling, and C. F. Krumm, "The AlInAs-GaInAs HEMT for microwave and millimeter-wave applications," *IEEE Trans. Microwave Theory Tech.*, vol. 37, no. 9, pp. 1279–1285, Sept. 1989.

[75] L. D. Nguyen, A. S. Brown, M. A. Thompson, and L. M. Jelloian, "50-nm self-aligned-gate pseudomorphic AlInAs/GaInAs high electron mobility transistors," *IEEE Trans. Electron Dev.*, vol. 39, no. 9, pp. 2007–2014, Sept. 1992.

[76] K. H. G. Duh, P. C. Chao, P. Ho, M. Y. Kao, P. M. Smith, J. M. Ballingall, and A. A. Jabra, "High performance InP-based HEMT millimeter-wave low-noise amplifiers," *1989 IEEE MTT-S Int. Microwave Symp. Dig.*, pp. 805–808.

[77] K. L. Tan, D. C. Streit, P. D. Chow, R. M. Dia, A. C. Han, P. H. Liu, D. Garske, and R. Lai, "140 GHz 0.1 μm gate length pseudomorphic $In_0.52Al_0.48As/In_0.60Ga_0.40As$/InP HEMT," *IEDM Tech. Dig.*, pp. 239–242, Dec. 1991.

[78] R. Isobe, C. Wong, A. Potter, L. Tran, M. Delaney, R. Rhodes, D. Jang, L. Nguyen, and M. Le, "Q- and V-band MMIC chip set using 0.1 μm millimeter-wave low noise InP HEMTs," *1995 IEEE MTT-S Dig.*, pp. 1133–1136.

[79] S. E. Rosenbaum, L. M. Jelloian, A. S. Brown, M. A. Thompson, M. Matloubian, L. E. Larson, R. F. Lohr, B. K. Kormanyos, G. M. Reibez, and L. P. B. Katehi, "A 213 GHz AlInAs/GaInAs/InP HEMT MMIC oscillator," *IEEE IEDM Tech. Dig.*, pp. 924–926, Dec. 1993.

[80] M. Matloubian, A. S. Brown, L. D. Nguyen, M. A. Melendes, L. E. Larson, M. J. Delaney, M. A. Thompson, R.A. Rhodes, and J.E. Pence, "20-GHz high-efficiency AlInAs-GaInAs on InP power HEMT," *IEEE Microwave Guided Wave Lett.*, vol. 3, no. 5, pp. 142–144, May 1993.

[81] M. Matloubian, L. M. Jelloian, A. S. Brown, L. D. Nguyen, L. E. Larson, M. J. Delaney, M. A. Thompson, R. A. Rhodes, and J. E. Pence, "V-band high-efficiency high-power AlInAs/GaInAs/InP HEMT's," *IEEE Trans. Microwave Theory Tech.*, vol. 41, no. 12, pp. 2206–2210, Dec. 1993.

[82] K. C. Hwang, P. Ho, M. Y. Kao, S. T. Fu, J. Liu, P. C. Chao, P. M. Smith, and A. W. Swanson, "W-band high power passivated 0.15 μm InAlAs/InGaAs HEMT device," *1994 Proceedings of International Conference on Indium Phosphide and Related Materials*, pp. 18–20.

[83] K. Y. Hur, R. A. McTaggart, B. W. LeBlanc, W. E. Hoke, P. J. Lemonias, A. B. Miller, T. E. Kazior, and L. M. Aucoin, "Double recessed AlInAs/GaInAs/InP HEMTs with high breakdown voltages," *1995 IEEE GaAs IC Symp. Dig.*, pp. 101–104.

[84] W. Lam, M. Matloubian, A. Igawa, C. Chou, A. Kurdoghlian, C. Ngo, L. Jelloian, A. Brown, M. Thompson, and L. Larson, "44-GHz high-efficiency InP-HEMT MMIC power amplifier," *IEEE Microwave Guided Wave Lett.*, vol. 4, no. 8, pp. 277–278, Aug. 1994.

[85] M. J. Vaughan and R. C. Compton, "Phase noise in arrays of mutually-coupled oscillators and its reduction via injection-locking and PLLs," *IEEE Trans. Microwave Theory Tech.*, Oct. 1995, submitted.

[86] R. A. York and R. C. Compton, "Measurement and modelling of radiative coupling in oscillator arrays," *IEEE Trans. Microwave Theory Tech.*, vol. MTT-41, pp. 438–444, March 1993.

[87] J. Lin and T. Itoh, "Two-dimensional quasi-optical power-combining arrays using strongly coupled oscillators," *IEEE Trans. Microwave Theory Tech.*, vol. 42, no. 4, pp. 734–741, April 1994.

[88] R. D. Martinez and R. C. Compton, "A quasi-optical oscillator/modulator for wireless transmission," *1994 MTT Symp.*, May 1994.

[89] R. D. Martinez and R. C. Compton, "High-Efficiency FET/microstrip-patch oscillators," *IEEE Antennas Propagat. Mag.*, vol. 36, no. 1, pp. 16–19, Feb. 1994.

[90] M. J. Vaughan, W. Wright, and R. C. Compton, "Active antenna elements for millimeter-wave cellular communications," presented at *Int. Symp. Signals Sys. Electron.*, San Francisco, CA, Oct. 1995.

[91] F. M. Gardner, *Phaselock Techniques,* Wiley, New York, 1979.

[92] C. E. Saavedra, M. J. Vaughan, and R. C. Compton, "An M-PSK modulator for quasi-optical wireless array applications," presented at the *1996 MTT-S Int. Microwave Symp.*, San Francisco, CA, June 1996.

[93] M. J. Vaughan and R. C. Compton, "28 GHz omni-directional quasi-optical transmitter array," *IEEE Trans. Microwave Theory Tech.*, vol. 43, no. 10, pp. 2507–2509, Oct. 1995.

[94] M. J. Vaughan and R. C. Compton, "28 GHz oscillator for endfire quasi-optical power combining arrays," *Electronics Lett.*, vol. 31, no. 17, pp. 1453–1455, Aug. 17, 1995.

[95] N. N. Fullilove, W. G. Scott, and J. R. Tomlinson, "The hourglass scanner—a new rapid scan, large aperture antenna," *1959 IRE National Convention Record,* New York, March 1959, pp. 190–200.

[96] M. J. Vaughan, R. C. Compton, and K. Y. Hur, "InP-Based 28 GHz integrated antennas for point-to-multipoint distribution," in *Proceedings of the IEEE/Cornell Conference on Advanced Concepts in High Speed Semiconductor Devices and Circuits,* Ithaca, NY, Aug. 1995, pp. 75–84.

[97] M. J. Vaughan and R. C. Compton, "Injection-Locked 28 GHz oscillator array with disk-cylinder reflector," *IEEE Trans. Microwave Guided Wave Lett.*

[98] D. R. Smith, *Digital Transmission Systems,* Van Nostrand Reinhold, New York, 1992.

[99] C. M. Thomas, J. E. Alexander, and E. W. Rahneberg, "A new generation of digital microwave radios for U.S. military telephone networks," *IEEE Trans. Commun.*, vol. COM-27, no. 12, pp. 1916–1928, Dec. 1979.

[100] R. G. Lyons, "The effect of a bandpass nonlinearity on signal detectability," *IEEE Trans. Commun.*, vol. COM-21, no. 1, pp. 51–60, Jan. 1973.

[101] D. Chakraborty and L. S. Golding, "Wide-band digital transmission over analog radio relay links," *IEEE Trans. Commun.*, vol. COM-23, no. 11, pp. 1215–1228, Nov. 1975.

[102] N. J. Kolias and R. C. Compton, "Thermal management for high-power active amplifier arrays," *IEEE Trans. Microwave Theory Tech.*, May 1995.

[103] P. F. Goldsmith, G. R. Huguenin, J. Kapitzky, and E. L. Moore, "Focal plane imaging systems for millimeter wavelengths," *IEEE Trans. Microwave Theory Tech.*, vol. 41, pp. 1664–1675, Oct. 1993.

[104] K. D. Stephan, P. H. Spooner, and P. F. Goldsmith, "Quasi-optical millimeter-wave hybrid and monolithic PIN diode switches," *IEEE Trans. Microwave Theory Tech.*, vol. 41, pp. 1791–1798, Oct. 1993.

[105] G. R. Huguenin, C.-T. Hsieh, J. E. Kapitzky, E. L. Moore, K. D. Stephan, and A. S. Vickery, "Contraband detection through clothing by means of millimeter-wave imaging," *SPIE Proc. 1942, Underground and Obscured Object Imaging and Detection,* Orlando, FL, April 1993, pp. 117–128.

[106] K. D. Stephan, Natalino Camilleri, and Tatsuo Itoh, "A quasi-optical polarization-duplexed balanced mixer for millimeter-wave applications," *IEEE Trans. Microwave Theory Tech.,* pp. 164–170, Feb. 1983 (also see correction, June 1983).

[107] V. D. Hwang and T. Itoh, "A quasi-optical HEMT self oscillating mixer," *IEEE MTT-S Int. Microwave Symp. Dig.,* pp. 1093–1096, 1988.

[108] R. Flynt, L. Fan, J. Navaro, and K. Chang, "Low cost and compact active integrated antenna transceiver for system applications," *IEEE MTT-S Int. Microwave Symp. Dig.,* pp. 953–956, 1995.

[109] C. W. Pobanz and T. Itoh, "Quasi-optical microwave circuits for wireless applications," *Microwave J.,* 1988, pp. 1093–1096.

[110] K. Cha, S. Kawasaki, and T. Itoh, "Transponder using self-oscillating mixer and active antenna," *IEEE MTT-S Int. Microwave Symp. Dig.,* 1994, pp. 425–428.

[111] C. W. Pobanz and T. Itoh, "A conformal retrodirective array for radar applications using a heterodyne phased scattering element," *IEEE MTT-S Int. Microwave Symp. Dig.,* pp. 905–908, 1995.

[112] E. F. Belohoubek, "Miniature hybrid ICs versus MMICs for phased array radars," in *1986 Military Microwave Conf. Dig.,* pp. 439–446.

[113] U. Dahlgren, J. Svedin, H. Johansson, O. J. Hagel, H. Zirath, C. Karlsson, and N. Rorsman, "An integrated millimeter wave BCB patch antenna HEMT receiver," *1994 IEEE MTT-S Dig.,* pp. 661–664.

[114] N. G. Alexopoulos, P. B. Katehi, and D. B. Rutledge, "Substrate optimization for integrated circuit antennas," *IEEE Trans. Microwave Theory Tech.,* vol. 31, no. 7, pp. 550–557, July 1983.

[115] L. M. Felton, "High yield GaAs flip-chip MMICs lead to low cost T/R modules," *1994 IEEE MTT-S Dig.,* pp. 1707–1710.

[116] H. Sakai, Y. Ota, K. Inoue, T. Yoshida, K. Takahashi, S. Fujita, and M. Sagawa, "A novel millimeter-wave IC on Si substrate using flip-chip bonding technology," in *1994 IEEE MTT-S Dig.,* pp. 1763–1766.

[117] H. Sakai, Y. Ota, K. Inoue, M. Yanagihara, T. Matsuno, M. Tanabe, T. Yoshida, Y. Ikeda, S. Fujita, K. Takahashi, and M. Sagawa, "A millimeter-wave flip-chip IC using micro-bump bonding technology," *1996 IEEE Int. Solid-State Circuits Conf.*

Index

Absolute power gain, 192, 223, 241
Active antennas, 85–128
Active arrays:
 amplifier arrays, *see* Antenna-array amplifiers
 oscillator arrays, *see* Coupled-oscillators
 versus grids, 17–20, 191–192
 see also Power combining
Active lens, 20, 192, 229–241
Angle diversity, 474–476
Antennas:
 active, *see* Active antennas
 for arrays, 21–25, 71–72, 193–204
 see also Planar antennas
Antenna arrays:
 active arrays, *see* Active arrays, power combining
 analysis techniques, 25–31
 characterization, *see* Figures-of-merit
 descriptive parameters, *see* Figures-of-merit
 scan blindness, 22
 tray vs. tile approach, 55–58
Antenna-array amplifiers:
 antenna elements for, 21–25, 193–204
 bandwidth, 192
 C-band folded-slot array, 210–213
 efficiency, 192
 feed, 190–192
 high-efficiency class-E, 208–210
 Ka-band quasi-monolithic, 216–219
 linear-to-circularly polarized, 206
 monolithic, 219–223, 223–226
 multilayer, 248–251, 263
 multiple-slot MMIC array, 213–214
 plane-wave fed, 205–226
 polarization-preserving, 205–208

 power-added efficiency (PAE), 209
 saturated class-A, 208, 216–219
 saturation, 207–209
 waveguide tapered-slot array, 214–216
Aperture efficiency, *see* Figures-of-merit
Applications, 2–3
 commercial, 246
 military, 246
 space communications, 481–482
Attenuation, atmospheric, 2

Bar-grid oscillator, 303–305
Beam amplifiers, *see* Antenna-array amplifiers
Beam control arrays:
 advantages and limitations, 387–394
 Fabry–Perot structures in, 385
 FETs in, 406
 insertion loss of, 389
 maximum power-handling capability of, 392
 mechanical limitations of, 392
 monolithic, 399–406
 other losses in, 391
 PIN diodes in, *see* PIN diodes
 varactors in, 390
 RF power limits of, 391
 see also Grid arrays, active
Beam forming, 469–470
Beam scanning/steering:
 with injection-locked oscillators, *see* Coupled oscillators
 scanning bandwidth, 468
 true time delay, 463–469
Beam switching, 470
Biasing, 17, 20, 196, 207, 209, 214, 230, 239

523

524 INDEX

Cavity combiner, see Power combining, circuit-based
Circuit-fed arrays, 10–11, 50–52
 circuit-fed/spatially combined, 68–82
 vs. spatial feeding, 58–61
 see also Coupled-oscillators; Antenna-array amplifiers
Combining efficiency, see Figures-of-merit
Constrained lenses, historical development, 227–229
Controlled-reflection surfaces, 406, 325
Controlled-transmission surfaces, 406
Conversion efficiency, 298
Corporate combiner, 5–7. See also Power-combining, circuit based
Coupled-oscillators, 135–183, 267–272
 broadside power combining, 156–163, 270–272
 coupling networks, 150–151, 267–269
 continuum modelling, 177–178
 dynamic analysis, 145–183
 extended resonance, 267–272
 noise analysis, 166–171
 scanning techniques, 171–181
 sensitivity analysis, 157
 stability analysis, 149–150
 transient response, 163–166

Devices, solid-state:
 comparison to vacuum tubes, 4
 power limitations, 3–4, 247–248
Direct-feedback grid oscillator, 302–305
Dual-frequency grid oscillator, 325

Effective isotropic power gain (EIPG), see Figures-of-merit
Effective isotropic radiated power (EIRP), see Figures-of-merit
Effective transmitted power, see Figures-of-merit
EMF method, see Induced EMF method
Extended resonance, 267–272

Fabry–Perot cavity, 20, 294–296
Feed networks, see Circuit-fed arrays
Figures-of-merit, 33–42
 aperture effiency, 34–37
 combining efficiency, 42, 189
 effective isotropic power gain (EIPG), 40–42, 190
 effective isotropic radiated power (EIRP), 39–40
 mode-coupling, 37–39
Finite-difference time-domain (FDTD) modeling, 28, 31, 112–115

Frequency-conversion:
 active antennas, 100–102
 grids, see Grid multipliers
Frequency selective surfaces (FSS), 18. See also Grid arrays, passive

Gain, see Figures-of-merit
Gate-feedback grid oscillator, 305–307
Gaussian beams, see Mode coupling
Grating lobes:
 in an array with a rectangular lattice, 467–470
 in an array with a triangular lattice, 467–470
Grid amplifiers:
 HBT, 335–336, 361–367
 HEMT, 342–361, 367–373
 modeling of, 337–342
Grid arrays, active:
 amplifiers, 331–372
 beam control, see Beam control arrays
 frequency conversion, see Grid multipliers
 oscillators, see Grid oscillators
 phase shifters, 386–388
 versus distinct component arrays, 61–63
Grid arrays, passive:
 analysis of, see Finite-difference time-domain (FDTD) modeling; Induced EMF analysis; Method-of-moments
 Jerusalem-cross pattern, 378, 384
 RADANT, 380–381
 mesh, 381–383
Grid mixer, see Grid multipliers; Sideband generator
Grid multipliers, 415–446
 doublers, 417–419, 422–446
 EMF analysis of passive array, 427–431
 harmonic-balance analysis, 444–446
 sideband generator, 446–449
 Schottky-diodes in, 423–425
 triplers, 420–422
Grid oscillators:
 analysis techniques, 310–320
  EMF analysis, 311–314
  full-wave analysis, 314–317
 cascaded grids, 322–327
 comparison, 307–309
 direct-feedback, 302–305
 feed, 459
 figures-of-merit, 296–298
 gate-feedback, 305–307
 power optimization, 320–321
 three-terminal, 301–307
 two-terminal, 299–301

INDEX   **525**

Heat removal, heat sinking, *see* Thermal management
Hybrid circuit construction, vs. monolithic, 63–65

Induced EMF analysis, 26–28, 311–314, 427–431. *See also* Grid arrays
Injection-locking, *see* Oscillator modeling; Coupled oscillators
Input/output isolation:
  ground-plane, 206
  in travelling-wave arrays, 215
  polarization, 332
Integrated antennas, *see* Active antennas
Isotropic conversion gain, *see* Active antennas

Kurokawa cavity combiner, *see* Power-combining, circuit-based

Lens amplifier arrays:
  focal point feed, 229, 462
  focal surface, 463, 469
  linear 7-element patch lens array, 229–231
  low-noise CPW slot array, 234–241
  receiving, 234–241, 470–471, 473–474
  transmitting, 231–233, 458–470
  two-dimensional patch array, 231–233
Local oscillator, 409–410. *See also* Grid multipliers
Losses:
  atmospheric, 2
  effect on combining efficiency, 6
  of common transmission media, 7
  in spatial combiners, 66–68

Measurements, *see* Antenna arrays, characterization
Method-of-moments, 194, 231
Mode coupling, *see* Figures-of-merit
Monolithic fabrication, vs. hybrid fabrication, 63–65
Multipath fading, 471–472
Multipliers:
  grids, *see* Grid multipliers
  waveguide, 411–415
Mutual, *see* Coupled oscillators
Mutual coupling, 20, 88–91

Noise reduction in arrays, 9. *See also* Phase noise; Coupled-oscillator
Noise sources, 234–236
Nonlinear dynamics, 31–33, 135–183
  injection-locking, 142–144
  mutual synchronization, 145–183

Nonlinear modeling, *see* Stability analysis; FDTD modeling

Oscillator arrays, *see* Grid Oscillators; Coupled-oscillators
Oscillator modeling, 138–145
  equivalent circuit, 139
  injection-locking, 142–144
  noise admittance, 144–145,
  stability, 141

Patch antenna, *see* Antennas; Active antennas; Planar antennas
Phase noise:
  in oscillators, *see* Oscillator modeling
  in oscillator arrays, *see* Coupled oscillators
  reduction, *see* Noise reduction in arrays
Phase shifter grids, *see* Grid arrays, active
Photonic band gap materials, 24
PIN diodes, in beam-control arrays, 380–387, 394–395, 479–481
Planar antennas, 21–25, 104–112, 193–204
  bandwidth, 24–25
  modeling, 112–115, 194, 231
  radiation efficiency, 22–24
  substrate modes, 22–24, 86–88
Planar quasi-optical combiners, 277–291
Polarization, input/output, 36–37
Polarizers:
  circular, 476–477
  need for, 34–37
Power-combining:
  active antenna arrays, 188–241
  circuit-based, 5–9
  device-level, 5, 189
  hierarchy, 19
  quasi-optical, 9–17
  spatial, 9–10, 49–82
  two-level, 251–263, 265, 458–463
  waveguide-based, 13
Power-combining efficiency, free-space, 189, 210
Power density, 15
  efficiency and unit cell, 16
Power grid oscillator, 320–321

Quasi-optical:
  definition of quasi-optical array, 12
  combining techniques, *see* Power combining, quasi-optical
  isolator, 478–479
  mixer:
    self-oscillating, 471–473
    subharmonic, 471–473

526 INDEX

Quasi-optical, mixer (*Continued*)
  see also Grid multipliers; Sideband generator
  phase modulator, 479–481
  receiver:
    dynamic range, 456
    noise figure, 456
    low-noise, 234–241
    tranceiver subsystem, 456–458
Quasi-optical amplifier gain, calibration, 190

Radiation efficiency, and grid unit cell, 20.
  See also Planar antennas

Scan blindness, *see* Antenna arrays
Scanning arrays, *see* Beam scanning
Schottky diodes, in grid doubler, 423–425
Sideband generator, 446–449
Simulation tools, 25–33, 85, 94–95
Slot antennas:
  antiresonant, 195
  broadband tapered, 201–204
  folded, 197–201
  loaded-folded, 201–202
  microstrip-fed, 196–197
  multiple-slot, 197–201
  off-center fed, 196
  see also Active antennas; Planar antennas
Smart antennas, 406. *See also* Active antennas
Spatial combining, *see* Power combining, spatial
Spatially fed array, 50–53. *See also* Power combining, spatial
Stability analysis:
  in active antennas arrays, 30–31
  in coupled-oscillator systems, *see* Coupled-oscillators
Substrate modes, *see* Planar antennas
Surface waves, *see* Planar antennas
System configurations, 53–66
  summary of design trade-offs, 65

Thermal management, 16
Three-dimensional grid oscillator, 325–327
Tray/tile approach, *see* Antenna arrays

Unit-cell:
  approximation, 310
  power density and efficiency, *see* Power density
  and thermal management, *see* Thermal management
  see also Antenna arrays, passive array analysis

Varactor diodes:
  in grid multiplers, 417–421
  in mode selection for grid oscillators, 324
  in phase shifter arrays, 390, 393, 399, 401–406
  in quasi-opticalVCOs, 322
  in waveguide multipliers, 411–414
Varistor, 412, 434
Voltage-controlled grid oscillator, 322–323

Wilkinson combiner, 7
Wire-grid polarizers, *see* Polarizers
Wireless applications, *see* Applications, wireless